SKELETAL & DEVELOPMENTAL ANATOMY

SECOND EDITION

Robert A. Walker, Ph.D.
New York Chiropractic College

C. Owen Lovejoy, Ph.D.
Kent State University

M. Elizabeth Bedford, Ph.D.
Los Angeles, California

William Yee, D.C., L.Ac.
Yonkers, New York

Linus
Publications, Inc.

Published by Linus Publications, Inc.

Deer Park, NY 11729

Copyright © 2007 Robert A. Walker, Ph.D., C. Owen Lovejoy, Ph.D., M. Elizabeth Bedford, Ph.D., & William Yee, D.C., L.Ac.

All Rights Reserved.

ISBN 1-934188-46-8

No part of this publications may be reproduced, stored in a retrieval system, or transmitted, in any form or by any means, electronic, mechanical, photocopying, recording, or otherwise, without the prior permission of the publisher.

Printed in the United States of America.

10 9 8 7 6 5 4 3 2 1

Table of Contents

Acknowledgements i
Preface iv
Preface to the Second Edition vi

Chapter - 1

Introduction: Anatomical Terminology

Anatomical Terminology 1

Chapter - 2

Bone: Anatomy, Growth, and Development

Functions of the Skeletal System 9
Structure of the Skeletal System 9
Skeletal Remodeling 24
Laws of Bone Growth and Deposition 26
Bone Formation and Growth 27
Basic Embryology of the Skeletal and Articular Systems 27
Two Types of Ossification 28
Typical Disorders of Bone Growth and Remodeling 37
References Cited 43

Chapter - 3

Development of the Vertebral Column

Embryological Development from Fertilization to Neurulation 47
Formation of the Vertebral Column 51

Development: Core Concepts 54

Molecular Control of Vertebral Column Formation 57

Hox Control of Vertebral Specification 60

References Cited 63

Chapter - 4

Bones of the Postcranial Skeleton: Vertebral Column

Bones of the Vertebral Column 75

Landmarks of the Vertebral Column 103

Ossification of the Vertebral Column 106

Introduction to Arthrology 109

Articulations of the Vertebral Column 111

Motions of the Joints of the Human Skeleton 123

Movements of the Vertebral Column 127

Variations and Disorders of the Vertebral Column 130

References Cited 134

Chapter - 5

Bones of the Postcranial Skeleton: Ribs and Sternum

The Ribs 135

The Sternum 139

Articulations of the Ribs 141

Articulations of the Sternum: 148

Ossification of the Ribs and Sternum 146

Disorders of the Ribs and Sternum 148

References Cited 149

Chapter - 6

Bones of the Postcranial Skeleton: Upper Limb

The Pectoral Girdle 152

Bones of the Upper Limb 155

Articulations of the Upper Limb 168

Ossification of the Upper Limb Bones 180

References Cited 185

Chapter - 7

Bones of the Postcranial Skeleton: Lower Limb

The Pelvic Girdle: The Os Coxae 187

Bones of the Lower Limb 197

Articulations of the Lower Limb 215

Ossification of the Lower Limb Bones 236

References Cited 240

Chapter - 8

Osteology of the Human Cranium

The Auditory Ossicles 242

The Hyoid Bone 242

The Skull 244

Ossification and Development of the Cranium, Mandible, and Hyoid 281

References Cited 292

Further Readings and Resources 293

Appendix:

Muscle Origins, Insertions, Innervations, Blood Supplies and Principal Actions

Major Muscles of the Trunk: The Back 297

Major Muscles of the Trunk: The Neck and Prevertebral Region 306

Major Muscles of the Trunk: The Thoracic Wall 309

Major Muscles of the Trunk: The Diaphragm 310

Major Muscles of the Trunk: The Abdomen 311

Upper Limb Musculature: The Shoulder 313

Upper Limb Musculature: The Pectoral Region 315

Upper Limb Musculature: The Arm 316

Upper Limb Musculature: The Forearm 318

Upper Limb Musculature: The Intrinsic Muscles of the Hand 323

Lower Limb Musculature: The Hip Joint Flexors 326

Lower Limb Musculature: The Gluteal Region 327

Lower Limb Musculature: Short Rotators of the Hip 328

Lower Limb Musculature: The Thigh 329

Lower Limb Musculature: The Leg 334

Lower Limb Musculature: Intrinsic Muscles of the Foot 337

Major Muscles of Mastication and of the Hyoid 341

References Cited 345

Acknowledgments

We would like to thank current and former members of the anatomy faculty of the New York Chiropractic College (NYCC) for their editorial comments and factual and typographic corrections: J. Donald Dishman, D.C.; Thomas M. Greiner, Ph.D.; Mike Lentini, D.C.; Raj Philomin, M.D, Ph.D.; Narayan Vijayashankar, M.D.; Maria Thomadaki, D.C.; Brigitte Tremblay, D.C.; Roger O. Walter, Ph.D.; and Michael Zumpano, Ph.D.

We thank Barry Yee for his assistance in the preparation of the illustrations. We would also like to thank Carl Jagos for his constant support.

The authors would like to thank the reviewers who read and reviewed the manuscript of the first edition: Barclay W. Bakkum, D.C., Ph.D.; Paul Barlett, Ph.D.; Craig L. Mekow, M.S.; Michael D. Reife, D.C.; Patricia A. Rogers, D.C.; and Carlos F. Soniera, M.D, M.Sc. Their many comments have improved the manuscript greatly.

Special thanks are due to Sue Firkins, M.S., formerly of the Department of Anatomy of New York Chiropractic College. Sue carefully proofread every version of this manual since its origins in the early 1990s at NYCC. Her many careful readings and suggestions for clarifications and corrections have produced invaluable insights that have greatly improved this manual.

Thomas Greiner, Ph.D., University of Wisconsin – La Crosse, and formerly of the NYCC anatomy department, has provided valuable input throughout the development of this book nearly from its inception.

Michael Zumpano, Ph.D., D.C., and NYCC Anatomy faculty member, provided commentary on the embryology and development of the craniofacial region.

Naryan Vijayashankar, M.D., and Raj Philomin, M.D., both also of the NYCC Anatomy faculty, provided many insights throughout the course of the preparation of the this manual.

Maria Thomadaki, D.C., has, through the years and multiple early versions of the this manual, kept it focused on the needs of future clinicians.

Fiona Jarrett-Thelwell, D.C., of the Radiology group of the New York Chiropractic College Diagnosis Department, contributed a great deal of useful information and checked the manuscript for consistency with the NYCC radiology instructional program.

The following reviewed in whole or in part the manuscript for the second edition: Joe Bernard, Northeast Ohio Universities Collge of Medicine, Barbara Brown, Ph.D., Northeast Ohio Universities College of Medicine; Steve Duray, Ph.D., Palmer Health Care University; Anthony Falsetti, Ph.D., University of Florida; Thomas M. Greiner, Ph.D., University of Wisconsin – La Crosse; Kenneth Parham, Ph.D., U.S. Army Natick Soldier Science Center;

Jerome C. Rose, Ph.D., University of Arkansas — Fayetteville; Tal Simmons, Ph.D., University of Central Lancashire; Linda Spurlock, Kent State University — Stark Campus; Robert Tague, Ph.D., Louisiana State University; and Steven Ward, Ph.D., Northeast Ohio Universities College of Medicine, Ph.D. Their valuable feedback has greatly improved this work.

The authors acknowledge and thank New York Chiropractic College, particularly the Division of Academic Affairs, for use of facilities, computer equipment, audiovisual assistance, secretarial support and encouragement. We specifically wish to thank Michael Mestan, Interim Executive Vice President for Academic Affairs at NYCC, and Frank Nicchi, President of NYCC.

Roberta Massey, our original development editor, drastically improved the flow of the original manuscript. She is deserving of all praise.

Finally, we would like to thank the many students who have used previous versions of this manual. Their comments and suggestions have aided greatly in the improvements to this manual.

Preface

This manual contains a detailed description of the osseous and ligamentous anatomy of the human skeleton, with special emphasis on the spine. While a number of texts describe human osteology in general terms, none treats osseous anatomy in the detail needed for clinical practitioners, basic sciences instructors in health care curricula, and serious students of human skeletal anatomy. Most such manuals generally ignore the ligaments of the axial skeleton and give, at best, cursory treatment to important ligaments of the appendicular skeleton. Most do not address the anatomy of the vertebral column in the detail provided here. We anticipate that this manual will be of use to health care professional students as well as advanced undergraduate and graduate students in programs where knowledge of human skeletal anatomy is essential.

Mention of craniometric points is kept to a minimum, as are detailed directions for osteometric measurements, indices and calculations such as stature estimation. Determination of sex is addressed where necessary as a consequence of function, but detailed descriptions of determination of sex and "race" from the skeleton are not addressed. References are given to the relevant literature which addresses forensic and paleodemographic analyses of human skeletal material, but these topics are not directly addressed in this work.

We have collectively taught human osteology, either as a stand-alone course or as part of a general human gross anatomy course, over 100 times. This includes courses taught to undergraduate and graduate students in biological anthropology and biomedical sciences, as well as to chiropractic, dental, medical, occupational therapy, paramedic, and emergency medical technician students. This text is based upon our experiences in teaching human osteology in all these contexts. In some areas, as a result, we take what we believe to be both a unique and useful approach to human osteology. The examination of the human cranium in Chapter 8, for example, approaches the cranium as a bony unit. The visible elements of the cranium are examined by standard view, rather than by individual bone: norma verticalis, norma frontalis, norma lateralis, normal occipitalis, norma basalis, and the interior of the cranium. Each is accompanied by a detailed description of relevant structures and landmarks to be identified in that view. We have found this to be a highly useful approach to the human cranium.

Preface to The Second Edition

With this second edition, we move to a new publisher. In general, the changes made from the first edition have been minor. The literature has been updated, new photographic illustrations have been added, and a few typographic errors from the first edition have been corrected.

But with the change in publisher, there has also been an important change in title. The first edition of this work, published by F.A. Davis, was titled *Skeletal and Developmental Anatomy for Students of Chiropractic*. With this new edition, we have dropped the second half of that title. The book is now simply *Skeletal and Developmental Anatomy, Second Edition*. While three of the four authors are, or were, affiliated with a chiropractic college, and one is indeed a chiropractor, we have always felt the book had a much broader audience and application. We feel the book is appropriate for all who have a need for, or are simply interested in, human osteology and arthrology, and we wrote the book with that audience in mind. The change in the title reflects that reality.

This manual is not intended as a substitute for the various forensic and archeological guides to the human skeleton. This manual is meant for those interested in all aspects of the biology and anatomy of the skeleton. Its principal function is not as a field guide for the skeleton, though it could be useful in that context as well. Our intention is to present a concise guide to osseous and ligamentous anatomy; bone growth, development, and microanatomy; basic arthrology and basic pathology of the human skeleton. With the focus on osseous landmarks and ligamentous anatomy, the musculature also becomes a central concern. Numerous muscles are referred to throughout the text. For that reason, a lengthy appendix outlines the origins, insertions, innervations, blood supplies, and principal actions of the major skeletal muscles.

Robert A. Walker

Seneca Falls, New York

May 1, 2007

CHAPTER One

Introduction: Anatomical Terminology

This text is a general overview of the gross anatomy of the human skeletal system. In particular, this manual concentrates on the axial skeleton, and treats its principal ligaments in great detail.

The vertebral skeleton is becoming progressively more important in clinical practice. It and its associated structures serve as the organizing axis of this manual. After a brief introduction to the vertebral column as a whole, we start a detailed description at the top of the cervical spine, and then proceed down the column to the thoracic, lumbar, sacral and coccygeal regions.

After examining the vertebral column, we view and discuss its closely associated structures: the ribs, costal cartilages and sternum. From here we proceed to the clavicles and scapulae, which attach the upper limb to the axial skeleton. After examining the osseous and ligamentous pectoral girdle, we then continue with an examination of the skeleton of the upper limb. Following this, we consider the pelvis and its ligaments, and then the skeleton of the lower limb. Following an examination of the postcranium, we examine the osteology of the cranium.

We suggest that you make frequent reference to the many detailed line drawings and photographs that accompany the text. The purpose of the drawings is to elucidate the structures described in the text. Often, a line drawing can give a clearer idea of the location and nature of structures, such as crests or protuberances, than a photograph. Where greater detail is needed to illustrate the structures discussed, photographs are provided.

Anatomical Terminology

Some basic understanding of anatomical terminology is necessary to understand the osteology of the human skeleton.

All human anatomical terminology is relative to the *anatomical position* (Figure 1-1). In the anatomical position, the subject is assumed to be standing. The legs are together, and the feet are flat on the surface, with the toes facing forward. The person's face and eyes are directed forward as well. The arms are allowed to hang at the subject's sides, with the palms facing forward and the hands held open with the fingers pointing downward and thumbs pointing away from the body. When describing parts of the body, they are always described *as if* the subject is standing in the position just described.

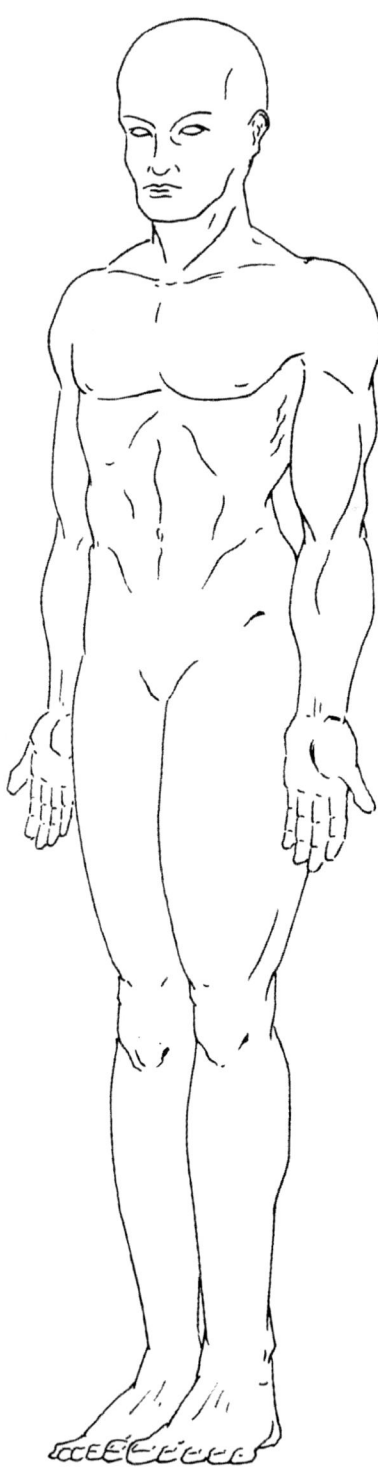

Figure 1-1 Person standing in the anatomical position.

Introduction: Anatomical Terminology

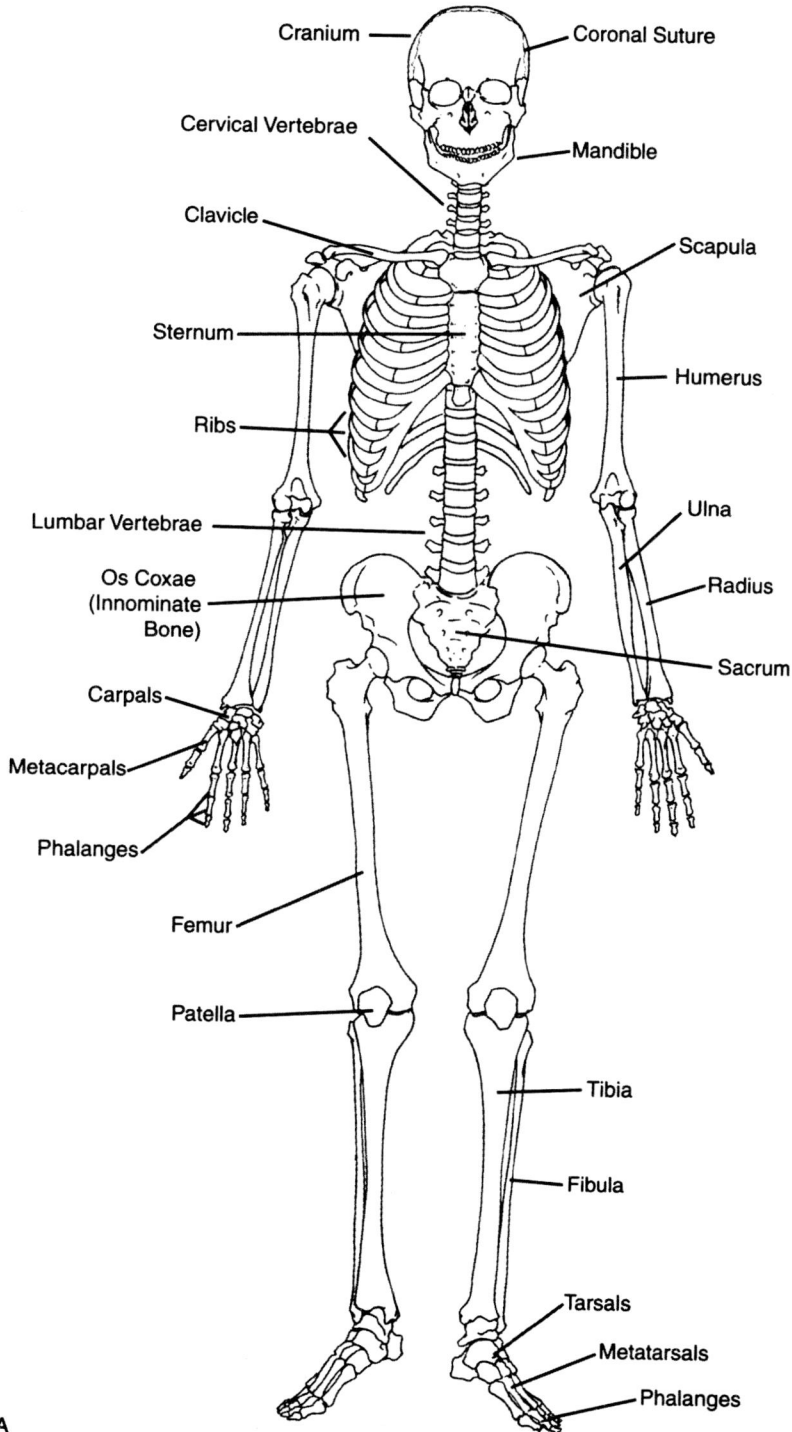

Figure 1-2 The human skeleton: (*A*) anterior.

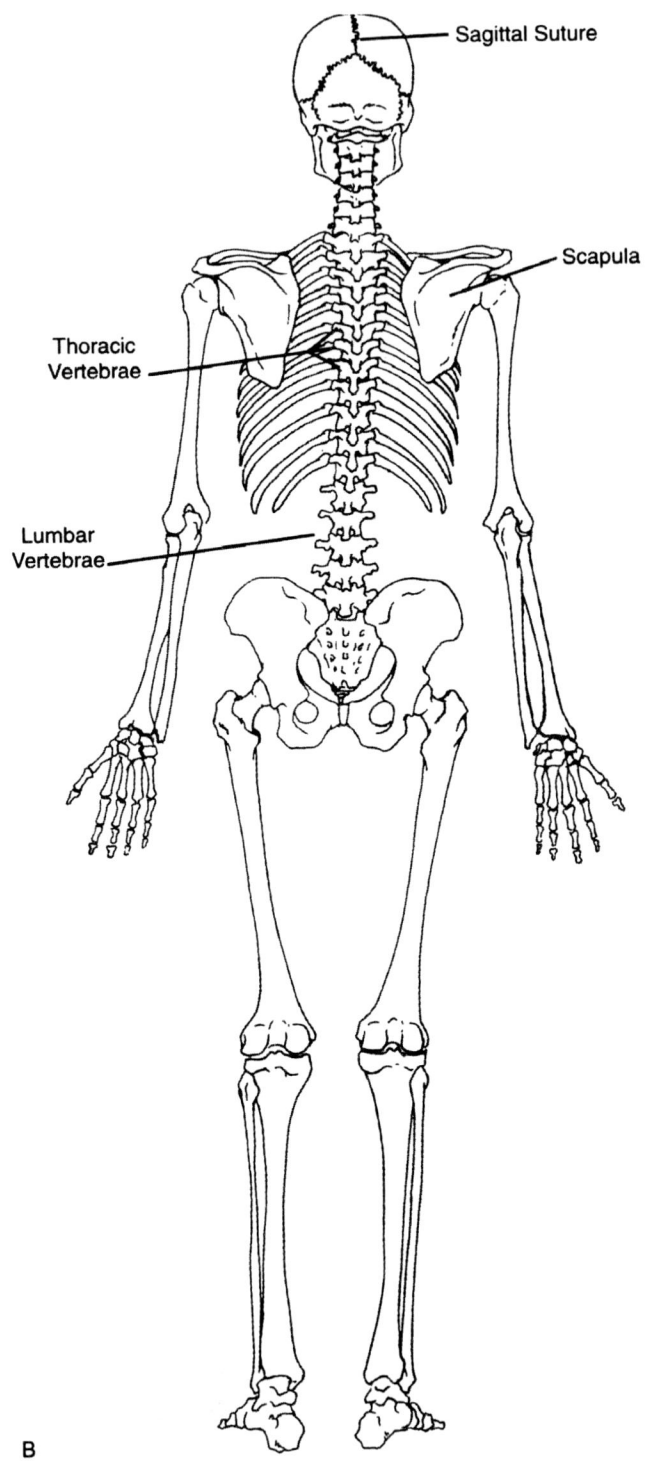

Figure 1-2 (*Continued*) (*B*) posterior.

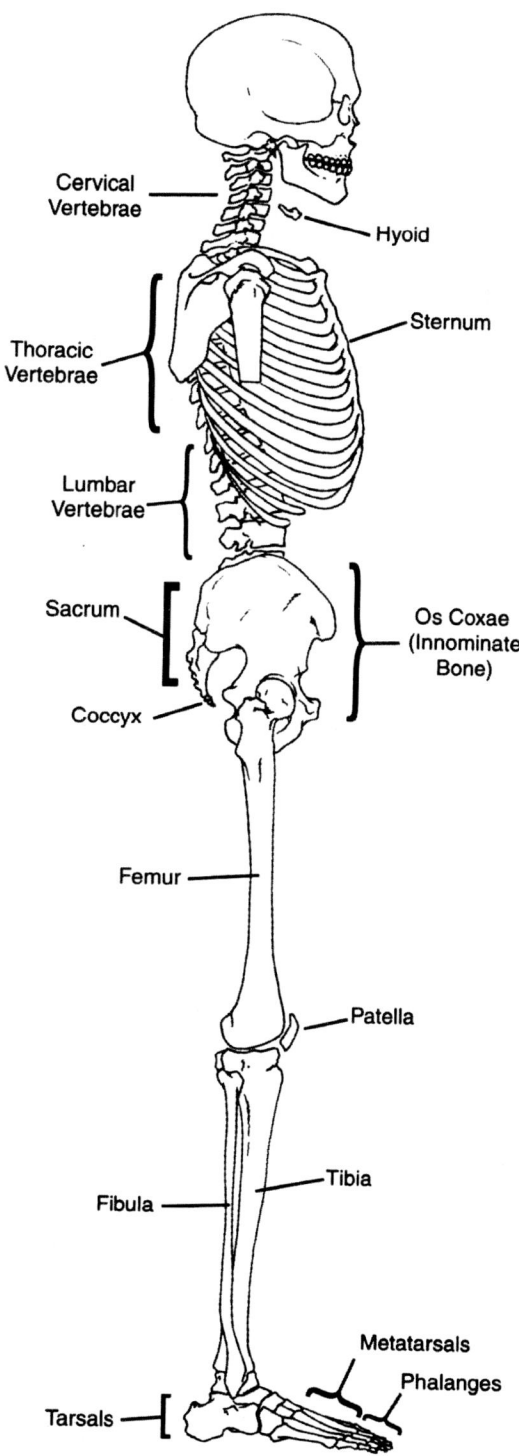

Figure 1-2 (*Continued*) (*C*) lateral.

The cardinal anatomical directions are based on the anatomical position as just defined.

With the subject in the anatomical position, the upward direction (toward the head) is described as *superior*. Toward the feet is defined as *inferior*. These directions are not absolute, but *relative* to other parts of the body. In the anatomical position, the knees are *superior* to the feet, but they are *inferior* to the chest. In human anatomy, a common synonym for superior is *cranial* (Latin: toward the head), and a common synonym for inferior is *caudal* (Latin: toward the tail). The terms cranial and caudal are more frequently encountered in embryology and comparative anatomy.

While superior and inferior describe one very important axis of the human body, the terms *anterior* and *posterior* define another very important axis. *Anterior*, in human anatomy, means toward the front side, or face side of the body. *Posterior* refers to the back side of the body. Refer to Figure 1-2. Note the locations of the *scapulae* ("shoulder blades") and *sternum* ("breast bone"). The scapulae are located posteriorly, while the body of the sternum is located anteriorly. The sternum is anterior to the thoracic vertebrae, while the thoracic vertebrae are posterior to the sternum. In human anatomy, a common synonym for anterior is *ventral* (the "belly" side), and a common synonym for posterior is *dorsal* (the "back" side). The terms ventral and dorsal, like cranial and caudal, are more frequently encountered in embryology and comparative anatomy.

This is a good point to introduce a basic difference between human anatomy and the anatomy

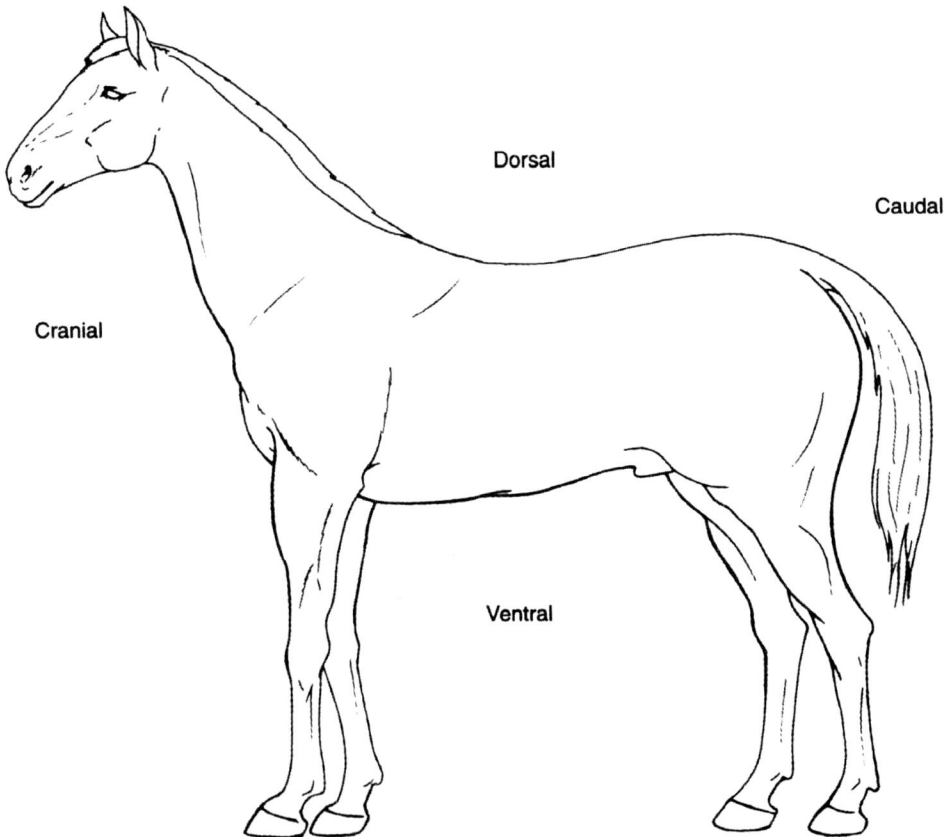

Figure 1-3 The horse, illustrating the anatomical position of a tetrapod, or four-legged vertebrate.

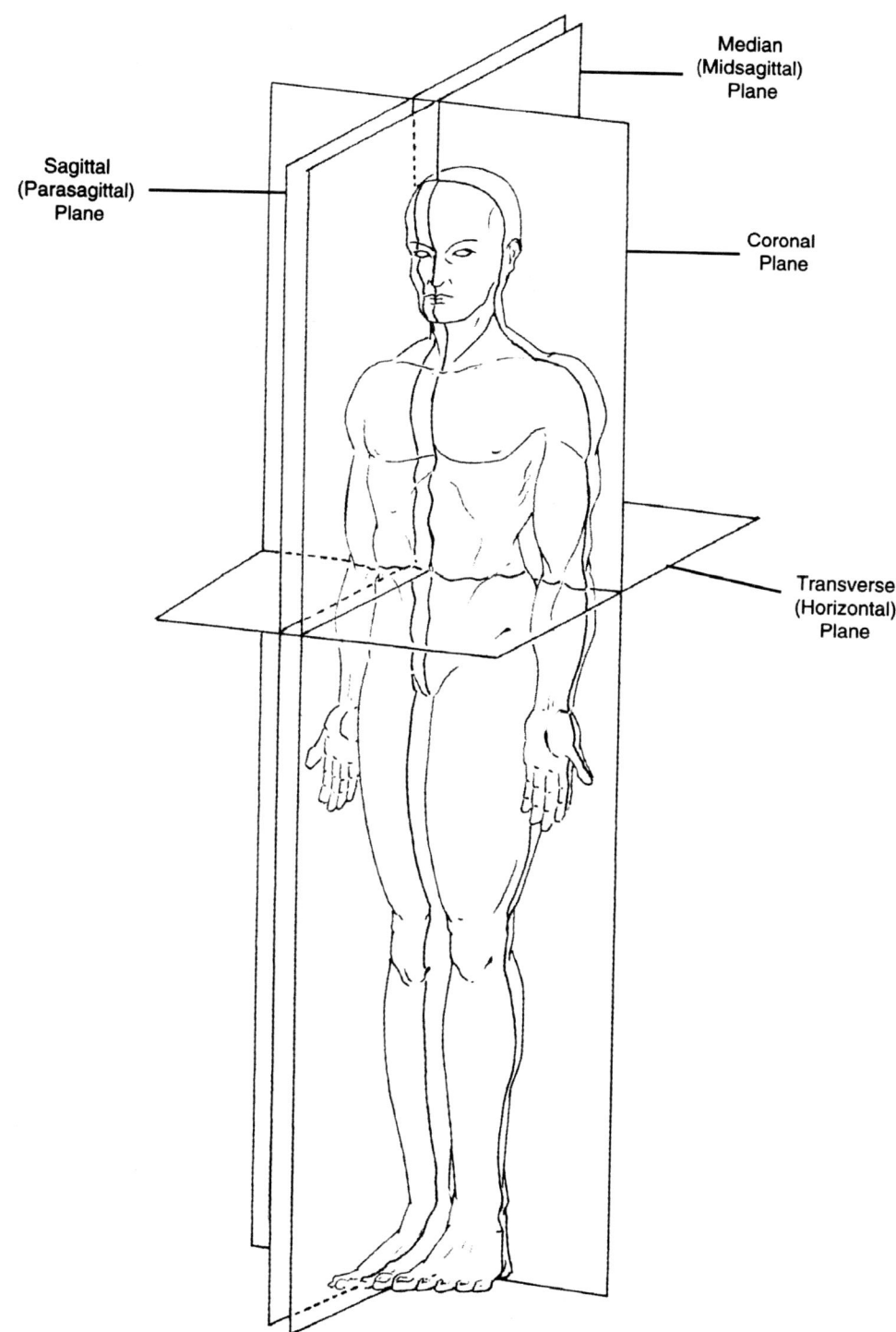

Figure 1-4 Basic anatomical planes.

of other terrestrial and amphibious vertebrates (tetrapods, or those vertebrates that have four limbs). Human beings are peculiar in a few respects, one of the most important being our obligatory bipedal style of locomotion (habitually moving by means of only the two hind limbs). As one consequence of this type of locomotion, human beings carry their trunks erect over the pelvis. As a result, the vertebral column is held perpendicular to the ground, whereas other tetrapods, supported by forelimbs and hindlimbs, carry the vertebral column, running along the long axis of the trunk, parallel to the ground (Figure 1-3). As a result, in a tetrapod, the cranial end of the body is anterior, rather than superior, and the caudal end is posterior, rather than inferior. Also, in tetrapods, inferior is ventral, while superior is dorsal (as in the dorsal fin of a fish or whale).

Two more important relative terms of direction are *medial* and *lateral*. When we describe a structure as being *medial*, we mean that it is closer to the midline or in the direction of the midline of the body. When we describe a structure as *lateral*, we mean that it lies further from the midline of the body. Again, these are relative terms. Refer back to Figure 1-1 representing the anatomical position. Note the thumbs and the other fingers of the hands. The thumbs are lateral to the other fingers, or conversely, the other digits are medial to the thumbs. Likewise, the ears are lateral to the cheeks, while the cheeks are lateral to the nose. Conversely, the cheeks are medial to the ears, but lateral to the nose. The term *median* refers to structures that lie in the midline of the body. Median is an absolute term. The nose is a median structure, as is the vertebral column.

Another pair of terms of relationship are *proximal* and *distal*, and are extensively used in the descriptions of the limbs. *Proximal* means closer to the point of origin or midline of the body, while *distal* refers to being further from the point of origin or midline of the body. Like most of the other terms we have examined, proximal and distal are relative terms. For example, the elbow joint is more proximal than the wrist joint, but distal to the shoulder joint.

In addition to these anatomical terms of relationship, there are some basic planes of anatomy that should be understood (Figure 1-4). There are three basic anatomical planes: *sagittal*, *coronal*, and *transverse* (or *horizontal*). If we return to the anatomical position, a *sagittal plane* divides the body into left and right portions. A sagittal plane is a plane that passes parallel to the plane of the sagittal suture of the cranium (refer to Figure 1-2). The sagittal plane that passes directly through the midline of the body and divides the body into equal left and right halves is called the *midsagittal plane* or *median plane*. Other sagittal planes that lie to one side or the other of the midline of the body, and divide the body into unequal left and right portions, are also referred to as *parasagittal planes*. A *coronal plane* is a plane that lies parallel to the plane of the coronal suture (refer to Figure 1-2) which runs from side to side across the top of the skull. Coronal planes divide the body into anterior and posterior portions. A *horizontal* or *transverse plane* divides the body into superior and inferior portions. The three anatomical planes are orthogonal to one another (they meet one another at 90 degree angles.) *Sections* through the body or a body part are generally defined according to one of these planes. An *oblique section* is a section which meets one or more of the anatomical planes at less than a 90 degree angle.

These are the principal anatomical terms needed to understand basic anatomical descriptions. Note that these terms may be combined, as in superoinferior, superolateral, anteroposterior, and so forth to describe directions through the body or portion of the anatomy.

CHAPTER Two

Bone: Anatomy, Growth, and Development

Functions of the Skeletal System

The skeletal system is closely related functionally to the *muscular system* and the *articulations* (joints). The skeletal and muscular systems are often referred to together as the *musculoskeletal system*. (The appendix lists origins and insertions of the principal muscles of the trunk, limbs, and masticatory apparatus.)

The skeletal system functions closely with other body systems in a variety of ways. It supports the soft tissues of the body and protects delicate structures such as the brain and heart. The skeletal system also acts as a system of levers for muscular action. Without relatively rigid levers for attachments, muscular contraction couldn't create movement or do work.

The skeleton also acts as a storage area for minerals, particularly calcium and phosphorous. Bone is highly vascular, allowing for a rapid exchange of minerals between bone and blood.

The skeletal system also provides a site for blood cell production (hematocytopoiesis). Red marrow is found in some bones, especially in children, and is one of the sites involved in blood cell production. By adulthood, red marrow is restricted to the vertebrae, ribs, sternum, scapulae, clavicles, pelvis and cranial bones, and the proximal femora and humeri. It is replaced by yellow marrow (composed principally of fat) in other bones of the adult.

Structure of the Skeletal System

There are many forms of *connective tissue*. They are found throughout the body and perform a number of functions. They are often broken down into two types: *loose* (meaning found extensively throughout the body and not demonstrating characteristic form; loose connective tissue has a very high proportion of cells and a minimal amount of extracellular materials), and *dense* (contributing to specific structures and generally having a very high proportion of extracellular material relative to the number of cells). The musculoskeletal system is generally composed of three specialized types of connective tissue: *bone* or *osseous tissue, cartilage*, and *fibrous tissue* (the primary element of tendons and ligaments). All of these involve cells whose primary role is to produce and maintain extracellular matrix, in which their cells become embedded. The cells are usually housed in small spaces called *lacunae*.

There are three principal differences between bone and cartilage. The first is that hydroxyapatite crystals are deposited preferentially in bone, but under most circumstances not in cartilage. The second difference is that the collagen fibers present in bone matrix are arranged in a highly organized, specific arrangement (except within primitive or woven bone, as will be described below) whereas in cartilage, collagen fibers are more often randomly arranged (except in some forms of fibrocartilage). Another important difference is that bone is a highly vascular tissue with a rich blood supply, whereas cartilage is avascular. This is clinically important because it allows bone to heal more readily than cartilage. The rich, vascular network within and around bone is a result of the ossification process (the formation of bone tissue) and a function of the highly organized lamellar ("layered") design of mature bone.

Characteristics of Cartilage

Different types of cartilage are distinguished by the kinds of fibers embedded in their extracellular matrices. As mentioned above, cartilage is avascular. It therefore receives nourishment primarily through osmosis and diffusion. There are three types of cartilage in the human body: *hyaline cartilage*, *fibrocartilage*, and *elastic cartilage* (Figure 2-1).

Hyaline cartilage (hyaline means "glassy") has no fibers which are visible to the naked eye or with the light microscope, and the cartilage appears clear and smooth. Hyaline cartilage does, however, contain very fine collagen fibers. Large amounts of hyaline cartilage are found associated with the skeletal system. As we shall see, the developmental models or *anlagen* (singular, *anlage*) for most bones of the skeleton are composed of hyaline cartilage. Furthermore, the *epiphyseal growth plates* of growing bones and *articular cartilages* of synovial joints are composed of *hyaline cartilage*.

Fibrocartilage contains large amounts of collagen fibers and is an important structural cartilage. Fibrocartilage does not form part of the bones, but is found in joints. The annulus fibrosus of the intervertebral disc is composed of fibrocartilage, and fibrocartilage is characteristic of the general class of joints called *symphyses*.

Elastic cartilage contains large amounts of a structural protein called *elastin*. As the name implies, elastin is a highly elastic protein and is thus capable of being stretched and distorted and without losing its original shape (it is said to have a low *elastic modulus*). Examples of elastic cartilage are the cartilage of the external ear and that surrounding the nostrils. Only a small amount of elastic cartilage is associated with the skeletal system. Most cartilage in the skeleton is either hyaline or fibrocartilage.

Characteristics of Bone

Bone matrix is unique and distinct from cartilage matrix in that it is both highly mineralized and, in mature bone, highly regular in its arrangement. Bone matrix consists of two portions, or *phases*: a *mineral phase* and an *organic phase*. The *mineral phase* makes up 65 to 70 percent of bone and consists primarily of calcium phosphate and calcium carbonate in the form of the mineral *hydroxyapatite*. This is embedded in the *organic phase* which makes up 30 to 35 percent of bone. This phase is composed of an *amorphous ground substance* made up of proteoglycans and glycoproteins, in which are embedded *collagen fibers* and smaller numbers of other fiber types. The organic phase, without the mineral phase, is very similar to fibrocartilage matrix (Cormack, 2001). See Avioli and Krane (1998) for a more detailed description of the mineral and organic phases of bone.

Classification of Bones

Bones can be classified by shape. There are various schemes, but probably the most commonly used classifies bones as long, short, flat, irregular, or sesamoid (Figure 2-2).

LONG BONES: The morphology of long bones is what is generally thought of as "typical". A long bone consists of a tubular shaft with expanded ends.

Bone: Anatomy, Growth, and Development

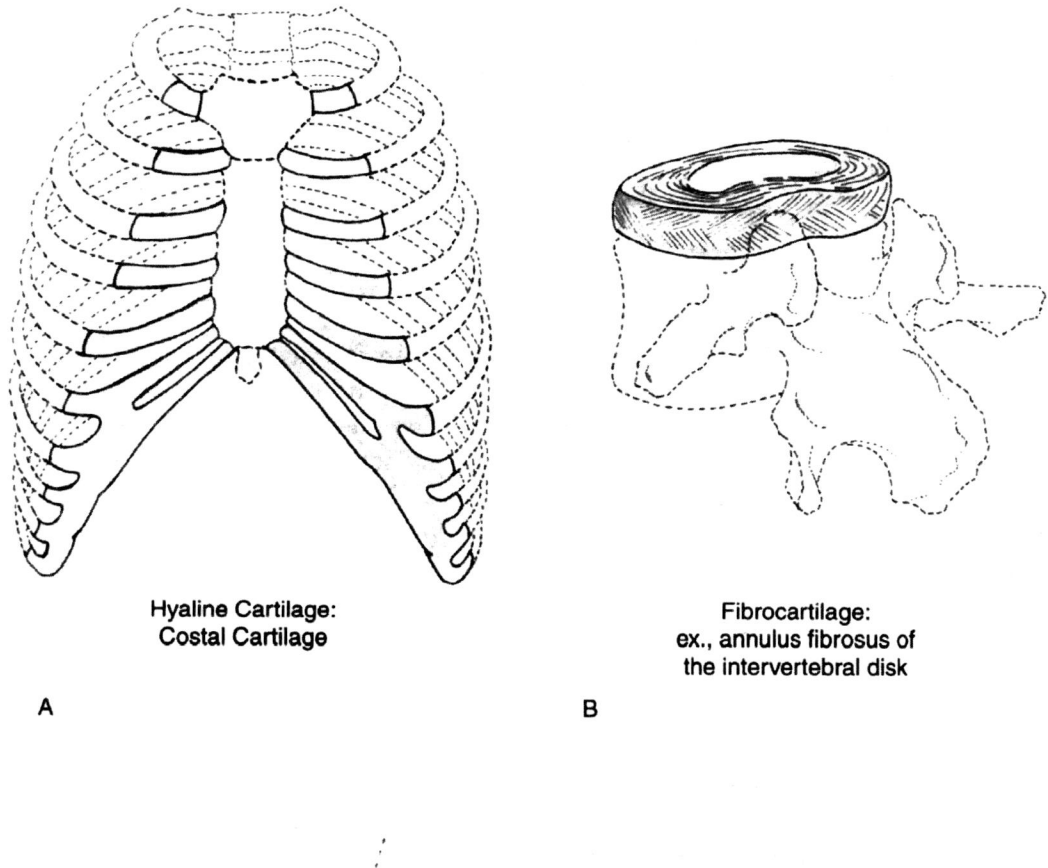

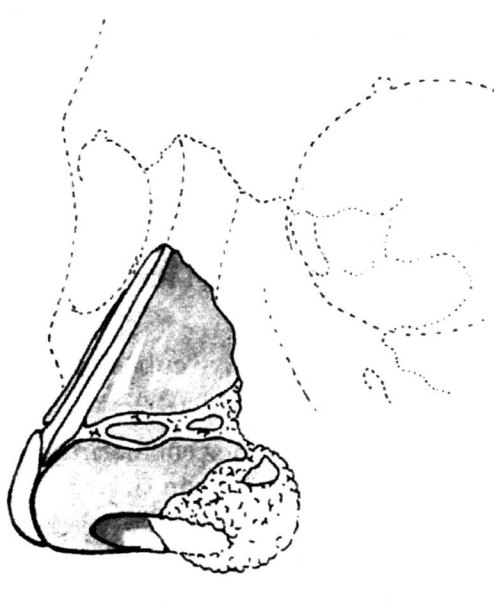

Figure 2-1 Three types of cartilage: (*A*) hyaline cartilage (costal cartilage), (*B*) fibrocartilage (annulus fibrosus of the intervertebral disc), and (*C*) elastic cartilage (cartilage of the nose).

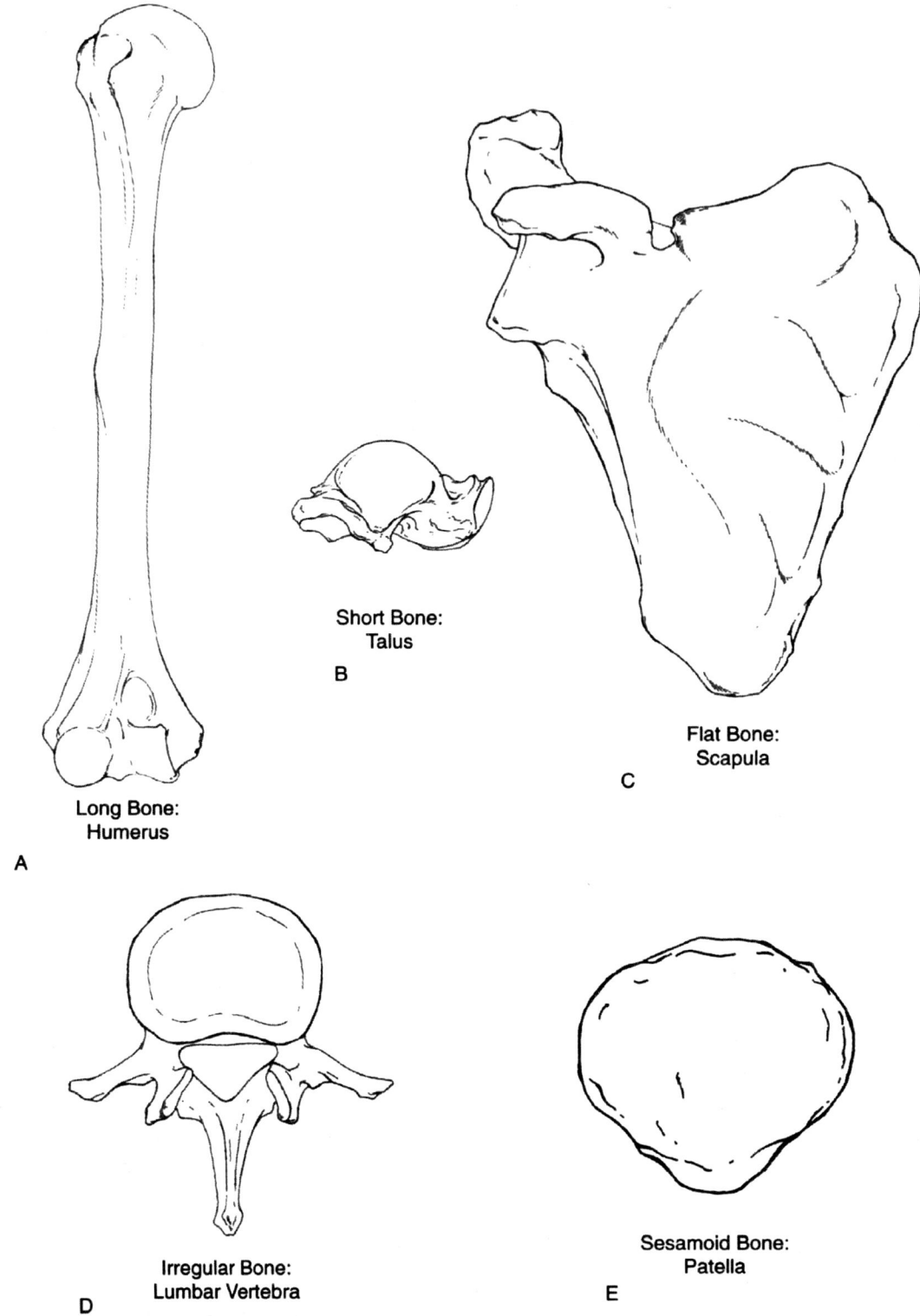

Figure 2-2 Classification of bones by shape: (*A*) long (humerus), (*B*) short (talus), (*C*) flat (scapula), (*D*) irregular (lumbar vertebra), and (*E*) sesamoid (patella).

SHORT BONES: Short bones are essentially "cubes" of bone, restricted to the *carpus* (wrist) and *tarsus* (ankle).

FLAT BONES: Flat bones are platelike bones consisting of a "sandwich" of two outer dense bone tables with sponge-like bone between them. The bones of the cranial vault that protect the brain are the best examples.

IRREGULAR BONES: Bones of irregular shape that don't fit the other categories. Vertebrae are good examples.

SESAMOID BONES: Sesamoid bones (i.e., resembling a sesame seed) are a special class of bones that develop in the tendons of muscles. The primary function of large, regularly occurring sesamoids is to move a tendon away from the center of a joint, and thus increase the moment arm of the muscle that acts across the joint. Current evidence suggests that most sesamoids are genetically determined, though it is possible that the development of some may be in response to stresses applied to a tendon as it crosses a bone. The *patella* ("knee cap") in the tendon of the quadriceps femoris muscle is the largest, and best, example of a sesamoid bone. Other sesamoid bones are frequently found in the gastrocnemius muscle of the calf (known as *fabellas*), at the ball of the foot in the flexor hallucis brevis muscle and elsewhere. (Refer to the appendix for muscle origins, insertions, and principal actions.)

Blood Supply and Innervation of Bone

Bone is a highly vascular tissue (it has a very rich blood supply). Every bone has at least one *nutrient artery*. As we shall see, the vascularization of the *anlage* or model of the forming bone is an essential step in bone formation. The nutrient artery (or arteries, as there can be more than one) passes into the *medullary (marrow) cavity* of the bone through an opening in the *compact bone* called the *nutrient foramen*. Within the medullary cavity, branches of the nutrient artery spread throughout the marrow and line the *inner (endosteal) surface* of the compact bone. There is a rich network of *subperiosteal vessels* that lie beneath the *periosteum* (the membrane that lines the outer surface of the compact bone) as well. These derive from the nutrient artery, from *metaphyseal arteries,* and from *epiphyseal arteries,* all of which are branches of systemic arteries that surround the bone, and from the muscles which invest the outer or fibrous layer of the periosteum. Within the substance of the bone matrix itself are a network of small channels, called *Haversian canals* (whose morphology we will explore further a little later) which contain small arterioles or capillaries. They are the ultimate blood supply to the bone tissue and its cells, called *osteocytes*. In a typical long bone, these Haversian canals run parallel to the long axis of the bone. They in turn are interconnected via transverse canals called *Volkmann's canals*. Volkmann's canals carry small arterial branches inward from the periosteal vessels and outward from endosteal vessels to *anastomose*, or network, with the small vessels in Haversian canals. Additional Volkmann's canals connect adjacent Haversian canals with one another and allow the vessels within the Haversian canals to anastomose with one anther. In addition, there are other vascular channels, called *primary osteons* or *nonhaversian canals* that transmit blood vessels. These vessels were originally subperiosteal vessels that became encased in bone matrix by subperiosteal deposition of bone during bone growth. Ultimately, in human bone, most of these primary osteons will be remodeled away and replaced by Haversian bone.

In some instances, the blood supply to a bone or part of the bone may be lost due to trauma. Fractures to the neck of the femur, for instance, can result in a loss of blood supply to the head of the femur. This loss of blood supply in such instances can lead to tissue death in the affected portion of the bone. This condition is referred to as *aseptic necrosis,* and is frequently seen in the femoral head. Veins accompany the arteries in all instances to provide drainage from the bone tissue.

Bones and joints receive innervation as well. Joints in general are innervated by the same nerves that innervate the muscles which cross these joints.

There is a wide network of sensory nerves in the periosteum of bone. Myelinated and nonmyelinated nerve fibers run with the nutrient artery and branch into the marrow and into the Haversian canals. Innervation of bones is particularly rich near the articular surfaces of long bones.

Morphology of Bone

The anatomy of a typical long bone will be described. Similar parts are present in all bones, though the relative shapes and sizes of the parts will vary from bone to bone (Figure 2-3).

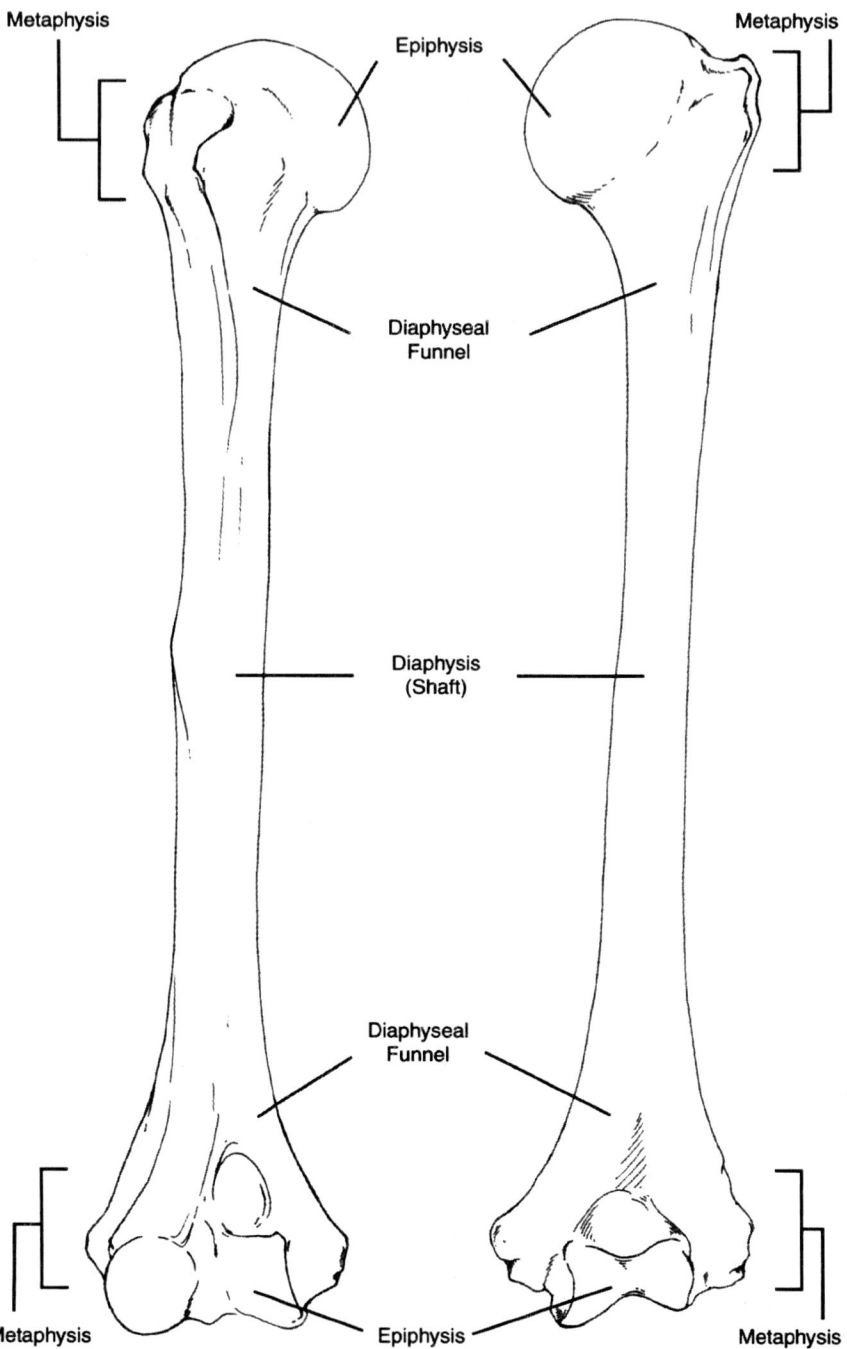

Figure 2-3 The morphology of a typical long bone.

The tubular shaft of long bones is the diaphysis. The broadened ends of the diaphysis adjacent to the metaphyses and epiphyses is referred to as the *diaphyseal funnel*. The *epiphyses* (sing., epiphysis) are the expanded proximal and distal ends of a long bone which have played a primary role in its development from the original anlage. Often there are multiple growth centers at one or both ends of long bones. Sometimes a distinction is made between those growth centers that take part in joint formation (*epiphyses*) and those which serve primarily as sites of muscle insertion (*apophyses* or "traction epiphyses"). The *metaphyses* are the transitional zone between the tubular shaft of the diaphysis and the expanded epiphyses and apophyses. In many long bones, an epiphysis is of considerably greater diameter than its diaphysis, and the metaphysis can be identified as the flared zone of the bone that joins the epiphysis with the diaphysis. The surfaces of the bones that take part in a synovial articulation are covered by *hyaline articular cartilage*. Joints between long bones, or between a long bone and another bone, such as the scapula, typically involve the epiphyses of these bones. The hyaline cartilage creates a very smooth surface, which is lubricated by the synovial fluid. This allows the joint to move smoothly and freely.

Except for the joint surfaces, the entire external surface of a bone is covered by a fibrous membrane, the *periosteum* (literally, "surrounding the bone"). The periosteum consists of two layers: an inner osteogenic layer, and an outer fibrous layer. Each has unique properties and functions.

The *osteogenic layer* is in contact with the external bone surface. It is referred to as the osteogenic layer because it is composed of *bone lining cells*, which derive from osteoprogenitor cells. These cells of the periosteum under some circumstances differentiate into *osteoblasts*, the cell type which forms bone matrix (Currey, 2002). The production of bone matrix deep to the osteogenic layer of the periosteum is referred to as *subperiosteal apposition*. It is through this subperiosteal apposition that bones grow in diameter. Any time the periosteum is pulled away from the surface of a bone, subperiosteal apposition will occur. It is in part the tearing away of the periosteum from a bone shaft during a fracture that triggers bone deposition and healing to occur.

Similarly, if an infection under the periosteum occurs (*periostitis*), the periosteum can be lifted away from the bone and bone will be deposited as healing occurs. This is referred to as a periosteal scab. This is frequently seen, particularly in archeological bone, on the tibias and other areas where the skin lies just above the bone.

Advanced infection of bones, which includes infection in the marrow cavity as well as beneath the periosteum, is known as *osteomyelitis*. Severe cases of osteomyelitis involve formation of massive amounts of bone callus beneath both the periosteum and the endosteum. The tibia is a frequent site of osteomyelitis partly because it lies so close to the skin and is frequently injured. Severe cases of osteomyelitis are very rare in the present day due to effective antibiotic treatment.

The outer layer of the periosteum is the *fibrous layer*. It consists of dense irregular connective tissue. Into the fibrous layer of the periosteum insert tendons of muscles. These fibers, which also extend into the underlying bone matrix, are known as *Sharpey's fibers*.

Within the diaphysis of a long bone, and extending into the epiphyses of mature bones, is the medullary cavity, or "marrow cavity." Many organs are described as having an outer cortex and an inner medulla. In the case of bones, the outer compact bone is the cortex, often described as *cortical bone*, while the *medulla* of the bone is comprised of the marrow. As a result, the cavity inside the diaphysis is referred to as the *medullary cavity* since it contains the bone's medulla.

The endosteum of a bone is a membrane very similar in composition to the osteogenic layer of the periosteum. The endosteum, however, lines the internal surface of the cortical bone, separating it from the bone marrow within the medullary cavity. Like the osteogenic layer of the periosteum, the bone lining cells of the endosteum help supply *osteoblasts* (bone forming cells) for the repair of fractures. The endosteum also serves as a site which provides osteoblasts for the formation of new *Haversian systems* (or *secondary osteons*) in the bone remodeling process (described later). Unlike the periosteum, the endosteum normally has no fibrous layer, or a poorly developed fibrous layer.

All bones will have one or more nutrient arteries, which are the principal blood supply to the bone. The principal nutrient artery invades the anlage of the bone early in the ossification process, and becomes surrounded by the forming bone. The passage left in the newly formed bone for the artery is termed the nutrient foramen. There is usually one nutrient foramen in the diaphysis of a long bone for the *primary center of ossification*, and one or more in each epiphysis for the *secondary centers of ossification* located there. Veins will accompany the nutrient arteries.

Microscopic Structure of Bone

Woven Bone

Woven bone consists of bone in which the collagen fibers of the bone matrix are arranged in a haphazard manner without a preferred orientation. It is found only in specific circumstances, such as the bone callus formed during fracture repair or in the centers of ossification in forming bones. Woven bone is also known as *primitive* or *immature* bone. Unlike trabecular and compact bone, woven bone is not lamellar. Whereas in lamellar bone the collagen fibers of the bone matrix are highly organized, in woven bone the collagen fibers have no specific fiber orientation.

A second type of immature bone is termed *"coarsely bundled bone"* (Cormack, 2001), or *parallel-fibered bone* (Enlow, 1969) in which thick collagen bundles are arranged parallel with one another and *osteocytes* ("resting" bone cells) occur between the collagen bundles.

Woven bone is known as immature bone as it is the first type of bone laid down in the cartilage or mesenchymal *bone model* (*anlagen*) when the process of *ossification* (or bone formation) begins. Woven bone is also the type of bone laid down in the *bone callus* that forms around a healing fracture. Ultimately, nearly all woven bone is remodeled away and replaced by lamellar bone. Small amounts of woven bone may remain, however, near cranial sutures, tooth sockets, tendon and ligament attachments, and in the bony labyrinth of the ear. Usually, however, in these locations the woven bone is present in patches surrounded by lamellar bone.

Lamellar Bone

Lamellar ("layered") *bone* is mature bone and takes the form of either compact bone or trabecular bone. *Compact bone* is the dense bone that makes up the cortex of the shaft of the diaphysis (*cortical bone*) and the thin layer of bone beneath articular cartilage (*subchondral bone*). *Trabecular bone* is generally found at the ends of long bones supporting the subchondral articular surface, in the bodies of the vertebrae, or between the bony plates forming the exterior of flat bones.

Trabecular Bone

Trabecular, or *cancellous*, bone is also known as spongy bone because it appears to be made up of a series of intersecting small pillars or beams (Latin: *trabeculae* means little beams) with a great deal of space between them, similar in appearance to a sponge. In contrast to compact bone, in which osseous tissue predominates over empty space per volume of bone, the situation is reversed in spongy bone.

Trabecular and compact bone are histologically very similar. Trabecular bone is very common under the thin shell of compact bone that underlies articular cartilage (subchondral bone). Although the shaft of a long bone is often subjected to bending and torsion, its articular ends are subjected almost exclusively to compression, allowing the subchondral bone to be supported by trabecular bone without a significantly increased probability of failure.

At the same time, the elastic modulus of cancellous bone is significantly lower than that of compact bone. This means that the cancellous bone is more easily deformed or has more "give." Hyaline cartilage cannot be repaired; once it is damaged, it loses some of its capacity to ensure friction-free motion within the joint. Therefore it is very important that it be protected from damage during the imposition of very high forces. The greater elasticity of cancellous bone therefore serves the very important function of preventing its overlying cartilage from being excessively compacted during the application of high compressive loads.

Trabeculae are generally so arranged that they maximize their capacity to resist the loads typically imposed upon the joint (Currey, 2002). This feature is largely the result of the sensitivity of the growth plate by which trabeculae are formed to the forces to which the growth plate is subjected during development. As noted, the articular surfaces at the ends of long bones are supported almost entirely by trabecular networks. Another example is the bodies of vertebrae, which are essentially thin-walled "drums" of bone filled with trabeculae.

Circumferential lamellar bone

Circumferential lamellar bone is a form of mature, compact bone that contains primary osteons (Figure 2-4). Both this type of bone and haversian bone have matrices that are composed of lamellae in which collagen fibers have a specific orientation. The principal differences between circumferential lamellar bone and haversian bone are the organization of their lamellae and their developmental origins. Circumferential lamellar bone is a form of mature bone in which the lamellae extend around the circumference of the shaft of a bone and lie in layers deep to its periosteum. In each individual lamella, most of the collagen fibers are organized parallel to one another. *Lacunae* (singular: *lacuna*), the small spaces in the bone matrix occupied by osteocytes, or "resting" bone cells, lie in the plane between adjacent lamellae.

Circumferential lamellar bone forms by the process of *subperiosteal apposition (deposition)*.

Throughout the period of growth, the bone grows in diameter by the addition of lamellae around the shaft. Each lamella is in some ways analogous with the growth rings of a tree. The analogy should not be carried too far: You cannot determine skeletal age simply by counting circumferential lamellae.

Osteoprogenitor cells in the osteogenic layer of the periosteum differentiate into osteoblasts, which then begin laying down bone matrix beneath the periosteum. Osteoblasts that become covered over by bone matrix which is formed by faster moving osteoblasts become encased in lacunae and become osteocytes. Again, these osteocytic lacunae come to lie in the plane between adjacent lamellae.

Beneath the periosteum is a rich network of small arteries. As new lamellae are laid down, some of these arteries become covered over by the advancing bone and hence come to lie in small osseous tunnels. A few concentric lamellae may be laid down around the blood vessel within these tunnels. The bony tunnels encasing these blood vessels are termed *primary osteons*, *primitive osteons* or "nonhaversian canals". These are termed *primary osteons* because they form as the bone itself is forming, as distinguished from *secondary osteons* (or *Haversian systems*) that form in already existing bone. The arteries within these primary osteons help supply blood to the osteocytes within their lacunae.

Since bone is a highly rigid material impermeable to osmosis and diffusion, the osteocytes must be able to receive oxygen and nutrients and dispose of waste products. Each osteocyte has a great number of hair like cellular processes that come into contact with other osteocytes. These cellular processes pass through minute channels in the bone matrix called *canaliculi* (Latin: "little channels"). The canaliculi and cellular processes of the osteocytes nearest a blood vessel communicate with it, and other osteocytes in turn communicate with that osteocyte and other osteocytes. As a result, all osteocytes are connected either with another osteocyte or with a blood vessel, and nutrients can reach the cells and waste products can be eliminated.

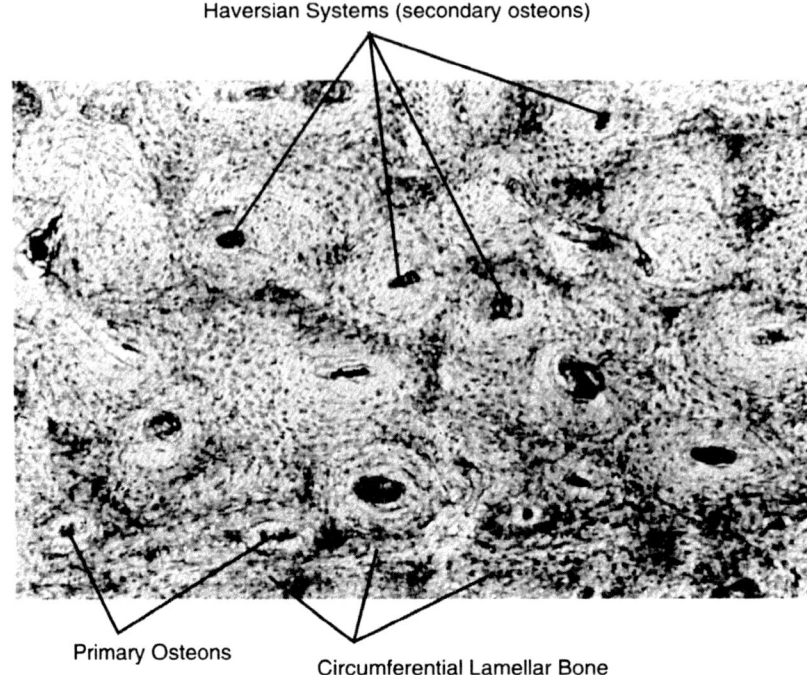

Figure 2-4 Photomicrograph of circumferential lamellar bone.

The orientation of *collagen fibers* in the lamellae of mature bone have important consequences for the mechanical properties of the bone. The mineral deposited in the bone matrix and around the collagen fibers gives bone its *compressive strength* (the ability to bear weight, or resist compression). The collagen fibers provide bone matrix with its *tensile strength* (ability to resist tension, or being pulled upon).

Collagen is a tough structural protein found throughout the skeleton and throughout connective tissue in general. There are many different subtypes, depending upon the tissue in which the collagen is located and its stage of development, but all have in common a high tensile strength. Within a typical long bone, many of its collagen fibers will lie parallel with the long axis of the bone shaft, and thus provide the bone with great tensile strength. Under a muscle, tendon, or ligamentous attachment, the collagen fibers in the bone are usually oriented along the directions in which those structures apply tension to the bone. (As noted before, the specialized collagen fibers that connect these structures to bone are called *Sharpey's fibers*.) While the overall arrangement of fibers gives overall tensile strength to the bone, the arrangement of fibers in the individual lamellae of circumferential lamellar bone, in combination with the hydroxyapatite crystals, also provide the bone with torsional strength.

Lamellar bone thus has design features to resist all the principal forces to which it is normally subjected: compression, tension, and torsion. When the design parameters of the bone are exceeded, the result is a fracture. Because of its random arrangement of collagen fibers, woven bone has less resistance to tension and torsion than does lamellar bone. (We will forgo a lengthy discussion of the mechanical properties of bone here. See Currey, 1984; Currey, 2002 and references cited therein for detailed discussions of the mechanical properties of bone.)

Haversian Bone

The basic unit of Haversian bone is the *secondary osteon* or *Haversian system* (Figures 2-5 and 2-6). Haversian bone is secondary lamellar bone. Circumferential lamellar bone forms as a bone grows in diameter. However, as one ages, microfractures

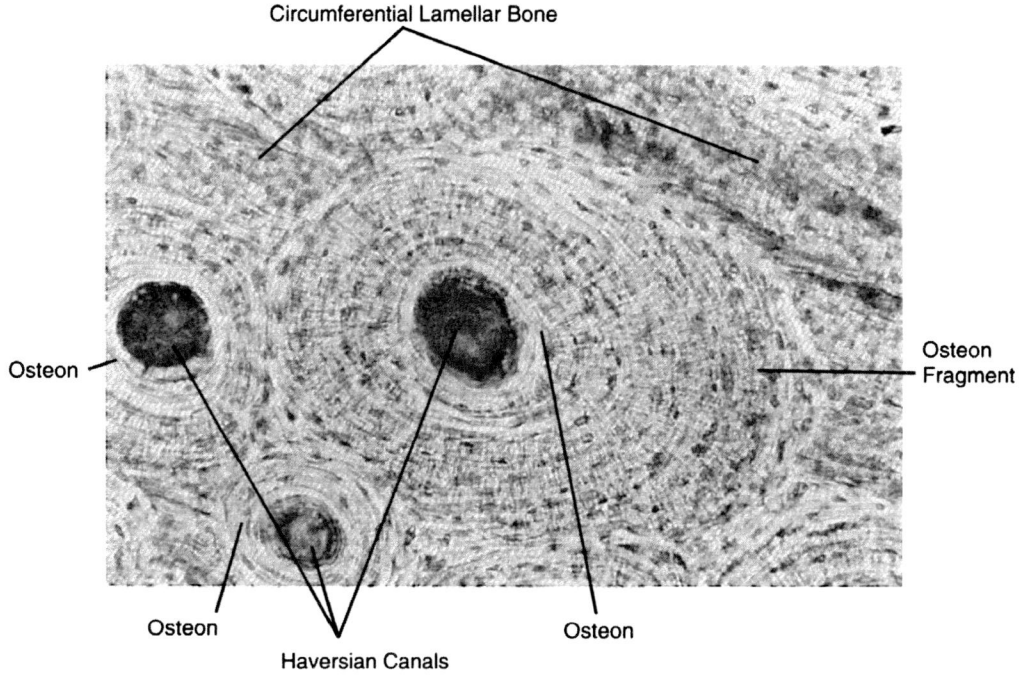

Figure 2-5 Photomicrograph of haversian bone.

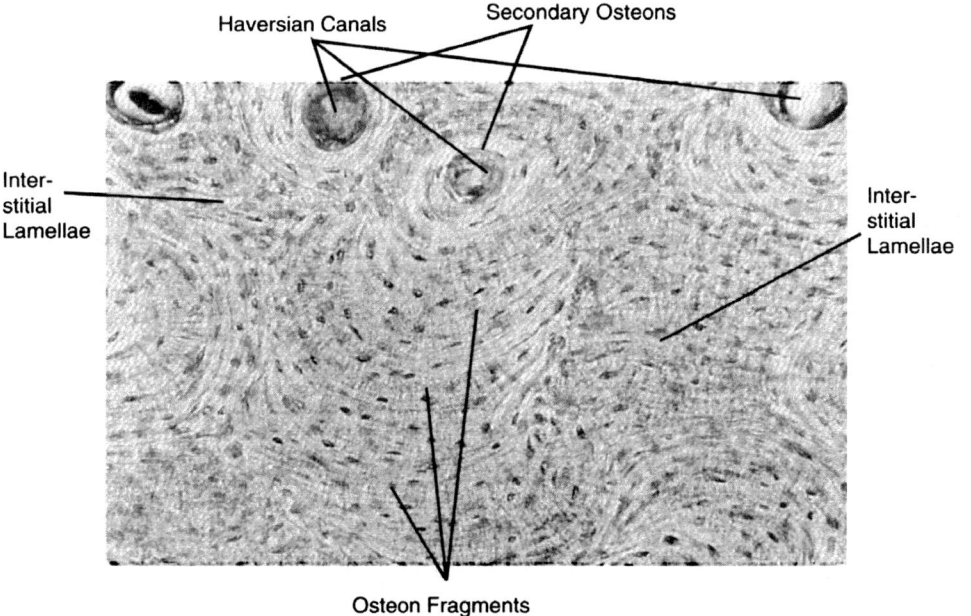

Figure 2-6 Photomicrograph of haversian bone, osteon fragments.

and other types of damage occur within that circumferential lamellar bone. If these microfractures and other insults to the integrity of lamellar bone were allowed to go unchecked, the bone would ultimately fail. As a result, such insults to lamellar bone must be repaired. The body responds by the formation of *secondary osteons*. They are referred to as *secondary osteons* because they form within bone that already exists, as opposed to *primary osteons* that are formed at the same time as the lamellar bone within which they exist (Figure 2-4). In essence, damaged bone is removed, stopping the spread of the crack from a microfracture, and the resulting cavity is filled by the formation of a secondary osteon. This is a widely accepted explanation for the formation and presence of secondary osteons (Burr, 1993; Currey, 2002).

A secondary osteon forms within a *resorption space* (Figure 2-7) carved out within the compact bone by *osteoclasts* (cells specialized for the destruction of bone matrix). Osteoclasts resorb the bone around the site of microfractures and other damage. In long bones, this forms a long narrow channel that generally follows the long axis of the diaphysis of the bone. Resorption spaces in cross section demonstrate scalloped edges. The scallops are known as *Howship's lacunae* (Figure 2-8) and represent the spaces occupied by the osteoclasts as they resorbed away the bone.

Obviously, if this process were allowed to continue unchecked, the remaining bone matrix would start to resemble Swiss cheese and the bone would become very much weakened. This, in fact, is precisely what happens in *osteoporosis*, in which the bone mass is substantially reduced and bones become easily subject to fracture. To prevent the bone matrix from becoming extremely weak, the resorption space formed at the site of a microfracture or other insult is eventually filled in with new lamellar bone formed by osteoblasts.

We have already noted that lamellar bone contains a rich blood supply, initially in the form of the arteries within *primary osteons* and the *subperiosteal vessels*, as well as branches of the *nutrient arteries*. Following the formation of the resorption space, blood vessels grow into the space, and bring with them osteoblasts. These osteoblasts then begin the process of forming lamellar bone inside the resorption space, filling it from the *outside inwards*. The lamellar bone that forms inside the resorption space eventually forms a *secondary osteon*

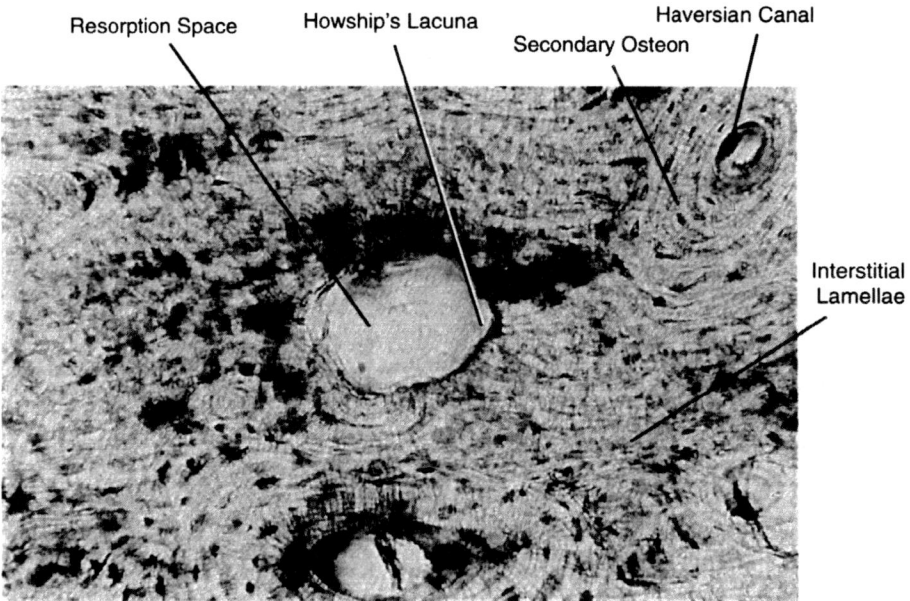

Figure 2-7 Photomicrograph of resorption space, osteons, and interstitial lamellae.

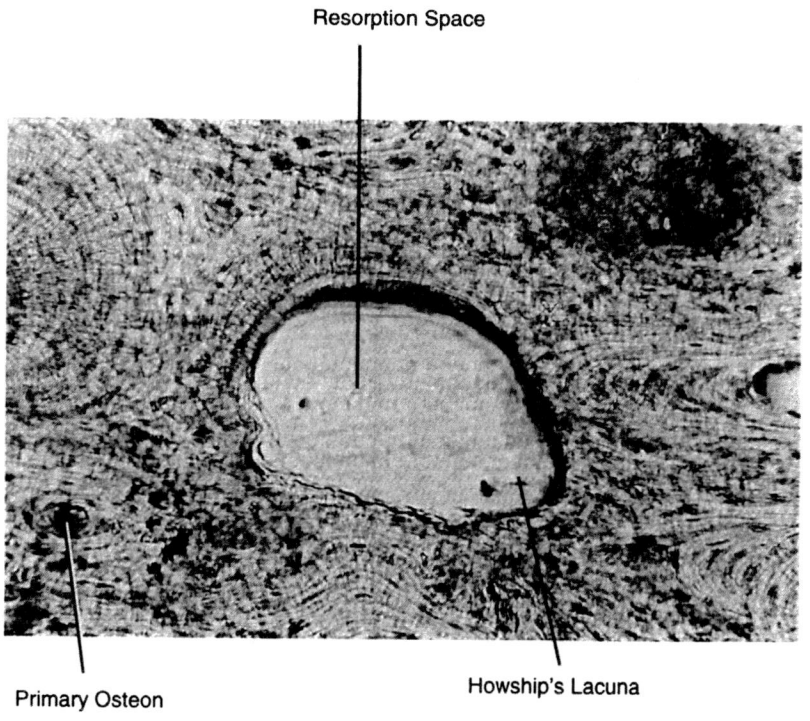

Figure 2-8 Photomicrograph of resorption space demonstrating Howship's lacunae.

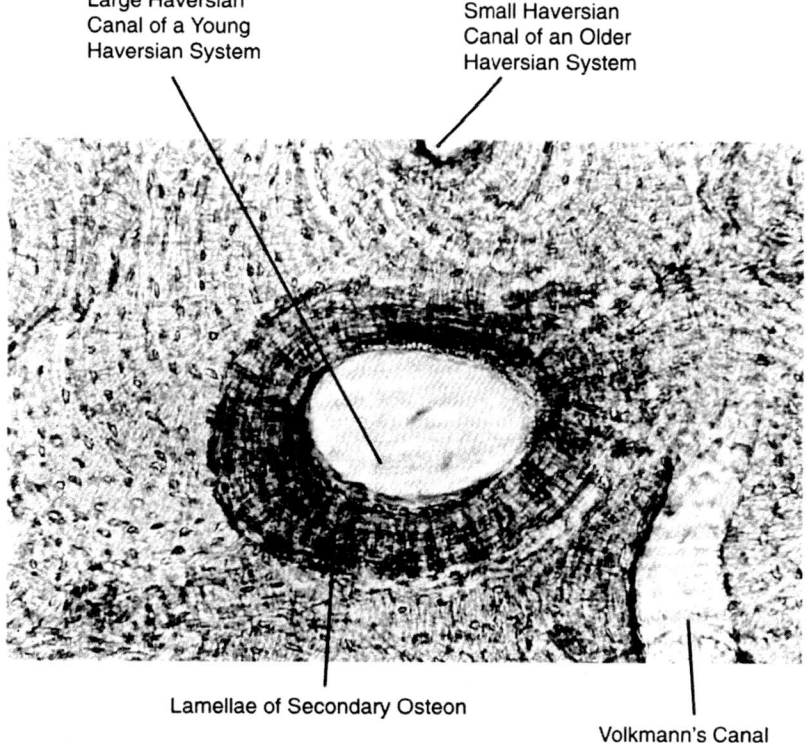

Figure 2-9 Photomicrograph of secondary osteon or haversian system.

or *haversian system* (Figure 2-9). The space that remains in the center contains an arteriole and a venule, or sometimes only a single capillary. There are no lymph vessels within Haversian systems; only periosteal tissue is supplied with lymphatics.

On average, human Haversian systems are approximately 300 microns (0.3 mm) in diameter, though their size can vary a great deal with age and location within the skeleton (Walker, 1989, 1993, 1998, Walker and Lovejoy, 1999; Walker et al., 1994).

With increased skeletal remodeling with age almost all circumferential lamellar bone will be converted to Haversian bone. While this slightly weakens the bone, it greatly increases its vascularity. In Haversian bone, there is, on average, an arteriole and venule every 0.3 mm in each Haversian canal. This increased vascularity of Haversian bone allows for a very rapid and efficient exchange of minerals, particularly calcium and phosphorous, between blood and bone. Individual Haversian systems also probably function to prevent the spread of cracks within the bone (Currey, 1984; 2002).

As circumferential lamellar bone is remodeled away and replaced by Haversian bone, some patches or islands of circumferential lamellar bone may remain between Haversian systems. Such isolated patches of circumferential lamellae are referred to as *interstitial lamellae* (see Figure 2-6). With increasing age and increasing amounts of remodeling, eventually older Haversian systems begin to be remodeled away and replaced by newer Haversian systems. Remnants of older Haversian systems are also a form of interstitial lamellae (see Figures 2-6 and 2-7). When it is apparent that these patches are remnants of partially eroded older Haversian systems, they are known as *fragmentary osteons* or *osteon fragments* (see Figure 2-6). The numbers of complete and fragmentary Haversian systems have a weak correlation with age at death, but appear to be more highly correlated with mechanical factors (Walker et al., 1994 and references therein).

Bone *cortex* or *compact bone* is formed by the dense bone of the shaft of the diaphysis. *Compact bone* is not limited to just the cortex of long bones, but it is most abundant here. The greatest mass of bone in the adult skeleton is compact bone. It can be either in the form of haversian bone or circumferential lamellar bone. Specific structures found within Haversian systems include the those described in the following paragraphs.

LAMELLAE: As with circumferential lamellar bone, the bone formed inside the Haversian system is lamellar (Figure 2-10; see also Figure 2-9), and is very similar in its organization to circumferential lamellar bone. However, they exist at different levels of magnitude. Whereas circumferential lamellae may surround the entire shaft of the femur, for example, the lamellae within a Haversian system will generally have a total circumference of far less than a millimeter. As with circumferential lamellae, the collagen fibers in any given lamella usually parallel to one another.

LACUNAE: As in circumferential lamellar bone, osteocytes reside in osteocytic lacunae (Figure 2-11; see also Figure 2-10) within secondary osteons. The lacunae are located at the interface between individual lamellae that make up the secondary osteon. Again, these represent osteoblasts that became entrapped within bone matrix by more rapidly advancing osteoblasts.

CANALICULI: Canaliculi (see Figures 2-10 and 2-11) join adjacent osteocytes with one another, and join the innermost osteocytes with the blood vessels in the Haversian canal. As with canaliculi in circumferential lamellar bone, these canaliculi contain cellular processes of the osteocytes which serve the function of passing oxygen and nutrients to, and draining metabolic wastes away from, osteocytes. In cross section, a secondary osteon somewhat resembles a spider web, with the Haversian canal at the center, surrounded by concentric rings formed by the planes between lamellae, with canaliculi radiating out from the central Haversian canal toward the outside perimeter of the osteon. Osteocytic lacunae are located along the planes of intersection between adjacent lamellae, and resemble flies caught in the spider's web.

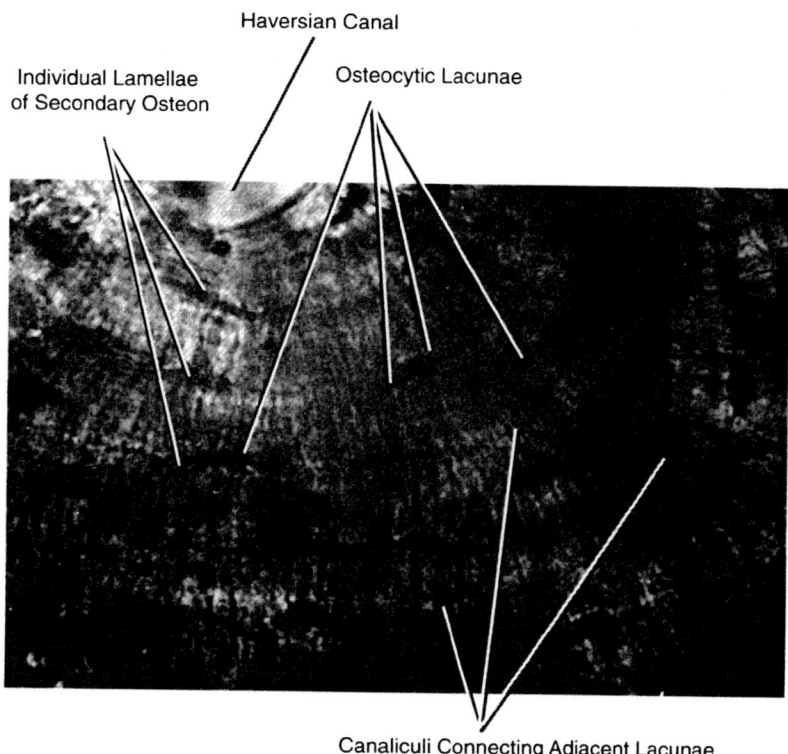

Figure 2-10 Photomicrograph of lamellae, lacunae, and canaliculi within a haversian system.

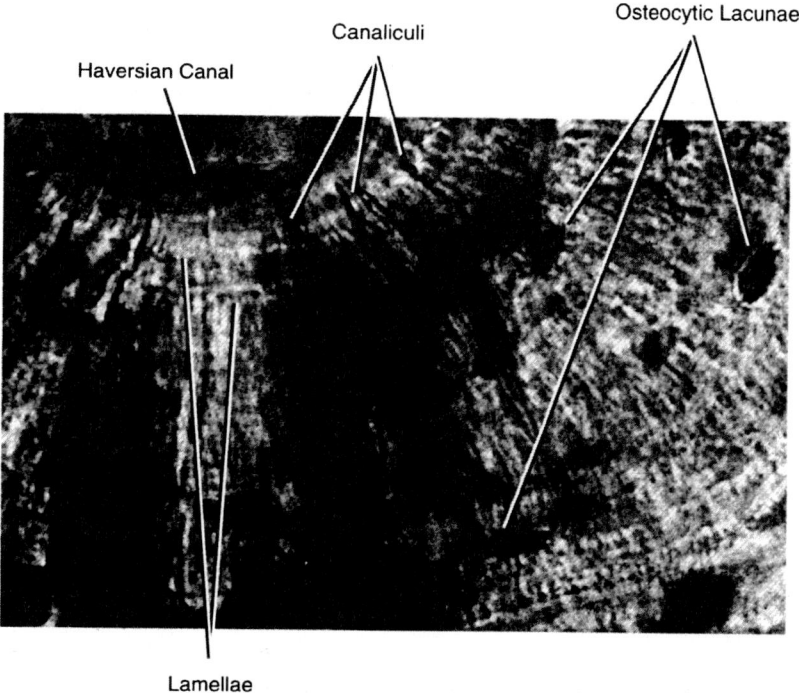

Figure 2-11 Photomicrograph of lamellae, lacunae, and canaliculi within a haversian system.

HAVERSIAN CANALS: As has been noted, the Haversian canals (see Figures 2-9 through 2-11) occupy the central space of the Haversian system and usually contain an arteriole and a venule.

VOLKMANN'S CANALS (see Figure 2-9): are similar to Haversian canals, but are oriented perpendicular to them. Volkmann's canals serve to transmit blood vessels and nerve fibers between adjacent Haversian canals, and serve to connect the vessels within the outermost Haversian canals to the subperiosteal blood vessels, and connect the innermost Haversian canals with endosteal blood vessels and branches of the nutrient arteries. Unlike Haversian canals, Volkmann's canals are not lined by concentric lamellae, and are oriented perpendicular to the long axis of the diaphysis rather than parallel to it as with Haversian canals.

Skeletal Remodeling

Skeletal modeling and remodeling are the processes by which bone undergoes changes in shape and size with growth and development during normal aging and in the course of disease processes. The formation of bone from the development of the anlage through ossification of the model is known as *modeling*. *Remodeling* is changes that occur secondary to that. For example, the development of primary osteons and circumferential lamellar bone are a result of the *modeling* process. The development of Haversian bone, by contrast, is the result of the *remodeling* process.

Both modeling and remodeling occur through the action of bone cells. As has already been alluded to, there are three types of bone cells. These are the *osteoblasts*, the *osteoclasts*, and the *osteocytes*. Each has specific role in the metabolism of bone. As we are more concerned with bone morphology than bone cell biology in this text, we will give here only a very cursory description of bone cell types. See any recent histology text or *Gray's Anatomy* for more detail regarding the histology and cell biology of the osteoblast, osteoclast, and osteocyte.

OSTEOBLASTS *(create bone)*: Osteoblasts are mononuclear cells that are found as a monolayer of cells on the surfaces of growing or remodeling bone. Osteoblasts arise from *osteoprogenitor cells* that are in turn derived from pluripotential stromal stem cells found throughout the bone marrow and in other connective tissues. As has already been noted, these osteoprogenitor cells are also found beneath the periosteum and endosteum. *Osteoblasts* synthesize, deposit, and mineralize bone matrix (Rodan and Rodan, 1984; Simmons and Grynpas, 1990).

OSTEOCYTES *("resting" bone cells in lacunae)*: Osteocytes are the principal type of cell in mature bone matrix. They derive from osteoblasts that have ceased or reduced their production of bone matrix and have become enclosed in bone matrix formed by other osteoblasts. They retain connections with each other and with blood vessels under the periosteum and endosteum through the network of canaliculi described earlier. The exact functions of osteocytes are not entirely clear, but it is assumed that they have an important role in the maintenance of bone. The deaths of the osteocytes within an area of bone matrix leads to the resorption of that matrix by osteoclasts.

Osteocytes probably are involved in the repair of fractured trabeculae by splinting them with microscopic amounts of bone callus. They may also act as sensors of the local mechanical and chemical state of bone, and initiate the resorption or deposition of bone matrix in the locality (Lanyon, 1993a, 1993b). They probably also play a role in mineral exchange with the bone at the level of the canaliculi and lacunae. Their activities are dependent upon a large number of factors, many of which are currently unknown, but which certainly include circulating levels of hormones such as calcitonin and parathyroid hormone.

OSTEOCLASTS *(remove bone matrix)*: Osteoclasts are large, polymorphous, polynucleated cells (often containing 15-20 or more nuclei). They are found on surfaces of bone undergoing active resorption, and reside in pits on the bone surface called *Howship's Lacunae* or *resorption bays*. Osteoclasts are responsible for the removal of bone matrix, but the exact mechanism by which they do so is not known. It is likely a combination of demineralization of the bone matrix and phagocytosis of the remaining matrix. Unlike osteoblasts and osteocytes, osteoclasts do not arise from osteoprogenitor cells, but from the fusion of mononuclear cells that originate in the bone marrow or other hematopoietic tissue. They show some similarity and relationship to the macrophages of the blood, which also arise in hematopoietic tissue. Osteoclasts respond to changes in parathyroid hormone and Vitamin D levels, as well as factors released by osteoblasts and other cell types. It has been speculated that when osteoclasts have completed their bone resorbing activities, they may dissociate back into mononuclear precursor cells.

BONE LINING CELLS: There is also a class of *bone lining cells*. These are flattened, epithelium like cells that are particularly abundant in adults. They are found lining surfaces of bone that are undergoing neither deposition nor resorption (Miller and Jee, 1987; 1992). Currey (2002) considers the bones of the osteogenic layers of the periosteum and endosteum to be bone lining cells. They form a continuous layer overlying the resting bone surface, and are in contact with adjacent osteocytes. They line the internal surface of the marrow cavity and line the walls of Haversian canals. These cells may be either once active osteoblasts that can potentially be reactivated, or may be a form of osteoprogenitor cells.

Bones grow and change their shape by two processes, which are mediated by the cell types just discussed. These two processes are *deposition* and *resorption*. The result of the combined action of these two processes is *skeletal modeling and remodeling*.

DEPOSITION *(appositional growth):* Deposition is the process whereby osteoblasts lay down new bone matrix on some preexisting surface. In mature bone, this is bone laid down on existing bone surfaces. In a growing bone, osteoblasts may deposit bone on calcified hyaline cartilage (during *endochondral ossification*) or in mesenchyme (during *intramembranous ossification*). This is referred to as *appositional growth*. This is in contrast to *interstitial growth* in which growth occurs from within the tissue.

In nonmineralized connective tissues, such as cartilage, the cells may produce new extracellular matrix that causes the matrix existing around the lacuna containing the cell to stretch and grow in volume. Because the extracellular matrix of cartilage and other connective tissues is usually unmineralized, this process can occur, and the matrix grows in volume essentially from the inside out. This is *interstitial growth*. Bone, on the other hand, has a highly mineralized matrix, preventing interstitial growth. Bone can therefore grow in mass only by the deposition of bone on preexisting surfaces (*appositional growth*).

RESORPTION *(absorption of old bone matrix):* Coupled with the process of bone deposition is *bone resorption*. As noted previously, bone resorption is carried out by specialized cells called *osteoclasts*. If, during growth and development, deposition were unaccompanied by resorption, a long bone would be as solid as a tree trunk, with no medullary cavity present. However, as a bone shaft expands in diameter by subperiosteal deposition, bone is at the same time removed from the endosteal surface of the compact bone, and the medullary cavity grows in conjunction with the growth in the diaphyseal diameter of the bone. In general, throughout most of life, deposition and resorption of bone remain coupled. When they become uncoupled, I.e. deposition outstrips resorption (*osteopetrosis*), or resorption outstrips deposition (*osteoporosis*), bone metabolic diseases are indicated.

The coupling of bone deposition and bone resorption throughout life results in *bone remodeling*. While a young bone is undergoing growth in length and diameter, subtle changes in shape are occurring as well. As the diaphysis grows in length, the epiphyses are pushed away from the center of the shaft of the bone. As a result, the diaphysis, particularly in the metaphyseal region, undergoes changes in shape while the bone grows in length.

Laws of bone growth and deposition

Bone growth can be described as following some fairly regular patterns. These patterns have been described as laws or principles. These laws are in reality more descriptions than proscriptions, but they are nonetheless so prevalent in discussions of bone tissue that the reader must be familiar with their general definitions and meaning. Two of the most important are Wolff's Law and the Heuter Volkmann Principle.

Wolff's Law

Wolff's law (Wolff, 1869; 1892), also referred to more grandiosely as the law of bone transformation is an old tenet of the study of bone growth that essentially states that *bone is laid down where needed to resist compressive and tensile forces and is removed where not needed in order to achieve the greatest economy of tissue.*

While this is more or less still a commonly held view by some researchers, more recent evidence suggests that a great deal of bone form is genetically determined and is not the result of local environmental factors and mechanical loading. Further, bone does not always remodel in a way that fits precisely with this "law" (Betram and Swartz, 1991; Swartz et al., 1998. See also Gefen and Seliktar, 2004; Pearson and Lieberman, 2004; Rubin et al., 2002; and Skedros and Baucom, 2007;). Bone is certainly known to be responsive to mechanical stimuli, and in the absence of loading, bone tissue is rapidly removed by altered bone metabolism. However, it is now clear that the "signal" that causes osteoblasts to synthesize new matrix or its removal by osteoclasts is highly complex and is not simply proportional to the loads to which the bone is subjected.

Strain studies of bone, in which gauges are placed upon the bone surface and the deformation of the bone is measured at different points along its surface show that bone metabolic activity and loading do not bear a simple relationship. In fact, it is not yet known whether the bone "signal" is mechanical (such as shear forces acting over cell surfaces or deformation of their bodies), hydrodynamic, electrical, or some complex combination of these factors.

The Heuter Volkmann Principle

The *Heuter Volkmann principle* states that *there is a direct relationship between static compressive forces parallel to the longitudinal axis of a long bone and the rate of growth of the epiphyseal cartilage.*

It has long been known from a variety of experiments in which bone shapes are intentionally altered (such as the removal of a triangular segment of its shaft) that differential growth within the growth plates of epiphyses and apophyses can result in a change in the bone's eventual shape. The angle in the human knee joint (usually referred to as "valgus") which causes the femoral shaft to angle superolaterally from the joint as it approaches the pelvis is probably the specialized result of habitual loading during bipedal walking, since this angle increases with age and does not develop normally in children with severe developmental insults such as the various forms of myodysplasia.

Simon and colleagues (Simon, 1983; Simon et al., 1984) demonstrated that increased, intermittent, compressive forces enhance limb bone growth in rats, initially using experimental bipedalism (and later chronic centrifugation) to simulate increased weight and compression. Using experimental bipedalism, growth plates that would normally have been closed or almost closed were open and with active columns of *chondrocytes* (cartilage cells). Low levels of chronic centrifugation significantly increased limb bone length. This research suggests that compressive forces may be important factors in regulating the growth of epiphyseal cartilage, and even long bone lengths, though a great deal of work remains to be done before we will have a more complete understanding of how mechanical signals are perceived and responded to by the cells of the growth plate.

migrate from the somites into the forming bud and become organized into dorsal and ventral muscle blocks. Shortly thereafter, each muscle block begins to split into smaller and smaller segments (which are, of course, at the same time rapidly expanding by growth). These eventually give rise to all the muscles of each limb. Shortly after this process begins, neurites from the developing spinal cord spread into the limb to innervate each of its muscles.

Mesenchyme is a diffuse network of cells forming embryonic mesoderm from which all connective tissues, including bone, develop. The anlagen of bones initially form as *mesenchymal condensations*, which may undergo direct, or *intramembranous* ossification, or may be transformed into cartilage, which then undergoes *endochondral* ossification. These mesenchymal cells have the ability to differentiate into other cell types, including *fibroblasts* that produce collagen and other fibers, *chondroblasts* that produce cartilage matrix, and *osteoblasts* that produce bone matrix. Some of the mesenchyme of the head region is derived from *neuroectoderm,* or *neural crest cells,* that give rise to cranial bones of *branchial arch* origin. The ossification of the cranium is discussed in Chapter 8.

Shortly after the anlagen of the bones of the limbs begin to form, they become progressively separated from one another at points where additional special signaling molecules are expressed. These are the presumptive *joints* (*articulations*, or *arthroses*) of the skeleton. *Synovial joints* differentiate from the *interzonal mesenchyme* between the anlagen of developing bones. Peripherally the mesenchymal cells become transformed into the fibrous joint capsule and the ligaments of the joint, while within, the cells disappear to form the joint cavity *(cavitation)*. Some mesenchymal cells give rise to the synovial membranes that line the internal aspects of the joint capsules of synovial joints. In *cartilaginous joints*, the interzonal mesenchyme differentiates into hyaline cartilage or fibrocartilage, but no joint cavity develops. In either case, the ends of the bones participating in the joint remain covered by hyaline cartilage. In the case of *fibrous joints* (also discussed in detail below), the interzonal mesenchyme differentiates into a dense fibrous connective tissue, forming ligaments that bind the bones together.

Two types of ossification

Ossification is the process whereby bone tissue is formed. This includes the formation of bone matrix, and in the case of lamellar bone, the highly organized network of collagen fibers in the bone matrix. Ossification also includes the deposition of calcium salts (calcification) in the form of hydroxyapatite in the matrix. Unlike calcification, ossification occurs only within the skeleton in normal circumstances. There are two types of ossification, differentiated by the type of tissue in which the bone is formed: *endochondral (intracartilaginous) ossification* (bone forms in a hyaline *cartilage anlage*), and *intramembranous (mesenchymal; dermal) ossification* (bone forms in a *mesenchymal membrane*).

Endochondral Ossification occurs throughout the postcranial skeleton and is therefore the more common or more typical type of ossification. We first endochondral ossification in detail first, and then examine intramembranous ossification. Intramembranous ossification occurs only in the sides and top of the skull and in the clavicles.

Endochondral Ossification

We first discuss the pattern of endochondral ossification as exemplified in a typical long bone with an epiphysis located at either extremity of the diaphysis (Figure 2-12). The pattern of endochondral ossification will be similar in other, more irregularly shaped bones such as vertebrae or the hip bones. Once the pattern is understood in a typical long bone, such as the tibia, it is much more readily understood in irregular bones.

Cartilage Model or Anlage

Endochondral ossification begins with the presence of a hyaline cartilage *anlage* (model). Such models

Bone Formation and Growth

Formation of bone is termed *ossification*. Ossification differs profoundly from simple *calcification*. Calcification is simply the deposition of calcium salts in a tissue. This occurs as one part of *ossification*, a term which implies actual bone formation, and includes the orderly arrangement of collagen fibers in the organic bone matrix in lamellae, as has already been discussed. Some tissues undergo calcification without ossification, in that calcium salts are simply deposited in the hyaline cartilage matrix of the thyroid cartilage, for example, without the presence of lamellar structures and the other features common to true bone tissue.

Basic Embryology of the Skeletal and Articular Systems

The skeleton and the articulations derive from *mesoderm*, or embryonic connective tissue. With the formation of the notochord and neural tube the mesoderm lateral to these two median embryonic structures thickens into two columns of *paraxial mesoderm* (For further details of embryology see Moore and Persaud, 2002; Sadler, 2006; Larson, 1993 or a similar general embryology text.). At about 20 days *in utero*, this paraxial mesoderm begins to organize itself into 42 to 44 segments.

Initially the segmented paraxial mesoderm is in the form of loose, whirl-like groups of cells termed somitomeres. With the exception of the first seven somitomeres that form in the cranial region, the somitomeres will further organize themselves into compact aggregates of cells termed somites. Externally somites appear as a series of pairs of beadlike elevations along the dorsal surface of the embryo. Each somite in turn becomes divided into a ventromedial part called the *sclerotome*, and a dorsolateral part called the *dermomyotome* (from which derive *myotomes* and *dermatomes*, which give rise in turn to most of the voluntary musculature and skin).

The *sclerotomal cells* soon surround the n tube and form the primordia, or anlagen, o vertebrae and ribs. The cells probably do not actu migrate, but because of differential growth of o structures, are shifted in position relative to the ne tube. The developing vertebral column replaces notochord (the stiffening dorsal rod of the emb characteristic of the animal phylum *Chordates*, which *vertebrates* belong) and takes on its supp function, while it also takes on the function of protecti the spinal cord.

The sclerotomal cells give rise to the vertebr column and its associated ligaments (see chapter 3) Each sclerotome divides, so that the superior half o one sclerotome unites with the inferior half of the adjacent sclerotome. For example, the inferior half of the T1 sclerotome unites with the superior half of the T2 sclerotome to form the T1 vertebrae. This results in the vertebrae being a half step off from the *meristic* (segmental) pattern of the body and thereby provides pathway for the spinal nerve and blood vessels so that they do not become encased within the bone.

Unlike somites, the cranial somitomeres are not perfectly segmental and do not cleanly divide into sclerotome, myotome and dermatome. The somitomeres are responsible, along with the top four "occipital" somites, for the formation of the neurocranium. This is done in a fashion very similar to the formation of the vertebrae, although they have to protect a portion of the central nervous system with a much larger diameter than the spinal cord: the brain.

Although the bones of the ribs and vertebral column develop from paraxial or somitic mesoderm, those of the limbs, pelvic girdle, and most of the pectoral girdle develop from *lateral plate mesoderm*. This tissue lines the lateral margins of the embryo and gives rise to paired limb buds. These buds emerge in response to a complex cascade of signaling molecules and are guided by a number of cell signaling regions which develop within each bud. All of the bones of the limbs and girdles develop in response to this complex cascade. Shortly after initiation of the limb bud, primitive muscle cells, or *promyoblasts*,

Bone: Anatomy, Growth, and Development

begin as vague condensations of mesenchyme. Soon thereafter cells around the periphery of the presumptive anlagen begin to express cell adhesion molecules and begin to form into a membrane which will eventually become the perichondrium. Cells in the interior of the presumptive anlagen continue to mitose and to synthesize collagen and other matrix molecules. Soon a hyaline cartilage anlage or model of the bone is thus formed.

Bone Collar Development

Soon after the appearance of the perichondrium that surrounds the anlage, it then differentiates into a

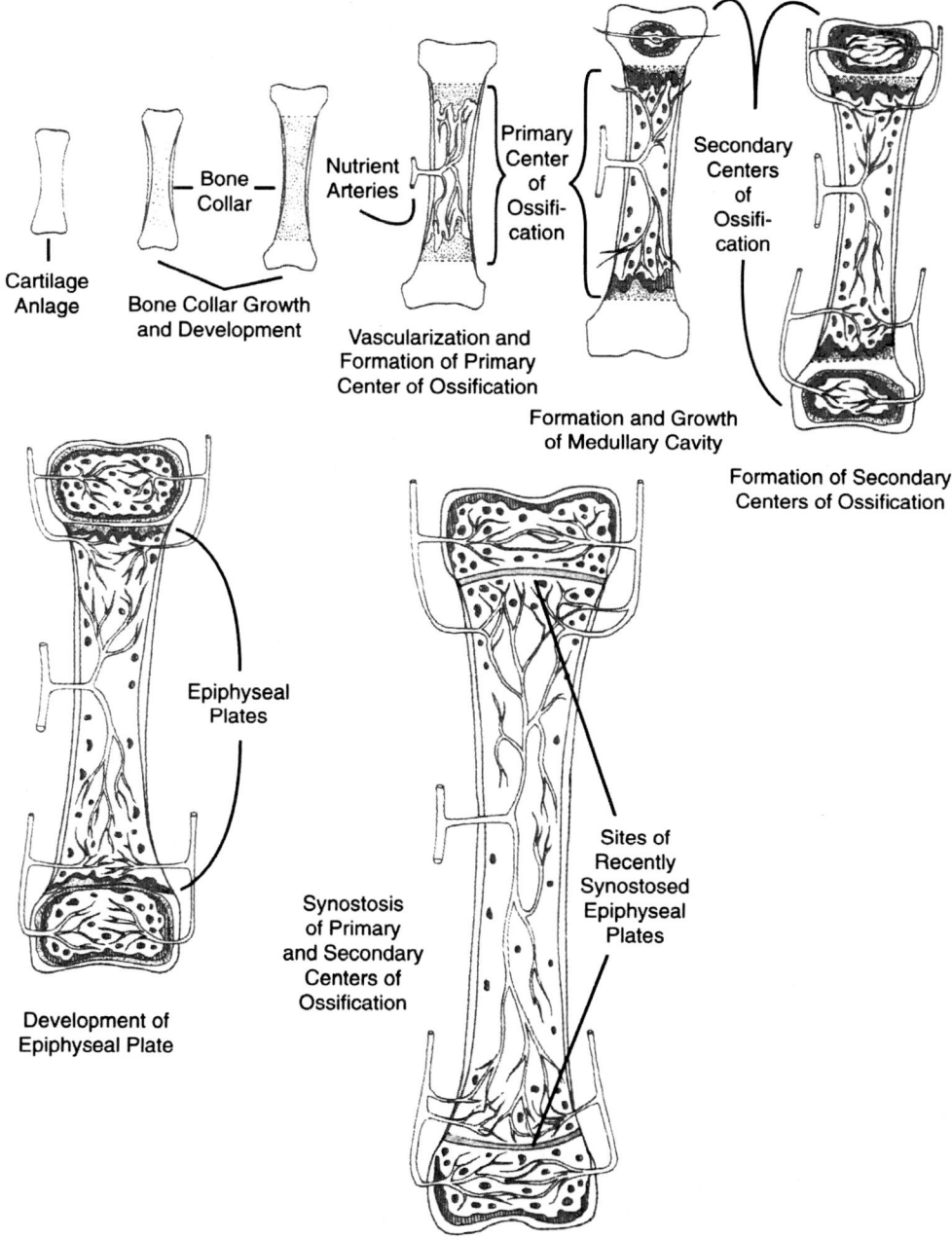

Figure 2-12 Steps in endochondral ossification.

periosteum (the cells of its deep layer metamorphose into osteoprogenitor cells which begin to synthesize osteoid rather than cartilaginous matrix). The activity of these new osteoblasts causes bone to be laid down around the periphery of the cartilage mode. This first bone that is laid down around the outside of the cartilage model is very similar to the intramembranous ossification that takes place elsewhere in the skeleton. Growing bone around the outside of the cartilage model in a long bone quickly takes the form of a *bone collar*. Osteogenic cells of the periosteum differentiate into osteoblasts that begin producing bone around the outside of the shaft of the anlage.

As the bone collar develops, mitosis within the center of the anlage continues and the cells which lie most deeply begin to suffer a loss of ready transport of materials (the anlage is avascular). This signals a change in these cells which undergo apoptosis and eventual calcification. At about this point in development, a branch of the arterial supply of the new periosteum migrates into the "nodule" of calcifying cartilage, carrying with it osteoprogenitor cells from the periosteum. This constitutes the nutrient artery of the bone diaphysis and its primary center of ossification.

Vascularization

As just noted, at about the same time that the bone collar begins to develop, the future nutrient artery invades the center of the anlage. Via this arterial route, osteoprogenitor cells are carried into the center of the anlage, where they will differentiate into osteoblasts and begin forming bone. As the cartilage model continues to grow in diameter and length by means of interstitial and appositional growth of the cartilage, chondrocytes (cartilage cells) in the center of the anlage continue to apoptose. As they do so they leave behind a matrix of hyaline cartilage. During the process of *apoptosis*, the chondrocytes in the center of the anlage hypertrophy, their lacunae coalesce, and compress the remaining cartilage matrix into a series of flattened plates, known as *spicules*, which are in reality the remaining walls of the chondrocytic lacunae. This remaining cartilage undergoes calcification through the deposition of calcium salts. It does not undergo ossification.

Formation of Primary Center of Ossification

Ossification is a process that is highly dependent upon vascularization (Cormack, 1987 and references therein). The vasculature carries in osteoprogenitor cells and osteoclasts to the developing center of ossification. Osteoclasts may be involved in carving a path through the bone collar, and chondroclasts in cutting through calcified cartilage matrix to create spaces through which the nutrient artery can pass. Once inside the diaphysis, branches of the nutrient artery ramify, forming *osteogenic buds*. These osteogenic buds consist of blind ended capillaries surrounded by osteoprogenitor cells and osteoclasts. The osteoprogenitor cells then differentiate into osteoblasts which begin the task of laying down woven bone on the *calcified cartilage spicules* left behind after the death of the hypertrophied chondrocytes. Once this process has begun in the center of the diaphysis, the *primary center of ossification* has been formed.

Branches of the nutrient artery continue to ramify through the center of the shaft of the anlage as it continues to grow. Subperiosteal deposition continues around the outside of the developing bone shaft, causing the bone continue to grow in diameter. By the processes just described the primary ossification center begins to spread throughout the center of the presumptive bone. The woven bone and the calcified cartilage matrix upon which it rests are a sort of temporary scaffolding that will be remodeled away as the diaphysis and the contained forming medullary cavity continue to grow. This occurs at about the same time as the initial formation of the bone collar and the invasion of the anlage by the nutrient artery.

Medullary Cavity Formation

The formation of the medullary (marrow) cavity is a consequence of the formation of the primary center of ossification. The primitive medullary cavity first forms when the lacunae of hypertrophied chondrocytes in the center of the anlage begin to coalesce. After the nutrient vasculature has invaded the anlage, and the primary center of ossification begins to form, the medullary cavity starts to grow. As we have seen, the first bone that is deposited within the primary center is woven bone laid down over calcified cartilage matrix. However, remodeling of this material begins almost immediately. As this bone is formed around the perimeters of the primitive marrow cavity, and more bone is laid down on the bone collar around the outside of the diaphysis, chondrocytes which lie around the periphery of the forming ossification center now begin to hypertrophy and apoptose, leaving behind more calcified cartilage matrix.

Osteoclasts and chondroclasts remove the older matrix and woven bone surrounding the early medullary cavity, and the process begins to repeat itself farther up and down the shaft of the bone. This process will continue until at some point the only remaining cartilage that has not undergone this process will be the epiphyseal plate. At the same time, lamellar bone is being laid down around the margins of the forming medullary cavity as well as beneath the periosteum. Later, when the epiphyseal plate has synostosed with the diaphysis, the medullary cavity will extend from the diaphysis into the epiphyses.

Thickening and Lengthening of the Bone Collar

While the primary center of ossification is enlarging and the medullary cavity is expanding, the subperiosteal bone collar continues to grow in length and in diameter. Of course, while the outside diameter of the shaft is increasing, osteoclastic activity continues to enlarge the medullary cavity so that it keeps pace in relative size with the enlarging bone collar. Generally, the epiphyses of a long bone are wider than the midshaft of the diaphysis. As the diaphysis approaches the epiphysis (or, as the bone collar approaches the epiphysis), the diaphysis widens to match the transverse diameter of the epiphysis. This broadened portion of the diaphysis is referred to as the *diaphyseal funnel*.

The *metaphysis* is at the broadest aspect of the diaphyseal funnel just subjacent to the epiphysis. As the bone collar lengthens and thickens, bone is constantly removed from its inner surface by osteoclastic action and redeposited on its periosteal surface to keep the bone collar growing in both diameter and length. It is of some interest that this scenario must be reversed in the region of the diaphyseal funnel. There the overall diameter of the shaft just below the metaphysis must actually be reduced as the bone lengthens in order to maintain a smooth transition between the diaphysis and metaphysis: that is the diaphyseal funnel must be moved farther from the center of the bone as it lengthens. Work in recent years on bone development has confirmed that much of the genetic information that is involved in bone formation is maintained and expressed within its growth plates (see below), and this fact accounts for the remarkable ability of a bone to make these kinds of adjustments as it grows in length.

Formation of Secondary Centers of Ossification in the Epiphyses

At some point well after the formation of the primary center of ossification, secondary centers of ossification begin to appear in the epiphyses. The process that occurs here is similar in all respects to the formation of the primary center. Each secondary center undergoes hypertrophy and apoptosis of its chondrocytes, is invaded by a nutrient artery, calcification of the cartilage matrix begins, and this is followed by the formation of true bone. A secondary center differs from a primary center only in the time of its occurrence, not in terms of its structure. In a typical long bone, there are one or more secondary centers at both the proximal and distal ends of the bone, or epiphyses. Eventually, all of the hyaline cartilage in the epiphysis will be replaced by bone by the mechanisms already described, with two exceptions. First, until growth in length of the bone ceases, the *epiphyseal plate* will remain cartilaginous. Second, a thin layer of hyaline cartilage will remain covering the surfaces of the epiphyses that participate in synovial joints. In a sense, the hyaline articular cartilage is the sole remnant of the cartilage anlage that survives into adulthood.

Development of the epiphyseal plate

The epiphyseal plate comes into existence after primary and secondary ossification centers have appeared. The epiphyseal plate develops from the plate of hyaline cartilage that remains between the primary and secondary ossification centers, and forms the *epiphyseal growth plate*. This growth plate has a distinct, layered, structure. From the epiphyseal side to the diaphyseal side of the epiphyseal plate four cartilaginous zones may be distinguished (Figure 2-13).

ZONE OF RESTING CARTILAGE: Also referred to as the zone of reserve cartilage, this is the layer of the epiphyseal plate closest to the epiphysis. The chondrocytes, or cartilage cells, in this layer are relatively quiescent, essentially being held in reserve for later growth of the epiphyseal plate. However, as the bottom of this layer, nearer the diaphysis, is approached, some cells are starting to undergo active mitosis. At this point, the first zone merges into the second zone. ZONE OF PROLIFERATING CARTILAGE: The second zone is a zone of actively mitosing chondrocytes, and is hence referred to as the zone of proliferating cartilage. These cells continue to produce extracellular matrix, and through the interstitial growth of the cartilage in this zone, the bone continues to grow in length. So long as the epiphyseal plate remains active and unossified, the bone is capable of continued growth in length. As the proliferative zone grows in thickness, the epiphysis is pushed further from the center of the diaphysis.

A few cells in the layer nearest the epiphysis remain constant as the zone of resting cartilage, while the remainder continue to mitose and transform. The mitosing cells form columnar arrangements throughout the epiphyseal plate, with resting cells nearest the epiphysis, and cells become increasingly active toward the diaphyseal side of the zone. Columns of extracellular matrix mark the spaces between columns of cells. Toward the bottom of this zone (nearer the diaphysis) is a transition to the third zone of the epiphyseal plate.

ZONE OF MATURING CARTILAGE: The third zone is referred to as the zone of maturing cartilage or as the zone of hypertrophying cartilage. Within this zone, the same sorts of changes are occurring as occur in the center of the primary and secondary centers of ossification. Chondrocytes within the maturing zone are growing in size, and undergo cell death. The lacunae of adjacent chondrocytes within the column begin to merge as the cells hypertrophy, and the columns of extracellular matrix between the lacunae become compressed. At the extreme diaphyseal side of the third zone, the remaining cartilage spicules begin to undergo calcification, just as was seen in the primary and secondary centers of ossification. This marks the transition to the fourth zone.

ZONE OF CALCIFYING CARTILAGE: The layer of the epiphyseal plate nearest the diaphysis is known as the zone of calcifying cartilage. Again, just as in the primary and secondary centers of ossification, once

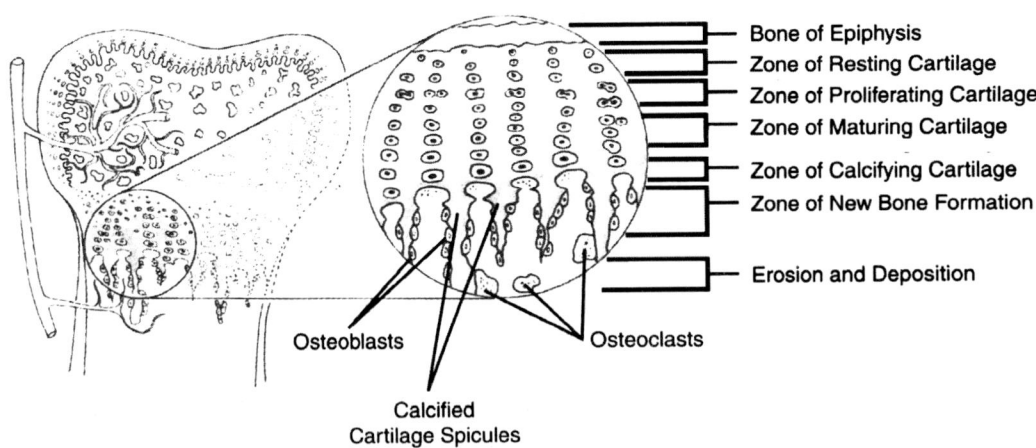

Figure 2-13 The four cartilaginous zones of the epiphyseal plate.

the hypertrophying cells enlarge, their lacunae coalesce to leave spicules of cartilage matrix. Once the chondrocytes undergo cell death, these remaining cartilage spicules become calcified.

At this point, we can say there is a fifth zone, a *zone of bone formation* or *zone of ossification,* wherein osteoblasts start laying down woven bone around the exposed calcified cartilage spicules. The epiphyseal plate does not remain static. As one side of the plate becomes ossified, the other side, the epiphyseal side, continues to produce new cartilage, so that the epiphysis is continually pushed away from the diaphysis, and the bone continues to grow in length as a result. What was once the zone of calcifying cartilage becomes ossified, while the previous layer of maturing cartilage becomes a new layer of calcifying cartilage, and so forth. The epiphyseal plate constantly rebuilds itself as it moves away from the center of the diaphysis.

At some genetically predetermined time — although it can be affected by nutritional status, disease, and other factors — the epiphyseal plate undergoes *synostosis,* during which it becomes solidly joined by bone to the diaphysis, and growth in bone length ceases. The times of appearance of primary and secondary centers of ossification, and the times of epiphyseal union differ from bone to bone, and between boys and girls. (See Ishcan and Krogman (1986) for a summary of these times.) Such patterns of appearance of ossification centers and union of epiphyses can, in fact, be used to estimate age in immature skeletal remains or to estimate biological age from radiographs of children.

Synostosis of Primary and Secondary Ossification Centers

As has been noted, the epiphyseal plate continues to exist until growth at the plate ceases and the epiphysis synostoses with the diaphysis. It should be remembered that the process of endochondral ossification is a continuous process throughout growth, and one phase blends into the next without sharp demarcation:

Nearly all epiphyseal plates synostose by 21 years of age. The last to fuse is generally the epiphyseal plate at the sternal end of the clavicle, which may remain open until 25 years of age or more, though there is also active bone growth within the pubic symphysis which may continue into the fourth decade of life (a small irregular epiphysis on the medial aspect of each pubic bone called the *ventral rampart* usually appears in the latter phases of the third decade of life and fuses in the early part of the fourth). Once all epiphyseal plates have fused, no further growth in stature or bone length can occur. (If the epiphyseal plate undergoes early synostosis, for example because of a fracture through the plate, the bone will not grow to its normal length.)

Not all bones formed by endochondral ossification follow so closely the model presented above. We have described the process in a typical long bone with a primary center of ossification in the diaphysis and secondary centers of ossification in the epiphyses. However, not all endochondral bones take this form. A typical vertebra, for example, has *three* primary centers of ossification, and *five* secondary centers. The *body* of a typical vertebra represents one of the primary centers, the *centrum,* while the *neural,* or *vertebral, arch* is formed by two primary centers that fuse in the midline to form the *spinous process* and fuse anteriorly with the body (Figure 2-14). In addition, a typical vertebra has secondary centers at the tips of each transverse process, one at the tip of the spinous process, and two ring-shaped annular epiphyses that cap the top and bottom of the cylinder of the body of the vertebra, for a total of *five* secondary centers. Each secondary center is separated from its primary center by an epiphyseal growth plate. The same processes occur there as are seen in a typical long bone, but the forms of the elements involved are substantially different. *Spina bifida* occurs when there is failure of the two halves of the neural arch to fuse with one another. As a result, the secondary center at the tip of the spinous process does not develop, the two halves of the neural arch do not meet, and the spinal cord is either exposed or only covered by soft tissue.

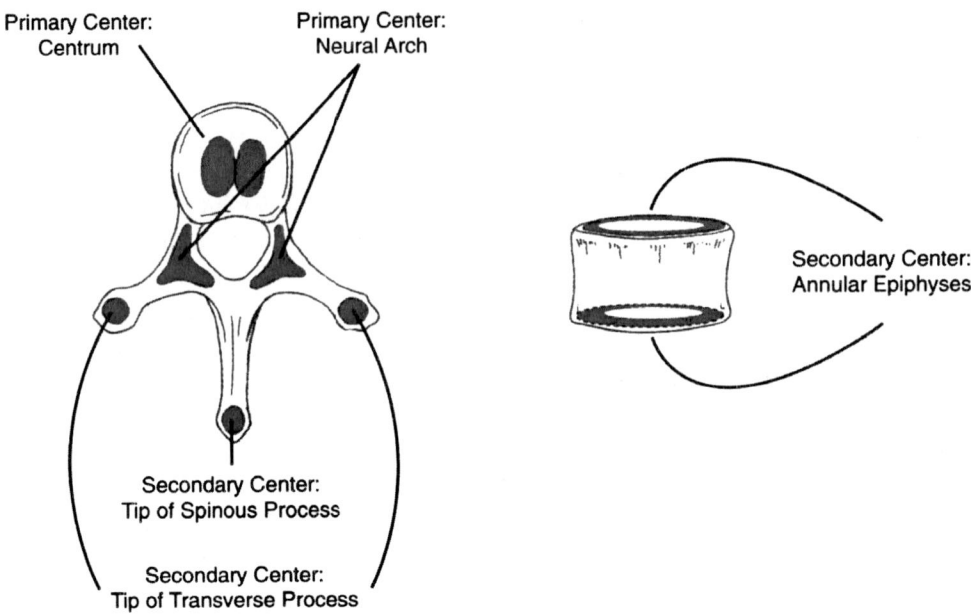

Figure 2-14 The primary and secondary centers of ossification of vertebrae.

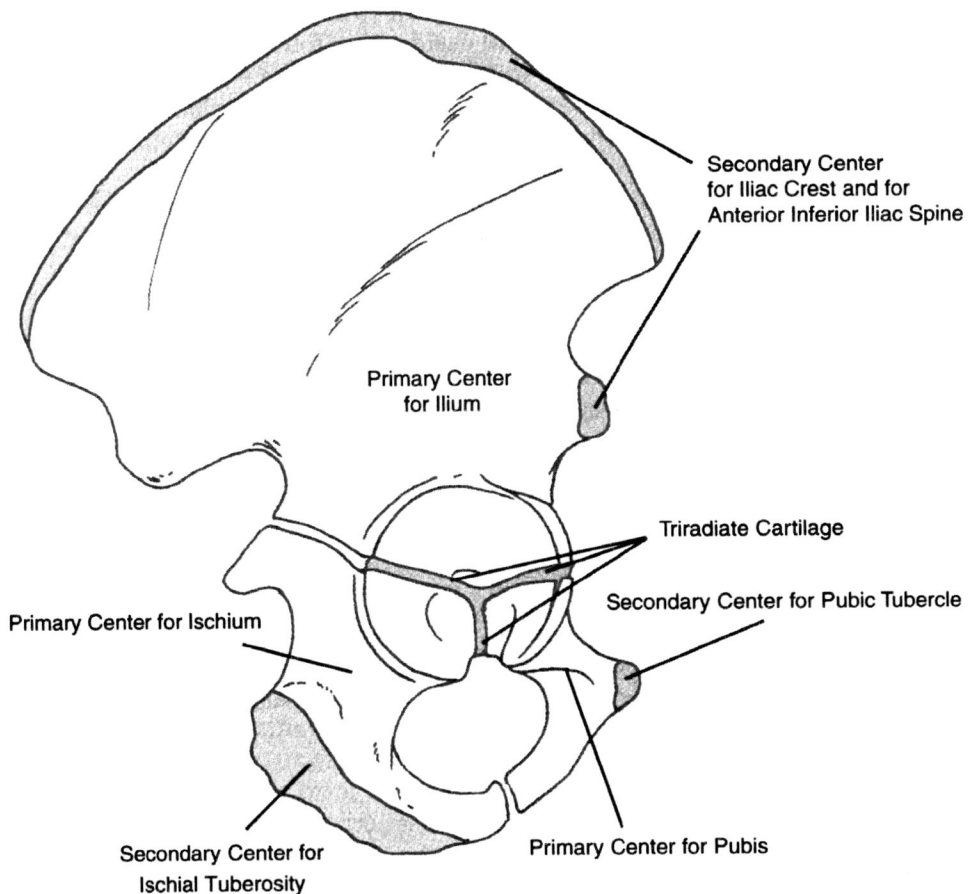

Figure 2-15 Centers of ossification of the os coxae.

Another good example of endochondral ossification in a less typical bone is the ossification of the *os coxae* (hip bone or innominate bone; Figure 2-15). The innominate is usually described as developing from three separate bones: the *ilium*, the *ischium*, and the *pubis*. In reality, these are three primary centers of ossification for the os coxae. They are separated from one another in the *acetabulum* (the socket for the head of the femur) by a **Y**-shaped epiphyseal plate. Again, this has the elements of an epiphyseal plate, but is greatly modified in its shape. In addition, the innominate has a number of secondary centers, all separated from the rest of the bone by epiphyseal growth plates. These include, but are not limited to, the *iliac crest*, the *ischial tuberosity*, and the ventral rampart of the *pubic symphyseal surface*, as just discussed above. Additionally, the *anterior inferior iliac spine* develops as a secondary center of ossification under the superior attachment of the rectus femoris muscle. Such a secondary center under a muscle attachment is known as a *traction epiphysis* or *apophysis*.

Intramembranous Ossification

Intramembranous ossification (*mesenchymal* or *dermal ossification*) is a somewhat less complicated form of ossification. It is restricted to most of the clavicle and the flat bones of the cranial vault. Intramembranous ossification is the direct ossification of highly vascular connective tissue, and spreads outward from a primary center of ossification.

Osteoprogenitor cells differentiate to osteoblasts in the highly vascular center of a mesenchymal membrane anlage. The mesenchymal membrane is free to continue to grow in size at its margins. Ossification, however, continues outward from the center of the bone reaching toward the margins of the membrane. Woven bone is laid down surrounding

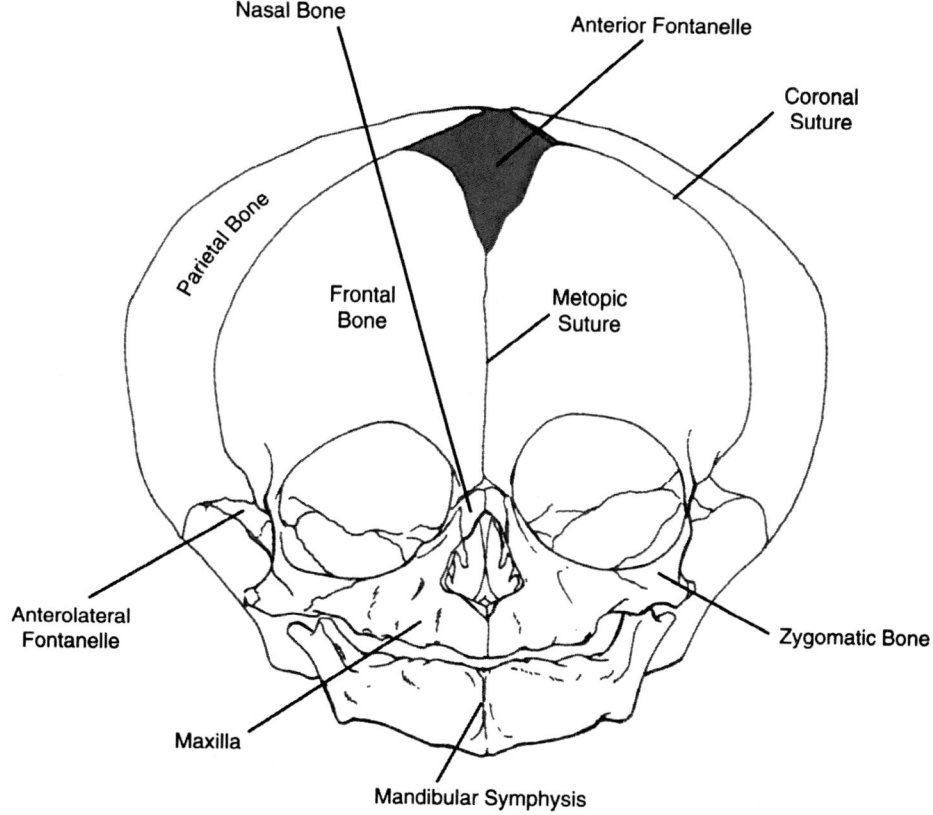

Figure 2-16 Bone development in the neonatal cranium, anterior view.

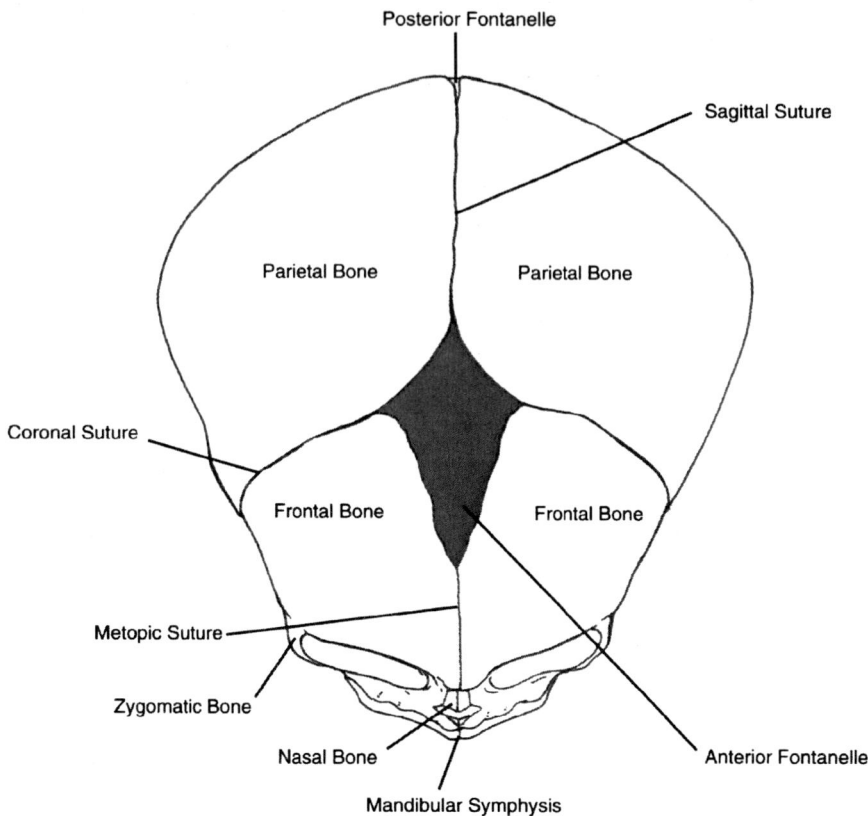

Figure 2-17   Bone development in the neonatal cranium, superior view.

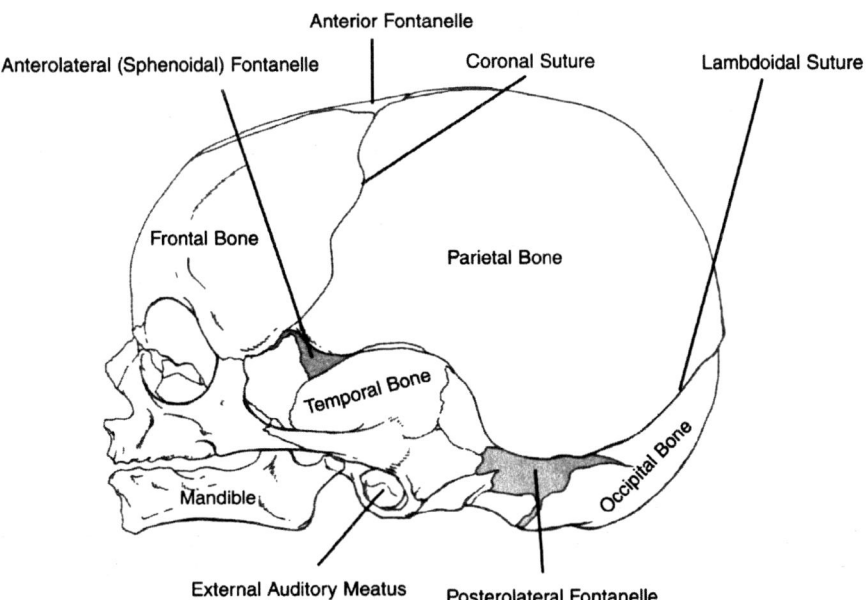

Figure 2-18   Bone development in the neonatal cranium, lateral view.

the vessels that enter the membrane and form primitive trabeculae. At the outer and inner surfaces of the forming membrane bone, these primitive trabeculae are joined together by woven bone to form the early compact cortical shell. Some blood vessels will become surrounded by bone in this newly formed cortex to form primary osteons, as was seen in endochondral ossification. Eventually, this early cortical bone will be remodeled and replaced by lamellar bone and Haversian systems. At birth, the margins of the bones of the cranial vault remain membranous. This enables the cranium to be somewhat compressed, which eases the passage of the fetal head through the birth canal. The fontanelles ("soft spots", Figures 2-16 through 2-18) are formed of remnant mesenchymal membrane. With increasing maturity, the cranial bones will become fully ossified and meet one another at sutures. Very often, especially after the age of 35, cranial sutures will synostose. This occurs with enough regularity that estimates of age at death can be obtained from the pattern of cranial suture closure (Meindl and Lovejoy, 1985).

Typical Disorders of Bone Growth and Remodeling

So far, we have outlined the normal progress of ossification and bone growth and development. Bone metabolism is subject to the effects of many factors, including nutrition, hormones, vitamins, mechanical loading and genetic disturbances. (For a good summary of bone metabolism, endocrinology, and metabolic bone disease, see Avioli and Krane, 1988.) *Some* of the problems related to bone growth, development and remodeling that are of clinical importance are briefly discussed here.

Rickets

Rickets is a disease of childhood in which the bone matrix that forms is not, or is incompletely, calcified. Uncalcified bone matrix is called *osteoid*. Osteoid is normally calcified rapidly after its production. In rickets, the osteoid does not calcify or is only poorly calcified at best. The resulting bone is very weak in compression, and a telltale characteristic of rickets in a young child is bowing of the femurs.

There are different varieties of rickets. The most common is *vitamin-D deficiency rickets*. Vitamin D (1,25 dihydroxycholicalciferol) is necessary for calcification of bone to occur. Adequate amounts of vitamin D are normally acquired through the diet or synthesized by the body. Ordinarily, vitamin D deficiency rickets responds to vitamin D supplementation.

Vitamin-D resistant rickets are forms of rickets in which the bone does not respond to vitamin D supplementation. These forms of rickets are due to problems with other steps in the metabolic pathways of calcification, rather than vitamin D deficiency.

Osteomalacia

Osteomalacia is sometimes referred to as the adult form of rickets. Its etiology is similar. It may be due to vitamin D deficiency or other causes. The net result is that the osteoid that is formed is not adequately mineralized. Of course, adults do not possess epiphyseal growth plates, so the manifestation of the disease will not be the same as in rickets. Most new bone formation that occurs in adults occurs during remodeling, and thus within Haversian systems. In osteomalacia, new Haversian systems form, but remain unmineralized or poorly mineralized. The result is that the newly formed bone is weaker than the surrounding healthy bone. This can be due to inadequate calcium in the diet, vitamin or hormonal problems, or due to pregnancy and lactation, where calcium is diverted to the fetus during growth of the fetal skeleton or to the infant through lactation.

Osteoporosis

Osteoporosis (Figures 2-19 through 2-21) is not a single disease, but a number of conditions that all result in a similar presentation. *Osteoporosis* can be most directly defined as *a condition in which the amount*

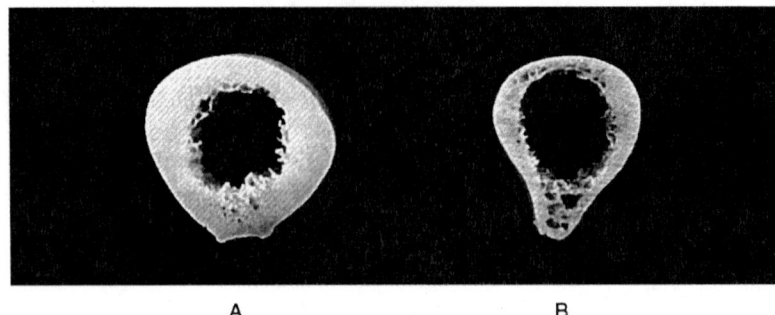

Figure 2-19 Osteoporosis. Cross-section of two femora at their midshafts, anterior at top. (A) A normal, healthy adult femur. (B) A femur with advanced osteoporosis. Note the thin cortex and increased porosity, particularly anteroposteriorly.

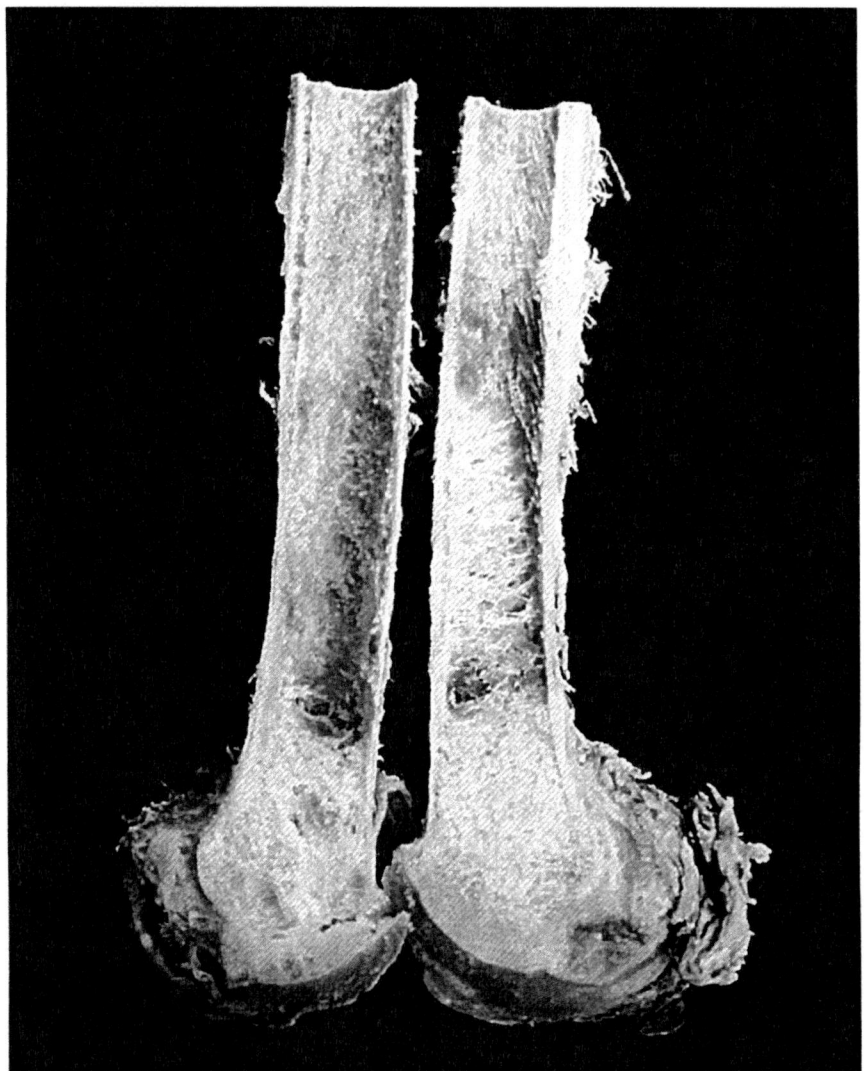

Figure 2-20 Disuse osteoporosis. These two femora are from the same individual. The femur on the left is from a leg that was amputated at the knee, whereas the femur on the right is from the patient's intact leg.

of actual bone matrix per unit volume is reduced. As the name implies, osteoporotic bone has greatly increased porosity. Spaces develop within the bone, and compact bone of the diaphyses of long bones becomes thinner (referred to as *cortical involution*). Not only does cortical bone become thinner, but the cortex that remains becomes increasingly porous (Walker, 1989 and references therein; Walker, 2000) and increasingly weak. It may become so thin and weak in advanced cases of osteoporosis, in fact, that it will undergo fracture under body weight alone. There are a number of subtypes of osteoporosis with different underlying causes.

Postmenopausal Osteoporosis

The most common, and generally the best known, type of osteoporosis is *postemenopausal osteoporosis*. Estrogens and androgens (i.e. testosterone) are both *anabolic* (promote tissue formation) for bone tissue. At menopause, there is a dramatic decrease in the production of, and circulating levels of, estrogen. In women who have low bone mass originally, low calcium intake, and/or who may have been less physically active prior to menopause, this can result in a dramatic decrease in bone mass (Gardsell et al., 1991). Postmenopausal osteoporosis appears to occur with higher frequencies among populations of European origin, though it can, and does, occur in people of all ancestries (Armelagos, 1969; Chalmers and Ho, 1970; Mangaroo et al., 1985; Ross et al., 1985; Martin et al., 1987; Slemenda and Johnston, 1990; Walker, 1991; Walker, 2000). It can be prevented, or at least the level of severity can be decreased, by adequate calcium intake and adequate exercise, both before and after menopause.

Senile Osteoporosis

Senile osteoporosis is the osteoporosis of old age. Senile osteoporosis affects both men and women. The effects in older women are worse than in older men, primarily because they suffer the combined effects of *both* postmenopausal and senile osteoporosis.

Disuse Osteoporosis

As the name implies, disuse osteoporosis (see Figure 2-20) is a form of osteoporosis that results from a lack of, or decrease in, applied external mechanical loads. Generally, there is a minimum mechanical signal ("minimum functional strain") which bone requires in order to be maintained at normal levels. Prolonged bed rest can result in significant reductions in bone mass, as can immobilization of a limb in a cast. Even the removal of the forces of gravity will result in a loss of bone mass. Space flight, even for relatively short periods, can result in a significant decrease in bone mass. Luckily, disuse osteoporosis is entirely reversible as long as it is not associated with paralysis or other conditions that prevent a body part from being loaded.

Idiopathic Osteoporosis

Idiopathic Osteoporosis is osteoporosis that does not have a readily observed cause. Idiopathic osteoporosis may strike anyone at any age. Loss of bone mass may result from a decrease in osteoblastic activity, from an increase in osteoclastic activity, or a combination of both osteoblastic and osteoclastic activity.

Different bones show different levels of susceptibility to osteoporosis. The vertebral column, especially in the thoracic region, shows an especially high susceptibility to osteoporosis. Vertebrae in older people, especially older women, may become so highly osteoporotic that they undergo compression fractures of the vertebral body. This will result in a severe *kyphosis* (anteriorly concave vertebral curvature) (Figure 2-22) of the thoracic vertebral column, sometimes referred to colloquially as a "dowager's hump." If the kyphosis is severe enough, it can compromise the ability of the heart and lungs to function properly.

Osteopetrosis

Osteopetrosis (marble bone disease) is almost the opposite of osteoporosis. In osteopetrosis, abnormally large amounts of bone matrix, and abnormally highly mineralized of bone matrix are produced. As a result, osteopetrotic bone appears almost entirely chalky white in radiographs. Because the amount of bone mineral is abnormally high relative to the amount of collagen fibers in the bone matrix, the bone is very brittle and easily fractured.

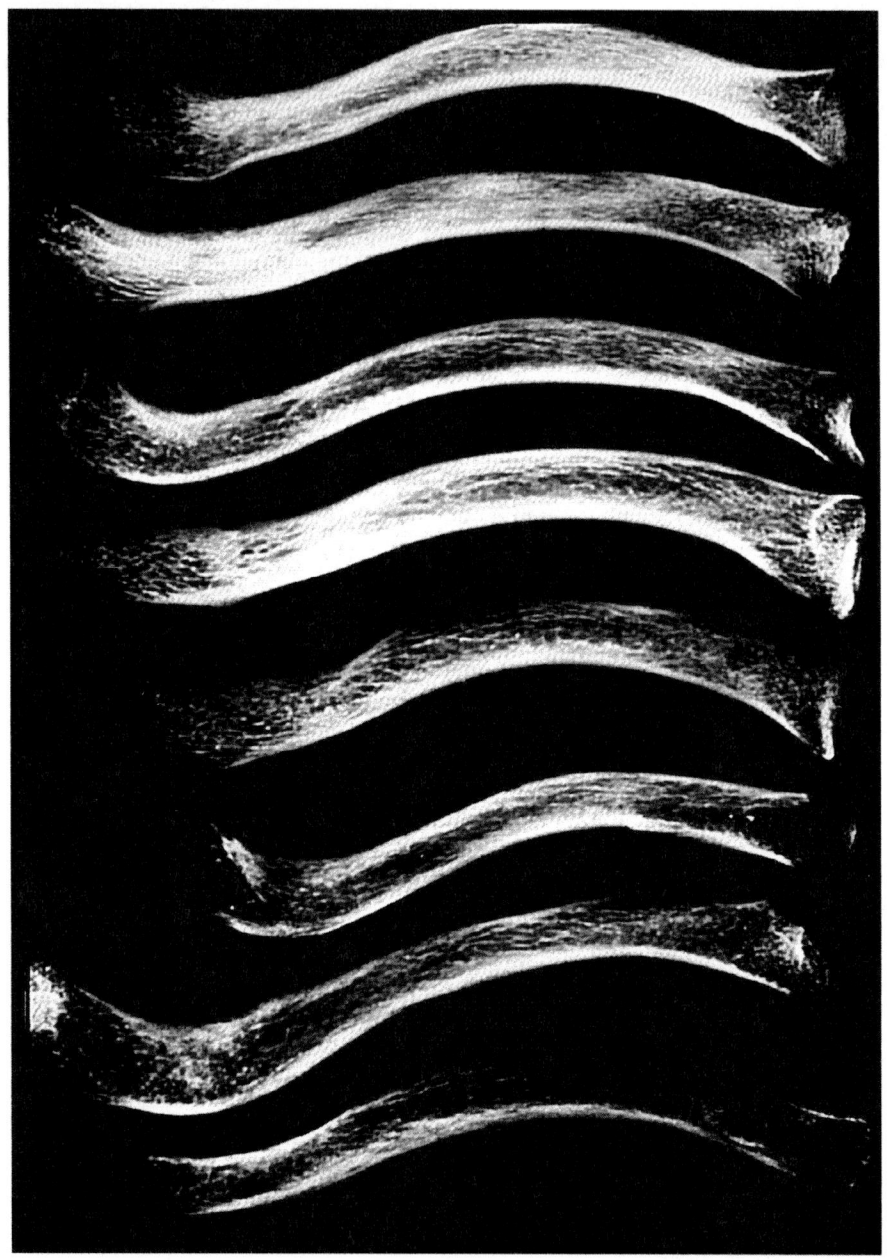

Figure 2-21 Osteoporosis in the clavicle. This figure represents through radiographs progressive bone loss in the clavicle with age. The youngest individual is represented at the top. (Source: Walker, RA, and Lovejoy, CO [1985]. Radiographic changes in the clavicle and proximal femur and their use in the determination of skeletal age at death. *American Journal of Physical Anthropology*, 68, 73. Copyright © 1985, Wiley-Liss, Inc. Reproduced by permission of Wiley-Liss, Inc., a subsidiary of John Wiley & Sons, Inc.)

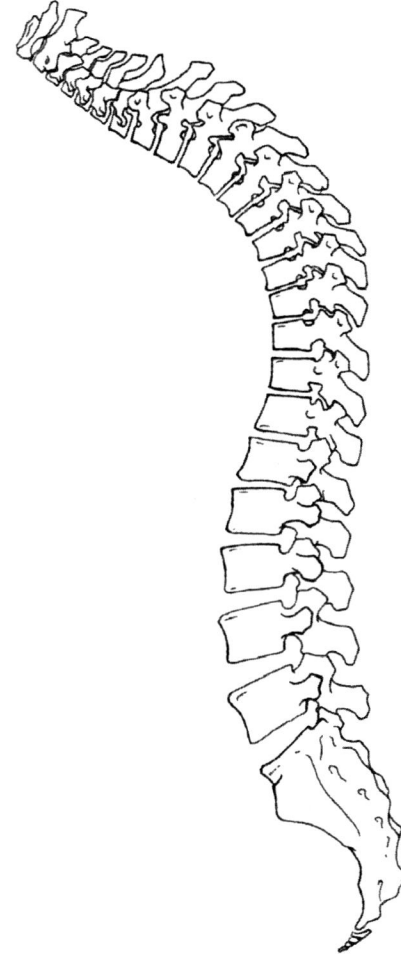

Figure 2-22 Kyphosis of the thoracic vertebral region. Anterior is to the left.

Osteopetrotic bones demonstrate very thick cortical bone, dense networks of cancellous bone, and large amounts of calcified cartilage. Experimentally, hypervitaminosis D in rats can present a similar condition. This condition is rare in humans, but is sometimes seen in infants with fatal anemia and as an inherited autosomal recessive trait.

Paget's Disease

Paget's Disease (Osteitis Deformans) has a presentation similar to osteopetrosis, but is a more localized phenomenon. Bone affected by Paget's disease shows thickened cortical bone and coarse, thickened trabeculae. It may be restricted to a single bone, or have multiple foci within a single individual. It is not, however, systemic, as is osteopetrosis.

Osteogenesis Imperfecta

The name of this disorder literally means "imperfectly formed bone." In osteogenesis imperfecta, the collagen of the bone matrix is affected. The bone as a result is soft and brittle. Osteogenesis imperfecta is inherited as an autosomal dominant trait.

Ankylosing Spondylosis

Bony *ankylosis* is an abnormal condition in which joints become immobilized by the production of bone across the joint, thus uniting the bones solidly to one another. The condition may be genetic, or induced by trauma, disease, or surgery (i.e. vertebral fusion). *Ankylosing spondylosis* is specifically the pathological ankylosis of the vertebral column.

Bone tissue responds to many hormonal influences (calcitonin, parathyroid hormone, and growth hormone to mention three), and disorders of hormonal equilibrium can result in changes in the skeletal system, including many of the conditions mentioned above. Other conditions may result as well. Decreased amounts of growth hormone results in dwarfism. Increased amounts of growth hormone during childhood (before the closure of epiphyseal growth plates) results in gigantism. Generally this is the result of disorders of the pituitary gland and is referred to as *pituitary gigantism*.

Increased amounts of growth hormone after the closure of the epiphyseal plates cannot result in any further growth in length of the bones, but it can result in increased amounts of subperiosteal bone deposition resulting in the coarsening of facial features and increased bone diameters. This disorder is known as *acromegaly*.

References Cited

Avioli, LV, and Krane, SM, eds. (1998) *Metabolic Bone Disease and Clinically Related Disorders.* Third Edition. Academic Press, New York.

Armelagos, G (1969) Disease in ancient Nubia. *Science* 163: 255-259.

Bertram JE, and Swartz, SM (1991) The 'law of bone transformation': a case of crying Wolff? Biol. Rev. Camb. Philos. Soc. 66: 245-73

Burr, DB (1993) Remodeling and the repair of fatigue damage. *Calcified Tissue International* 53(Supplement 1):
S75-81

Ash, MM (1984) *Wheeler's Dental Anatomy, Physiology, and Occlusion.* WB Saunders Company, Philadelphia.

Chalmers J, and Ho KC (1970) Geographical variations in senile osteoporosis. *Journal of Boneand Joint Surgery* 52B: 667-675.

Cormack, D.H., ed. (1987) *Ham's Histology*, 9th Edition. Lippincott, Philadelphia.

Cormack, D.H. (2001) *Essential Histology,* 2nd Edition. Lippincott-Raven, Philadelphia.

Currey, JD (1984) *The Mechanical Adaptations of Bone.* Princeton University Press, Princeton,New Jersey.

Currey, JD (2002) *Bones: Structure and Mechanics.* Princeton University Press, Princeton,New Jersey.

Enlow, DH (1969) The bone of reptiles. In: *Biology of the Reptilia*, vol. 1, pp. 45-80. C. Gans,ed. Academic Press, New York.

Frankel, VH and Nordin, M (1980) *Basic Biomechanics of the Skeletal System.* Lea and Febiger, Philadelphia.

Gardsell P, Johnell O, Nilsson BE, and Sernbo I (1991) Bone mass in an urban and a rural population: A comparative, population-based study in southern Sweden. *Journal of Bone and Mineral Research* 6: 67-75.

Gefen A, and Seliktar, R (2004) Comparison of the trabecular architecture and the isostatic stress flow in the human calcaneus. Medical Engineering and Physics 26: 119-29

Krogman, W.M. and M. Y. Iscan (1986) *Human Skeleton in Forensic Medicine*, 2nd Edition. Charles C. Thomas, Springfield, Illinois.

Lanyon, LE (1993a) Osteocytes, strain detection, bone modelling and remodelling. *Calcified Tissue International (Supplement 1)*: S102-S107.

Lanyon, LE (1993b) Skeletal responses to physical loading. In: Mundy, GR, and Martin, TJ (eds.) *Physiology and Pharmacology of Bone.* Springer-Verlag, Berlin, pp. 485-505.

Larson, WJ (1993) *Human Embryology,* Third Edition. Churchill-Livingstone, New York.

Mangaroo J, Glasser JH, Roht LH, and Kapadia AS (1985) Prevalence of bone demineralization in the United States.*Bone* 6: 135-139.

Martin DL, Magennis AL, and Rose JC (1987) Cortical bone maintenance in an historic Afro-American Cemetery Sample from Cedar Grove, Arkansas. *American Journal of Physical Anthropology* 74: 255-264.

Meindl, RS, and C.O. Lovejoy (1985) Ectocranial suture closure: A revised method for the determination of skeletal age at death based on the lateral-anterior sutures. *American Journal of Physical Anthropology 68(1):* 57-66.

Miller, SC, and Jee, WSS (1987) The bone lining cell: A distinct phenotype? *Calcified Tissue International,* 41: 1-5.

Miller, SC, and Jee, WSS (1992) Bone lining cells. In: Hall, BK, ed., Bone. Volume 4. Bone Metabolism and Mineralization. CRC Press, Boca Raton, Florida, pp. 1-19.

Moore, KL and Persuad, TVN (2002) *The Developing Human: Clinically Oriented Embryology*, 7th Edition. Saunders.

Pearson, OM and Lieberman, DE (2004) The aging of Wolff's "law": Ontogeny and responses to mechanical loading in cortical bone. *American Journal of Physical Anthropology*, Supplement 39: 63-99.

Rodan, GA, and Rodan, SB (1984) Expression of the osteoblastic phenotype. In: Peck, WA, ed., *Bone and Mineral Research Annual 2*. Elsevier, Amsterdam. Pp. 244-285.

Ross PD, Wasnich RD, and Davis JW (1990) Fracture prediction models for osteoporosis prevention. *Bone* 11: 327-331.

Rubin C, Judex S, and Hadjiargyrou, M (2002) Skeletal adaptation to mechanical stimuli in the absence of formation or resorption of bone. *Journal of Musculoskeletal and Neuronal Interactions* 2: 264-267.

Sadler, TW (2006) *Langman's Medical Embryology,* 10th Edition. Lippincott, Williams and Wilkins, Baltimore.

Skedros, JG, and Baucom, SL (2007) Mathematical analysis of trabecular "trajectories" in apparent trajectoral structures: the unfortunate historical emphasis on the human proximal femur. *Journal of Theoretical Biology* 244: 15-45.

Slemenda, CW, and Johnston, CC Jr (1990) Osteoporotic fractures. In: Simmons, DJ, ed., *Nutrition and Bone Development.* Oxford University Press, New York and Oxford, pp. 131-147.

Simmons, DJ, ed. (1990) *Nutrition and Bone Development.* Oxford University Press, New York and Oxford.

Simmons, DJ, and Grynpas, MD (1990) Mechanisms of bone formation *in vivo*. In: Hall, BK, ed., *Bone, Volume 1: The Osteoblast and the Osteocyte.* The Telford Press, Caldwell, New Jersey. Pp. 193-302.

Simon, MR (1983) The effect of dynamic loading on the growth of the epiphyseal growth plate of the rat. *Acta Anatomica* 101: 176-183.

Simon, MR, Holmes, KR, and Olsen, AM (1984) The effects of quantified amounts of increased intermittent compressive forces for 30 and 60 days on the growth of limb bones in the rat. *Acta Anatomica* 120: 173-179.

Swartz, SM, Parker A, and Huo C (1998) Theoretical and empirical scaling patterns and topological homology in bone trabeculae. Journal of Exp. Biol. 201: 573-90.

Walker, RA (1989) *Assessment of Cortical Bone Dynamics and Skeletal Age at Death from Femoral Cortical Histomorphology.* Doctoral Dissertation submitted to the School of Biomedical Sciences, Kent State University, Kent, Ohio.

Walker, RA (1991) Variation in femoral cortical thickness in three human populations. Paper presented at the 60th Annual Meeting of the American Association of Physical Anthropologists, Milwaukee, Wisconsin, April 4, 1991. (Abstract published in *American Journal of Physical Anthropology*, Supplement 12: 179 (1991))

Walker, RA (1993) Relationship of cross-sectional biomechanical properties of cortical bone to histomorphology in white Americans over 50. Paper presented at the 62nd Annual Meeting of the American Association of Physical Anthropologists, Toronto, Ontario, April 15, 1993. (Abstract published in *American Journal of Physical Anthropology,* Supplement 16: 202 (1993)).

Walker, RA (1997) Intraindividual variation in cortical thickness and histomorphology of the midshaft femur. Paper presented at the 66th Annual Meeting of the American Association of Physical Anthropologists, St. Louis, Missouri, April 3, 1997. (Abstract published in *American Journal of Physical Anthropology,* Supplement 24: 234 (1997)).

Walker, RA (1998) A survey of remodeling in the mammalian skeleton: a pilot study. Paper presented at the 67th Annual Meeting of the American Association of Physical Anthropologists, Salt Lake City, Utah, April 2, 1998. (Abstract published in *American Journal of Physical Anthropology*, Supplement 26: 224-225 (1998)).

Walker, RA (1999) A survey of remodeling in the vertebrate skeleton, part II. Paper presented at the 68th Annual Meeting of the American Association of Physical Anthropologists, April 29, 1999. (Abstract published in *American Journal of Physical Anthropology,* Supplement 28: 272 (1999)).

Walker, R.A. (2000) Properties of the cortex of the midshaft femur: Variation in three human populations (2000) *Homo: Journal of Comparative Human Biology* 51: 180-199.

Walker, RA, and Lovejoy, CO (1985) Radiographic Changes in the clavicle and proximal femur and their use in the determination of skeletal age at death. *American Journal of Physical Anthropology* 68:67-78.

Walker, RA, Lovejoy, CO, and Meindl, RS (1994) The histomorphological and geometric properties of human femoral cortex in individuals over 50: Implications for histomorphological determination of age-at-death. *American Journal of Human Biology* 6: 659-667.

CHAPTER Three

Development of the Vertebral Column

By B. Rosenman*, C.O. Lovejoy*, and M. McCollum**

Clinicians concerned with the functioning of the human body often deal solely with adult anatomical complexes. This is especially true of the spinal column because many of its pathologies occur only in adulthood. To fully understand the etiology of many ailments and therefore better diagnose them, however, a basic awareness is required of the processes by which the vertebral column is assembled. This chapter reviews early vertebral column development, with an emphasis on the pattern and processes involved in its creation.

Embryological Development From Fertilization To Neurulation

How does a single cell develop into a complex, multicellular organism? This process involves a multitude of steps, all of which are precisely determined and coordinated.

In this section, we will trace development from the fertilized egg to the creation of the germ layers that give rise to the tissues and structures of the body, including the vertebral column.

Before fertilization, the egg is surrounded by an acellular layer called the *zona pellucida*. This layer serves as one of the barriers to multiple fertilizations by normally allowing the entrance of no more than one sperm. Approximately 1 day after fertilization, the fertilized egg (zygote) initiates a series of cell divisions called *cleavage*. Because they are bounded by the zona pellucida, the resultant daughter cells, or *blastomeres*, become smaller with each subsequent *mitosis*, or cell division. By the 32-cell stage (the fourth day postconception), these blastomeres take on the appearance of a mulberry, hence their designation as the *morula* (Latin: morum means "mulberry"). The morula gives rise to the embryo, the *extraembryonic membranes*, and the placenta.

Up to this stage, *embryogenesis* (the process of embryo formation) is roughly similar in all vertebrates. However, from this point on, the process in mammals begins to differ significantly from that in other animals, especially the chick, whose development has been most intensively studied. Until recently, many of the detailed events of mammalian development were poorly understood. Many medical texts continue to use bird development as a general outline in discussing early embryology; this is no longer necessary, however, because current research has greatly expanded our knowledge of mammalian

---

* School of Biomedical Sciences and Department of Anthropology, Kent State University,
** Department of Cell Biology, University of Virginia School of Medicinee

embryogenesis. Here we touch on only some of the principal events in mammalian development.

About two thirds of the cells of the morula lie on its external surface. These are called the *trophectoderm*, also known as the *trophoblast*. The remaining one third of the cells are internal and are called the *inner cell mass*. Collectively, both structures are called the *blastocyst*. Cells of the trophectoderm begin pumping fluid into the interior of the blastocyst, causing formation of a fluid-filled vesicle called the *blastocoele*. The fluid pushes the inner cell mass against one end of the trophectoderm termed the *embryonic pole*.

At the same time, the inner cell mass differentiates into two distinct regions. The outer region, which is in contact with the trophectoderm, is called the *epiblast*, whereas the inner region, called the *primitive endoderm*, borders the blastocoele (Figure 3-1). These events begin at day 3 1/2 and are completed by day 4 ½ when the zona pellucida ruptures, freeing the blastocyst to interact directly with the uterine wall.

At implantation, day 5 1/2, some of the trophectoderm gives rise to giant cells that invade the uterus. The remainder of the trophectoderm contributes to the *placenta*, which supplies nutrition to the embryo. At the same time, the primitive endoderm spreads over the internal surface of the trophectoderm that is not in contact with the epiblast and forms the *parietal endoderm*. The portion of the primitive endoderm that remains in contact with the epiblast is now called the *visceral endoderm*.

By day 6 the epiblast has assumed a cuplike shape with an internal cavity (U-shaped in cross-section). It is now composed of about 1000 cells, and its external surface is visceral endoderm. At day 6 1/2

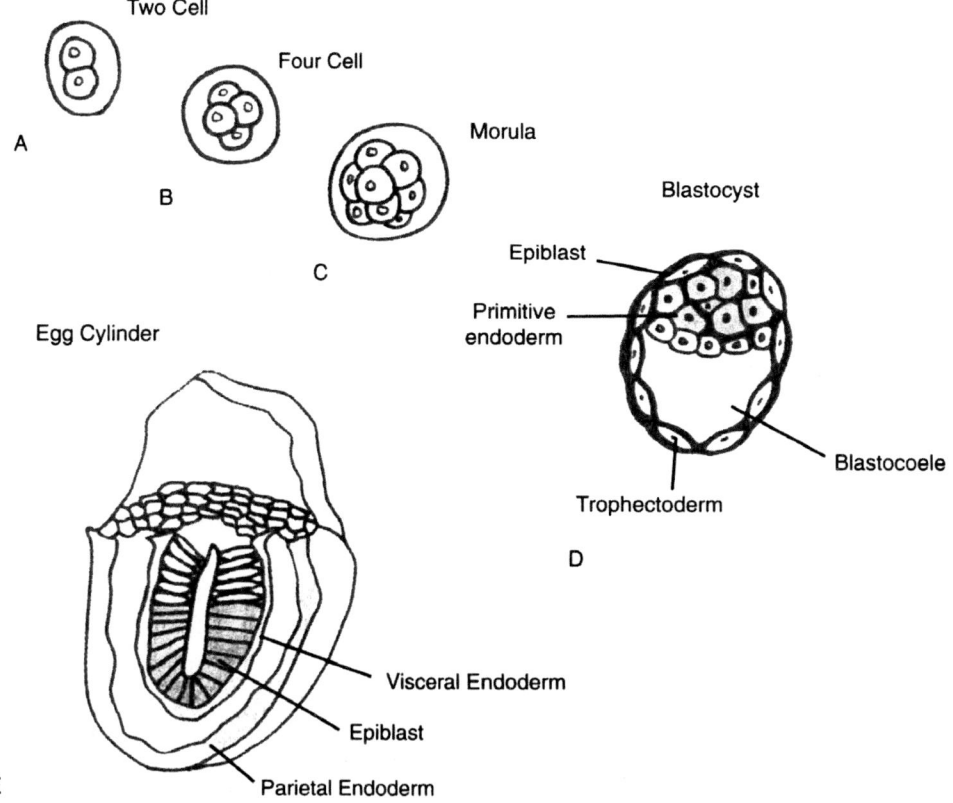

Figure 3-1 Early embryogenesis. The fertilized egg enters a series of stages: (*A*) two cell, (*B*) four cell, (*C*) morula, (*D*) blastocyst, and (*E*) egg cylinder.

the process of *gastrulation* begins. This is the process during which the three primary germ layers (*ectoderm, mesoderm,* and *endoderm*) are formed.

At a point on the circumference of the cup-shaped epiblast, at the future posterior end of the embryo, a localized thickening of cells appears. This is called the *primitive streak.* Cells of the epiblast proliferate and migrate through those of the primitive streak. As they do so, they spread anteriorly and laterally between the epiblast (which soon becomes the *embryonic ectoderm*) and the visceral endoderm. These cells are the *embryonic mesoderm.* Some epiblast cells penetrate the visceral endoderm layer and isolate it from the remainder of the epiblast. These new epiblast-derived cells form the *embryonic endoderm* (Figure 3-2). While these events are occurring, the primitive streak elongates toward the future anterior end of the embryo.

By this point, the simple "cup" of the epiblast has been transformed into one cup with three distinct layers. The internal surface of this cup (the ectoderm) is the dorsal surface of the embryo, and its external surface is the endoderm. Once gastrulation is complete, the formation of the basic body plan is well on its way. The body's primary tissues have been defined, as have its primary axes (craniocaudal, mediolateral [midline to right and left], and dorsoventral). Each germ layer forms different organ systems and tissues: the ectoderm forms epidermis, glands, hair, nails, tooth enamel, neural tissue, and much of the skull. The mesoderm forms the postcranial skeleton; connective tissue; striated muscle; and the urinary, reproductive, and circulatory systems. The endoderm forms the digestive tract, lungs, and several internal organs.

In the final stages of gastrulation, a series of complex foldings takes place that internalizes the embryonic endoderm and externalizes the embryonic ectoderm. In essence, the cup turns inside out. Many of the internal movements are complex, however; for example, the presumptive heart and liver first form anterior to the future head,

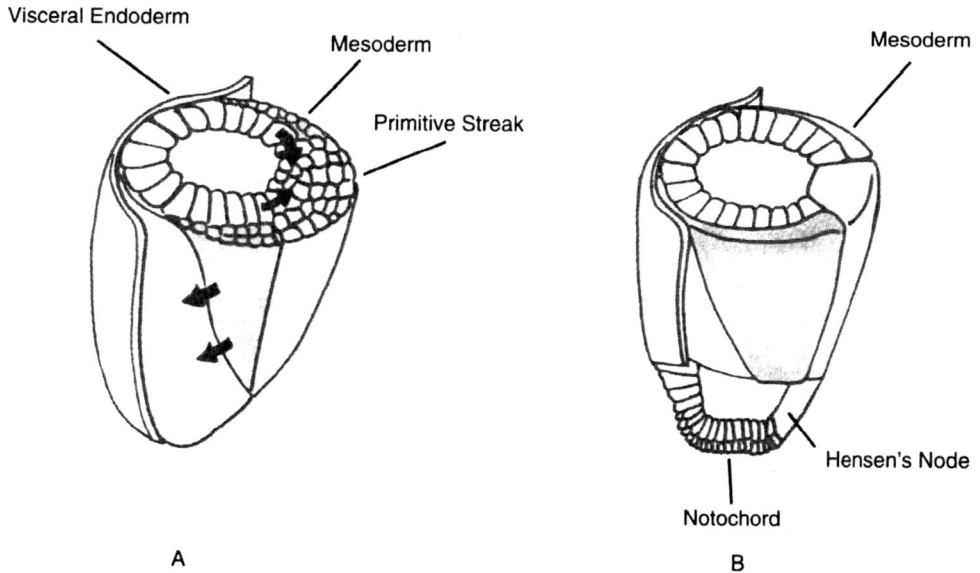

Figure 3-2 Gastrulation: (*A*) The primitive streak forms at the future posterior midline of the developing embryo. Epiblast cells pass through the primitive streak and migrate anteriorly and laterally between the epiblast and visceral endoderm to form the mesoderm. Some epiblast cells penetrate and displace the visceral endoderm to become the embryonic endoderm. (*B*) Hensen's node is formed at the anterior end of the primitive streak. Cells passing through Hensen's node form the notochord, which plays a central role in later development of the neuraxis and vertebral column.

but the tissues that form the head also fold. As they do so, they move the presumptive heart and liver to their more central location in the developing embryo.

Recall that the primitive streak first forms at the posterior end of the embryo and elongates anteriorly. At its anterior termination is found a condensation of cells traditionally called Hensen's

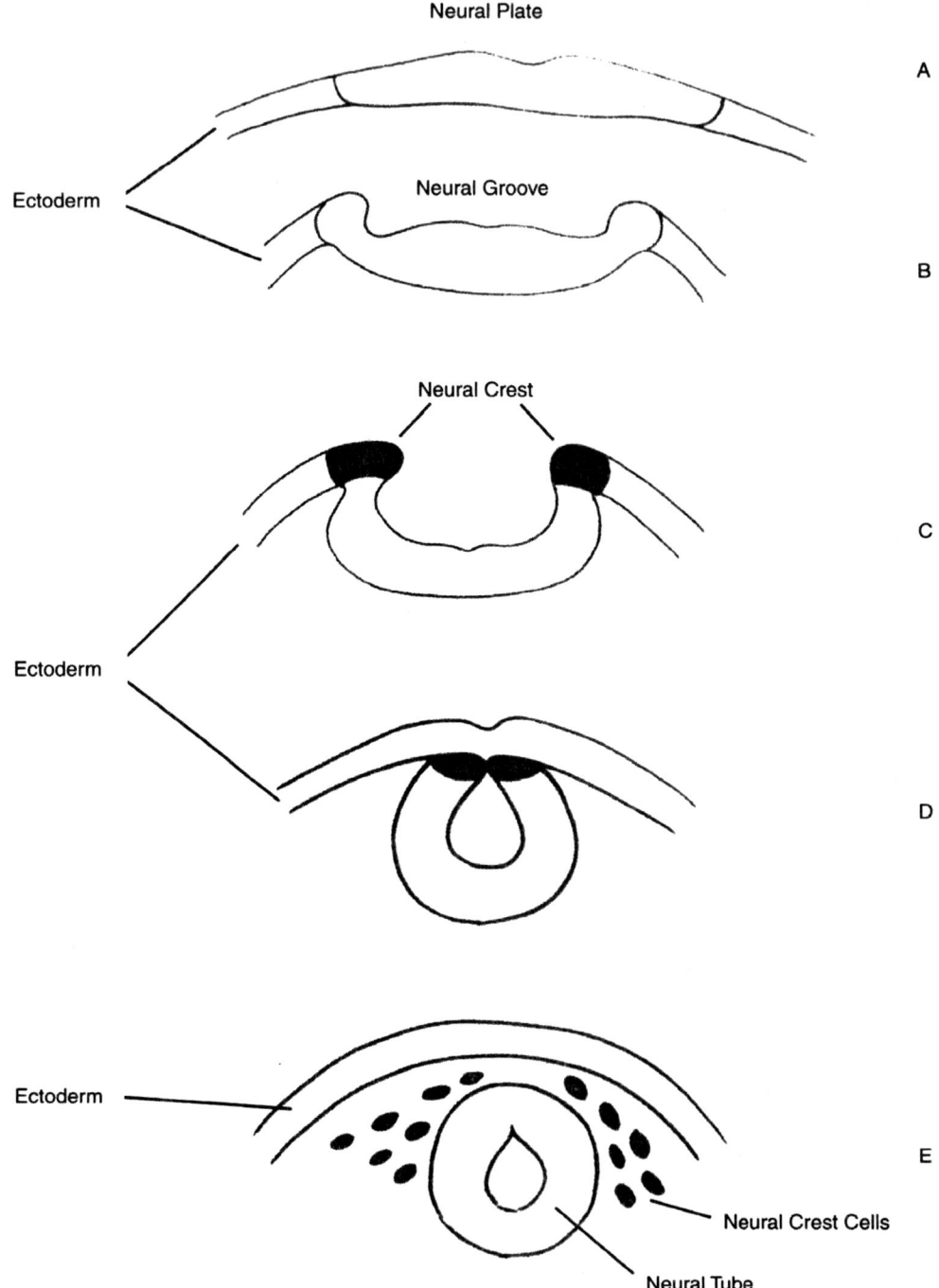

Figure 3-3 The process of neurulation: (A) neural plate, (B) neural groove, (C) appearance of neural crest, (D) completion of neural tube, and (E) formation of neural crest.

node. Epiblast cells that pass through this node and migrate anteriorly become organized into a solid rod called the *notochord*. The first cells to do so form its anteriormost, or most cranial, portion. The notochord passively elongates posteriorly, or caudally, as these new cells are added.

Once established, the notochord induces (causes to occur by interaction with) overlying ectodermal cells to form a layer called the *neural plate* (Figure 3-3). As with the notochord, formation of the neural plate occurs in a cranial to caudal sequence. Cranially, the neural plate is comparatively broad and gives rise to the three primary components of the developing brain: the prosencephalon (forebrain), mesencephalon (midbrain), and rhombencephalon (hindbrain). Caudal to the rhombencephalon the neural plate forms the spinal cord.

Beginning in the region of the future midbrain and spreading both cranially and caudally, the neural plate begins to fold along its midline. Its two sides, the *neural folds*, both move dorsally. The process is called *neurulation*, and the point of flexure is called the *neural groove*. The two sides then meet and fuse to form the neural tube. Upon fusion, the neural groove becomes the *neural canal*. Eventually the neural tube becomes isolated from surface ectoderm and is completely encased in mesoderm.

During neurulation, cells lying along the crests of the neural folds separate and migrate ventrally in the developing embryo. These *neural crest cells* eventually give rise to many diverse structures, including the ciliary and pupillary muscles of the eye, facial skeleton, parts of the developing teeth, melanocytes of the skin, connective and muscular tissue of the vessels around the heart, adrenal medulla, and various ganglia of the nervous system and their supportive cells. With respect to the vertebral column, neural crest cells form the sensory cell bodies of the dorsal root ganglia.

Formation of The Vertebral Column

As the notochord is forming during gastrulation (and elongating caudally), the mesoderm flanking the developing neural tube begins to differentiate (Christ and Ordahl, 1995). The mesoderm lying adjacent to either side of the developing *neuraxis*, the presumptive central nervous system and vertebral column, becomes the *paraxial mesoderm* (Figure 3-4). This gives rise to a series of tissue blocks called *somites* (Tam and Trainor, 1994). Lateral to the paraxial mesoderm lies a second sheet of mesodermal cells called the *lateral plate mesoderm*. It gives rise to much of the body wall as well as to the skeleton and connective tissues of the limbs, although not to the

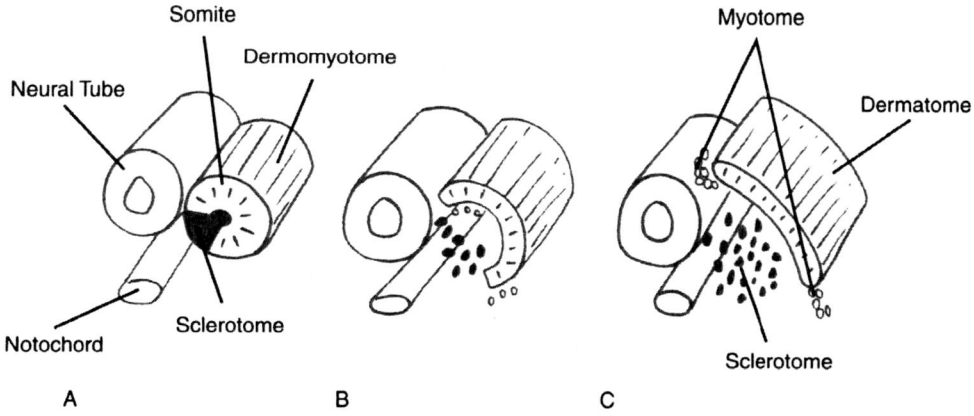

Figure 3-4 The somite: (A) A newly formed epithelial somite, before dissociation. (B) The sclerotome is beginning to lose its epithelial character, whereas the dermomyotome remains epithelial. (C) Sclerotomal cells migrate around the notochord and the dermomyotome has differentiated into its constituent parts, the dermatome and the myotome.

limb musculature, which arises from somitic mesoderm and migrates into the limbs (Chevallier, 1978; Chevallier et al., 1977). Finally, lying between the paraxial mesoderm and the lateral plate mesoderm on each side is a band of intermediate mesoderm. This gives rise to important structures of the reproductive and urinary systems.

Somite formation proceeds in a strict craniocaudal direction (Christ et al., 2000; Gossler and Hrabe de Angelis, 1998; McGrew and Pourquie, 1998). At anyone time, mature somites may be found at the cranial end of the embryo, with newly formed somites budding off from the paraxial mesoderm more caudally. Early in this process electron microscopy can detect swirls of mesenchymal cells in the paraxial mesoderm (Meier, 1979). These "presomites" are called *somitomeres* and are essentially the first stage in somite formation. Somites form when unsegmented paraxial mesenchyme undergoes a process called *epithelialization*. This process is facilitated by the expression of cell adhesion molecules within each somitomere (N-cadherin and fibronectin [Christ and Wilting, 1992; Horikawa et al., 1999; Lash et al., 1984; Linask et al., 1998]).

The early somite is a sphere of epithelial cells surrounded by a thick basement membrane (Solursh et al., 1979). At the center of each somite lies a lumen, termed the *somitocoele*, which contains loose mesenchymal cells that contribute to the formation of several structures, including the intervertebral discs (Huang et al., 1994, 1996). Each somite next undergoes a series of subdivisions (Christ and Wilting, 1992). It is first divided into a cranial and a caudal half by the *intrasegmental fissure* (called *von Ebner's fissure* in older literature). A second division separates each somite into dorsolateral and ventromedial portions. The ventromedial portion becomes a structure called the *sclerotome*, the precursor to the vertebrae, intervertebral discs, and ribs (Christ et al., 1998; Hall, 1977; Huang et al., 2000b). Separated by the intrasegmental fissure, cells in the cranial half of the sclerotome are less dense than those in the caudal half, and the developing axons and neural crest cells therefore migrate through only the cranial half of each sclerotome.

The sclerotome next subdivides into dorsolateral and ventromedial components, each of which further divides to become specific components of vertebral structure. The dorsolateral halves of the sclerotome give rise to the vertebral arch and the ribs. On each side of the future column, dorsolateral sclerotomal cells migrate around the neural tube to reach each other at the dorsal midline, making a complete neural arch around the developing spinal cord. At the same time, the more laterally placed sclerotomal cells expand laterally. In future thoracic vertebrae, these extensions form ribs. In nonthoracic vertebrae, they are smaller and form part of the transverse processes.

As the neural arches take shape, the notochord, which lies ventral to the neural arch, is surrounded only by extracellular matrix. The ventromedial group of sclerotomal cells migrates ventromedially to completely surround the notochord, encasing it in a *perichordal tube*, or *sheath* (Balling et al., 1996). Unlike the cells of the dorsolateral sclerotome, which maintain their segmental identity via the intrasegmental fissures, the perichordal tube loses its segmental borders and becomes one continuous mass of cells. Later, condensations appear in the perichordal tube (Bundy et al., 1998). These then develop into the intervertebral discs, whereas the less dense sections of the tube develop into the vertebral bodies. Whether these condensations are segmental or intersegmental (i.e., whether the vertebra of a specific level corresponds to the original somite level) has been one of the most controversial topics in vertebral column development; therefore, we must examine the theory of *resegmentation* (also known as "Neugliederung"; Figure 3-5).

Remak first proposed resegmentation in the middle of the last century (Verb out, 1976, 1985). He argued that a vertebra is made not from a single somite but from the caudal half of one somite and the cranial half of the somite just caudal to it. Therefore, one vertebra is the union of two adjacent somites. There are some exceptions to this "rule." The first four somites form part of the occipital bone instead of a vertebra (Kant and Goldstein, 1999). In addition, the body of the atlas (C1) has become incorporated into the axis during development, forming the *odontoid process*, or *dens*.

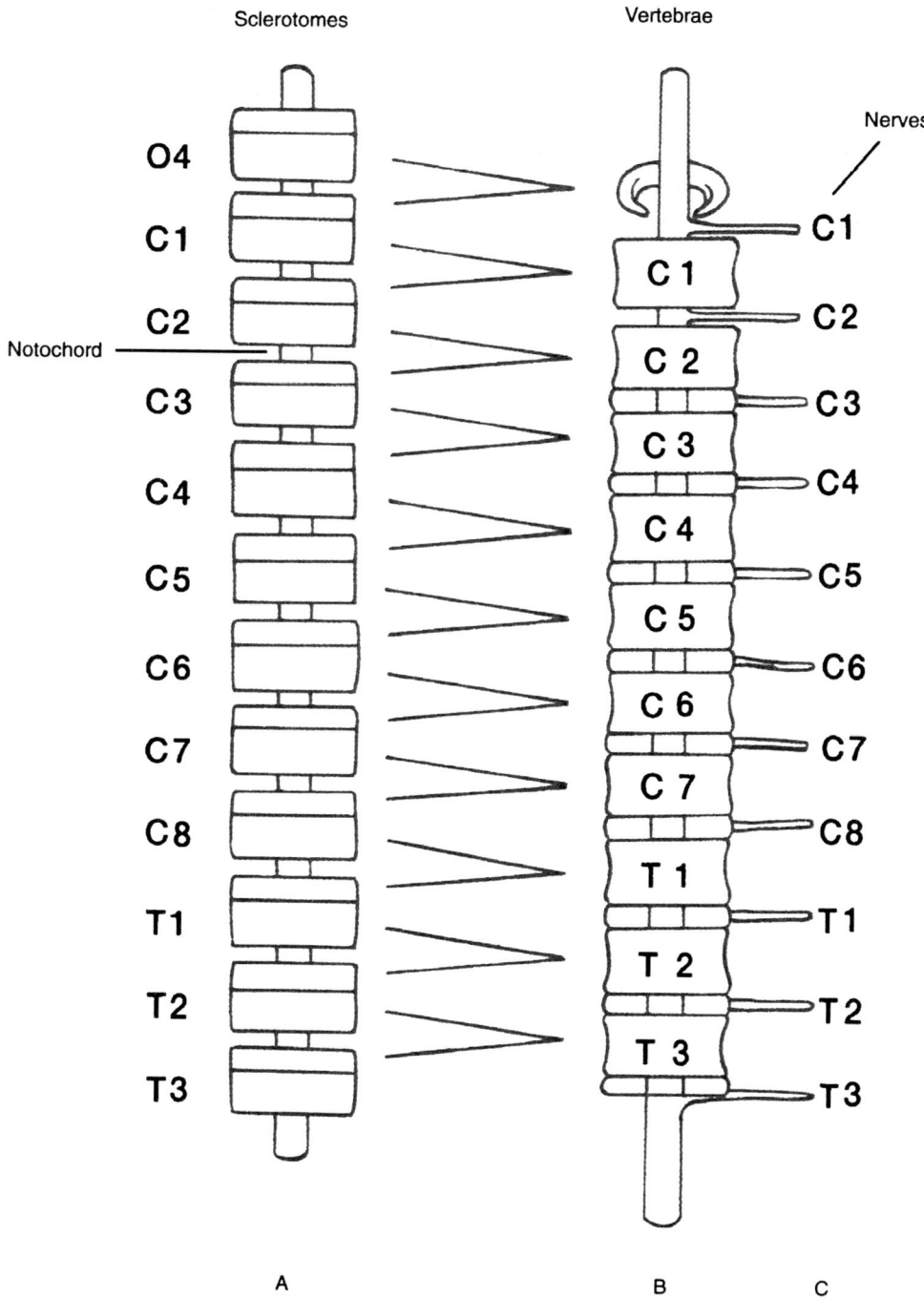

Figure 3-5 Resegmentation: (A) Sclerotomes O4 (fourth occipital sclerotome, which contributes to the occipital bone of the skull) through T3 (third thoracic sclerotome) are illustrated. Each sclerotome has surrounded the notochord and has become divided by intrasegmental fissures. (B) Vertebrae C1 through T3 after segmentation. Note that each centrum is formed by the union of half of two adjacent sclerotomes, with vertebra C1 being formed by the caudal half of sclerotome C1 and the cranial half of sclerotome C2, and that the notochord persists to eventually become the nucleus pulposus of each vertebral disc. (C) The vertebral nerves, each of which supplies a number of structures in the limb and body wall (C5 through T1 contribute to the brachial plexus).

The greatest criticism against resegmentation arises from the fact that no one has directly witnessed the splitting of a somite. However, labeling and quail transplantation studies of somites (Bagnall, 1992a, 1992b; Bagnall et al., 1988a, 1988b, 1989; Ewan and Everett, 1992; Stern and Keynes, 1987) provide evidence that supports resegmentation. The recent work of Huang and colleagues (2000a) definitively shows that the vertebral body, neural arch, pedicles, and transverse processes (including the distal ribs) are created by the fusion of two somites. Only the head of the rib is derived from a single somite.

As the perichordal sheath envelops the notochord, the notochordal cells in each of the presumptive vertebra undergo programmed cell death, or *apoptosis*. Those cells in the presumptive intervertebral disc, however, do not die (Walmsley, 1953) but contribute to the formation of the nucleus pulposus, the watery, jellylike deformable center of the intervertebral disc. By birth it appears that cells from the outer annulus fibrosis have replaced most of the notochordal-derived cells of the nucleus pulposus.

The dorsolateral part of the somite remains epithelial. It is termed the *dermomyotome*. It is really two structures: a dorsal *dermatome*, which gives rise to the dermis of the skin of the back, and the underlying *myotome*, which supplies all the muscle cells for the spinal column, body wall, and limbs, but not most of those for the muscles of the head (these cells are derived from the neural crest). Myotome formation is complex and not yet fully understood. Opinions as to the source of myogenic precursor cells differ, and two recent reviews (Kalcheim et al., 1999; Venters et al., 1999) indicate that there may be species-specific differences in the details of myotome formation.

Although it has long been commonly assumed that the epaxial and hypaxial portions of the body's musculature arise from a single muscle cell pool in the myotome, recent studies indicate that this is not the case (Denetclaw and Ordahl, 2000; Ordahl, 1993; Ordahl and Le Douarin, 1992; Williams and Ordahl, 1997). Using chick-quail chimeras and transplantation studies, workers have found that the epaxial muscles originate solely in the medial half of the somite. Furthermore, those myogenic cells that eventually migrate from the somite, such as those that invade the limb bud, are found exclusively in the lateral somite.

Development: Core Concepts

Before discussing the molecular controls underlying the processes described previously, we must first address several key embryological concepts and review some general developmental molecular mechanisms. The phenomenon of *induction* occurs when a group of cells influences the fate of another. Signals emanating from an initial cell population can launch a developmental program in a secondary group and determine the latter's fate. Cells affected by an inducing signal become specified to a particular cell type; after a time, the targeted cells are unable to return to their initial state or differentiate into another cell type. An important corollary to induction is that the affected cells must be *competent* to respond to an inducing signal. Competence is often time dependent (i.e., cells from an earlier or later developmental stage do not react to an inducer). The competence of cells usually depends on previous inductive interactions. Induction is achieved primarily through short-range signaling molecules (discussed later). Without induction, increasing embryonic complexity is impossible.

Other important developmental concepts are *pattern formation* and *positional information* (Neumann and Cohen, 1997; Wolpert, 1978, 1996; Woychik et al., 1997). Pattern formation is the process by which an embryo is constructed in space; positional information directs this assembly. Positional information enables a cell to differentiate into its fated type based on its placement in a three-dimensional coordinate system; for example, positional information dictates which cell develops into a chondroblast of a developing vertebra and which develops into a myoblast of a developing erector spinae muscle.

Pattern formation, mediated through positional information, occurs in several steps (Wolpert, 1978). First, cells interpret their position relative to other cells within the embryo; that is, communication among cells allows each to "understand" exactly what portion of its genetic code it should express. That expression causes the cell to become a particular type. Cells receive information from other cells because either they express specific receptors for extracellular signaling molecules that they and other cells can synthesize and secrete or they can make bonds with one another and use these bonds as avenues of expression. These signaling molecules include members of the Hedgehog, Bmp, Wnt, Fgf, and Notch families, as well as retinoic acid.

Once their positional information is established, the cells then carry out specific programs of genetic expression by activating *transcription factors*. These are nuclear proteins that act upon their own DNA to cause production of other proteins, some of which may also be transcription factors. Obviously, the most fundamental transcription factors are those expressed earliest in development. In the development of the spine, one group of transcription factors of crucial importance are the *homeobox genes*.

Homeobox genes are found throughout the animal kingdom in organisms as diverse as fruit flies, chicks, and mice (Figure 3-6; Krumlauf, 1994). They specify much of the embryonic anteroposterior axis, as well as secondary axes such as those in the limbs. Their name derives from the *homeotic* genes of the fly — genes that, when mutated, replace one body part with another, a phenomenon called *homeosis*. One subset of these genes, the *Hox* genes, is essential to axis formation. All *Hox* genes contain a specific region of DNA, the *homeobox* (hence the name), which is highly conserved among diverse taxa. The homeobox is approximately 180 base pairs long and codes for a 60-amino acid region of the gene's protein product called the *homeodomain*. The homeodomain binds to specific regions of DNA to control its target gene's transcription, either negatively or positively (Duboule, 1994).

In fruit flies there is only one set of 8 homeobox genes, but in vertebrates there are four sets (*Hoxa*, *Hoxb*, *Hoxc*, *Hoxd*), each located on a different chromosome. The additional three sets in vertebrates have arisen by two duplications of the original set. Therefore, many *Hox* genes share sequence similarities with those in the other three complexes. These are called paralogs. As an example, *Hox12* paralogs share a degree of similarity not found in other paralogous groups (such as *Hox11* or *Hox13*). Although each chromosome possesses 13 possible slots, not every *Hox* complex contains all 13 genes (there are only 38 *Hox* genes out of a possible 52 [4 groups X 13 paralogous slots = 52 genes]). The missing genes have been lost during evolution.

Hox genes are ordered sequentially on a chromosome, with the group 1 paralogs located at the 3' area of the DNA master strand and the group 13 paralogs at the 5' region (Duboule, 1994). Genes located at the 3' end are expressed in more anterior parts of the embryo than those in the 5' end, which are expressed more posteriorly; for example, group 3 paralogs are expressed in the cervical region, whereas group 11 paralogs are expressed in the presumptive sacrum.

This correlation between a *Hox* gene's placement on the chromosome and its expression along the embryonic anteroposterior axis is termed *spatial colinearity*. Additionally, the 3' genes are expressed earlier in development than those in the 5' region, a phenomenon called *temporal colinearity*. If either spatial or temporal colinearity is disrupted during development, a partial or full homeosis (transformation into a vertebra of another type) may occur.

The upstream regulation of *Hox* expression is poorly understood. Data have been culled primarily from studies of *Hox*-related fruit fly genes and vertebrate brain development (Charite et al., 1998; Hanson et al., 1999; Manzanares et al., 1999). These studies identify the genes *cdx* and *kreisler* as important guides in assigning the positional information relevant to *Hox* expression in the developing brain and spinal cord (Subramanian et al., 1995). Several other genes have been found to affect *Hox* gene expression in the developing spinal column. They include *Mll* and *Bmi1*, which are related to the fruit fly gene families Trithorax and Polycomb-group (Hanson et al., 1999). Manipulations of these genes result in vertebral homeoses. Retinoic acid, a derivative of vitamin A (Durston et al., 1997; Marshall et al., 1996), is also a potent regulator of *Hox* expression (Kessel, 1992; Kessel and Gruss, 1991). A recent discovery is growth and differentiation factor 11 (*Gdf11*). *Gdf11* mutants show altered *Hox* expression that, as with retinoic acid, results in vertebral homeoses, indicating that it may be upstream of *Hox* (McPherron et al., 1999). Although these preliminary studies give us important clues as to the control of *Hox*, further work must be completed in the search of more universal regulators of *Hox* expression.

Although *Hox* genes are the primary regulators of embryonic patterning, other homeobox genes also contribute. *Pax* genes are another well-studied homeobox gene family (Dahl et al., 1997; Gruss and Walther, 1992). *Pax* genes also contain a highly conserved region of DNA, called the paired box (hence the name). Nine *Pax* genes are known, *Pax1* through *Pax9*. *Pax* genes have been implicated in many areas of development and are primarily, but not exclusively, expressed in such ectodermally derived structures as the eye and brain. Our interest lies in *Pax1*, *Pax3*, and *Pax9*, because their expression in the mesoderm is necessary for normal vertebral column development.

The signaling molecule *Hedgehog (Hh)* was also first discovered in the fruit fly. It is a secreted protein that mediates cell-to-cell interactions (Fietz et al., 1994). Its effect is exerted over a distance of only a few cell diameters (Hammerschmidt, 1997; Johnson and Tabin, 1995), and in fly development it helps determine cell polarity (i.e., resulting in a tissue axis) and fate. Genes related to *Hh* have been discovered in vertebrates and show high homology in both their genetic sequence and signaling pathways (Goodrich et al., 1996). Of the *Hh* homologues identified in vertebrates, the best known are *Sonic hedgehog (Shh), Indian hedgehog (Ihh),* and *Desert hedgehog (Dhh)* (Hammerschmidt, 1997). *Dhh* is involved in the development of the reproductive, nervous, and cardiovascular systems, whereas Ihh is involved in the digestive system and bone growth plates. Shh is integral to several key organizing centers of the embryo and is used several times in the developing vertebral column.

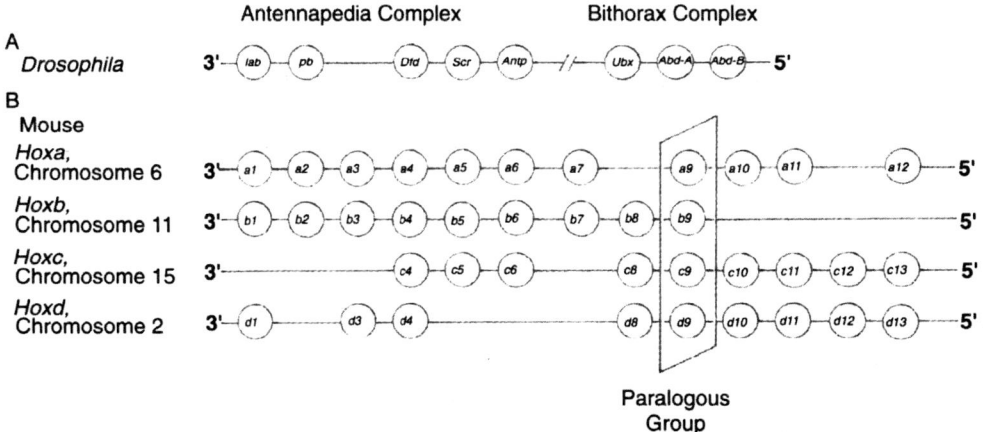

Figure 3-6 The mammalian *Hox* gene complex, homologous with (*A*) the fruit fly genes. (*B*) Each mammalian complex (*Hoxa, Hoxb, Hoxc, Hoxd*) is located on a different chromosome. Each cluster of genes is ordered sequentially, from *Hox1* to *Hox13*. *Hoxa9*, *Hoxb9*, *Hoxc9*, and *Hoxd9* are classified together as an example of a paralogous group. Note that no chromosome possesses the full complement of genes. The missing genes were deleted during evolution.

Among the most important discoveries is the fact that vertebrate *Hedgehog* genes are coexpressed with *bone morphogenetic protein (Bmp)* genes at different sites in the embryo, indicating that *Bmp*s may be downstream targets of *Hh* expression. This situation is analogous to that found in the fly, in which the *Bmp*-related gene decapentaplegic is a downstream target of *Hh* (Bitgood and McMahon, 1995).

*Bmp*s received their name because the first gene discovered in this family produced bone when it was applied experimentally to ectopic tissue sites. The name is misleading, however, because *Bmp*s are used in many situations that are unrelated to the development of the skeletal system. *Bmp*s are members of the larger *Tgf-ß* (transforming growth factor-ß) superfamily and, like *Hh*, are secreted proteins. With more than 20 members, *Bmp*s are involved in cell proliferation, differentiation, and determination (Hogan, 1996). *Bmp*s help control a variety of embryonic processes, such as mesoderm specification, neuronal patterning, bone development (including the vertebral column), and formation of several organs and organ systems (Hogan, 1996).

*Wnt*s (pronounced "wints") are yet another family of secreted cell-to-cell signaling molecules involved in many aspects of embryonic patterning (Cadigan and Nusse, 1997; Loureiro, 1999; Moon et al., 1997). *Wnt*s are secreted glycoproteins, which, unlike *Shh*, can be found several cell diameters away from their initial source. Like *Bmp*s, *Wnt*s are involved in cell polarity and neuronal development, and they are found in a variety of adult tissues. Uncontrolled Wnt expression has been implicated in several cancers. As is the case for the other signaling molecules, *Wnt*s are also implicated in vertebral column morphogenesis. *Fibroblast growth factors* (*Fgf*s) are a group of multifunctional secreted factors essential for proper embryonic and postembryonic development (Goldfarb, 1996). As in the other signaling molecules discussed here, *Fgf*s control the proper formation of the lungs, hair, and limbs. They have the ability to initiate the growth of entire structures, such as the limbs. They act by several mechanisms, primarily by inducing growth (hence the name *growth factor*). Alternately, they may prevent growth while inducing a cell lineage to differentiate. *Fgf*s and *Bmp*s are likely downstream targets of *Hox* genes.

The Notch signaling pathway is quite complex, and its importance in embryogenesis is confirmed by its ubiquitous presence in the animal kingdom (Artavanis-Tsakonas et al., 1999; Nye, .1999). First discovered in the fruit fly Drosophila, Notch proteins are cell receptors; cells communicate through the Notch pathway. Nye stated in 1999 that "[cells] that receive a Notch signal assume one cell fate, whereas those that do not receive it assume another fate." Many genes are involved in transducing the signal received by Notch cell receptors, including members of the Deltalike (*Dll*) family, *Jagged*, and *lunatic fringe*. The Notch pathway is used in a variety of developmental situations.

The interplay between *Hox* and *Pax* genes and the signaling molecules discussed here is quite complex. In the next section we find that these relationships are used repeatedly and in different combinations in forming a vertebra from undifferentiated paraxial mesoderm.

Molecular Control of Vertebral Column Formation

A number of the genes previously described play fundamental roles in the development of the vertebral column. It is difficult to say that one is more important than another, because all must properly interact for normal formation to proceed. The effects of some gene mutants are very striking, however; for example, *Wnt3a* mutants form ectopic neural tubes and fail to show any paraxial mesodermal markers (Yoshikawa et al., 1997; for additional effects of Wnt3a, see also Ikeya and Takada, 2001).

Wnt3a guides paraxial mesoderm formation through its effect on the expression of the T-box gene *T (Brachyury)* and other T-box family genes, such as Tbx6 (Chapman et al., 1996; Yamaguchi et al., 1999); mutations in these genes result in severe abnormalities. *Tbx6* mouse mutants show only a few irregularly shaped cervical somites, with numerous other malformations as well (Chapman and Papaioannou, 1998).

One of the most interesting problems in spinal development is the determination of segment number (Cooke, 1975; Cooke and Zeeman, 1976; Kerszberg and Wolpert, 2000; Schnell and Maini, 2000). How does an embryo determine the proper species specific number of somites? When an embryo is experimentally reduced in size, it still forms the same number of somites as in a full-sized specimen. How is this accomplished? One proposal is called the clock-and-wavefront model. In this model, cells are assumed to have clocklike control devices that cause them to cycle at regular intervals until they receive a signal that leads to further differentiation. During spinal development that signal is sent down the column as a "wavefront" (Cooke, 1998).

The clock-and-wavefront theory had little supporting evidence before the analysis of the avian gene *c-hairy1*, which is related to a Drosophila segmentation gene (Palmeirim et al., 1997). The *c-hairy1* gene is expressed dynamically in the paraxial mesoderm of the chick, rising to high and then sinking to undetectable levels with regular periodicity.

The expression of another gene, called *lunatic fringe*, also regularly cycles in the paraxial mesoderm before somitogenesis, or somite formation, and may be a target of *c-hairy1* (Aulehla and Johnson, 1999; Dale and Pourquie, 2000; Forsberg et al., 1998; Holley et al., 2000; McGrew et al., 1998;). Mutants of *lunatic fringe* fail to form boundaries between and within somites, in addition to undergoing other changes. Several genes within the Notch pathway are also essential in border maintenance, as well as synchronization of the segmentation "clock" (Jiang et al., 2000; Jouve et al., 2000).

Notch1 mutants exhibit delayed and disorganized somitogenesis (Conlon et al., 1995). *Dll1* is required for normal polarity in somites (Beckers et al., 2000; Hrabe de Angelis et al., 1997), and *Dll3* mutants lack segment polarity because epithelialization fails to occur (Kusumi et al., 1998). In these mutants, the descendants of the somite, the sclerotome and dermomyotome, do form but are disorganized. Analysis of these and other genes has shown that segmentation and somitogenesis are different processes and that epithelial somites are required for the proper patterning of the sclerotome and dermomyotome.

Genes not directly involved with the Notch pathway also playa role in somite boundary formation. The gene Paraxis is a factor that controls epithelialization of somites (Barnes et al., 1997; Burgess et al., 1996; Correia and Conlon, 2000; Johnson et al., 2001; Sosic et al., 1997; see also Wilson-Rawls et al., 1999) and is required for the proper development and patterning of the body segments.

Mutations in other genes can also have dramatic effects on the formation of the spine. These genes, and the signals they convey, are expressed in two separate spinal organizing centers. The first such center is the notochord and the ventral part of the neural tube known as the floorplate. The second center is the dorsal part of the neural tube known as the roofplate and the adjoining lateral plate mesoderm. The first center primarily determines the dorsoventral patterning of the vertebral column; the second dictates mediolateral patterning. There is, however, interplay between these two centers. We examine dorsoventral patterning first.

If a notochord or a floorplate is experimentally moved to a site in the embryo where it is not normally found (e.g., dorsal to the neural tube), dramatic alterations occur in the adult form. The parts of the somite that normally form the dermomyotome instead take on the identity of a secondary sclerotome, resulting in a vertebral column with an excess of bone but a scarcity of muscle. Shh is produced by the notochord and floorplate (Chiang et al., 1996; Fan et al., 1995; Fan and Tessier-Lavigne, 1994; Johnson et al., 1994; Marcelle et al., 1999) and ultimately controls sclerotomic determination.

A downstream target of Shh is *Pax1* (Balling et al., 1996; Barnes et al., 1996a, 1996b; Ebensperger et al., 1995; Furumoto et al., 1999; Peters et al., 1999; Smith and Tuan, 1995). *Pax1* is expressed in the more ventral parts of the sclerotome, those parts that become the vertebral bodies, intervertebral discs, pedicles, and proximal ribs (Deutsch et al., 1988; Koseki et al., 1993; Wallin et al., 1994). Mutations in *Pax1*, which are known from the mouse mutant family *undulated* (Balling et al., 1988; Chalepakis et al., 1991; Deutsch et al., 1988; Dietrich and Gruss, 1995; Wallin et al., 1994), result in severe disorganization or lack of vertebral bodies and intervertebral discs. The neural arches remain relatively unaffected, whereas the proximal ribs are missing or deformed.

Although Shh controls much of the development of the bony parts of the vertebrae, other factors also come into play. Mutations in the gene *Uncx4.1* produce vertebrae that lack pedicles and transverse processes (Leitges et al., 2000; Mansouri et al., 2000). The gene *Bagpipe* appears to control proliferation of the sclerotome and its ability to interact with the notochord (Lettice et al., 1999; Tribioli and Lufkin, 1999). The development of the spinous process is controlled through local signaling by the roofplate and overlying ectoderm. These tissues express *Bmp4*. *Bmp4* in turn regulates the expression of the homeobox-containing genes *Msx1* and *Msx2*. Increased expression of these genes results in larger spinous processes, whereas their removal produces unfused neural arches lacking spinous processes, a condition resembling spina bifida (Monsoro-Burq et al., 1994, 1996).

Signals from the second organizing center, the roofplate, surface ectoderm, and lateral plate mesoderm, control dorsoventral patterning of the somite, often by counteracting the effects of one another. *Wnt1* and *Wnt3a* are expressed in the dorsal roofplate of the neural tube (Capdevila et al., 1998; Fan et al., 1997; Ikeya and Takada, 1998; Reshef et al., 1998; Stern et al., 1995). In this context, Wnts participate in two processes.

First, *Wnts* may prevent the dorsal expansion off Shh and therefore aid in forming the sclerotome (Borycki et al., 2000; Capdevila et al., 1998). Second, *Wnt1* and *Wnt3a* regulate medial dermomyotome formation by counteracting the effect of the lateral plate, which produces *Bmp4* (Pourquie et al., 1996; Tonegawa et al., 1997). *Wnts* accomplish this by their production of a *Bmp* antagonist (a molecule that counteracts the effect of another molecule), *Noggin* (Hirsinger et al., 1997; Marcelle et al., 1997; McMahon et al., 1998). Correct specification of the mediolateral axis in the developing somite is consequently dependent on the proper interplay between *Wnts*, *Noggin*, and *Bmp4*.

A major target of *Bmp4*, which is produced in the lateral plate, is *Pax3*. *Pax3* is first expressed widely in presomitic tissues, but as the somite matures, its expression is restricted to the dermomyotome (Goulding et al., 1994; Maroto et al., 1997; Williams and Ordahl, 1994). *Pax3* regulates the onset of myogenesis, the formation of muscle tissue from more primitive cells. *Pax3* mutants lack ventral body wall musculature, indicating that the gene aids in the long-range migration of progenitor muscle cells (Tremblay et al., 1998). Mouse *splotch* mutants, who have a mutation in *Pax3*, show rib truncations, probably because of *Pax3*'s effect on intercostal muscles (Dickman et al., 1999; Henderson et al., 1999).

We should note that other axial structures have been implicated in the control and maintenance of myogenesis. These studies have implicated the notochord, floorplate, and roofplate (Buffinger and Stockdale, 1995; Munsterberg and Lassar, 1995; Pownall et al., 1996; Spence et al., 1996; Stern and Hauschka, 1995). The specific role of each of these structures in myogenesis is still debated.

The role of *Pax3* appears to be in its mediation of signals involved in the activation of myogenic differentiation factors, or Mdfs (Dietrich, 1999; Maroto et al., 1997). Myogenic differentiation factors include *myoD*, *myogenin*, *myf5*, and *mrf4*. They are expressed at different times and in different cell lineages (Currie and Ingham, 1998). Regardless, different members of this family share much functional redundancy, as evidenced by the fact that deletion of one gene often has no detectable phenotypic effect. Both *myoD* and *mrf4* have redundant functions (Rawls et al., 1998), as do *myf5* and *myogen* in (Wang and Jaenisch, 1997).

Although *myoD* and *myf5* appear necessary for initiation of muscle formation, downstream products such as *mrf4*, myogenin, and Mef2 (from another family of myogenic regulators) are required for maintenance of myogenesis (Brand-Saberi and Christ, 1999). *Growth and differentiation factor 8 (Gdf8)*, also known as myostatin, determines muscle mass by regulating mitosis in myoblasts. Mutants of this gene produce double the muscle mass of the wildtype (McPherron et al., 1997), as seen most spectacularly in the double-muscled European cattle breeds Belgian Blue and Piedmontese (McPherron and Lee, 1997).

Hox control of vertebral specification

Although we have thoroughly addressed the molecular processes in vertebral assembly, we have not yet discussed the molecular controls that determine vertebral identities. This role is clearly attributable to *Hox* genes. *Hox* genes encode positional information for vertebral identity before somitogenesis (Figure 3-7).

When presumptive thoracic paraxial mesoderm is transplanted ectopically to the cervical region, cervical vertebrae with ribs are formed (Kieny et al., 1972; for an update of this experiment, see Nowicki and Burke, 2000 and Burke and Nowicki, 2001). *Hox* genes obey spatial colinearity and are expressed in increasing numbers the farther caudally one moves down the vertebral column. Targeted gene deletions* have been instrumental in determining the functions of *Hox* genes in vertebral identity.

*The literature of *Hox* mutants and the vertebral column is voluminous. References may be found in the categories that follow.

General: Burke, 2000; Burke et al., 1995; Gaunt, 1991, 1994, 2000, 2001; Gaunt and Strachan, 1994; Gaunt et al., 1999; Kessel, 1992; Kessel and Gruss, 1990, 1991; Medina-Martinez et al., 2000; Richardson et al., 1998; Subramanian et al., 1995. *Hox3*: Condie and Capecchi, 1993, 1994; Manley and Capecchi, 1997. *Hox4*: Boulet and Capecchi, 1996; Folberg et al., 1999; Horan et al., 1994; Horan et al., 1995a, 1995b; Kostic and Capecchi, 1994; Ramirez-Solis et al., 1993; Saegusa et al., 1996. *Hox5*: Aubin et al., 1998; Jeannotte et al., 1993. *Hox6*: Jegalian and De Robertis, 1992. *Hox7*: Chen et al., 1998; Kessel et al., 1990. *Hox8*: Belting et al., 1998; Charite et al., 1994; Charite et al., 1995; Le Mouellic et al., 1992; Pollock et al., 1992; Shashikant and Ruddle, 1996; van den Akker et al., 2001. *Hox9*: Chen and Capecchi, 1997; de la Cruz et al., 1999; Fromental-Ramain et al., 1996; Suemori et al., 1995. *Hox10*: Rijli et al., 1995; Wahba et al., 2001. *Hox11*: Beckers et al., 1996; Davis and Capecchi, 1994; Davis et al., 1995; Favier et al., 1995; Gerard et al., 1996; Hostikka and Capecchi, 1998; Small and Potter, 1993; Zákány et al., 1996.

Both posterior and anterior vertebral transformations occur in mutants. Moreover, the severity of the mutant phenotype appears to be dose dependent; for example, *Hoxd3* mutants show anterior transformations of parts of the atlas into the occipital bone and of the axis into atlaslike structures. With the addition of *Hoxa3* mutants, however, the dysmorphology increases. Double homozygous mutants for *Hoxa3* and *Hoxd3* show a complete loss of the atlas, with only a cartilage rudiment in its place (Condie and Capecchi, 1994). Furthermore, these studies show that paralogous genes may have redundant functions. The functional compensation of paralogous genes can rescue the loss of function of the targeted deletions and can hinder easy interpretation of results.

Kessel and Gruss (1991) proposed that individual vertebrae are specified by a particular combination of *Hox* genes expressed at a vertebral level-a *Hox* code. Duboule (1995) as well as Condie and Capecchi (1993, 1994) have suggested instead that combinations of *Hox* genes do not specify particular vertebrae but determine rates of proliferation in the mesenchymal vertebral precursors. Alterations in overall growth rates within different parts of a vertebra could explain the discrepancy in size and shape between cervical and lumbar vertebrae; for example, Condie and Capecchi (1994) suggest that the lack of an axis in *Hoxd3* and *Hoxa3* mutants results because the cell condensations forming this vertebra failed to proliferate.

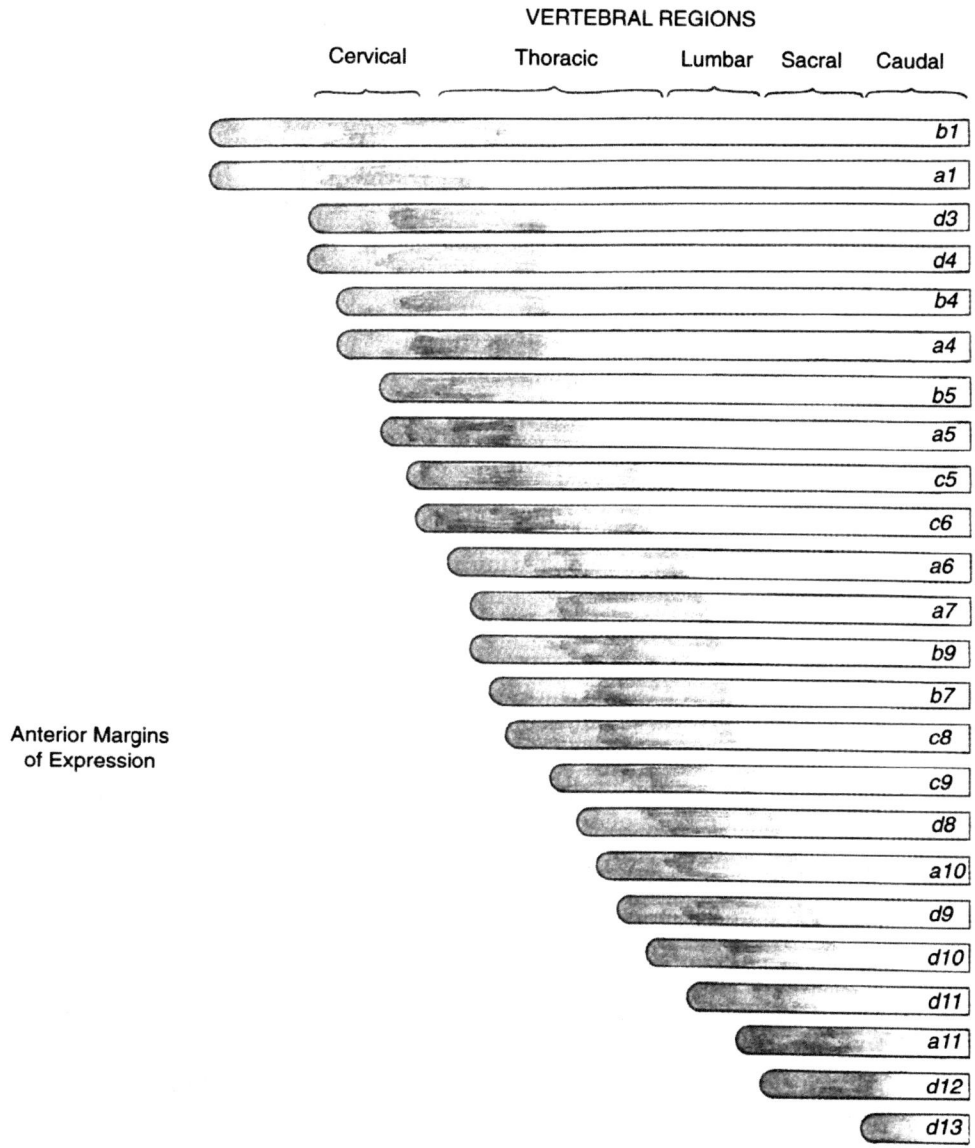

Figure 3-7 *Hox* gene expression along the developing spinal column. The darker banding indicates the strongest and most anterior limits of each gene's expression. The expression of each gene becomes weaker as the distance increases between the point of its initial expression and the most posterior extent of its domain.

Burke and colleagues (1995) proposed two equally possible mechanisms by which *Hox* genes specify vertebrae. First, *Hox* genes may score an invariant coordinate system along the craniocaudal axis that does not vary with respect to segment number. Changes in vertebral type would be specified by changes in the targets of *Hox*. There would be no correlation between the *Hox* position along the craniocaudal axis and a particular vertebral type.

Alternatively, vertebral morphology could change in concert with differences in *Hox* expression patterns up and down the body axis. *Hox* genes would therefore directly control changes in vertebral type. Differences in the timing and expression domains of *Hox* genes would cause shifts in type. Burke and colleagues (1995) and Belting and colleagues (1998) conclusively support the latter model. In a comparison of *Hox* expression between chicks and mice, these researchers found that specific genes are always associated with a change of one vertebral type to another. Important associations are found in *Hoxc5* and *Hoxc6*, which are connected to the cervicothoracic transition; *Hoxa9*, *Hoxb9*, and *Hoxc9*, which are associated with the end of the thoracic vertebrae; and *Hox10* paralogs, which are expressed in the first sacral vertebra.

Furthermore, Belting and colleagues (1998) noted a correlation between *Hoxc8* and the brachial plexus. They also found that the regulatory circuitry of this gene differed only very slightly between the mouse and the chick. Therefore, minor and subtle changes in the regulation of where and when *Hox* genes are expressed may have large downstream effects on adult morphology.

Finally, Chen and colleagues (1998) have noted that alterations caused by *Hox* mutants are often centered on transitional zones from one type of vertebra to another, even when the anterior limits of the affected genes do not correspond to these points. Transition zones appear to be more sensitive to *Hox* mutations than are vertebrae within a particular segmental zone.

Mutated genes with an anterior expression domain within a particular vertebral type often elicit their effects in the nearest caudal transition. This observation led the authors to suggest that "vertebral types are made in blocks and that the temporal phasing of these blocks is mediated by the ordered dosage of *Hox* gene expression within each region" (Chen et al., 1998, p. 55). Although this is an interesting hypothesis, further work is needed to support it.

Our understanding of vertebral column development has burgeoned considerably over the past decade. Progress has been made on many fronts, from the clarification of descriptive models of vertebral development, such as resegmentation, to the discovery of the genes that control the initiation, specification, and maintenance of vertebral morphogenesis. Future work will further expand our knowledge of the processes underlying the creation of the vertebral column.

References Cited

Artavanis-Tsakonas, S, Rand, MD, and Lake, RJ (1999). Notch signaling: Cell fate control and signal integration in development. *Science*, 284, 770-776.

Aubin, J, Lemieux, M, Tremblay, M, Behringer, RR, and Jeannotte, L (1998). Transcriptional interferences at the *Hoxa4/Hoxa5* locus: Importance of correct *Hoxa5* expression for the proper specification of the axial skeleton. *Developmental Dynamics*, 212, 141-156.

Aulehla, A, and Johnson, RL (1999). Dynamic expression of *lunatic fringe* suggests a link between notch signaling and an autonomous cellular oscillator driving somite segmentation. *Developmental Biology*, 207, 49-61.

Bagnall, KM (1992a). Segmentation and compartments in the vertebrate embryo. In R Bellairs, EJ Sanders, and JW Lash (Eds), *Formation and Differentiation of Early Embryonic Mesoderm* (pp 133-147). Plenum Press, New York.

Bagnall, KM (1992b). The migration and distribution of somite cells after labelling with the carbocyanine dye, DiI: The relationship of this distribution to segmentation in the vertebrate body. *Anatomy and Embryology*, 185, 317-324.

Bagnall, KM, Higgins, SJ, and Sanders, EJ (1988). The contribution made by a single somite to the vertebral column: Experimental evidence in support of resegmentation using the chick-quail chimaera model. *Development*, 103, 69-85.

Bagnall, KM, Higgins, SJ, and Sanders, EJ (1989). The contribution made by cells from a single somite to tissues within a body segment and assessment of their integration with similar cells from adjacent segments. *Development*, 107, 931-943.

Bagnall, KM, Sanders, EJ, Higgins, SJ, and Learn, H (1988b). The effects of somite removal on vertebral formation in the chick. *Anatomy and Embryology*, 178, 183-190.

Balling, R, Deutsch, U, and Gruss, P (1988). *undulated*, a mutation affecting the development of the mouse skeleton, has a point mutation in the paired box of *Pax 1*. *Cell*, 55, 531-535.

Balling, R, Neubuser, A, and Christ, B (1996). *Pax* genes and sclerotome development. *Seminars in Cell and Developmental Biology*, 7, 129-136.

Barnes, GL, Alexander, PG, Hsu, CW, Mariani, BD, and Tuan, RS (1997). Cloning and characterization of chicken Paraxis: A regulator of paraxial mesoderm development and somite formation. *Developmental Biology*, 189, 95-111.

Barnes, GL, Hsu, CW, Mariani, BD, and Tuan, RS (1996a). Chicken *Pax-1* gene: Structure and expression during embryonic somite development. *Differentiation*, 61, 13-23.

Barnes, GL, Mariani, BD, and Tuan, RS (1996b). Valproic acid-induced somite teratogenesis in the chick embryo: Relationship with *Pax-1* gene expression. *Teratology*, 54, 93-102.

Beckers, J, Gerard, M, and Duboule, D (1996). Transgenic analysis of a potential *Hoxd-11* limb regulatory element present in tetrapods and fish. *Developmental Biology*, 180, 543-553.

Beckers, J, Schlautmann, N, and Gossler, A (2000). The mouse rib mutation disrupts anterior-posterior somite patterning and genetically interacts with a *Delta1* null allele. *Mechanisms of Development*, 95, 35-46.

Belting, HG, Shashikant, CS, and Ruddle, FH (1998). Modification of expression and cis-regulation of *Hoxc8* in the evolution of diverged axial morphology. *Proceedings of the National Academy of Sciences of the United States of America*, 95, 2355-2360.

Bitgood, MJ, and McMahon, AP (1995). *Hedgehog* and *Bmp* genes are co-expressed at many diverse sites of cell-cell interaction in the mouse embryo. *Developmental Biology*, 172, 126-138.

Borycki, AG, Brown, AMC, and Emerson, CP (2000). Shh and Wnt signaling pathways converge to control Gli gene activation in avian somites. *Development*, 127, 2075-2087.

Boulet, AM, and Capecchi, MR (1996). Targeted disruption of *hoxc-4* causes esophageal defects and vertebral transformations. *Developmental Biology*, 177, 232-249.

Brand-Saberi, B, and Christ, B (1999). Genetic and epigenetic control of muscle development in vertebrates. *Cell and Tissue Research*, 296, 199-212.

Buffinger, N, and Stockdale, FE (1995). Myogenic specification of somites is mediated by diffusible factors. *Developmental Biology*, 169, 96-108.

Bundy, J, Rogers, R, Hoffman, S, and Conway, SJ (1998). Segmental expression of aggrecan in the nonsegmented perinotochordal sheath underlies normal segmentation of the vertebral column. *Mechanisms of Development*, 79, 213-217.

Burgess, R, Rawls, A, Brown, D, Bradley, A, and Olson, EN (1996). Requirement of the Paraxis gene for somite formation and musculoskeletal patterning. *Nature*, 384, 570-573.

Burke, AC (2000). *Hox* genes and the global patterning of the somitic mesoderm. *Current Topics in Developmental Biology*, 47, 155-181.

Burke, AC, Nelson, CE, Morgan, BA, and Tabin, C (1995). *Hox* genes and the evolution of vertebrate axial morphology. *Development*, 121, 333-346.

Burke, AC, and Nowicki, JL (2001). *Hox* genes and axial specification in vertebrates. *American Zoologist*, 41, 687-697.

Cadigan, KM, and Nusse, R (1997). Wnt signaling: A common theme in animal development. *Genes and Development*, 11, 3286-3305.

Capdevila, J, Tabin, C, and Johnson, RL (1998). Control of dorsoventral somite patterning by Wnt-1 and ß-catenin. *Developmental Biology*, 193, 182-194.

Chalepakis, G, Fritsch, R, Fickenscher, H, Deutsch, D, Goulding, M, and Gruss, P (1991). The molecular basis of the *undulated/Pax-1* mutation. *Cell*, 66, 873-884.

Chapman, DL, Agulnik, I, Hancock, S, Silver, LM, and Papaioannou, VE (1996). *Tbx6*, a mouse T-box gene implicated in paraxial mesoderm formation at gastrulation. *Developmental Biology*, 180, 534-542.

Chapman, DL, and Papaioannou, VE (1998). Three neural tubes in mouse embryos with mutations in the T-box gene *Tbx6*. *Nature*, 391, 695-697.

Charite, J, de Graaff, W, Consten, D, Reijnen, MJ, Korving, J, and Deschamps, J (1998). Transducing positional information to the *Hox* genes: Critical interaction of cdx gene products with position-sensitive regulatory elements. *Development*, 125, 4349-4358.

Charite, J, de Graaff, W, and Deschamps, J (1995). Specification of multiple vertebral identities by ectopically expressed *Hoxb-8*. *Developmental Dynamics*, 204, 13-21.

Charite, J, de Graaff, W, Shen, S, and Deschamps, J (1994). Ectopic expression of *Hoxb-8* causes duplication of the ZPA in the forelimb and homeotic transformation of axial structures. *Cell*, 78, 589-601.

Chen, F, and Capecchi, MR (1997). Targeted mutations in *hox*a-9 and *hox*b-9 reveal synergistic interactions. *Developmental Biology*, 181, 186-196.

Chen, F, Greer, J, and Capecchi, MR (1998). Analysis of *Hoxa7/Hoxb7* mutants suggests periodicity in the generation of the different sets of vertebrae. *Mechanisms of Development*, 77, 49-57.

Chevallier, A (1978). Etude de la migration des cellules somitiques dans Ie mesoderme somatopleural de l'ebauche de l'aile. *Wilhelm Roux's Archives*, 184, 57-73.

Chevallier, A, Kieny, M, and Mauger, A (1977). Limb-somite relationship: Origin of the limb musculature. *Journal of Embryology and Experimental Morphology*, 41, 245-258.

Chiang, C, Litingtung, Y, Lee, E, Young, KE, Corden, JL, Westphal, H, and Beachy, PA (1996). Cyclopia and defective axial patterning in mice lacking Sonic hedgehog gene function. *Nature*, 383, 407-413.

Christ, B, Huang, R, and Wilting, J (2000). The development of the avian vertebral column. *Anatomy and Embryology*, 202, 179-194.

Christ, B, and Ordahl, CP (1995). Early stages of chick somite development. *Anatomy and Embryology*, 191, 381-396.

Christ, B, Schmidt, C, Huang, R, Wilting, J, and Brand-Saberi, B (1998). Segmentation of the vertebrate body. *Anatomy and Embryology*, 197, 18.

Christ, B, and Wilting, J (1992). From somites to vertebral column. *Annals of Anatomy*, 174, 23-32.

Condie, BG, and Capecchi, MR (1993). Mice homozygous for a targeted disruption of *Hoxd-3* (*Hox-4.1*) exhibit anterior transformations of the first and second cervical vertebrae, the atlas and axis. *Development*, 119, 579-595.

Condie, BG, and Capecchi, MR (1994). Mice with targeted disruptions in the paralogous genes *hoxa-3* and *hoxd-3* reveal synergistic interactions. *Nature*, 370, 304-307.

Conlon, RA, Reaume, AG, and Rossant, J (1995). *Notch 1* is required for the coordinate segmentation of somites. *Development*, 121, 1533-1545. Cooke, J (1975).

Control of somite number during morphogenesis of a vertebrate, *Xenopus laevis*. *Nature*, 254, 196-199. Cooke, J (1998). A gene that resuscitates a theory-Somitogenesis and a molecular oscillator. *Trends in Genetics*, 14, 85-88.

Cooke, J, and Zeeman, EC (1976). A clock and wavefront model for control of the number of repeated structures during animal morphogenesis. *Journal of Theoretical Biology*, 58, 455-476.

Correia, KM, and Conlon, RA (2000). Surface ectoderm is necessary for the morphogenesis of somites. *Mechanisms of Development*, 91, 19-30.

Currie, PD, and Ingham, PW (1998). The generation and interpretation of positional information within the vertebrate myotome. *Mechanisms of Development*, 73, 321.

Dahl, E, Koseki, H, and Balling, R (1997). Pax genes and organogenesis. *Bioessays*, 19, 755-765.

Dale, KJ, and Pourquie, 0 (2000). A clock-work somite. *BioEssays*, 22, 72-83.

Davis, AP, and Capecchi, MR (1994). Axial homeosis and appendicular skeleton defects in mice with targeted disruption of *hox*d-11. *Development*, 120, 2187-2198.

Davis, AP, Witte, DP, Hsieh-Li, HM, Potter, SS, and Capecchi, MR (1995). Absence of radius and ulna in mice lacking *hox*a-11 and *hox*d-11. *Nature*, 375, 791-795.

de la Cruz, CC, Der-Avakian, A, Spyropoulos, DD, Tieu, DD, and Carpenter, EM (1999). Targeted disruption of *Hoxd9* and *Hoxd10* alters locomotor behavior, vertebral identity, and peripheral nervous system development. *Developmental Biology*, 216, 595-610.

Denetclaw, WF, and Ordahl, CP (2000). The growth of the dermomyotome and formation of early myotome lineages in thoracolumbar somites of chicken embryos. *Development*, 127, 893-905.

Deutsch, U, Dressler, GR, and Gruss, P (1988). *Pax 1*, a member of a paired box homologous murine gene family, is expressed in segmented structures during development. *Cell*, 53, 617-625.

Dickman, ED, Rogers, R, and Conway, SJ (1999). Abnormal skeletogenesis occurs coincident with increased apoptosis in the *Splotch (Sp^{2H})* mutant: Putative roles for *Pax3* and *PDGFR* in rib patterning. Anatomical Record, 255, 353-361.

Dietrich, S (1999). Regulation of hypaxial muscle development. *Cell and Tissue Research*, 296, 175-182.

Dietrich, S, and Gruss, P (1995). undulated phenotypes suggest a role of *Pax-1* for the development of vertebral and extravertebral structures. *Developmental Biology*, 167, 529-548.

Duboule, D (Ed) (1994). *Guidebook to the Homeobox Genes*. Oxford University Press, Oxford.

Duboule, D (1995). Vertebrate *Hox* genes and proliferation: An alternative pathway to homeosis? *Current Opinion in Genetics and Development*, 5, 525-528.

Durston, AJ, van der Wees, J, Pijnappel, WWM, Shilthuis, JG, and Godsave, SF (1997). Retinoid signaling and axial patterning during early vertebrate embryogenesis. *Cellular and Molecular Life Sciences*, 53, 339-349.

Ebensperger, C, Wilting, J, Brand-Saberi, B, Mizutani, Y, Christ, B, Balling, R, and Koseki, H (1995). *Pax-1*, a regulator of sclerotome development, is induced by notochord and floor plate signals in avian embryos. *Anatomy and Embryology*, 191, 297-310.

Ewan, KBR, and Everett, AW (1992). Evidence for resegmentation in the formation of the vertebral column using the novel approach of retroviral-mediated gene transfer. *Experimental Cell Research*, 198, 315-320.

Fan, CM, Lee, CS, and Tessier-Lavigne, M (1997). A role for WNT proteins in induction of dermomyotome. *Developmental Biology*, 191, 160-165.

Fan, CM, Porter, JA, Chiang, C, Chang, DT, Beachy, PA, and Tessier-Lavigne, M (1995). Long-range sclerotome induction by Sonic hedgehog: Direct role of the amino-terminal cleavage product and modulation of the cyclic AMP signalling pathway. *Cell*, 81, 457-465.

Fan, CM, and Tessier-Lavigne, M (1994). Patterning of mammalian somites by surface ectoderm and notochord: Evidence for sclerotome induction by a hedgehog homolog. *Cell*, 79, 1175-1186.

Favier, B, Le Meur, M, Chambon, P, and Dolle, P (1995). Axial skeleton homeosis and forelimb malformations in *Hoxd-11* mutant mice. *Proceedings of the National Academy of Sciences of the United States of America*, 92, 310-314.

Fietz, MJ, Concordet, JP, Barbosa, R, Johnson, R, Krauss, S, McMahon, AP, Tabin, C, and Ingham, PW (1994). The *hedgehog* gene family in *Drosophila* and vertebrate development. *Development Supplement*, 43-51.

Folberg, A, Kovacs, EN, Huang, H, Houle, M, Lohnes, D, and Featherstone, MS (1999). *Hoxd4* and *Rarg* interact synergistically in the specification of the cervical vertebrae. *Mechanisms of Development*, 89, 65-74.

Forsberg, H, Crozet, F, and Brown, NA (1998). Waves of mouse *Lunatic fringe* expression, in four-hour cycles at two-hour intervals, precede somite boundary formation. *Current Biology*, 8, 1027-1030.

Fromental-Ramain, C, Warot, X, Lakkaraju, S, Favier, B, Haack, H, Birling, C, Dierich, A, Dolle, P, and Chambon, P (1996). Specific and redundant functions of the paralogous *Hoxa-9* and *Hoxd-9* genes in forelimb and axial skeleton patterning. *Development*, 122, 461-472.

Furumoto, T, Miura, N, Akasaka, T, Mizutani-Koseki, Y, Sudo, H, Fukuda, K, Maekawa, M, Yuasa, S, Fu, Y, Moriya, H, Taniguchi, M, Imai, K, Dahl, E, Balling, R, Pavlova, M, Gossler, A, and Koseki, H (1999). Notochord-dependent expression of *MFH1* and *PAX1* cooperates to maintain the proliferation of sclerotome cells during vertebral column development. *Developmental Biology*, 210, 15-29.

Gaunt, SJ (1991). Expression patterns of mouse *Hox* genes: Clues to an understanding of developmental and evolutionary strategies. *BioEssays*, 13, 505-513.

Gaunt, SJ (1994). Conservation in the *Hox* code during morphological evolution. *International Journal of Developmental Biology*, 38, 549-552.

Gaunt, SJ (2000). Evolutionary shifts of vertebrate structures and *Hox* expression up and down the axial series of segments: A consideration of possible mechanisms. *International Journal of Developmental Biology*, 44, 109-117.

Gaunt, SJ (2001). Gradients and forward spreading of vertebrate *Hox* gene expression detected by using a *Hox/lacZ* transgene. *Developmental Dynamics*, 221, 26-36.

Gaunt, SJ, Dean, W, Sang, H, and Burton, RD (1999). Evidence that *Hoxa* expression domains are evolutionary transposed in spinal ganglia, and are established by forward spreading in paraxial mesoderm. *Mechanisms of Development*, 82, 109-118.

Gaunt, SJ, and Strachan L (1994). Forward spreading in the establishment of a vertebrate *Hox* expression boundary: The expression domain separates into anterior and posterior zones, and the spread occurs across implanted glass barriers. *Developmental Dynamics*, 199, 229-240.

Gerard, M, Chen, JY, Gronemeyer, H, Chambon, P, Duboule, D, and Zakany, J (1996). In vivo targeted mutagenesis of a regulatory element required for positioning the *Hoxd-11* and *Hoxd-10* expression boundaries. *Genes and Development*, 10, 2326-2334.

Goldfarb, M (1996). Functions of fibroblast growth factors in vertebrate development. *Cytokine and Growth Factor Reviews*, 7, 311-325.

Goodrich, LV, Johnson, RL, Milenkovic, L, McMahon, JA, and Scott, MP (1996). Conservation of the *hedgehog / patched* signaling pathway from flies to mice: Induction of a mouse patched gene by Hedgehog. *Genes and Development*, 10, 301-312.

Gossler, A, and Hrabe de Angelis, M (1998). Somitogenesis. *Current Topics in Developmental Biology*, 38, 225-287.

Goulding, M, Lumsden, A, and Paquette, AJ (1994). Regulation of *Pax-3* expression in the dermomyotome and its role in muscle development. *Development*, 120, 957-971.

Gruss, P, and Walther, C (1992). *Pax* in development. *Cell*, 69, 719-722.

Hall, BK (1977). Chondrogenesis of the somitic mesoderm. *Advances in Anatomy, Embryology and Cell Biology*, 53, 1-50.

Hammerschmidt, M (1997). The world according to *hedgehog*. *Trends in Genetics*, 13, 14-21.

Hanson, RD, Hess, JL, Yu, BD, Ernst, P, van Lohuizen, M, Berns, A, van der Lugt, NMT, Shashikant, CS, Ruddle, FH, Seto, M, and Korsmeyer, SJ (1999). Mammalian Trithorax and polycomb-group homologues are antagonistic regulators of homeotic development. *Proceedings of the National Academy of Sciences of the United States of America*, 96, 14372-14377.

Henderson, DJ, Conway, SJ, and Copp, AJ (1999). Rib truncations and fusions in the Sp2H mouse reveal a role for *Pax3* in specification of the ventrolateral and posterior parts of the somite. *Developmental Biology*, 209, 143-158.

Hirsinger, E, Duprez, D, Jouve, C, Malapert, P, Cooke, J, and Pourquie, 0 (1997). *Noggin* acts downstream of *Wnt* and *Sonic hedgehog* to antagonize *BMP4* in avian somite patterning. *Development*, 124, 4605-4614.

Hogan, BLM (1996). Bone morphogenetic proteins: Multifunctional regulators of vertebrate development. *Genes and Development*, 10, 1580-1594.

Holley, SA, Geisler, R, and Nusslein-Volhard, C (2000). Control of herl expression during zebrafish somitogenesis by a Delta-dependent oscillator and an independent wavefront activity. *Genes and Development*, 14, 1678-1690.

Horan, GSB, Kovacs, EN, Behringer, RR, and Featherstone, MS (1995a). Mutations in paralogous *Hox* genes result in overlapping homeotic transformations of the axial skeleton: Evidence for unique and redundant function. *Developmental Biology*, 169, 359-372.

Horan, GSB, Ramirez-Solis, R, Featherstone, MS, Wolgemuth, DJ, Bradley, A, and Behringer, RR (1995b). Compound mutants for the paralogous *hoxa-4*, *hoxb-4*, and *hoxd-4* genes show more complete homeotic transformations and a dose-dependent increase in the number of vertebrae transformed. *Genes and Development*, 9, 1667-1677.

Horan, GSB, Wu, K, Wolgemuth, DJ, and Behringer, RR (1994). Homeotic transformation of cervical vertebrae in *Hoxa-4* mutant mice. *Proceedings of the National Academy of Sciences of the United States of America,* 91, 12644-12648.

Hostikka, SL, and Capecchi, MR (1998). The mouse *Hoxc-11* gene: Genomic structure and expression pattern. *Mechanisms of Development*, 70, 133-145.

Hrabe de Angelis, M, McIntyre, J, and Gossler, A (1997). Maintenance of somite borders in mice requires the Delta homologue *Dll1*. *Nature*, 386, 717-721.

Huang, R, Zhi, Q, Brand-Saberi, B, and Christ, B (2000a). New experimental evidence for somite resegmentation. *Anatomy and Embryology*, 202, 195-200.

Huang, R, Zhi, G, Neubuser, A, Muller, TS, Brand-Saberi, B, Christ, B, and Wilting, J (1996). Function of somite and somitocoele cells in the formation of the vertebral motion segment in avian embryos. *Acta Anatomica*, 155, 231-241.

Huang, R, Zhi, Q, Schmidt, C, Wilting, J, Brand-Saberi, B, and Christ, B (2000b). Sclerotarnal origin of the ribs. *Development*, 127, 527-532.

Huang, R, Zhi, Q, Wilting, J, and Christ, B (1994). The fate of somitocoele cells in avian embryos. *Anatomy and Embryology*, 190, 243-250.

Ikeya, M, and Takada, S (1998). Wnt signaling from the dorsal neural tube is required for the formation of the medial dermomyotome. *Development,* 125, 4969-4976.

Ikeya, M, and Takada, S (2001). Wnt-3a is required for somite specification along the anteroposterior axis of the mouse embryo and for regulation of cdx-1 expression. *Mechanisms of Development*, 103, 27-33.

Jeannotte, L, Lemieux, M, Charron, J, Poirier, F, and Robertson, EJ (1993). Specification of axial identity in the mouse: Role of the *Hoxa-5* (*Hox1.3*) gene. *Genes and Development,* 7, 2085-2096.

Jegalian, BG, and De Robertis, EM (1992). Homeotic transformations in the mouse induced by overexpression of a human *Hox3.3* transgene. *Cell*, 71, 901-910.

Jiang, YJ, Aerne, B, Smithers, L, Haddon, C, Ish-Horowicz, D, and Lewis, J (2000). Notch signalling and the synchronization of the somite segmentation clock. *Nature*, 408, 475-479.

Johnson, J, Rhee, J, Parsons, SM, Brown, D, Olson, EN, and Rawls, A (2001). The anterior/posterior polarity of somites is disrupted in Paraxis-deficient mice. *Developmental Biology*, 229, 176-187.

Johnson, RL, Laufer, E, Riddle, RD, and Tabin, C (1994). Ectopic expression of Sonic hedgehog alters dorsal-ventral patterning of somites. *Cell,* 79, 1165-1173.

Johnson, RL, and Tabin, C (1995). The long and short of hedgehog signaling. *Cell*, 81, 313-316.

Jouve, C, Palmeirim, I, Henrique, D, Becker, J, Gossler, A, Ish-Horowicz, D, and Pourquie, O (2000). Notch signalling is required for cyclic expression of the hair-like gene HES1 in the presomitic mesoderm. *Development*, 127, 1421-1429.

Kalcheim, C, Cinnamon, Y, and Kahane, N (1999). Myotome formation: A multistage process. *Cell and Tissue Research*, 296, 161-173.

Kant, R, and Goldstein, RS (1999). Plasticity of axial identity among somites: Cranial somites can generate vertebrae without expressing *Hox* genes appropriate to the trunk. *Developmental Biology*, 216, 507-520.

Kerszberg, M, and Wolpert, L (2000). A clock and trail model for somite formation, specialization and polarization. *Journal of Theoretical Biology*, 205, 505-510.

Kessel, M (1992). Respecification of vertebral identities by retinoic acid. *Development*, 115, 487-501.

Kessel, M, Balling, R, and Gruss, P (1990). Variations of cervical vertebrae after expression of a *Hox-1.1* transgene in mice. *Cell*, 61, 301-308.

Kessel, M, and Gruss, P (1990). Murine developmental control genes. *Science*, 249, 374-379.

Kessel, M, and Gruss, P (1991). Homeotic transformations of murine vertebrae and concomitant alteration of *Hox* codes induced by retinoic acid. *Cell*, 67, 89-104.

Kieny, M, Mauger, A, and Sengel, P (1972). Early regionalization of the somitic mesoderm as studied by the development of the axial skeleton of the chick embryo. *Developmental Biology*, 28, 142-161.

Koseki, H, Wallin, J, Wilting, J, Mizutani, Y, Kispert, A, Ebensperger, E, Herrmann, BG, Christ, B, and Balling, R (1993). A role for Pax-1 as a mediator of notochordal signals during the dorsoventral specification of vertebrae. *Development*, 119, 649-660.

Kostic, D, and Capecchi, MR (1994). Targeted disruptions of the murine *Hoxa-4* and *Hoxa6* genes result in homeotic transformation of components of the vertebral column. *Mechanisms of Development*, 46, 231-257.

Krumlauf, R (1994). *Hox* genes in vertebrate development. *Cell*, 78, 191-201.

Kusumi, K, Sun, ES, Kerrebrock, AW, Bronson, RT, Chi, DC, Bulotsky, MS, Spencer, JB, Birren, BW, Frankel, WN, and Lander, ES (1998). The mouse pudgy mutation disrupts *Delta* homologue *Dll3* and initiation of early somite boundaries. *Nature Genetics*, 19, 274-278.

Lash, JW, Seitz, AW, Cheney, CM, and Ostrovsky, D (1984). On the role of fibronectin during the compaction stage of somitogenesis in the chick embryo. *Journal of Experimental Zoology*, 232, 197-206.

Leitges, M, Neidhardt, L, Haenig, B, Herrmann, BG, and Kispert, A (2000). The paired homeobox gene Uncx4.1 specifies pedicles, transverse processes and proximal ribs of the vertebral column. *Development*, 127, 2259-2267.

Le Mouellic, H, Lallemand, Y, and Brulet, P (1992). Homeosis in the mouse induced by a null mutation in the *Hox-3.1* gene. *Cell*, 69, 251-264.

Lettice, LA, Purdie, LA, Carlson, GJ, Kilanowski, F, Dorn, J, and Hill, RE (1999). The mouse bagpipe gene controls development of the axial skeleton, skull, and spleen. *Proceedings of the National Academy of Sciences of the United States of America*, 96, 9695-9700.

Linask, KK, Ludwig, C, Han, MD, Liu, X, Radice, GL, and Knudsen, KA (1998). Ncadherin/catenin-mediated morphoregulation of somite formation. *Developmental Biology* 202, 85-102.

Loureiro, JJ (1999). The *Wnts*. *Current Biology*, 9, R4.

Manley, NR, and Capecchi, MR (1997). *Hox* group 3 paralogous genes act synergistically in the formation of somitic and neural crest-derived structures. *Developmental Biology*, 192, 274-288.

Mansouri, A, Voss, AK, Thomas, T, Yokota, Y, and Gruss, P (2000). *Uncx4.1* is required in the formation of the pedicles and proximal ribs and acts upstream of *Pax9*. Development, 127, 2251-2258.

Manzanares, M, Trainor, PA, Nonchev, S, Ariza-McNaughton, L, Brodie, J, Gould, A, Marshall, H, Morrison, A, Kwan, CT, Sham, MH, Wilkinson, DG, and Krumlauf, R (1999). The role of *kreisler* in segmentation during hindbrain development. *Developmental Biology*, 211, 220-237.

Marcelle, C, Ahlgren, S, and Bronner-Fraser, M (1999). In vivo regulation of somite differentiation and proliferation by Sonic Hedgehog. *Developmental Biology*, 214, 277-287.

Marcelle, C, Stark, MR, and Bronner-Fraser, M (1997). Coordinate action of BMPs, Wnts, Shh and Noggin mediate patterning of the dorsal somite. *Development*, 124, 3955-3963.

Maroto, M, Reshef, R, Munsterberg, AE, Koester, S, Goulding, M, and Lassar, AB (1997). Ectopic *Pax-3* activates *MyoD* and *Myf-5* expression in embryonic mesoderm and neural tissue. *Cell*, 89, 139-148.

Marshall, H, Morrison, A, Studer, M, Popperl, H, and Krumlauf, R (1996). Retinoids and *Hox* genes. *FASEB Journal*, 10, 969-978.

McGrew, MJ, Dale, JK, Fraboulet, S, and Pourquie, O (1998). The *lunatic fringe* gene is a target of the molecular clock linked to somite segmentation in avian embryos. *Current Biology*, 8, 979-982.

McGrew, MJ, and Pourquie, O (1998). Somitogenesis: Segmenting a vertebrate. *Current Opinion in Genetics and Development*, 8, 487-493.

McMahon, JA, Takada, S, Zimmerman, LB, Fan, CM, Harland, RM, and McMahon, AP (1998). Noggin-mediated antagonism of BMP signalling is required for growth and patterning of the neural tube and somite. *Genes and Development*, 12, 1438-1452.

McPherron, AC, Lawler, AM, and Lee, SJ (1997). Regulation of skeletal muscle mass in mice by a new TGF- superfamily gene. *Nature*, 387, 83-90.

McPherron, AC, Lawler, AM, and Lee, SJ (1999). Regulation of anterior/posterior patterning of the axial skeleton by growth/differentiation factor 11. *Nature Genetics*, 22, 260-264.

McPherron, AC, and Lee, SJ (1997). Double muscling in cattle due to mutations in the myostatin gene. *Proceedings of the National Academy of Sciences of the United States of America*, 94, 12457-12461.

Medina-Martinez, O, Bradley, A, and Ramirez-Solis, R (2000). A large targeted deletion of *Hoxb1-Hoxb9* produces a series of single-segment anterior homeotic transformations. *Developmental Biology*, 222, 71-83.

Meier, S (1979). Development of the chick embryo mesoblast: Formation of the embryonic axis and establishment of the metameric pattern. *Developmental Biology*, 73, 25-45.

Monsoro-Burq, AH, Bontoux, M, Teillet, MA, and Le Douarin, NM (1994). Heterogeneity in the development of the vertebra. *Proceedings of the National Academy of Sciences of the United States of America*, 91, 10435-10439.

Monsoro-Burq, AH, Duprez, D, Watanabe, Y, Bontoux, M, Vincent, C, Brickell, P, and Le Douarin, N (1996). The role of bone morphogenetic proteins in vertebral development. *Development*, 122, 3607-3616.

Moon, RT, Brown, JD, and Torres, M (1997). WNTs modulate cell fate and behavior during vertebrate development. *Trends in Genetics*, 13, 157-162.

Munsterberg, AE, and Lassar, AB (1995). Combinatorial signals from the neural tube, floor plate and notochord induce myogenic *bHLH* gene expression in the somite. *Development*, 121, 651-660.

Neumann, C, and Cohen, S (1997). Morphogens and pattern formation. *Bioessays*, 19, 721-729.

Nowicki, JL, and Burke, AC (2000). *Hox* genes and morphological identity: Axial versus lateral patterning in the vertebrate mesoderm. *Development*, 127, 4265-4275.

Nye, JS (1999). The Notch proteins. *Current Biology*, 9, R118. Ordahl, CP (1993). Myogenic lineages within the developing somite. In M. Bernfield (Ed), *Molecular Basis of Morphogenesis* (pp 165-176). Wiley-Liss, New York.

Ordahl, CP, and Le Douarin, NM (1992). Two myogenic lineages within the developing somite. *Development*, 114, 339-353.

Palmeirim, I, Henrique, D, Ish-Horowicz, D, and Pourquie, O (1997). Avian *hairy* gene expression identifies a molecular clock linked to vertebrate segmentation and somitogenesis. *Cell*, 91, 639-648.

Peters, H, Wilm, B, Sakai, N, Imai, K, Maas, K, and Balling, R (1999). *Pax1* and *Pax9* synergistically regulate vertebral column development. *Development*, 126, 5399-5408.

Pollock, RA, Jay, G, and Bieberich, CJ (1992). Altering the boundaries of *Hox3.1* expression: Evidence for antipodal gene regulation. *Cell*, 71, 911-923.

Pourquie, O, Fan, CM, Coltey, M, Hirsinger, E, Watanabe, Y, Breant, C, Francis-West, P, Brickell, P, Tessier-Lavigne, M, and Le Douarin, NM (1996). Lateral and axial signals involved in avian somite patterning: A role for BMP4. *Cell*, 84, 461-471.

Pownall, ME, Strunk, KE, and Emerson, CP (1996). Notochord signals control the transcriptional cascade of myogenic *bHLH* genes in somites of quail embryos. *Development*, 122, 1475-1488.

Ramirez-Solis, R, Zheng, H, Whiting, J, Krumlauf, R, and Bradley, A (1993). *Hoxb-4* (*Hox2.6*) mutant mice show homeotic transformation of a cervical vertebra and defects in the closure of the sternal rudiments. *Cell*, 73, 279-294.

Rawls, A, Valdez, MR, Zhang, W, Richardson, J, Klein, WH, and Olson, EN (1998). Overlapping functions of the myogenic bHLH genes *MRF4* and *MyoD* in double mutant mice. *Development*, 125, 2349-2358.

Reshef, R, Maroto, M, and Lassar, AB (1998). Regulation of dorsal somitic cell fates: BMPs and Noggin control the timing and pattern of myogenetic regulator expression. *Genes and Development*, 12, 290-303.

Richardson, MK, Allen, SP, Wright, GM, Raynaud, A, and Hanken, J (1998). Somite number and vertebrate evolution. *Development*, 125, 151-160.

Rijli, FM, Matyas, R, Pellegrini, M, Dierich, A, Gruss, P, Dolle, P, and Chambon, P (1995). Cryptorchidism and homeotic transformations of spinal nerves and vertebrae in *Hoxa-10* mutant mice. *Proceedings of the National Academy of Sciences of the United States of America*, 92, 8185-8189.

Saegusa, H, Takahashi, N, Noguchi, S, and Suemori, H (1996). Targeted disruption in the mouse *Hoxc-4* locus results in axial skeleton homeosis and malformation of the xiphoid process. *Developmental Biology*, 174, 55-64.

Schnell, S, and Maini, PK (2000). Clock and induction model for somitogenesis. Developmental *Dynamics*, 217, 415-420.

Shashikant, CS, and Ruddle, FH (1996). Combinations of closely situated cis-acting elements determine tissue-specific patterns and anterior extent of early *Hoxc8* expression. *Proceedings of the National Academy of Sciences of the United States of America*, 93, 12364-12369.

Small, KM, and Potter, SS (1993). Homeotic transformations and limb defects in *Hox A11* mutant mice. *Genes and Development*, 7, 2318-2328.

Smith, CA, and Tuan, RS (1995). Functional involvement of *Pax-1* in somite development: Somite dysmorphogenesis in chick embryos treated with *Pax-1* paired-box antisense oligodeoxynucleotide. *Teratology*, 52, 333-345.

Solursh, M, Fisher, M, Meier, S, and Singley, CT (1979). The role of extracellular matrix in the formation of the sclerotome. *Journal of Embryology and Experimental Morphology*, 54, 75-98.

Sosic, D, Brand-Saberi, B, Schmidt, C, Christ, B, and Olson, EN (1997). Regulation of paraxis expression and somite formation by ectoderm and neural tube-derived signals. *Developmental Biology*, 185, 229-243.

Spence, MS, Yip, J, and Erickson, CA (1996). The dorsal neural tube organizes the dermomyotome and induces axial myocytes in the avian embryo. *Development*, 122, 231-241.

Stern, CD, and Keynes, RJ (1987). Interactions between somite cells: The formation and maintenance of segment boundaries in the chick embryo. *Development*, 99, 261-272.

Stern, HM, Brown, AM, and Hauschka, SD (1995). Myogenesis in paraxial mesoderm: Preferential induction by dorsal neural tube and by cells expressing *Wnt-1*. *Development*, 121, 3675-3686.

Stern, HM, and Hauschka, SD (1995). Neural tube and notochord promote in vitro myogenesis in single somite explants. *Developmental Biology*, 167, 87-103.

Subramanian, V, Meyer, BI, and Gruss, P (1995). Disruption of the murine homeobox gene *Cdx1* affects axial skeletal identities by altering the mesodermal expression domains of *Hox* genes. *Cell*, 83, 641-653.

Suemori, H, Takahashi, N, and Noguchi, S (1995). *Hoxc-9* mutant mice show anterior transformation of the vertebrae and malformation of the sternum and ribs. *Mechanisms of Development*, 51, 265-273.

Tam, PPL, and Trainor, PA (1994). Specification and segmentation of the paraxial mesoderm. *Anatomy and Embryology*, 189, 275-305.

Tonegawa, A, Funayama, N, Veno, N, and Takahashi, Y (1997). Mesodermal subdivision along the mediolateral axis in chicken controlled by different concentrations of BMP4. *Development*, 124, 1975-1984.

Tremblay, P, Dietrich, S, Mericskay, M, Shubert, FR, Li, Z, and Paulin, D (1998). A crucial role for *Pax3* in the development of the hypaxial musculature and the long-range migration of muscle precursors. *Developmental Biology*, 203, 49-61.

Tribioli, C, and Lufkin, T (1999). The murine *Bapx1* homeobox gene plays a critical role in embryonic development of the axial skeleton and spleen. *Development*, 126, 5699-5711.

van den Akker, E, Fromental-Ramain, C, de Graaf, W, Le Mouellic, H, Brulet, P, Chambon, P, and Deschamps, J (2001). Axial skeletal patterning in mice lacking all paralogous group 8 *Hox* genes. *Development*, 128, 1911-1921.

Venters, SJ, Thorsteinsdóttir, S, and Duxson, MJ (1999). Early development of the myotome in the mouse. *Developmental Dynamics*, 216, 219-232.

Verbout, AJ (1976). A critical review of the "Neugliederung" concept in relation to the development of the vertebral column. *Acta Biotheoretica*, 25, 219-258.

Verbout, AJ (1985). The development of the vertebral column. *Advances in Anatomy, Embryology and Cell Biology*, 90, 11-19.

Wahba, GM, Hostikka, SL, and Carpenter, EM (2001). The paralogous *Hox* genes *Hoxa10* and *Hoxd10* interact to pattern the mouse hindlimb peripheral nervous system and skeleton. *Developmental Biology*, 231, 87-102.

Wallin, J, Wilting, J, Koseki, H, Fritsch, R, Christ, B, and Balling, R (1994). The role of *Pax-1* in axial skeleton development. *Development*, 120, 1109-1121.

Walmsley, R (1953). The development and growth of the intervertebral disc. *Edinburgh Medical Journal*, 60, 341-364.

Wang, Y, and Jaenisch, R (1997). Myogenin can substitute for *Myf5* in promoting myogenesis but less efficiently. *Development*, 124, 2507-2513.

Williams, BA, and Ordahl, CP (1994). *Pax-3* expression in segmental mesoderm marks early stages in myogenic cell specification. *Development*, 120, 785-796.

Williams, BA, and Ordahl, CP (1997). Emergence of determined myotome precursor cells in the somite. *Development*, 124, 4983-4997.

Wilson-Rawls, J, Hurt, CR, Parsons, SM, and Rawls, A (1999). Differential regulation of epaxial and hypaxial muscle development by *Paraxis*. *Development*, 126, 5217-5229.

Wolpert, L (1978). Pattern formation in biological development. *Scientific American*, 239, 154-164.

Wolpert, L (1996). One hundred years of positional information. *Trends in Genetics*, 12, 359-364.

Woychik, R, Hogan, B, Bryant, S, Eichele, G, Kimelman, D, Noden, D, Shoenwolf, G, and Wright, C (1997). Pattern formation. *Reproductive Toxicology*, 11, 339-344.

Yamaguchi, TP, Takada, S, Yoshikawa, Y, Wu, N, and McMahon, AP (1999). T (Brachyury) is a direct target of Wnt3a during paraxial mesoderm specification. *Genes and Development*, 13, 3185-3190.

Yoshikawa, Y, Fujimori, T, McMahon, AP, and Takada, S (1997). Evidence that absence of *Wnt-3a* signalling promotes neuralization instead of paraxial mesoderm development in the mouse. *Developmental Biology*, 183, 234-242.

Zákány, J, Gerard, M, Favier, B, Potter, SS, and Duboule, D (1996). Functional equivalence and rescue among group 11 *Hox* gene products in vertebral patterning. *Developmental Biology*, 176, 325-328.

CHAPTER | Four |

Bones of the Postcranial Skeleton: Vertebral Column

The adult human skeleton (Figure 4-1) is composed of 206 bones, which can be divided into an *appendicular* and an *axial skeleton*. The *appendicular skeleton* consists of the bones of the upper and lower limbs and their associated girdles. The *axial skeleton* consists of the *cranium (*or *skull)*, the *vertebrae,* and associated structures:

Cranium	28 bones.
Hyoid	1 bone.
Vertebral	Column 26 bones.
Ribs	24 bones.
Sternum	1 bone.

The *appendicular skeleton* consists of the *bones of the upper and lower limbs,* including the *pectoral and pelvic girdles:*

Upper Limbs	60 bones.
Pectoral Girdle	4 bones.
Lower Limbs	60 bones.
Pelvic Girdle	2 bones.

This chapter will consider the osteology and articulations of the vertebral column, and the next chapter will cover the ribs and sternum. The following two chapters will discuss the appendicular skeleton, and the last chapter will be devoted to the cranium.

Bones of the vertebral column

The Vertebral Column in General

The human *vertebral column* is composed of 33 segments or *vertebrae.* Some are fused in adult life, so that the total number of individual bones is only 26. Between each vertebral segment is a fibrocartilaginous *intervertebral disc.*

The vertebrae may be broken into five groups, cranially to caudally (Figure 4-2):

Cervical	7 bones.
Thoracic	12 bones.
Lumbar	5 bones.
Sacral	5 segments fused into one bone.
Coccyx	4 segments typically fused into one bone.

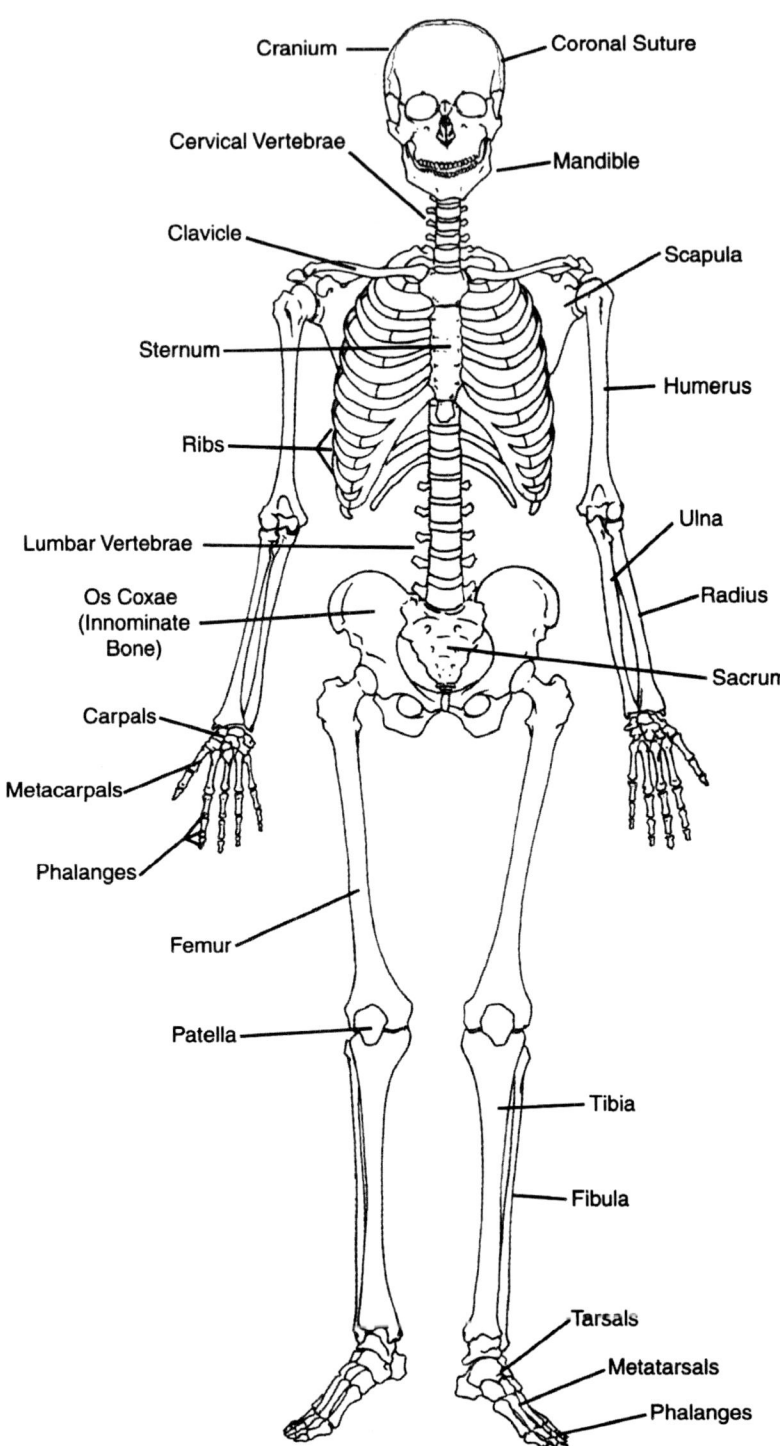

Figure 4-1 The human skeleton: (A) anterior.

Bones of the Postcranial Skeleton: Vertebral Column

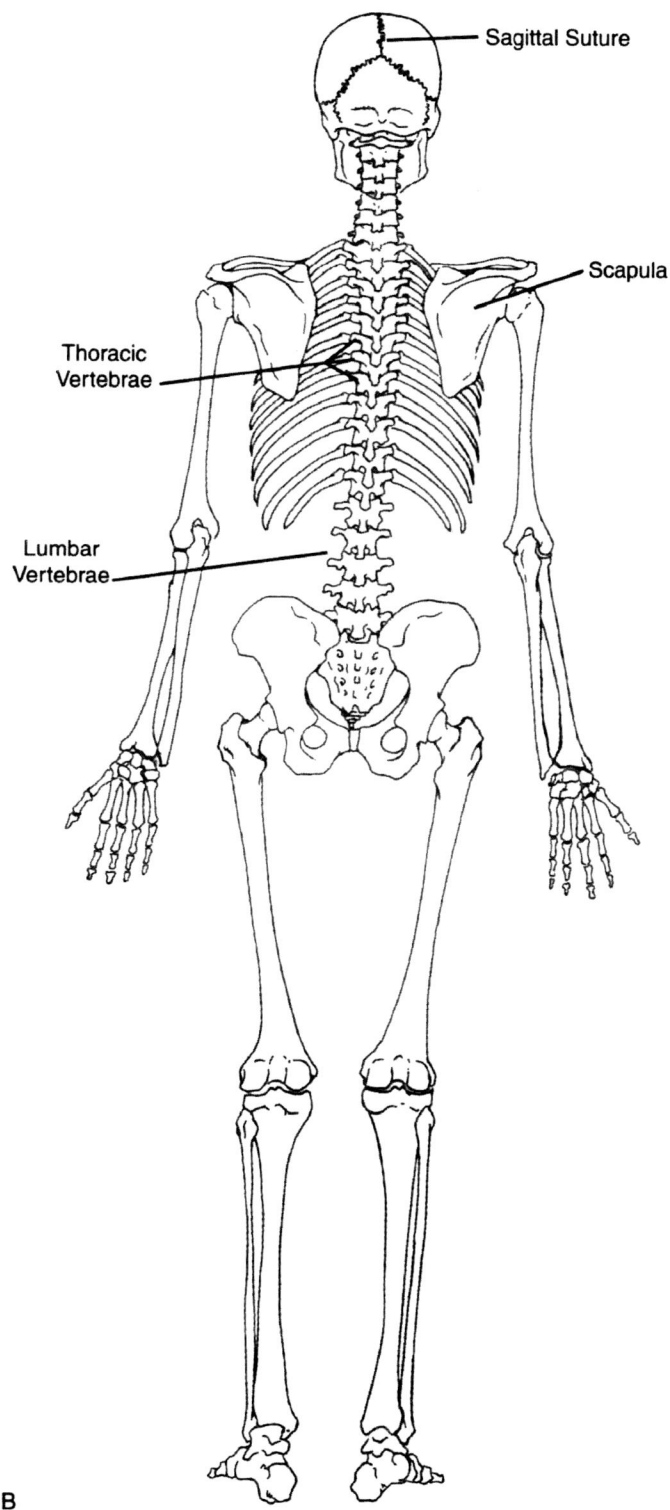

B

Figure 4-1 *(Continued)* *(B)* posterior.

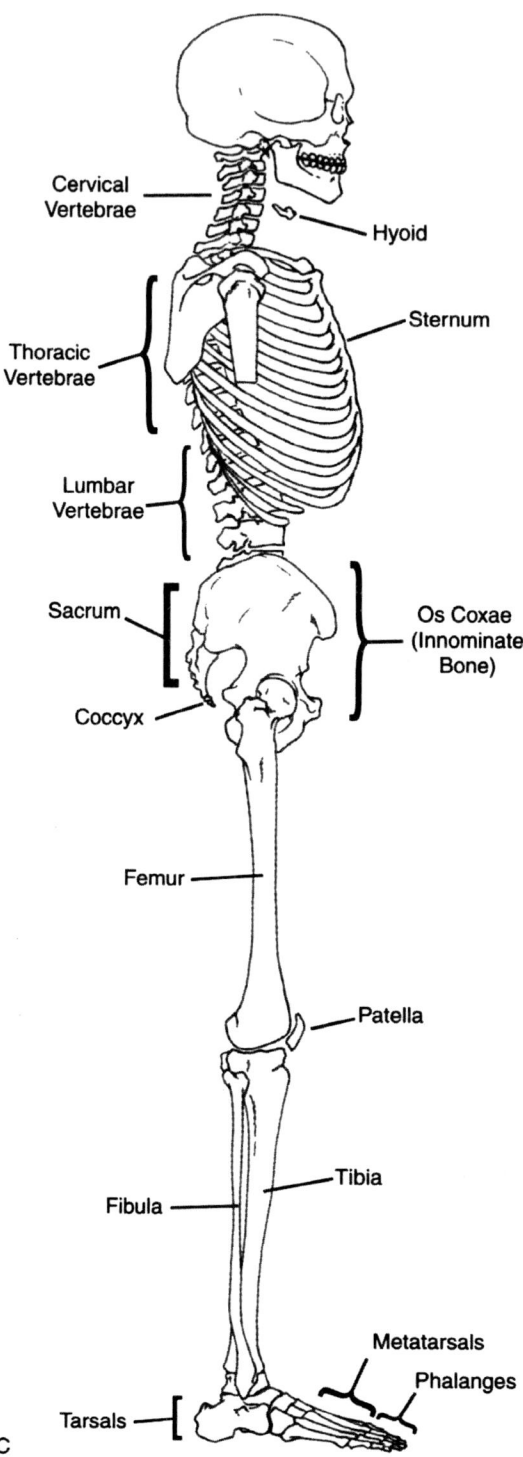

Figure 4-1 (*Continued*) (*C*) lateral.

It should be noted that these are the most typical numbers of vertebrae in each segment, but that there can be some variation. There is rarely (in fact, virtually never) any variation in the number of cervical vertebrae, but it is fairly common to find one more or one less segment in the lumbar, sacral, and coccygeal segments of the vertebral column. The probability that an atypical number of vertebrae will be encountered increases caudally.

The vertebral column serves three primary functions: (1) to support the trunk and transmit its weight to the pelvis and lower limb; (2) to house and protect the spinal cord and its membranes; (3) to provide a central axis for the thorax (each rib articulates with a vertebra and coordinated motion of the ribs is a critical aspect of normal breathing) and protect its contents. (The thorax also serves as the attachment point of the forelimb). All vertebrae are

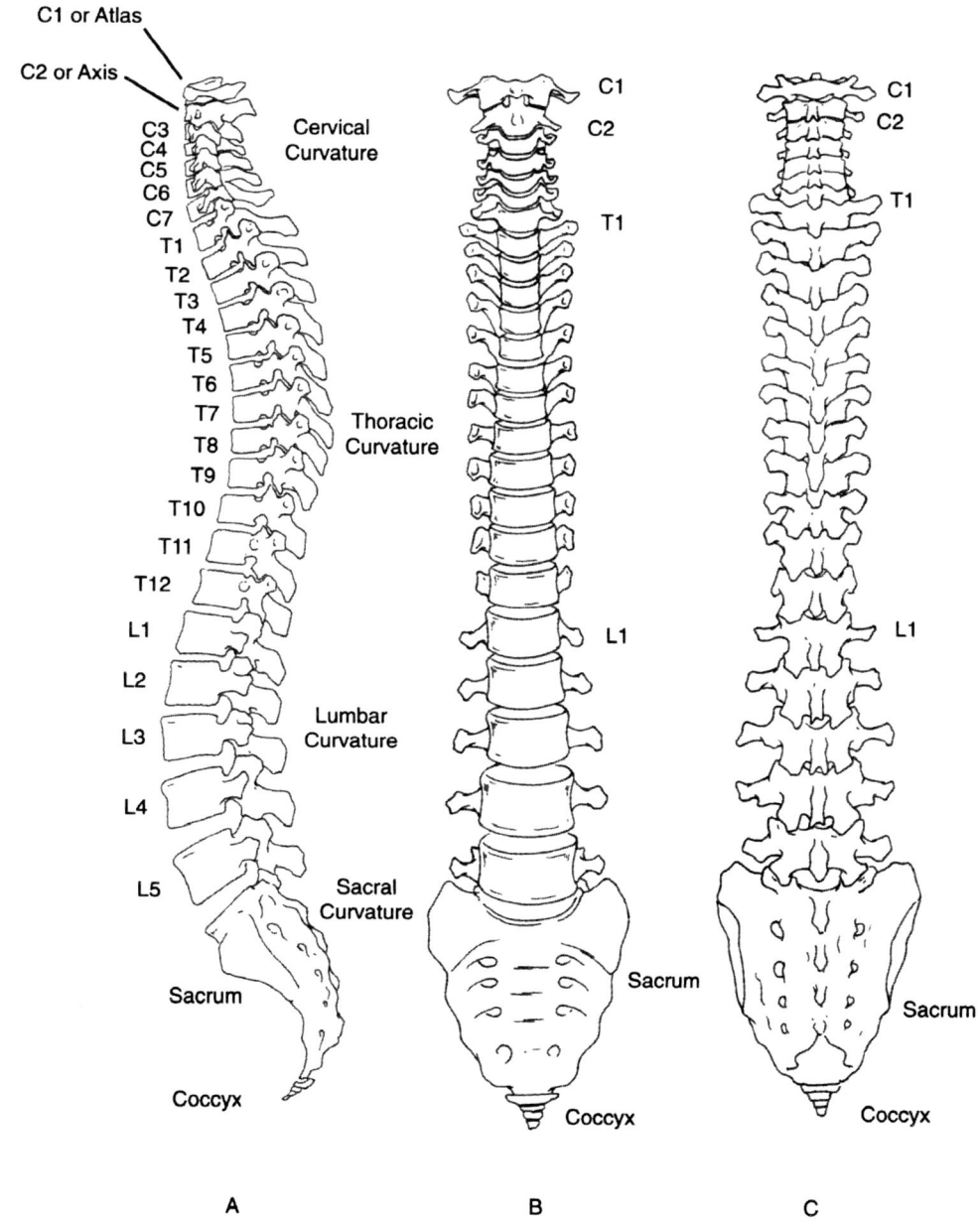

Figure 4-2 The vertebral column: (A) left lateral, (B) anterior, and (C) posterior aspects.

formed along a common plan, but vary according to their positions and functions in the spinal column. These parts may be examined in a typical vertebra.

Regions of the Vertebral Column

As mentioned above, the vertebral column can be divided into five groups or regions, cervical (C1-C7), thoracic (T1-T12), lumbar (L1-L5), sacral (S1-S5), and coccygeal (no individual designation). Vertebrae of each region demonstrate specific characteristics that are diagnostic for that region. All 33 vertebral segments and their associated intervertebral discs taken together constitute the *vertebral column.*

In the human being, the vertebral column demonstrates a series of *curvatures* associated with these different regions (see Figure 4-2). There are *four curvatures* in the human vertebral column, the *cervical curvature*, the *thoracic curvature*, the *lumbar curvature*, and the *sacral curvature.* The curve of the coccyx continues along the sacral curvature. In most terrestrial vertebrates, the vertebral column demonstrates a single horizontal arc that is convex dorsally, and is supported cranial and caudally by the pectoral and pelvic limbs. Human beings, however, are unique in our *bipedal* form of walking and in that we habitually balance all of the weight of the trunk over the pelvic limb and girdle alone, with the vertebral column held more or less vertically over the sacrum.

Embryologically, there is present a single dorsoconvex arc in the human vertebral column, just as is true of other terrestrial vertebrates. The thoracic and sacral curvatures of the human adult vertebral column are convex dorsally. They are called *primary curvatures*, as they were present embryologically and are the adult remnants of this embryological curvature. The cervical and lumbar curvatures begin to appear before birth, but are not really obvious until later. They are referred to as *secondary curvatures*. The cervical curvature becomes apparent when the infant begins to hold its head up. It can be obliterated by flexing the neck, but it is present in normal standing, sitting, and walking. The thoracic curvature is formed by the articulation of the twelve thoracic vertebrae. It is permanent. The lumbar curvature, which becomes pronounced as a child begins to walk, is generally greater in women and less in men, while the sacral curvature is generally slightly greater in men than in women, though the sexes overlap a great deal in this feature.

Osteological Features of Vertebrae in General

As examples of typical vertebrae, Figure 4-3 A illustrates a thoracic vertebrae viewed superiorly and Figure 4-3 B, a lumbar vertebrae viewed laterally. The most anterior portion of a typical vertebra is the *body*. The *body* is the largest portion of a vertebra and is approximately cylindrical in shape. Its *cranial (superior)* and *caudal (inferior) surfaces* are roughened and give attachment to the *intervertebral discs*. Located around the circumferences of the cranial and caudal surfaces of the body are slightly elevated rims. The diameters of the cranial and caudal surfaces of the vertebral bodies are somewhat greater than the diameter of the center portion of the body so that it appears constricted or "waisted".

The posterior surface of the body faces the *vertebral foramen*, an opening for the spinal cord, and contains one or more large, irregular openings, the *foramina vasculare (basivertebral foramina)*, for the passage of the basivertebral veins draining from the interior of the vertebral bodies to the internal vertebral plexus, a venous plexus that surrounds the spinal cord and receives venous drainage form both the vertebral column and spinal cord.

The bodies of the vertebrae transmit the weight of a person's body, and muscular and other compressive forces (e.g., reaction forces from walking, running, jumping, etc.). The vertebrae are separated from one another by fibrocartilaginous intervertebral discs, which transmit forces from one vertebra to the next. Posterior to the vertebral body is the *vertebral arch* , formed by the pedicles and laminae (see Figure 4-3A)

The *neural arch* is the developmental structure that contributes to both the vertebral arch and to part of the vertebral body. It is sometimes used, incorrectly, as a synonym for vertebral arch. The vertebral arch, along with the posterior surface of the vertebral body, encloses an opening, the *vertebral foramen*. The spinal cord and its associated membranes pass through the *vertebral canal*, formed by the vertebral foramina of all the vertebrae. The vertebral arches do not bear weight, but rather protect the spinal column, and are connected to one another by strong ligaments.

As mentioned, the vertebral arch consists of a pair of *pedicles* and a pair of *laminae* (singular *lamina*). These two portions of the arch together support seven processes: four *articular processes*, two *transverse processes*, and one *spinous process*. Each vertebral arch can be divided into right and left halves posteriorly by the spinous process. Laterally, each half can be divided into anterior and posterior portions by the transverse process. The portion of the arch between the body anteriorly and the transverse process posteriorly is the *pedicle*. It is by the pedicle that the vertebral arch is attached to the vertebral body. It is short, strong, and rounded.

The portion of the vertebral arch between the transverse process and the spinous process is the

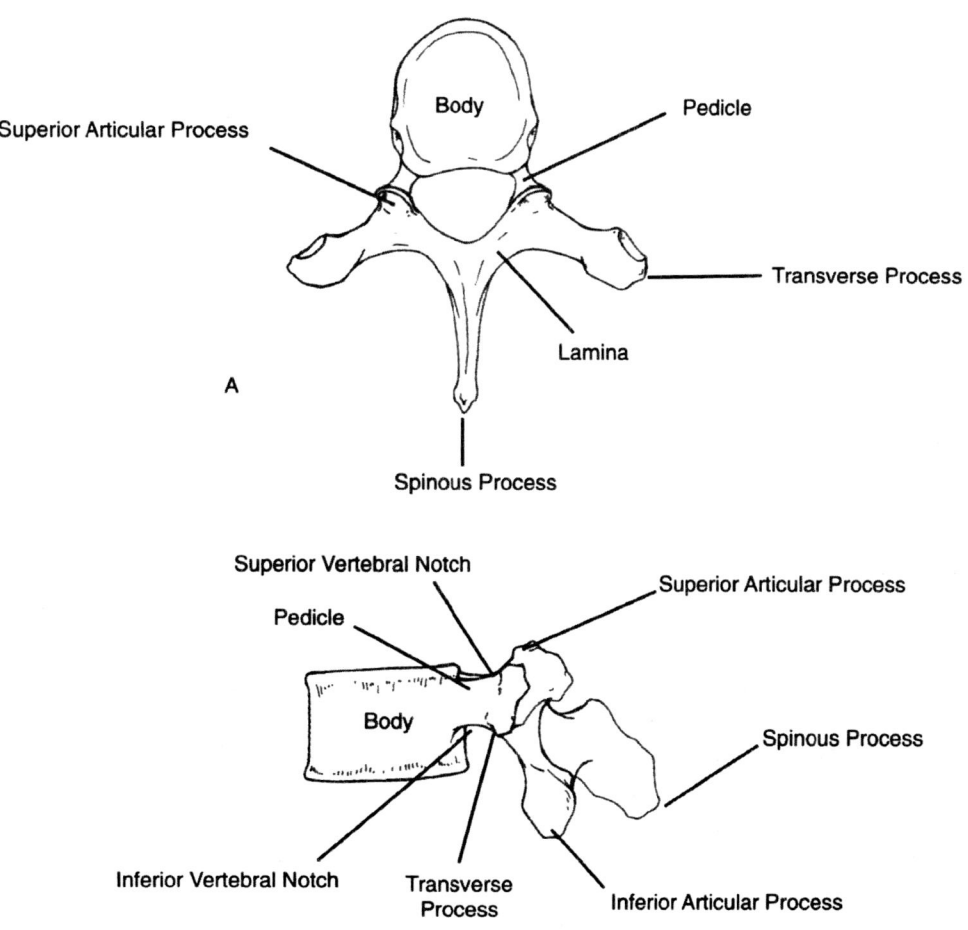

Figure 4-3 Features of typical vertebrae: (*A*) superior view of a midthoracic vertebra, and (*B*) left lateral view of a lumbar vertebra.

lamina. The lamina is flat and platelike. Arches of adjacent vertebrae articulate with one another by means of *superior and inferior articular processes.* The articular processes are also known as *zygapophyses* (singular *zygapophysis*). On each side of the vertebral arch, the superior articular process is located superior to the transverse process, while the inferior articular process is located inferior to the transverse process. Each inferior articular process articulates with the superior articular process of the vertebra below it. These form small synovial joints, and each articular process bears a small smooth *articular facet.*

Because the craniocaudal height of the pedicles is less than the height of the bodies of the vertebrae to which they are attached, there exist "notches" or depressions on the superior and inferior surfaces of the pedicles. These are the *superior vertebral notches* above the pedicles and the *inferior vertebral notches* below them (see Figure 4-3 B). The inferior vertebral notch is much deeper than the superior because the pedicle attaches nearer the superior surface of the vertebral body than the inferior surface of the body. The superior and inferior vertebral notches of adjacent vertebrae form the *intervertebral foramen.* It is through these intervertebral foramina that the spinal nerves exit the spinal column to reach their peripheral distributions.

In addition to the body, the pedicles, the laminae, the transverse processes, the spinous process, and the articular processes, associated with each vertebra is a *costal element* (Figure 4-4). The costal element takes

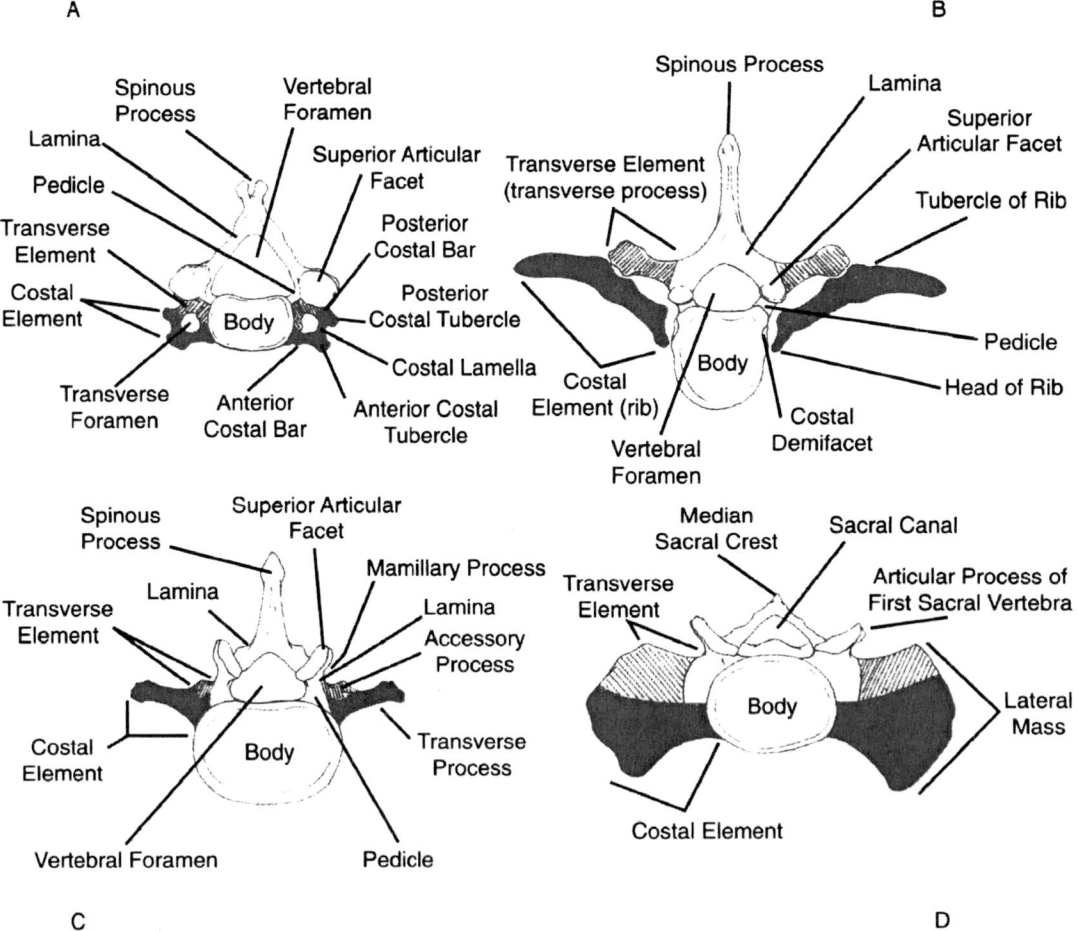

Figure 4-4 *Costal elements* (stippled) and *transverse elements* (hatched) of vertebrae: (*A*) cervical, (*B*) thoracic, (*C*) lumbar, and (*D*) sacral vertebrae, diagrammatic representations, superior views.

a different form in different regions of the spinal column. *Costal* means "rib" (Latin: *costa*). In the thoracic region, the costal element is not a part of the thoracic *vertebra*, but is present as a separate bone, the *rib*. In cervical, lumbar, and sacral vertebrae, the costal element is incorporated into the vertebrae. In cervical vertebrae, the costal element is present as the *anterior and posterior costal tubercles,* the *anterior costal bar,* and the *costal lamella* (described in detail with the cervical vertebrae), all parts of the transverse process. These define the anterior and lateral borders of the characteristic *transverse foramen* of cervical vertebrae.

In lumbar vertebrae, the costal element also forms the bulk of the transverse process the "true" transverse process being represented by the *accessory process,* a posterior protrusion that serves as a muscle attachment. In the sacral vertebrae the costal elements become incorporated into the *lateral masses* of the sacrum. The coccygeal vertebrae consist primarily of degenerate bodies and lack costal elements. The specific details of each type of vertebrae are discussed in the following sections.

Regional Osteology of the Vertebral Column

The Cervical Vertebrae

There are seven cervical vertebrae (C1-C7). They are located in the neck between the base of the skull and the thoracic vertebrae. The identifying characteristic of cervical vertebrae is the presence of the transverse foramen in the transverse process. The transverse foramina of the first six cervical vertebrae transmit the (paired) vertebral arteries and their accompanying venous and sympathetic plexuses to the head. The vertebral arteries do not pass through the transverse foramina of the seventh cervical vertebra. The third through sixth cervical vertebrae share a common morphological pattern (Figure 4-5), while the first, second, and seventh possess unique morphological details. These three vertebrae also have unique names. The first is called the atlas, the second the axis, and the seventh is the vertebra prominens.

A *typical cervical vertebra* (see Figure 4-5) has a small but relatively broad *body*. The *vertebral foramen* is large and triangular in shape. The pedicles project dorsolaterally. The laminae then curve dorsomedially from them. The large vertebral foramina of the cervical vertebrae house the cervical enlargement of the spinal cord. The superior and inferior vertebral notches are of almost equal depth. The laminae are long and narrow with a thin upper border and are roughened ventrally. The spinous process is short and bifid. The distribution of bifid spinous processes is variable from population to population. Frequently cervical spinous processes end in a single, undivided apex (Duray et al., 1999).

The superior and inferior articular processes of a single cervical vertebra are at opposite ends of an *articular pillar*. The transverse process is centered around the transverse foramen. The anterior and posterior costal bars form the anterior and posterior margins of the transverse process. Connecting these two elements and forming the lateral border of the transverse foramen is the costal lamella. The most lateral structures on the transverse process are the anterior and posterior costal tubercles, which are the tips of the anterior and posterior costal bars, respectively. The costal lamella and the more lateral portions of the costal bars represent the final morphology of the developmental costal element of the cervical vertebrae. The more medial part of the posterior costal bar is the only part of the cervical transverse process that arose from the same origin or element as the transverse process of thoracic vertebrae (developmental transverse process).

Occasionally *cervical ribs*, which are separate costal elements, may be present in the lower cervical vertebrae. Cervical ribs are not a normal condition When a cervical rib is present, there is no transverse foramen associated with the transverse process. The cervical rib may compress the roots of the nerves comprising the brachial plexus, leading to potential neurological problems in the upper limb and other structures innervated by branches of the brachial plexus.

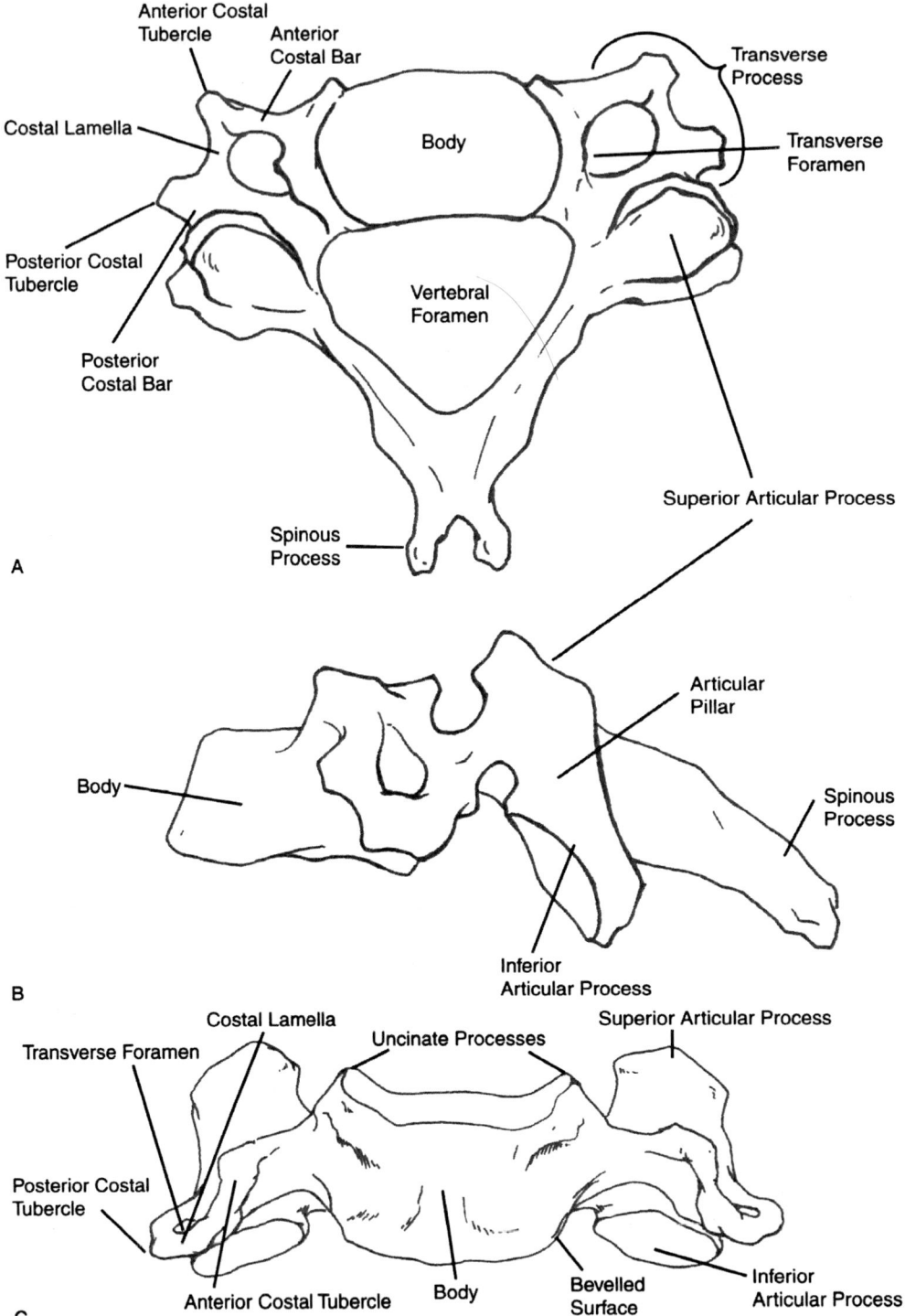

Figure 4-5 Features of cervical vertebrae 3 through 6: (A) superior view, (B) lateral view, (C) anterior view showing the uncinate processes.

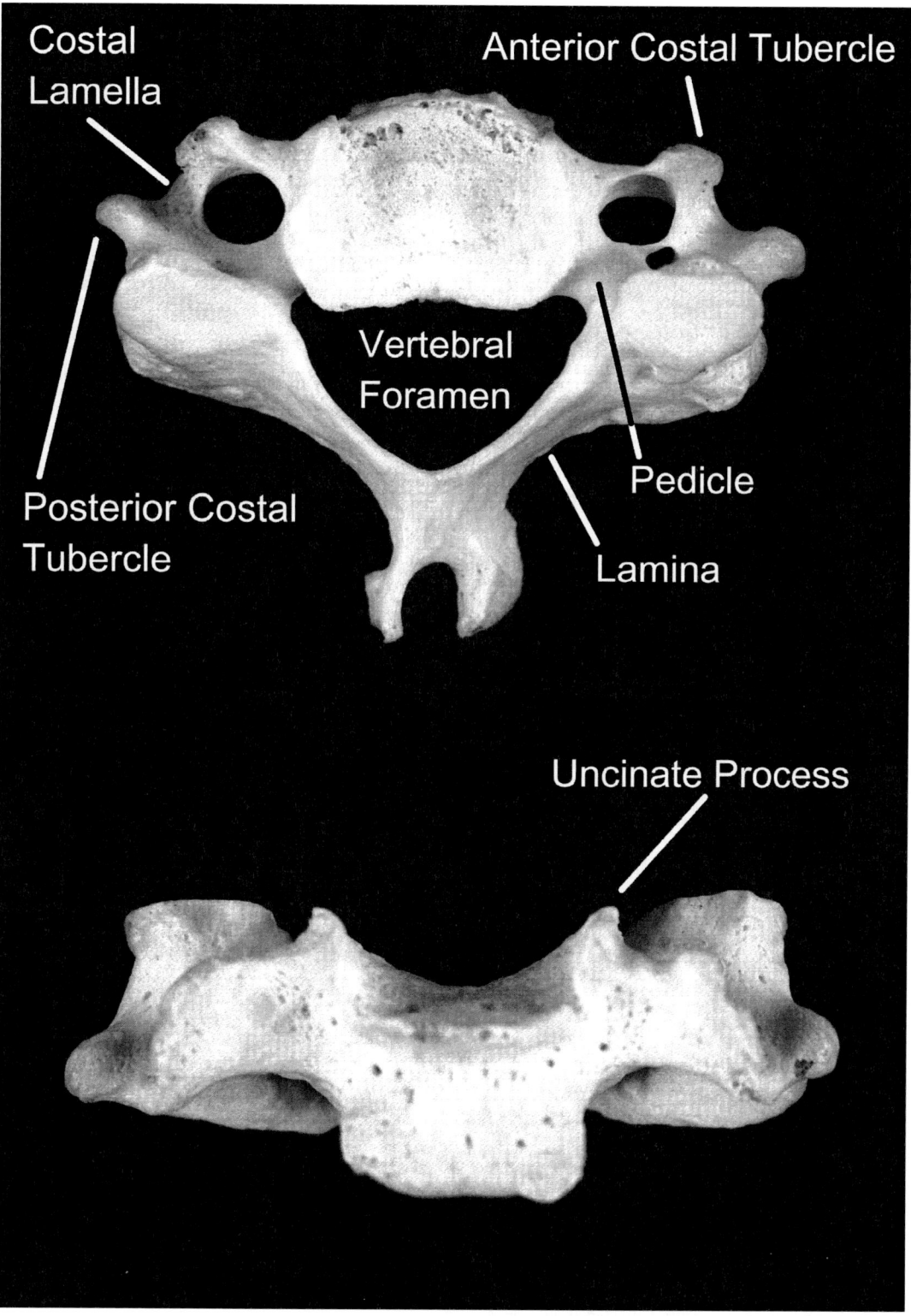

Figure 4-5 *(Continued)* Photographs of superior and anterior views of a typical cervical vertebra. See accompanying line drawing for identification of additional landmarks.

The Atlas

The first cervical vertebra (Figure 4-6) is called the *atlas* because it supports the globe of the head. It is unique among the vertebrae in that it lacks a body. Its body instead takes the unique form of the *dens*, a "process" or projection of the second cervical vertebra around which the atlas rotates. The atlas also lacks a spinous process and consists primarily of two lateral masses connected by a short anterior arch and a longer, more curved, posterior arch.

The *anterior arch* of the atlas composes about one-fifth of the circumference of the ring of the atlas. On the ventral surface of the anterior arch is a swelling called the *anterior tubercle*. On the dorsal surface of the anterior arch is a small articular facet, the *fovea dentis*. The dens of the axis articulates here.

The *posterior arch* forms about two-fifths of the ring of the atlas. Located on the dorsal surface of the posterior arch is the *posterior tubercle* of the atlas. The posterior tubercle represents the spinous process of the atlas. Located on the cranial surface of the posterior arch, between the posterior tubercle and the lateral masses, are the *grooves for the vertebral artery and first cervical spinal nerve*. This groove may occasionally be converted to a foramen called the *arcuate foramen*. This occurs by the formation of a bony strut passing over the nerve and artery called the *posterior ponticle* passing over the nerve and artery. Inferiorly are the shallow inferior vertebral notches.

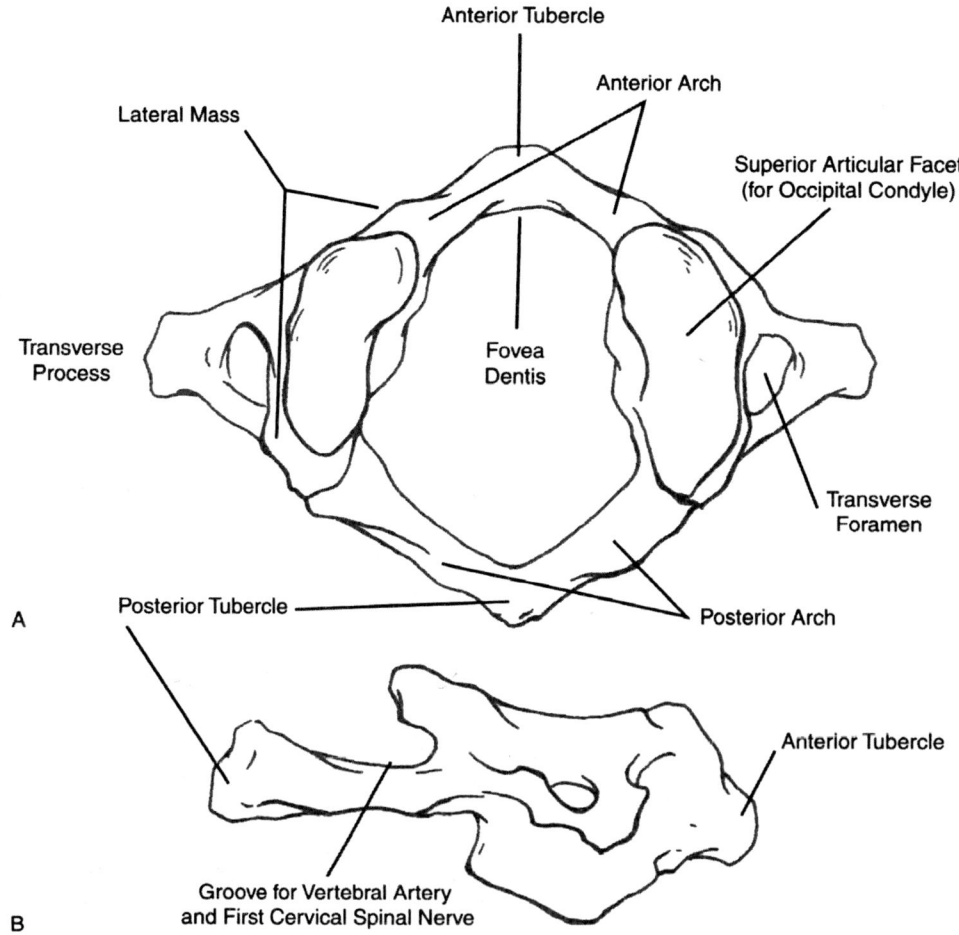

Figure 4-6 The atlas (first cervical vertebra): (A) superior view, (B) right lateral view.

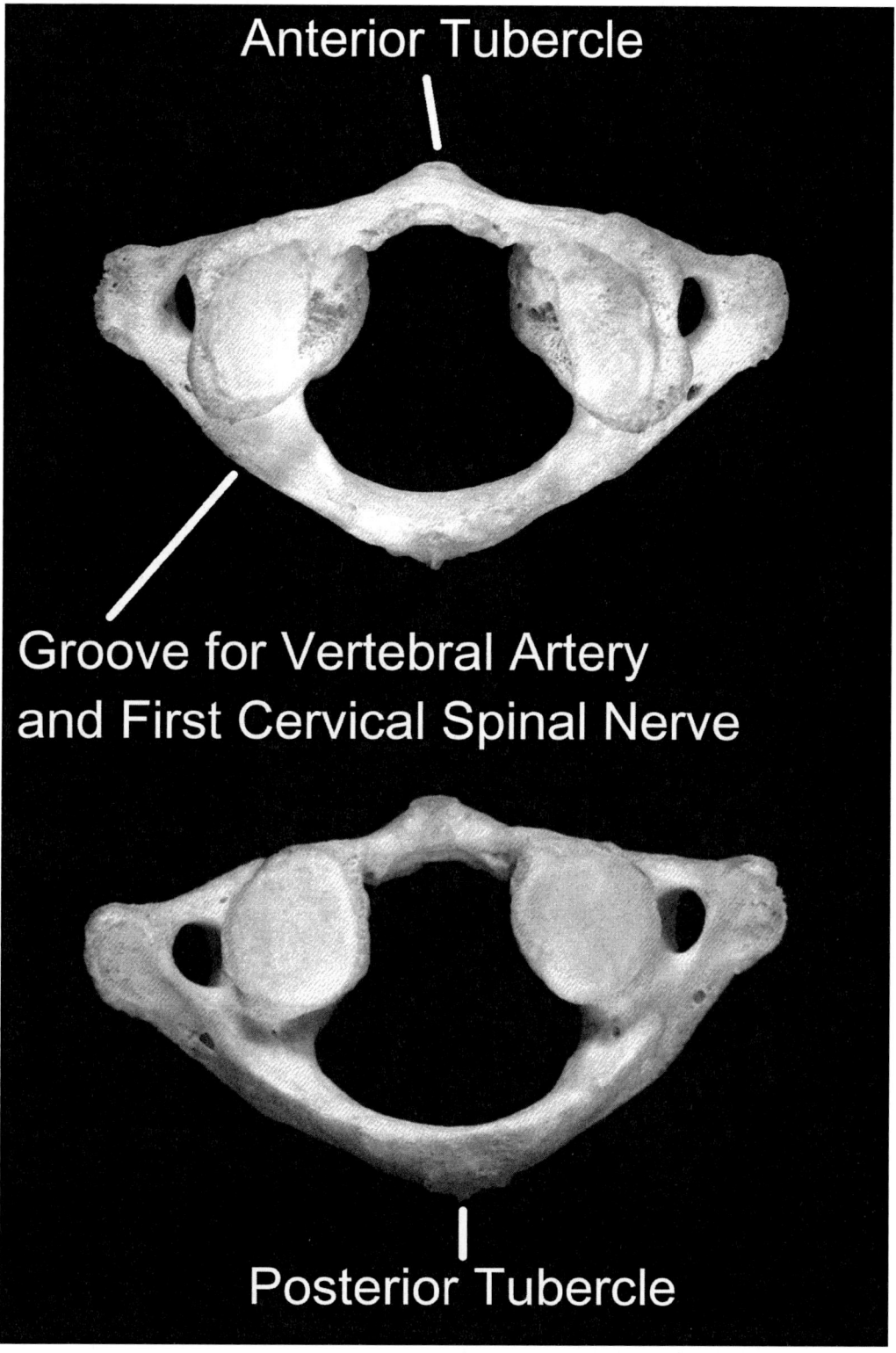

Figure 4-6 *(Continued)* Photographs of superior (top) and inferior views of the atlas (first cervical vertebra). See accompanying line drawing for identification of additional landmarks.

The *lateral masses* (see Figure 4-6 B) are masses of bone on the lateral aspects of the atlas that bear the superior and inferior articular facets. The superior articular facets are large and ovoid. They articulate with the occipital condyles of the skull. On the medial surfaces of the masses are tubercles for the attachment of the transverse atlantal ligament. The inferior articular facets project posteriorly and slightly medially. They are almost circular and slightly concave. They articulate with the superior articular facets of the axis. The transverse processes project laterally and inferiorly from the lateral masses. The transverse processes are very long, and the atlas is the widest of the cervical vertebrae with the exception of the seventh. The transverse process ends in a single tubercle homologous to the posterior costal tubercle of other cervical vertebrae. A very small vestigial anterior tubercle may occasionally be present on the lateral mass. The transverse foramen is oval-shaped.

The first cervical spinal nerve emerges from the spinal cord within the vertebral canal to exit between the skull and the atlas. Note that this is not a true intervertebral foramen because it is not formed between two vertebrae. The boundaries of the passage for the first cervical spinal nerve are:

Anterior: The lateral mass

Posterior: The posterior atlantooccipital membrane (described later)

Superior: The edges of an opening in the posterior atlantooccipital membrane

Inferior: The groove for the vertebral artery and first cervical spinal nerve

The Axis

The second cervical vertebra (Figure 4-7) is called the *axis* (or *epistropheus*) because it forms the pivot upon which the atlas, carrying the head, rotates. Its characteristic feature is the *dens* (or *odontoid process*), a superior projection of the body that articulates with the anterior arch of the atlas. The dens has a slight constriction or neck where it joins the body of the axis. It was at one time considered to be the displaced body of the first cervical vertebra, but more recent research does not entirely support this view. The dens has an anterior *articular facet* for articulation with the anterior arch of the atlas. There is a groove on the posterior aspect of the dens for the transverse atlantal ligament. This ligament keeps the dens applied to the anterior arch of the atlas.

The *body of the axis* is small, and is continuous with the dens. The vertebral foramen is triangular in shape. It is large, but it is smaller than the vertebral foramen of the atlas. The transverse processes are smaller and shorter than those of the atlas. The transverse processes are located anterior to the inferior articular processes. They each end in a single posterior costal tubercle (there is no anterior tubercle on the axis). The pedicles are broad and strong. They are covered cranially by the superior articular processes. The laminae are thick and strong as well. The superior vertebral notches are shallow and lie dorsal to the articular processes. The inferior vertebral notches lie ventral to the articular processes, as in other cervical vertebrae. The spinous process is large and bifid. The superior articular facets are round, slightly convex, directed cranially and laterally, and are supported by the body, the pedicles, and the transverse processes. The inferior articular facets face ventrally, similar to those in other cervical vertebrae. The intervertebral foramen between the atlas and axis is not well defined. It allows the exit of the second cervical spinal nerve. Its boundaries are:

Anterior: Lateral atlantoaxial joint capsule (described later)

Posterior: Posterior atlantoaxial ligament (described later)

Superior: Inferior vertebral notch of the posterior arch of the atlas

Inferior: Superior surface of the lamina of the axis

The Third to Sixth Cervical Vertebrae

The third to sixth cervical vertebrae (see Figure 4-5) are morphologically similar and exhibit the typical features of cervical vertebrae. Their bodies are oval in outline, broader mediolaterally than anteroposteriorly, and uniform in height posteriorly and anteriorly. The pedicles are short and the laminae are thin and long. The spinous process is short and bifid, and the two divisions are often unequal in size. The superior and inferior articular processes are fused to form an articular pillar.

The superior articular facets are directed superiorly and dorsally, and are at about a 45 degree angle to the horizontal plane. The inferior articular facets are directed inferiorly and ventrally. The superior and inferior vertebral notches are formed by constrictions of the pedicles. The anterior tubercle of the sixth cervical vertebra is relatively large, and is known as the *carotid tubercle*. It is possible for the common carotid artery to be compressed between this tubercle and the body of the vertebra, hence the name.

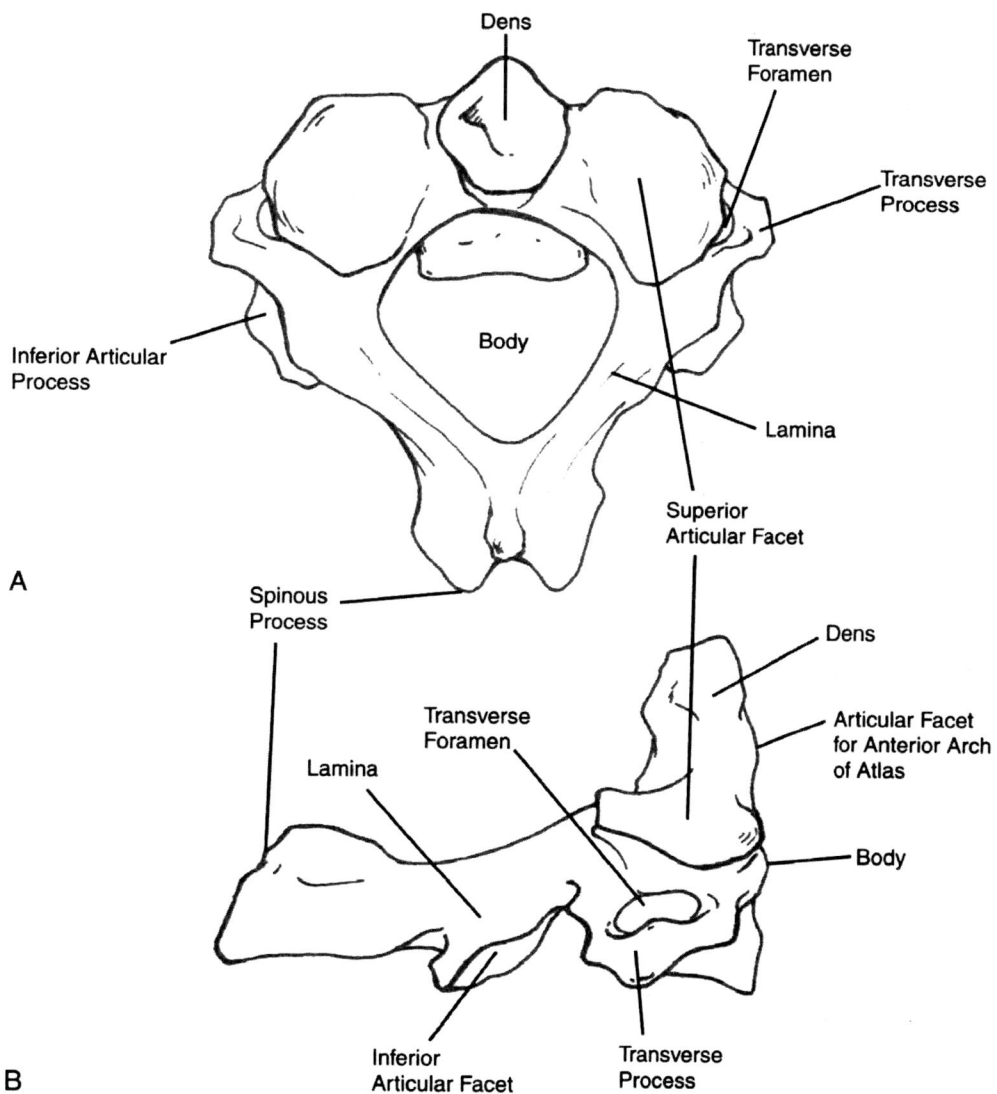

Figure 4-7 The axis (second cervical vertebra): (A) superior view, (B) right lateral view.

A specialization of the vertebral bodies of the caudal five cervical and the first, and sometimes second, thoracic vertebrae is the *uncinate process* (*Lip of Luschka*) (see Figure 4-5 C), a liplike projection on the superior lateral surface of the vertebral bodies. This uncinate process articulates

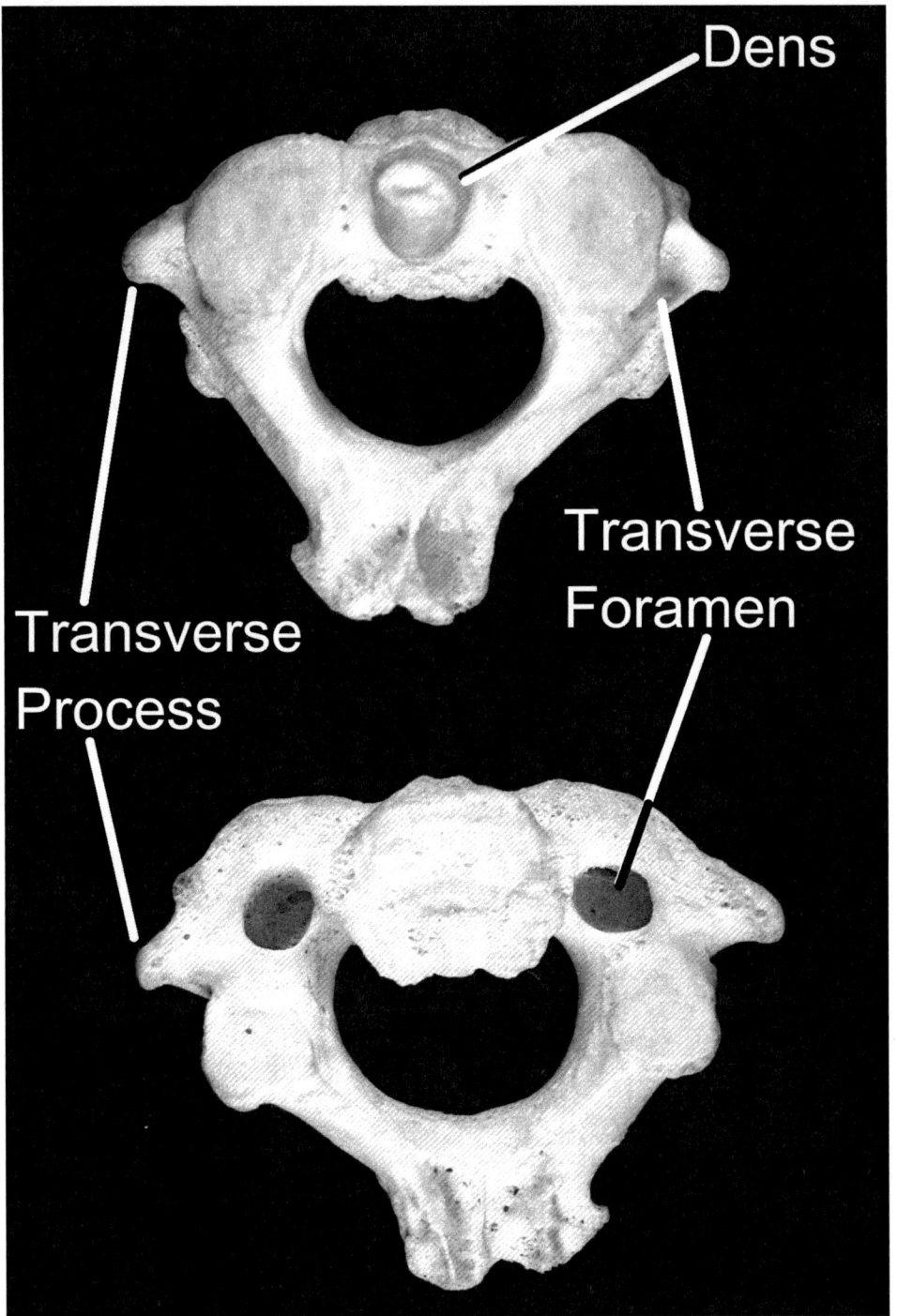

Figure 4-7 *(Continued)* Photographs of superior (top) and inferior views of the axis (second cervical vertebra). See accompanying line drawing for identification of additional landmarks.

with a *beveled surface* on the lateral aspect of the inferior surface of the vertebra above it. This beveled surface is found on the second through seventh cervical vertebrae, and sometimes on the first thoracic. The articulation of the uncinate process with the beveled surface on the vertebra above is the *uncovertebral joint*, or *Joint of Luschka*.

The *intervertebral foramina* are formed similarly among the lower cervical vertebrae. The boundaries of each intervertebral foramen among these vertebrae are as follows:

Superior: Inferior vertebral notch of the cranial vertebra

Inferior: Superior vertebral notch of the caudal vertebra

Posterior: Joint capsule of the articular processes

Anterior: Uncovertebral joint (joint of Luschka)

The Seventh Cervical Vertebra

The *seventh cervical vertebra* (Figure 4-8) is also called the *vertebra prominens* (literally, "the most prominent vertebra"). It is a typical cervical vertebrae, but is unique in one feature: the spinous process is very long and prominent and the tip of the spinous process is not bifid as it often is in the other cervical vertebrae. The palpable prominence of the spinous process of the seventh cervical vertebra makes it a useful landmark for counting vertebral levels. The transverse process of C7 usually has a prominent posterior costal tubercle, but the anterior costal tubercle

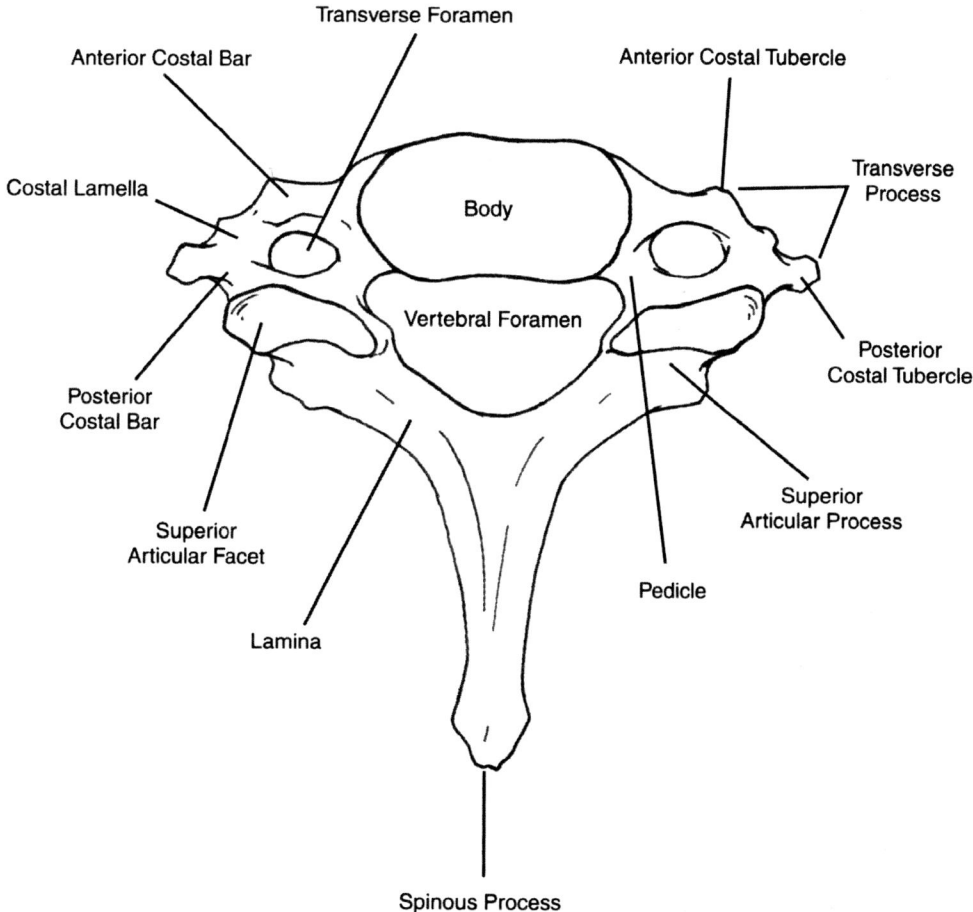

Figure 4-8 The vertebra prominens (seventh cervical vertebra): superior view.

is rudimentary. In addition, the transverse foramen may be small, or double, or absent. As mentioned before, the transverse foramen of the seventh cervical vertebra does not transmit the vertebral artery.

The Thoracic Vertebrae

There are twelve *thoracic vertebrae* (T1-T12; refer to Figures 4-9 and 4-10 throughout the discussion of the thoracic vertebrae). They are of intermediate in size and located between the cervical and lumbar vertebrae. The thoracic vertebrae are unique in that they articulate with the ribs. They are easily distinguished by the presence of *costal facets* on the sides of the bodies for articulation with the heads of the ribs, and facets on their transverse processes for articulation with the tubercles of the ribs. Some of the thoracic vertebrae bear complete, rounded *costal*

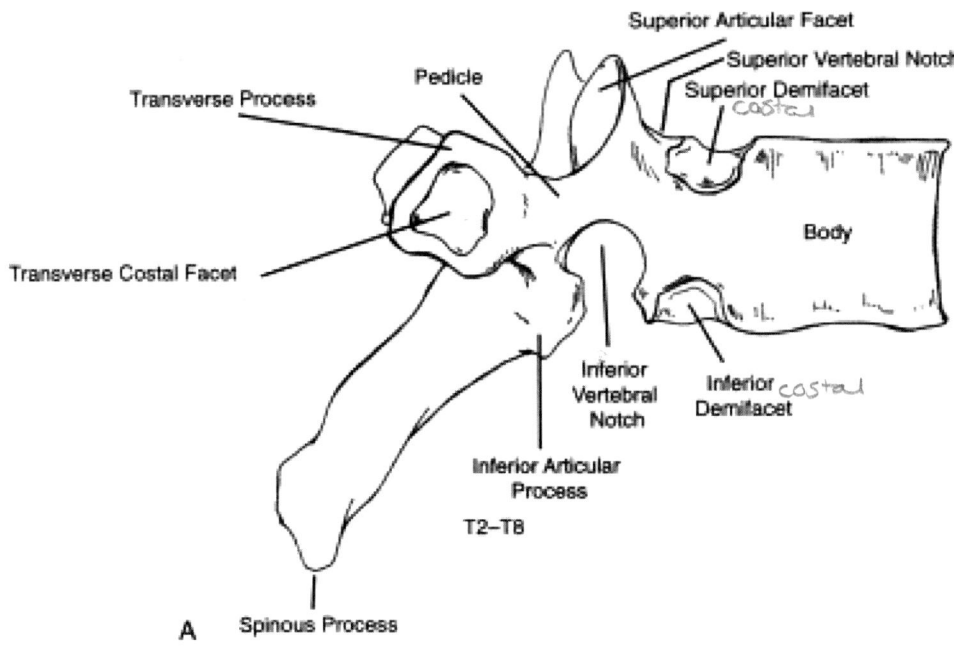

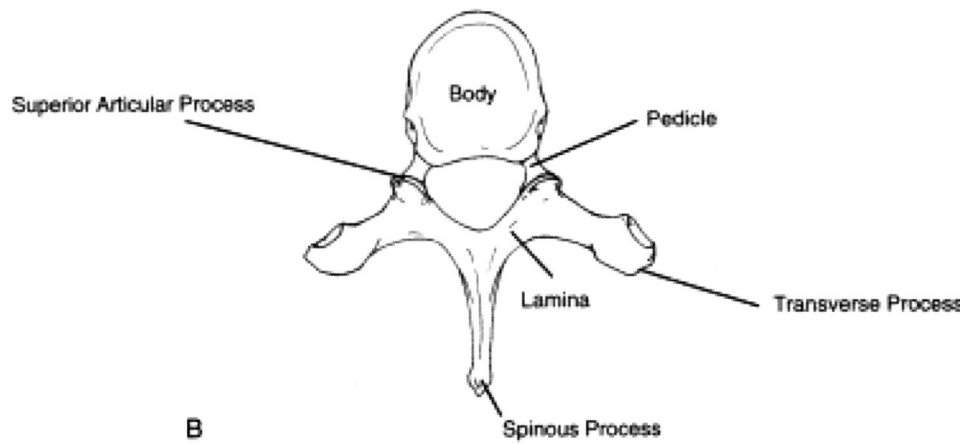

Figure 4-9 Features of a typical thoracic vertebra: (A) right lateral view and (B) superior view.

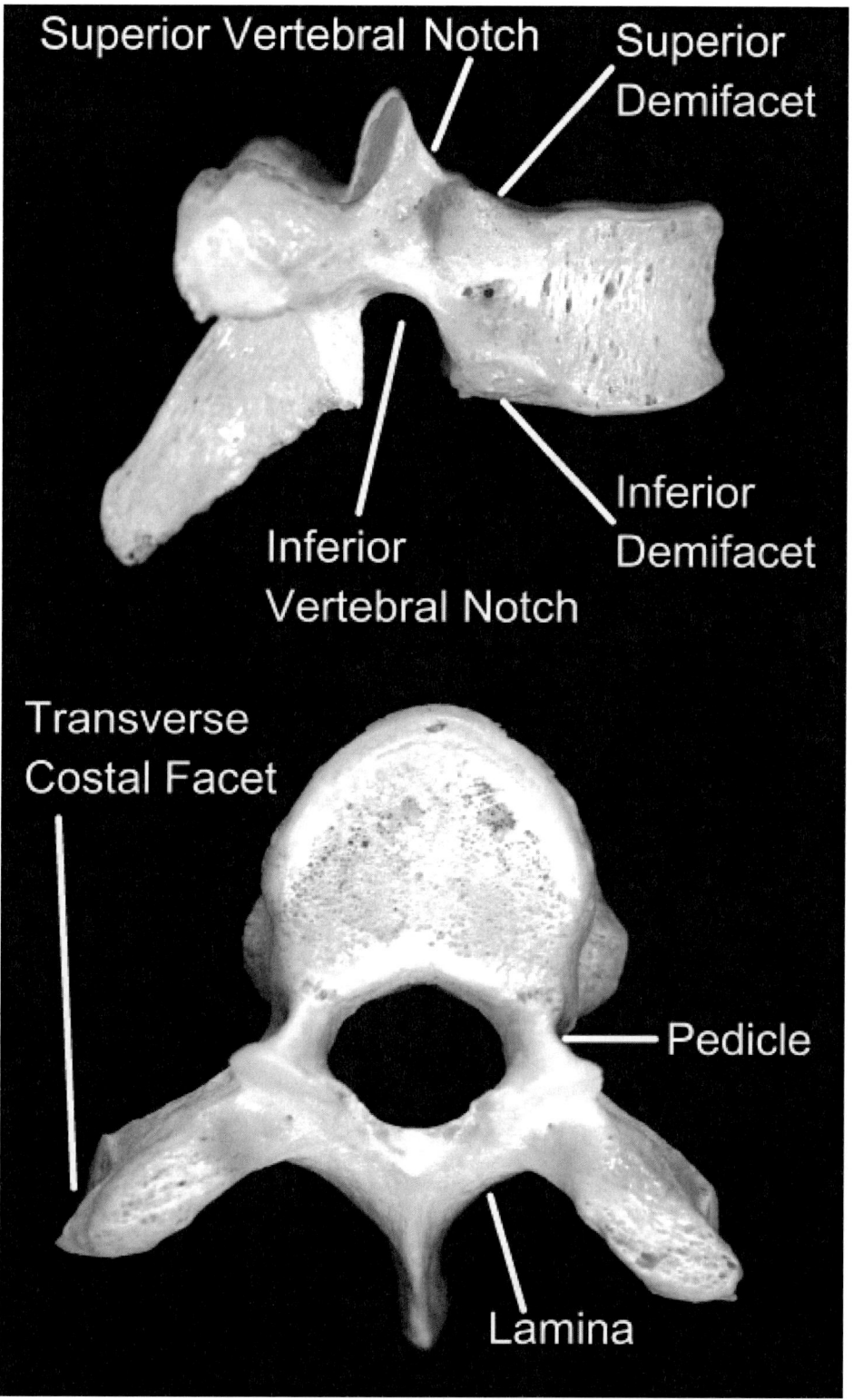

Figure 4-9 *(Continued)* Photographs of the right lateral (top) and superior views of a typical thoracic vertebra. See accompanying line drawing for identification of additional landmarks.

facets (T1 and T10-T12), by which a rib articulates with a single vertebral body.

In most of the thoracic vertebrae (inferior portion of T1, T2-T8, and superior portion of T9) each side of the body bears both a superior and inferior semicircular shaped costal facet, called a *demifacet*. The inferior demifacets of the more cranial vertebrae combine with the superior demifacets of the adjacent caudal vertebra to form a cuplike cavity that receives the head of the

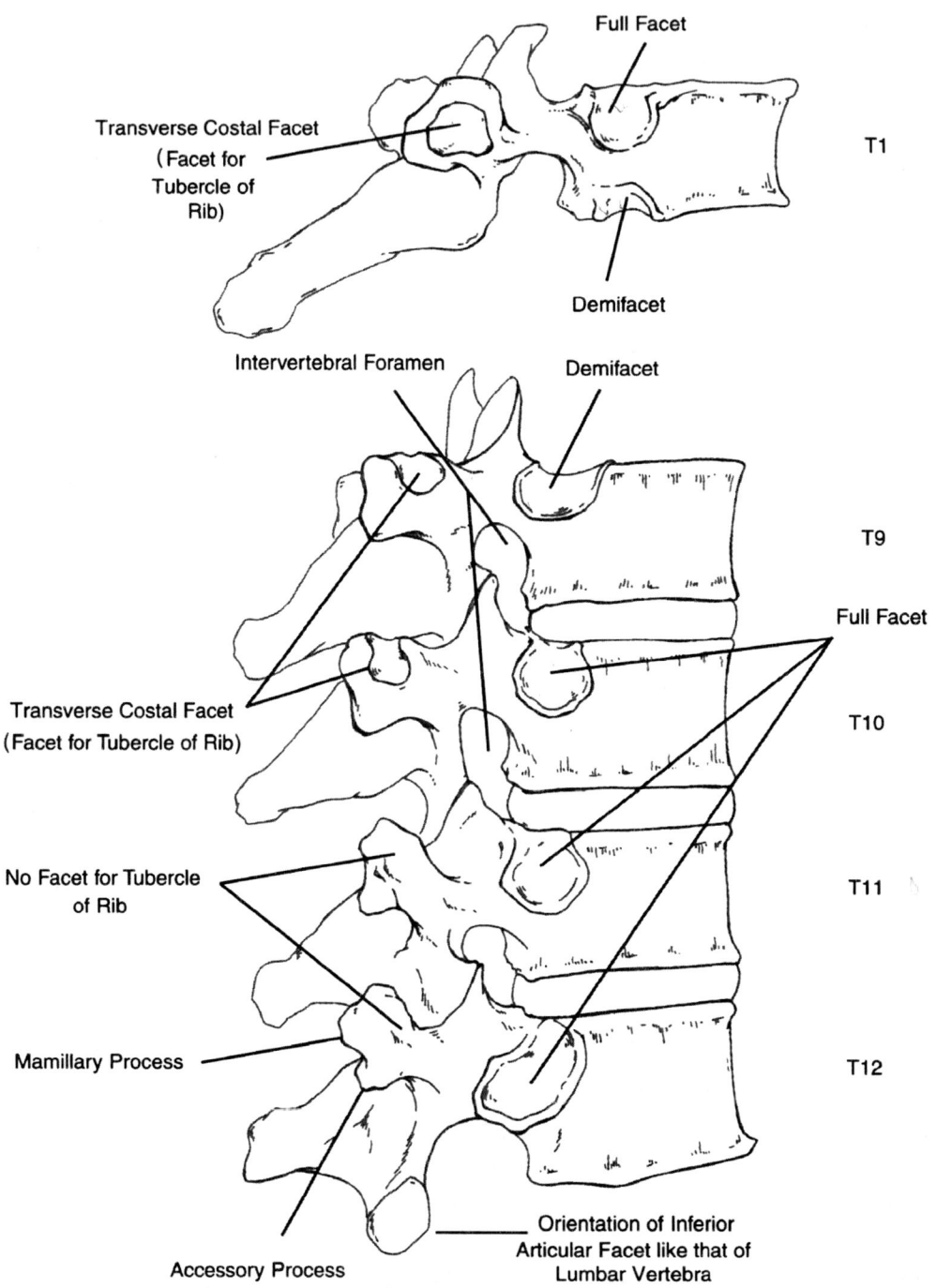

Figure 4-10 Features of atypical thoracic vertebrae, right lateral view.

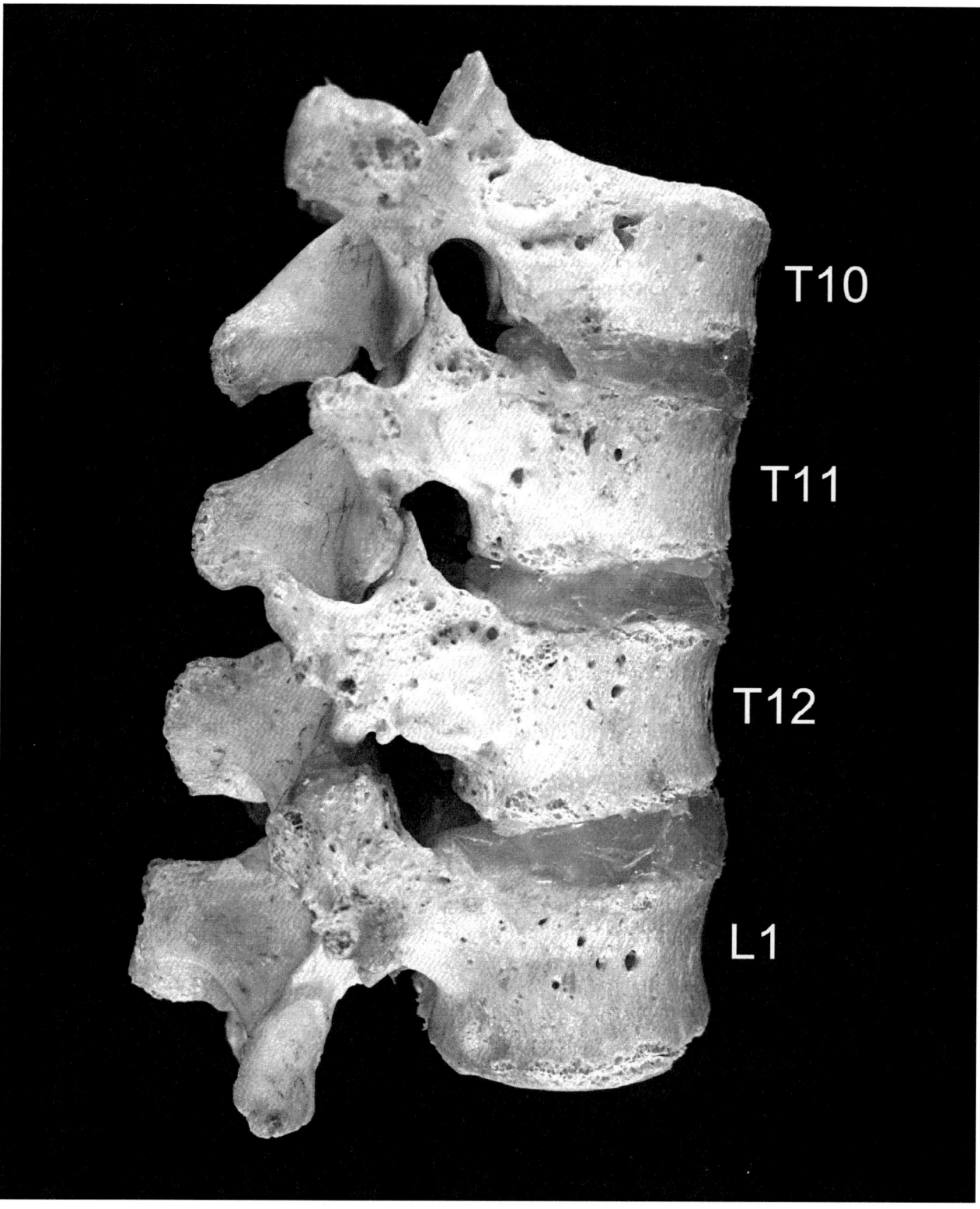

Figure 4-10 *(Continued)* Photograph of the right lateral view of T9 (top) to L1 articulated. See accompanying line drawing for identification of additional landmarks.

rib. The articulations between the heads of the ribs and the costal facets form the *costovertebral joints*. The *costotransverse joints* are formed by the articulation of the tubercles of the ribs with the transverse processes of the thoracic vertebrae.

The bodies of the thoracic vertebrae, in the midthoracic region, are heart shaped, and as broad anteroposteriorly as mediolaterally. The body is slightly thicker dorsally than ventrally, contributing to the thoracic curvature of the spine. Superior demifacets are located just anterior to the pedicles, while the inferior demifacets are located just anterior to the inferior vertebral notch. The pedicles are directed slightly cranially. The superior vertebral notches are very shallow and scarcely recognizable. The inferior vertebral notches are large, and deeper than in other regions of the spine. The laminae are broad and thick and, unlike other regions of the column, overlap those of the vertebrae below. The vertebral foramen is small and circular.

The spinous processes of thoracic vertebrae are long, and directed obliquely inferior. The thoracic spinous process ends in a single tubercle. In T1 and T2, the process is more horizontal, and the interspinous spaces are large. In T3 and T4, they are more oblique but the interspinous spaces are still large. The spinous processes of T5-T8 are very oblique and overlap to a greater extent than do the spinous processes of more cranial vertebrae and the interspinous spaces become smaller. The processes of T9 and T10 are oblique with small interspinous spaces, while those of T11 and T12, like those of the lumbar vertebrae, are horizontal.

The superior articular processes are thin and platelike, and project up from the junctions of the pedicles and laminae. The articular facets are practically flat and face dorsally and laterally. The inferior articular processes are fused to a considerable extent with the laminae, and project only slightly beyond the inferior border of the laminae. The inferior articular facets are oriented at nearly 90 degrees to the horizontal and are angled to the coronal plane such that they face slightly medially.

The transverse processes arise from the vertebral arch just posterior to the superior articular processes and pedicles. They are long, thick and strong with an obliquely posterior and lateral direction. On the ventral surface of the lateral extremity of the transverse process is the *transverse costal facet*, which is the facet for articulation with the tubercle of the rib.

In general, the body of first thoracic vertebra will bear an entire costal facet for the first rib, and an inferior demifacet for the superior portion of the head of the second rib. The body of the first thoracic vertebra is transversely broad, like a cervical vertebra. The spinous process is long, thick, and almost horizontal.

The second through eighth thoracic vertebrae are the most typical. They bear two demifacets on each side, a superior and an inferior.

The ninth thoracic vertebra usually bears only a superior demifacet and no inferior demifacet. In some persons, however, it may bear both a superior and a small inferior demifacet.

The 10th thoracic vertebra usually bears one large circular articular facet on each side, placed partly on the pedicle and partly on the body. In the case of the ninth thoracic vertebra bearing a small inferior demifacet, the tenth then bears one large demifacet on each side.

The 11th thoracic vertebra approaches the lumbar vertebrae in size and form. It bears only a single costal facet on each side, which is large and placed upon the lateral aspect of the pedicle. The pedicles are thicker and stronger in the eleventh and twelfth thoracic vertebrae than in any other part of the column. In both the eleventh and twelfth thoracic vertebrae, transverse processes are short and bear no transverse costal facets.

The 12th thoracic vertebra is very similar to the eleventh. It may be distinguished, however, by its inferior articular facets. They are convex and directed laterally like those of the lumbar vertebrae. As noted above, it bears no transverse costal facets on its transverse processes. In many ways, the twelfth thoracic vertebra is a transitional vertebra between the thoracic and lumbar regions, just as the first represented a transition between the cervical and thoracic regions. T12 will exhibit accessory processes which are very small rough elevations on the inferolateral surfaces of the transverse processes. They are also found occasionally on T11 and T10. The *mamillary processes* are small round enlargements on the superior aspect of the transverse processes of T12 and possibly T11 and T10. Both accessory and mamillary processes are characteristic of lumbar vertebrae though their locations are slightly different in thoracic and lumbar regions, probably due to the developmental origins of thoracic and lumbar transverse processes from transverse and costal elements, respectively.

The intervertebral foramina in the thoracic region have the following borders:

Superior: Inferior vertebral notch of cranial vertebra

Inferior: Superior vertebral notch of caudal vertebra

Posterior: Joint capsule of articular process

Anterior: Body of cranial vertebra, articular capsule of costovertebral joint

The Lumbar Vertebrae

There are normally five lumbar vertebrae (Figure 4-11). Aside from the fused upper sacral segments, the lumbar vertebrae are the largest vertebrae. They are characterized by large bodies, the presence of mamillary processes and accessory processes, and a lack of transverse costal facets or transverse foramina.

The body is large, wider transversely, and a little thicker ventrally. It is flattened or slightly concave superiorly and inferiorly. The sides of the body are concave. The vertebral foramen is triangular, but smaller than in the cervical region. The transverse processes are anterior to the articular processes (instead of posterior to them as in the thoracic vertebrae) and posterior to the inferior vertebral notch. They are long, slender, and horizontal in the cranial three lumbars, while they incline a little caudally in the caudal two lumbars. The accessory processes are small, rough elevations on the dorsal surfaces at the bases of the transverse processes.

The spinous process is thick, broad, and quadrilateral in shape. It is thickest caudally. The spinous process of the fifth lumbar vertebrae is small for the lumbosacral angle.

The superior articular processes possess slightly concave, vertical facets which face medially and dorsally. On the posterolateral borders of the superior articular processes are found round enlargements, the mamillary processes. The inferior articular processes possess convex, vertical facets which face laterally and ventrally.

The pedicles unite with the cranial part of the body. As a result, the inferior vertebral notches are very deep, while the superior vertebral notches are shallow. The laminae are tall superoinferiorly, narrow mediolaterally, and robust.

The intervertebral foramina allow the passage of lumbar spinal nerves to exit the vertebral canal. In the lumbar region, the borders of the foramina are:

Superior: Inferior vertebral notch of cranial vertebrae

Inferior: Superior vertebral notch of caudal vertebrae

Posterior: Zygapophyseal joint capsule

Anterior: Bodies of cranial and caudal vertebrae and their intervertebral disc

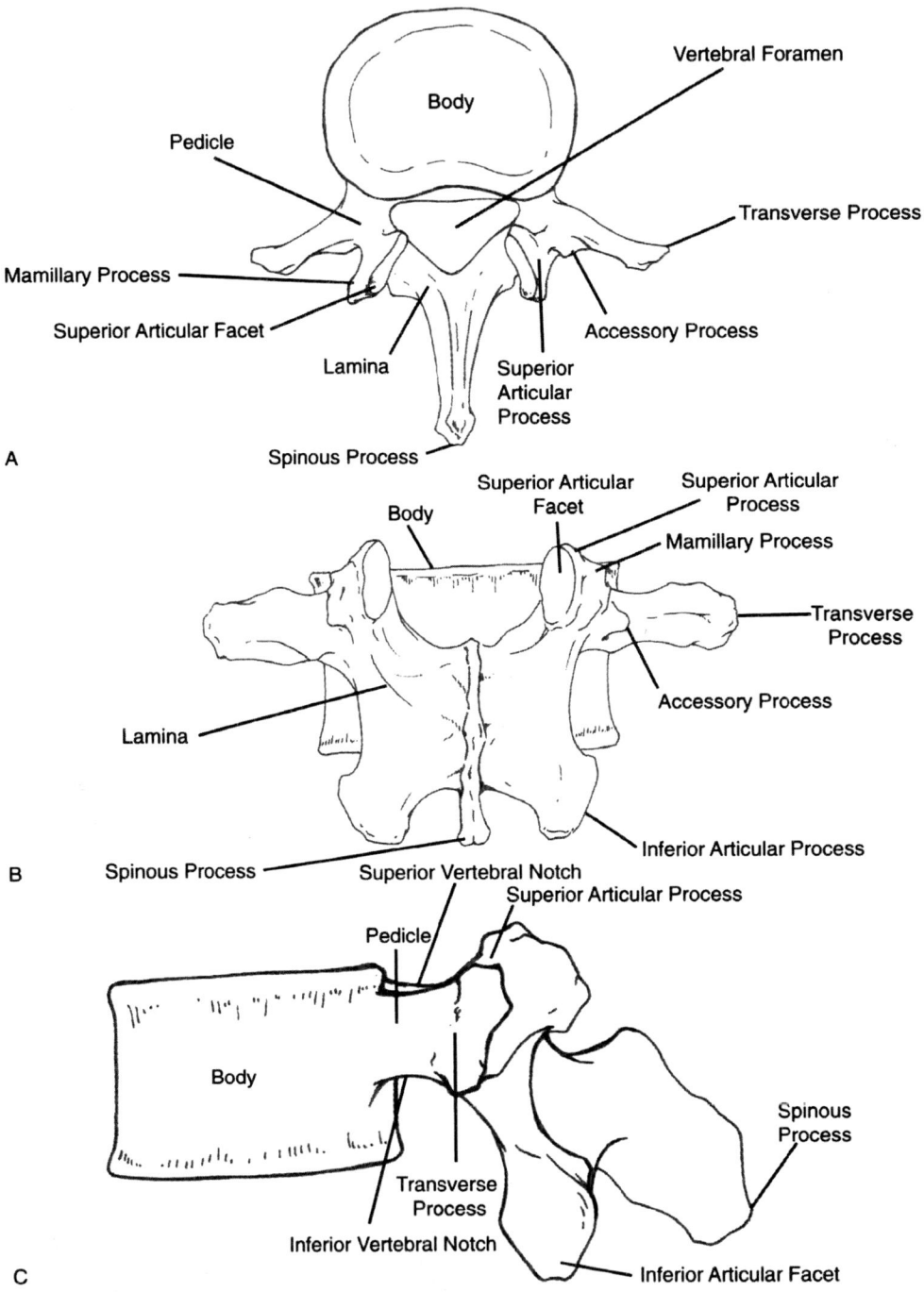

Figure 4-11 A lumbar vertebra: (A) superior, (B) posterior, and (C) left lateral views.

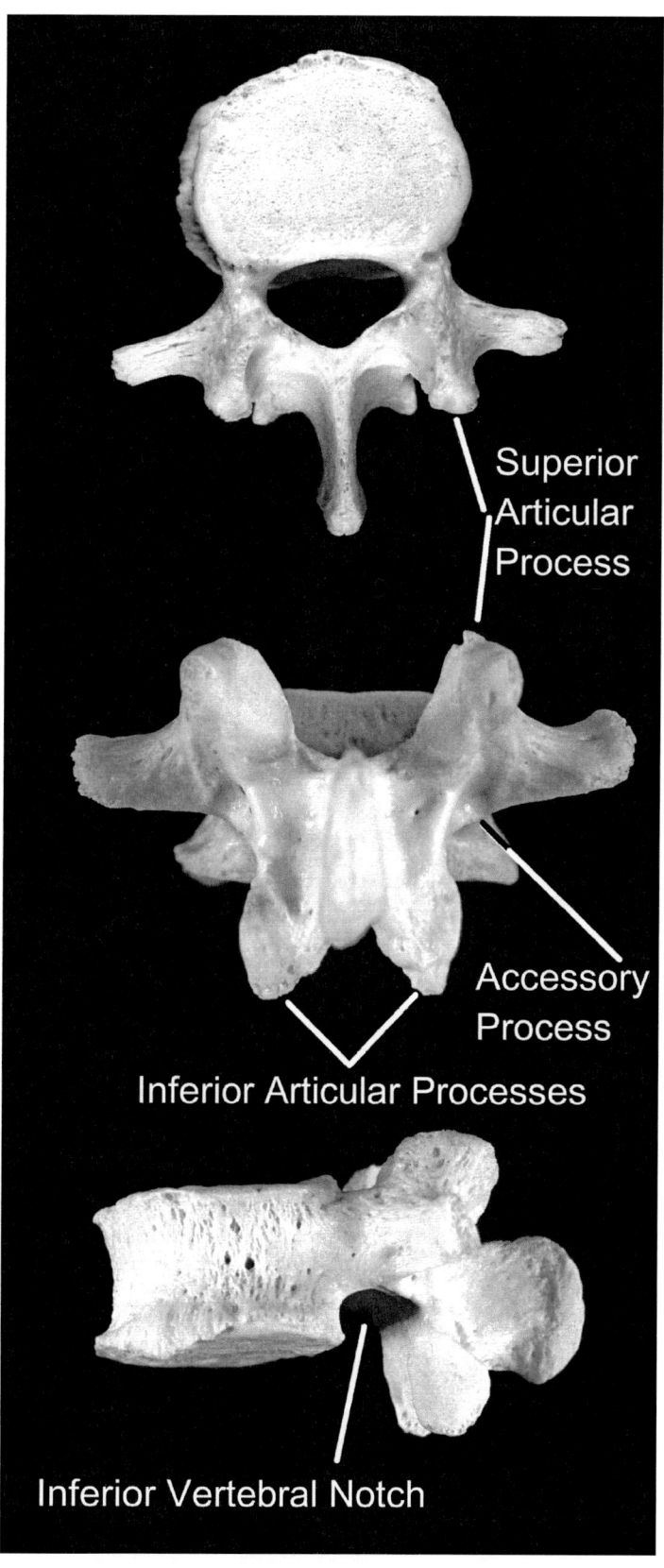

Figure 4-11 *(Continued)* Photographs of the superior (top), posterior, and left lateral views of a lumbar vertebra. See accompanying line drawing for identification of additional landmarks.

The Sacrum

The sacrum (S1-S5; refer to Figures 4-12 through 4-15 throughout the discussion of the sacrum) is a large triangular mass formed by the fusion of five vertebrae. It forms the posterosuperior wall of the pelvic cavity, being wedged between the auricular surfaces of the two os coxae. It presents for examination a base, a pelvic (ventral) surface, a dorsal surface, two lateral surfaces, and an apex. The base of the sacrum is superior and articulates with the 5th lumbar vertebrae, while its apex is inferior and articulates with the coccyx. Because the sacrum is formed by the fusion of five (on average) separate vertebrae, all the typical features of a vertebra are present in the sacrum, though in modified form.

The *base* is the superior surface of the body of the first sacral vertebral segment. The base projects ventrally to form the lumbosacral angle. The *promontory* is the ventral border of superior surface of the body of S1. It is the most ventral portion of the sacrum and helps define the border of the pelvic inlet. The *sacral canal* is formed by the five vertebral foramina of the sacral vertebrae. The *alae* (Latin: "wings"; singular: *ala*) are the lateral processes extending from the bodies of S1 and S2. They are formed by the fusion of the costal elements of the sacral vertebrae. They support the auricular surfaces. The *superior articular processes* of the sacrum articulate with the inferior articular facets of the 5th lumbar vertebra. The articular facets face dorsomedially.

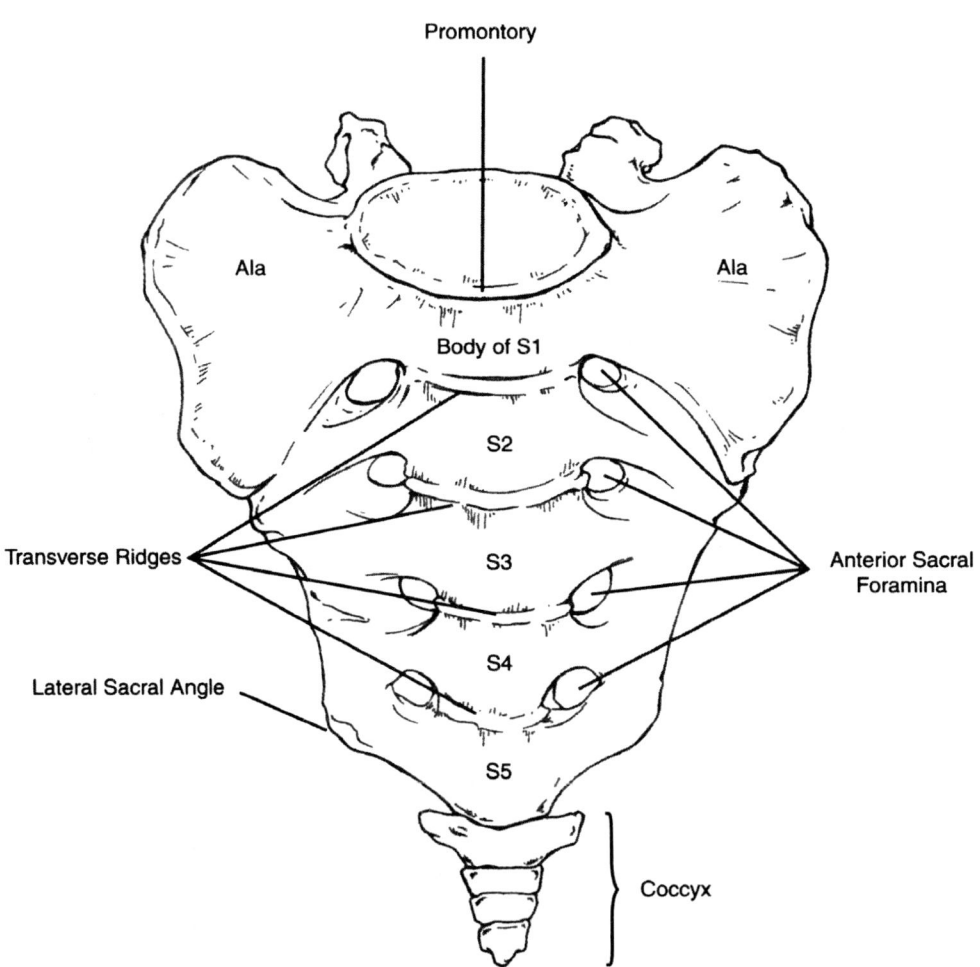

Figure 4-12 The sacrum and coccyx, anterior view (pelvic surface).

Bones of the Postcranial Skeleton: Vertebral Column

The *pelvic surface* is concave in shape. It is characterized by four *transverse ridges*, which indicate the site of fusion of the five sacral vertebrae. Also on the pelvic surface are four pairs of *anterior sacral foramina*. They communicate with the sacral canal by means of the intervertebral foramina and allow the passage of the ventral rami of the first four sacral spinal nerves, and the lateral sacral arteries and veins.

The posterior (dorsal) surface is convex and narrower than the pelvic surface. In the midline, the *median sacral crest* is surmounted by three or four tubercles, the rudimentary spinous processes of the first three or four sacral segments. The *sacral groove* is a shallow groove on either side of the median sacral crest. It is formed by the union of the laminae of the sacral vertebrae. The *sacral hiatus* is a defect in the floor of the sacral groove inferiorly. It is due to the nonfusion of the laminae of S5 with those of S4.

On either side of the dorsal surface, on the lateral aspect of the sacral groove, is the *intermediate sacral crest*. It is a linear series of tubercles produced by the fusion of the articular processes of the sacral vertebrae. At the distal ends of the intermediate crests are the two *sacral cornua*. These are the inferior articular processes of S5 and articulate via ligaments with the coccyx. The *lateral sacral crest* is formed by the fusion of three or four transverse tubercles,

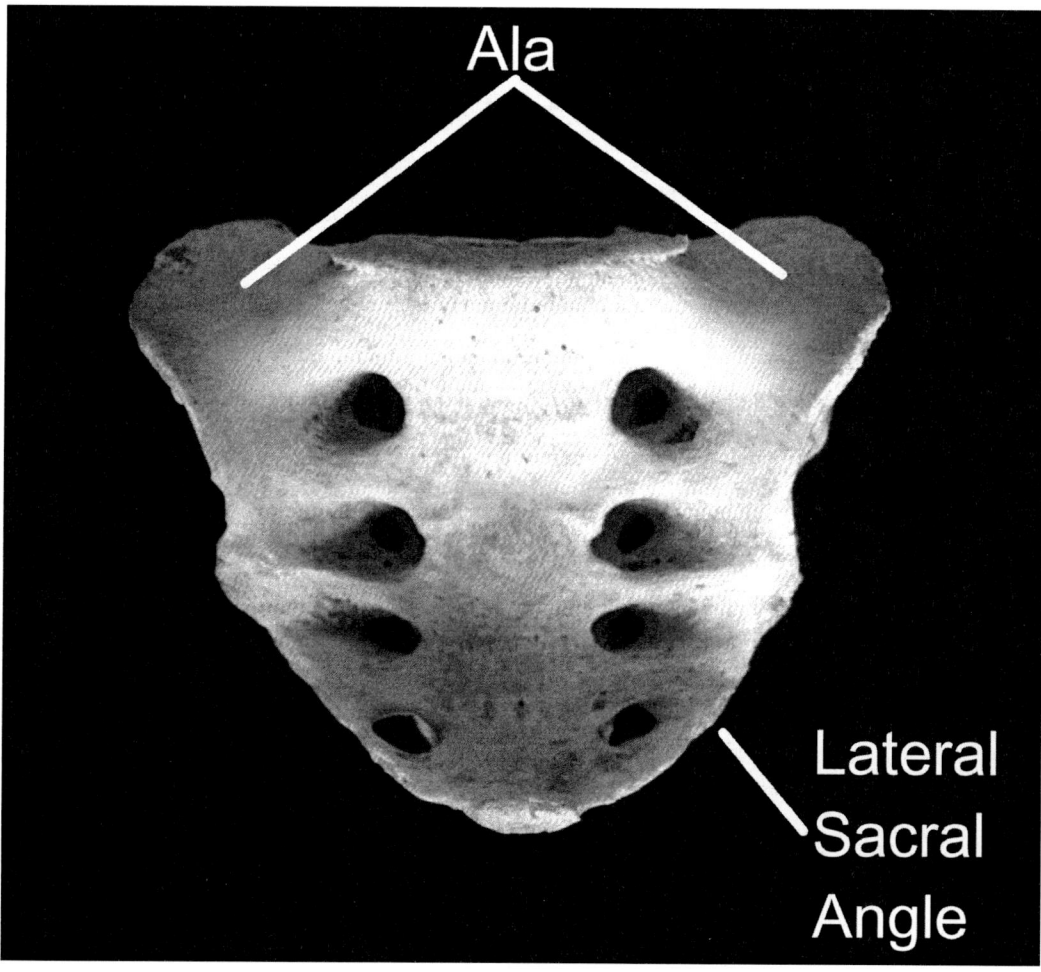

Figure 4-12 *(Continued)* Photograph of the anterior surface of the sacrum. See accompanying line drawing for identification of additional landmarks.

which are rudimentary transverse processes of the sacral vertebrae. There are four *posterior sacral foramina* between the intermediate and lateral sacral crests. These allow the exit of the dorsal rami of the sacral spinal nerves from the sacral canal.

The lateral surface of the sacrum is dominated by the *auricular surface*. This surface articulates with the similarly shaped auricular surface of the ilium. The *sacral tuberosity* is the rough area between the auricular surface and the lateral sacral crest. It has deep impressions for the attachments of the interosseous sacroiliac ligaments. Caudal to the auricular surface the lateral surface is nonarticular. More caudally the lateral surface curves medially to the body of the fifth sacral vertebra at the *lateral sacral angle*.

The *apex* of the sacrum is its caudal end. It presents a *facet* for articulation with the coccyx. This is the inferior surface of the body of the fifth sacral vertebra.

In order to increase the size of the pelvic cavity for childbirth, the female sacrum is wider and relatively shorter, and its pelvic surface faces more caudally

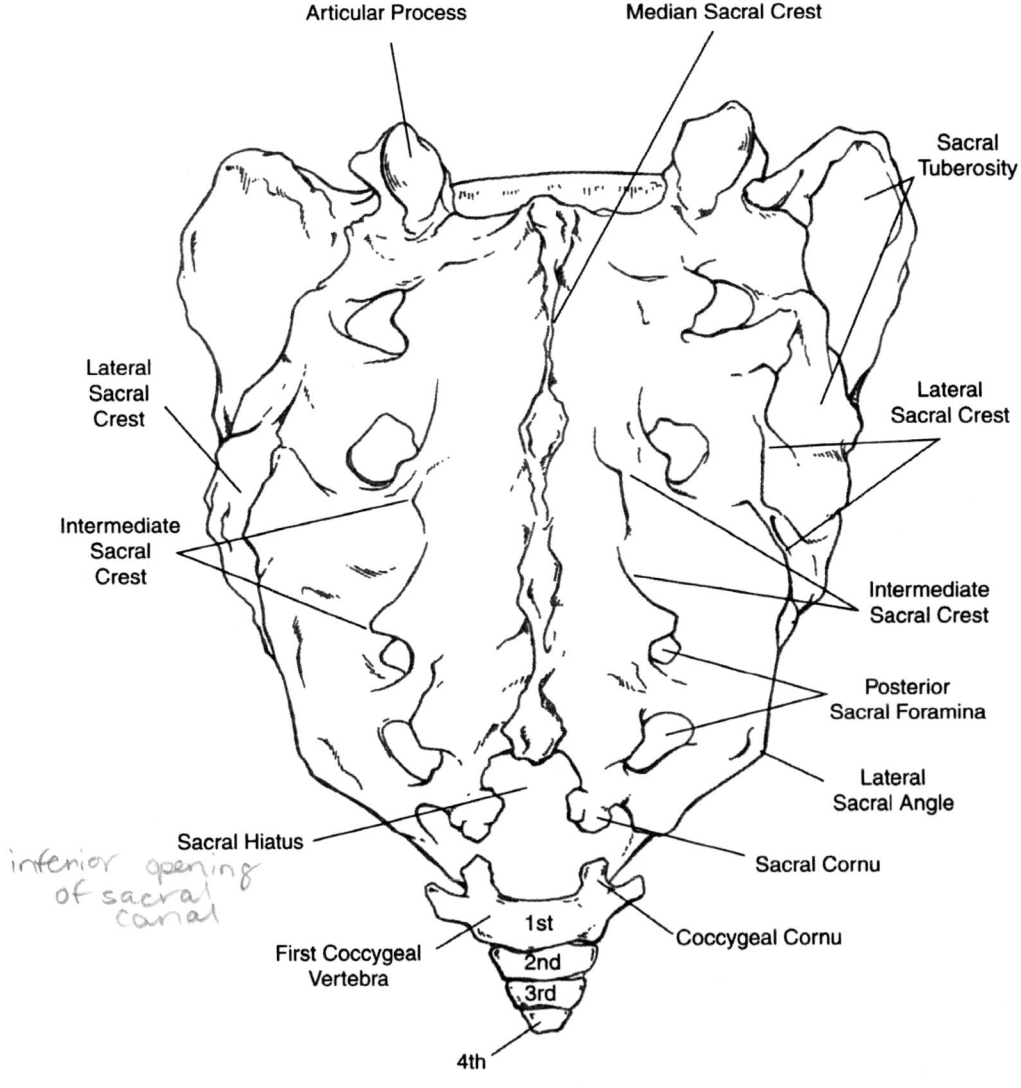

Figure 4-13 The sacrum and coccyx, posterior view (dorsal surface).

than in the male because of the specialized growth of the ilium that decreases the acuity of the greater sciatic notch.

The Coccyx

The *coccyx* (refer to Figures 4-12 through 4-15 throughout the discussion of the coccyx) is the most caudal portion of the vertebral column. It is formed by the fusion of (usually) four rudimentary vertebral bodies. As a rule, only the first coccygeal vertebra possesses rudiments of transverse processes and superior articular processes. These articular processes are termed the *cornua* of the coccyx. The coccyx is as a site of muscle attachment for the pelvic floor.

Landmarks of The Vertebral Column

Specific landmarks of the vertebral column and back may be projected to the surface of the body and palpated. Such landmarks are important in recognizing the vertebral level of interest in a living person.

The *external occipital protuberance* (see Chap. 8, Fig. 8-19) is the prominence on the back of

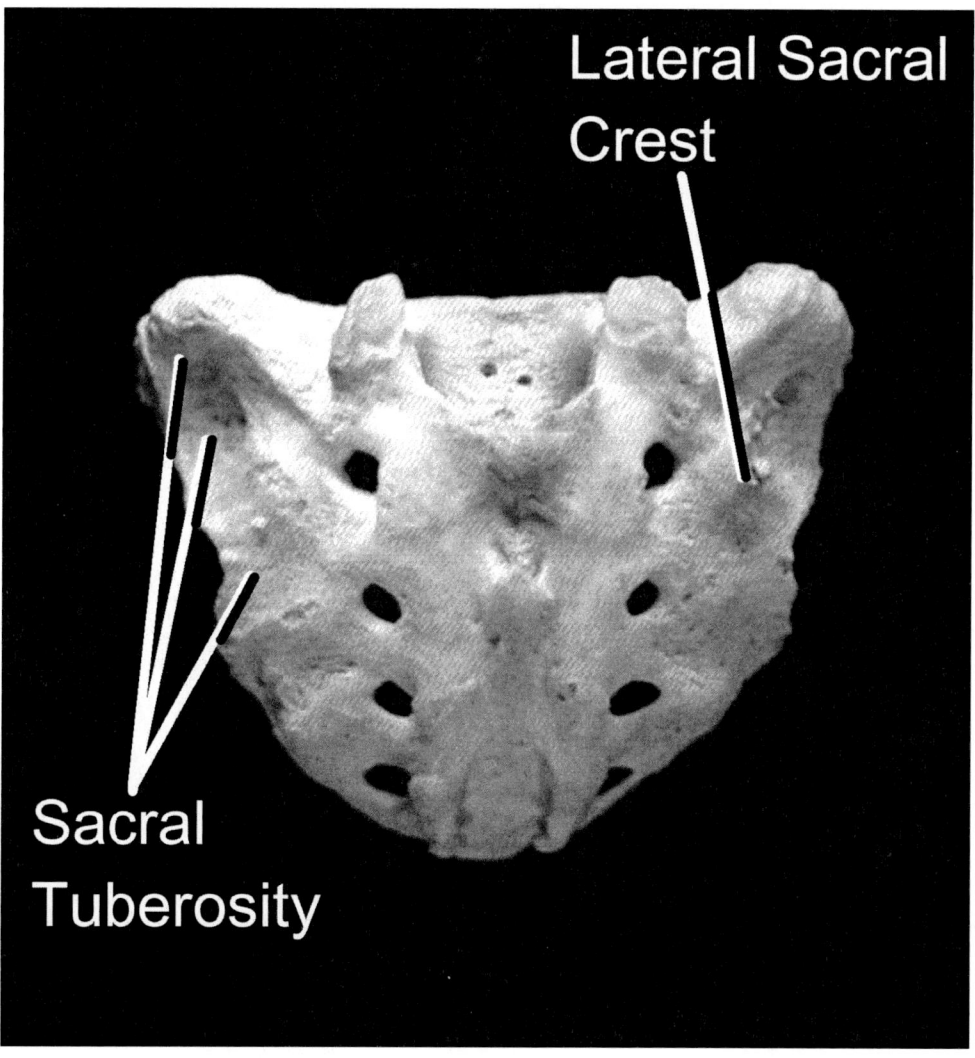

Figure 4-13 *(Continued)* Photograph of the posterior surface of the sacrum. See accompanying line drawing for identification of additional landmarks.

the cranium that marks the superior attachment of the ligamentum nuchae (the ligamentous structure that passively supports the head) and the cervical musculature. If the finger is drawn down the midline of the neck from the external occipital protuberance, the first bony structure that can be felt is the spinous process of the second cervical vertebra. The posterior tubercle of the atlas is not palpable.

At the base of the neck, inferior to the external occipital protuberance, is the vertebra prominens. We have already referred to this structure. This is the

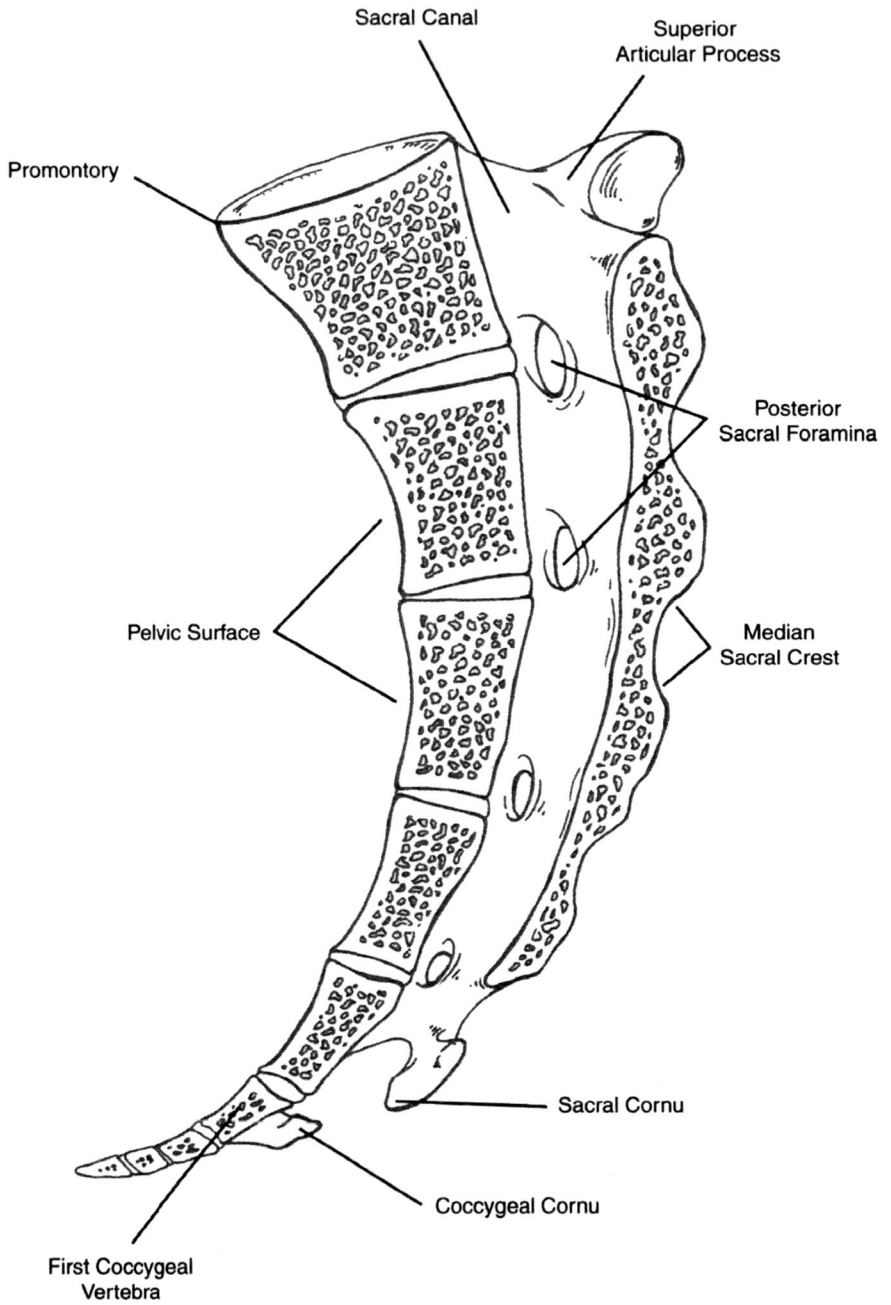

Figure 4-14 The sacrum and coccyx, median section, left lateral view.

spinous process of the seventh cervical vertebra. It projects considerably farther posteriorly than the spinous processes of the upper six cervical vertebrae and marks the junction of the cervical and thoracic portions of the vertebral column.

The *scapulae* (see Chap. 6, Fig. 6-14) are important guides to vertebral levels as well. In a person in the anatomical position, the *spines of the scapulae* lie in the same plane as the spinous process of the third thoracic vertebra. Similarly, in the anatomical position, the *inferior angles of the scapulae* lie at the level of the spinous process of the seventh thoracic vertebra.

Points on the bony pelvis serve as guides to lower vertebral levels. The *crest of the ilium* (the highest point of the os coxae, or innominate bone; see Chap. 7, Fig. 7-2) lies at the level of the spinous process of the fourth lumbar vertebra. The *posterior superior*

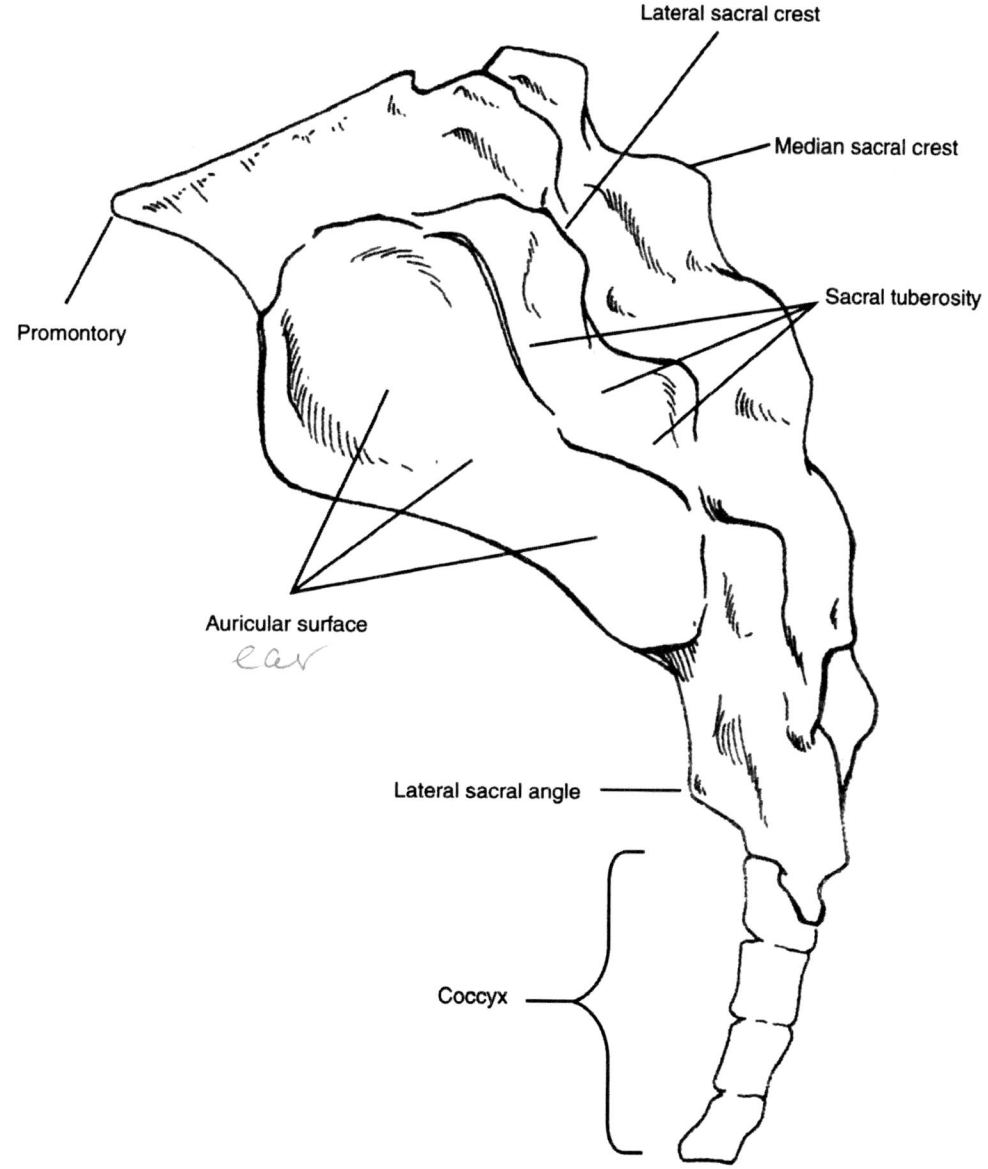

Figure 4-15 The sacrum and coccyx, left lateral view.

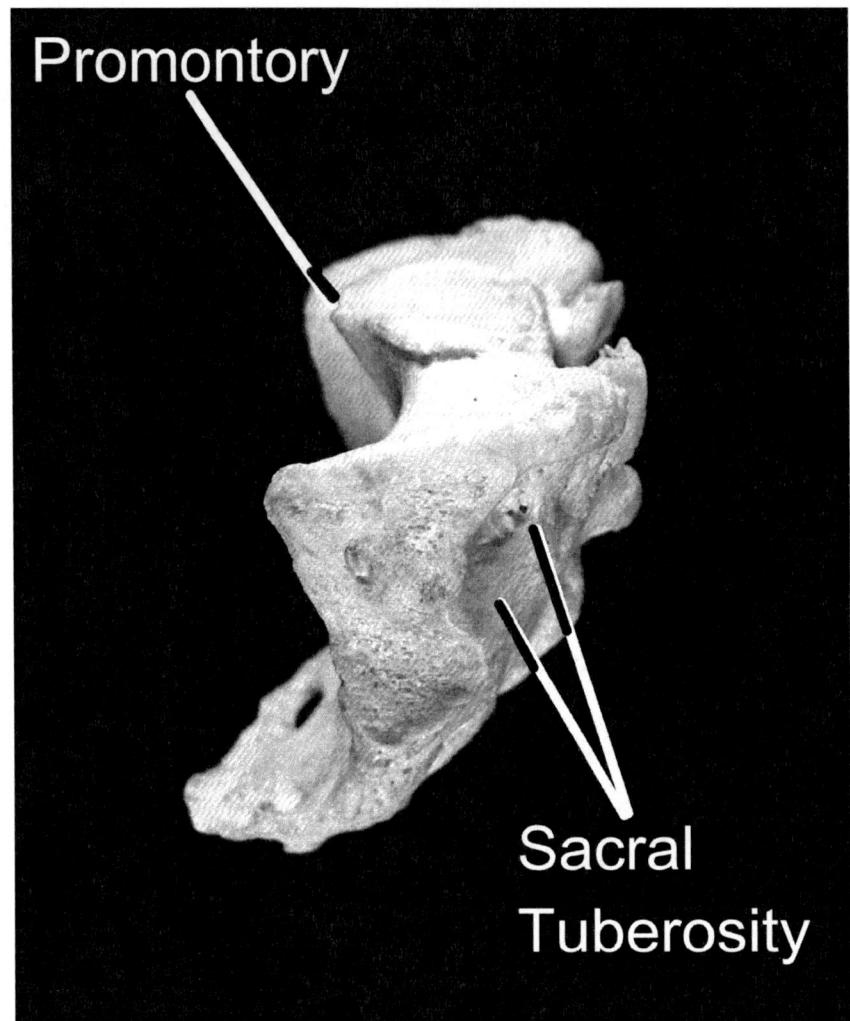

Figure 4-15 *(Continued)* Photograph of an oblique view of the left lateral surface of the sacrum. See accompanying line drawing for identification of additional landmarks.

iliac spine (a posterior projection of the os coxae, or innominate bone) is at the same level as the second sacral vertebra. This level corresponds to the obvious *sacral dimples* on the overlying skin.

Bony landmarks of the thorax (see Chap. 5, Figs. 5-1 and 5-5) can be projected posteriorly to vertebral levels. Anteriorly, the *sternal angle* marks the junction of the manubrium and the body of the sternum. It also marks the location of the *junction of the costal cartilage of the second rib with the sternum*. Projected directly posterior, this marks the location of the *intervertebral disc between T4 and T5* in a supine person. The *jugular notch* of the manubrium lies at the level of the lower part of the second thoracic vertebra. The *xiphisternal junction* between the body and xiphoid process of the sternum lies at the level of the 10th thoracic vertebra.

Ossification of the Vertebral Column

A typical vertebra has *three primary centers of ossification, and five s*econdary centers. The *body* of a typical vertebra represents one of the primary centers (the *centrum;* plural: *centra*), while the neural (vertebral) arch is formed by two primary centers that fuse in the midline to form the spinous process and fuse anteriorly with the body (Figure 4-16 A). In addition, a typical vertebra has secondary centers at the tips of each transverse process, one at the tip of the spinous process, and two ring-shaped annular epiphyses that cap the top and bottom of the cylinder of the body of the vertebra, for a total of *five* secondary centers. Each secondary center is separated from the primary centers by an epiphyseal growth plate.

The primary centers for the centra appear first in the lower thoracic and upper lumbar regions at nine or ten weeks in utero and proceed cranially and

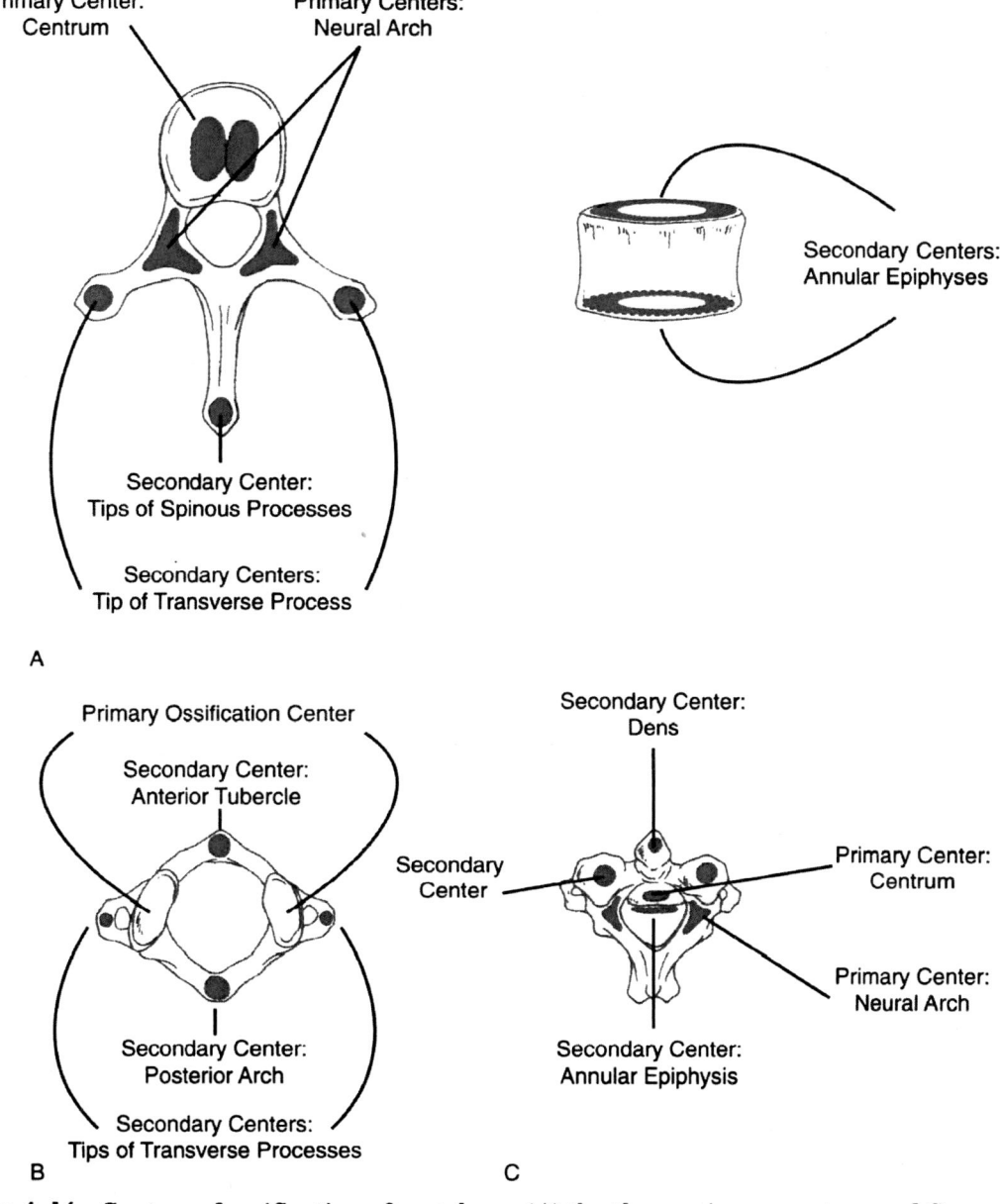

Figure 4-16 Centers of ossification of vertebrae: (A) the three primary centers and five secondary centers of a typical vertebra, (B) centers of ossification in the atlas, and (C) centers of ossification in the axis.

caudally from there. At the eighth or ninth week in utero the centers for the neural arches appear in the cervical region and upper two thoracic segments, T1 and T2 ossifying before some of the cervical levels, and then ossification proceeds caudally.

At birth all three primary centers for each vertebrae are still distinct, but are joined by cartilage. The halves of the neural arches fuse with each other during the first year, starting in the lumbar region. Following the same pattern, the arches then fuse with the bodies of the vertebrae a few years later. The costal element of the vertebra articulates with the primary center for the neural arch, not the primary center for the body (the centrum). Thus, the *centrum* is not strictly synonymous with the *body* of the vertebrae. For example, in thoracic vertebrae, the costal demifacets are described as part of the body of the vertebrae, but developmentally they are part of the primary center of ossification for the neural arch.

The five secondary centers appear after puberty, and fuse with the primary centers by the age of 21 (Breathnach, 1965; Gray, 1985, 1995). In U.S. males, the superior and inferior annular epiphyses of the centrum fuses at 17 or 18 at the earliest, and are always completed by age 25 (McKern and Stewart, 1957). The presacral vertebral column is completely ossified by the twenty-fifth year, with T4 and T5 representing the final elements to complete ossification. The sacrum is complete by age 23, though there may be a gap between the centra of S1 and S2 as late as age 32 (McKern and Stewart, 1957; Krogman, 1962). Some individual vertebrae or vertebral segments exhibit some differences from this general arrangement.

Atlas

The atlas (see Figure 4-16 B), or first cervical vertebrae, differs from other vertebrae in its lack of a body or spinous process and its possession of anterior and posterior arches. This is reflected in its pattern of ossification. It has two primary ossification centers, a left and a right, from which the bulk of the posterior arch and lateral mass of each side develops. The two halves of the posterior arch fuse at about age four, while the anterior arch forms and joins the two lateral masses between ages five and nine. Secondary centers have been noted at the tips of the transverse processes, for the posterior tubercle and for the anterior tubercle (Breathnach, 1965; Gray, 1985, 1995).

Axis

The axis (see Figure 4-16 C) has all the typical centers noted for other vertebrae, plus two for the dens, which some ascribe to being the displaced body of the atlas. The centrum begins to ossify in the third or fourth month *in utero*, with the centers for the dens appearing in the fourth or fifth month *in utero*. It has two centers of ossification, placed laterally, which unite before birth. The centers for the neural arch appear by the ninth week *in utero*. The bone consists of four centers at birth: the dens, the true centrum, and each half of the neural arch. By age six or seven the primary centers will have fused, but the center of the union between the dens and the centrum may remain cartilaginous until middle age. An annular epiphysis for the inferior surface of the body appears at about puberty (Breathnach, 1965; Gray, 1985, 1995).

Sacrum

The body of each sacral vertebrae develops from a primary center for the centrum, and two epiphyseal plates, superior and inferior. The vertebral arches form from the typical two primary centers of ossification as elsewhere in the vertebral column. Two more additional ossification centers appear laterally for each of the upper three or four sacral segments for the costal elements. On each lateral surface of the sacrum, two secondary centers of ossification develop, one for the auricular surface, another for the lower, thinner part of the lateral edge of the bone. At eight or nine weeks in utero, the body of the first segment begins to ossify. The second and third follow rapidly, but the fourth and fifth do not ossify until between five and eight fetal months.

Between 6 and 8 months in utero, the centers for the costal elements and the centers for the neural arches begin to appear. Between the second year and the fifth or sixth year, the neural arches join with the bodies of the sacral vertebrae, the lower segments fusing first, and the upper segments fusing last. Around age 16 the annular epiphyses for the bodies appear, and at about age 18 to 20, the lateral epiphyses appear. Intervertebral discs separate the sacral vertebral bodies, but synostosis of the discs begins at about age 18 between S4 and S5, and then proceeds cranially so that by age 25 to 30, all sacral segments will be fused solidly together (Breathnach, 1965; Gray, 1985, 1995). McKern and Stewart (1957) found that, among U.S. Army males killed in the Korean war, by age 24 all but the disc between S1 and S2 had synostosed, and by age 33, S1 and S2 had synostosed in all cases.

Coccyx

The coccyx has four primary centers of ossification, one for each segment. Embryos show the rudiments of at least six coccygeal segments, but the distal ones fuse, so that there are only four cartilaginous segments at birth. The primary ossification center for the first segment appears soon after birth, and between ages 5 and 10 in the second segment. The third segment ossifies between 10 and 15 years, and the fourth after puberty, between 14 and 20 years. With age, there is a tendency for the segments to fuse with one another, after age 25 or 30 for the first and second segments, and later for the others. The coccyx has a tendency to fuse with the sacrum, especially among women (Breathnach, 1965; Gray, 1985, 1995).

Introduction to Arthrology

Before discussing the joints or articulations of the vertebral column, it will be useful to discuss at this point some basics about the joints of the body. This will serve as a general introduction to arthrology for all remaining discussions of articulations in later chapters.

Any joint between two bones is known as an *articulation* or *arthrosis*. The study of the joints is the science of *arthrology*. *Joints*, or arthroses, are defined as junctions between bones. They may take a number of different forms, based upon the types of connective tissue that join the bones together and the shapes of the surfaces of the bones that are articulating. In some instances joints serve to provide mobility between body segments, as in the joints of the limb. In other instances they serve to stabilize and unify bony elements, as for example at the cranial sutures. Throughout the descriptions that follow of the different elements of the skeleton, we will be making frequent reference to the joints of the body, and so here we offer a short introduction to the descriptive terminology of the joints that we will employ.

Joints or articulations may be defined in a number of ways. One of the most useful schemes is a classification of joints according to the manner in which the bones involved are joined to one another. Broadly they can be divided into diarthroses (synovial joints), and synarthroses. Synarthroses may further be subdivided into fibrous joints and cartilaginous joints.

Diarthroses

Diarthroses or *synovial joints* are characterized by the presence of cavitated connective tissue between bones. In other words, these joints have a joint cavity. Diarthroses are usually highly moveable, and have opposing articular surfaces covered in hyaline articular cartilage with the presence of a synovial cavity between them. They are also characterized by the presence of an articular capsule, composed of an outer fibrous layer and an inner synovial membrane, surrounding the joint. Synovial joints are commonly categorized by the shapes of the articular surfaces and motions allowed. One common schema lists the synovial joints according to the following classifications:

PLANE SYNOVIAL JOINT: A plane synovial joint, or gliding joint, is characterized by opposition of almost flat surfaces with pure translatory (gliding) motion. Examples include zygapophyseal joints and joints between carpal bones. In many instances, the surfaces of plane joints may have a small degree of curvature. This is true of the articular surfaces of zygapophyseal joints, as well as the joints between carpal bones and the joints between tarsal bones.

GINGLYMUS JOINT: A ginglymus, or hinge, joint is uniaxial; that is, it has 1 degree of freedom of motion. It is usually not pure hinge motion but more often is coupled with a slight degree of rotation. An example is the humeroulnar (elbow) joint.

TROCHOID JOINT: A trochoid, or pivot, joint is uniaxial. It is an osseous pivot in which one bone rotates about an axis. In the vertebral column, the joint between the dens of the axis and the anterior arch of the atlas is an example of a trochoid joint: As the head turns, the atlas rotates relative to the fixed dens of the axis. Another good example of a trochoid joint is the proximal radioulnar joint: The round head of the radius is bound to the ulna by the annular ligament, which forms a "collar" around the head of the radius. In pronation and supination of the forearm, the radial head spins within the circular confines formed by the annular ligament and radial notch of the ulna.

BICONDYLAR JOINT: A bicondylar, or condyloid, joint, is largely uniaxial, with some rotatory movement. It consists of two convex condyles rigidly fixed to the same bone and articulating with two concave joint surfaces. Examples include the joint between the femoral condyles and the tibial plateau (knee joint), the temporomandibular joint, and the atlanto-occipital joints.

ELLIPSOID JOINT: An ellipsoid joint is biaxial; an oval convex surface articulates with an oval concave surface. It allows movement on two orthogonal axes (i.e., flexion/extension in one axis and abduction/adduction in the other). The best example is the joint occurring between the distal end of the radius and the scaphoid and lunate.

SELLAR JOINT: A sellar, or saddle, joint is multiaxial. Two concavoconvex (saddle shaped) surfaces articulate with one another. Examples include the pollical carpometacarpal joint between the base of the thumb and the carpal bones, as well as the sternoclavicular joint.

SPHEROIDAL JOINT: A spheroidal joint, or ball-and-socket joint, is multi axial. It allows 3 degrees of freedom of motion. The only examples in the human skeleton are the glenohumeral (shoulder) joint and the hip joint (between the acetabulum and the head of the femur).

Synarthroses

Synarthroses are the second broad category of articulations. They are characterized by the presence of solid connective tissue between the bones involved. They are usually categorized by the types of materials with which the bones are joined. Movements at synarthroses are usually more restricted than those at synovial joints. They are described below.

Fibrous Joints

Fibrous joints are characterized by a union of the bones involved by fibrous connective tissue. They can be further subdivided into sutures, gomphoses, and syndesmoses.

Sutures

Sutures are joints between bones of the cranium. The bones are joined with one another at the sutures by short, strong, collagenous *intersutural ligaments*. For all practical mechanical purposes, they effectively block motion at the sutures and unify the cranium into a rigid structural unit that protects the delicate brain. (There is a school of thought in chiropractic and osteopathic medicine that physiologically important amounts of motion continue at the sutures and that adjustment of the sutures can affect cerebrospinal fluid (Adams et al., 1992; Heisey and Adams, 1993). See Rogers and Witt (1997) for a contrary view.)

Gomphoses

Gomphoses (singular: *gomphosis*) are joints between the teeth and alveolar processes of the mandible and maxilla. Gomphoses are unusual in that bone is present on one side of the joint only. In a gomphosis, the bone of the alveolar process of the maxilla or mandible is joined to the cementum (a bone-like substance) of the tooth's root by a short strong *periodontal ligament*. (Teeth are composed of three types of calcified tissue: the cementum of the root, the enamel of the external surface of the tooth crown, and the dentin at the core of the crown and root. Like bone, they are calcified connective tissue, but differ from bone in their microscopic structure and materials properties. See Ash (1984) or Bhaskar (1990) for more information about teeth.)

Syndesmoses

Syndesmoses (singular: *syndesmosis*). Bonding of bones by collagenous (and sometimes elastic) interosseous ligaments, cords, or membranes. *Ligaments* are bands of connective tissue connecting bones together. Most ligaments are examples of syndesmoses. Exceptions are the intersutural ligaments and the periodontal ligaments, discussed previously. *Interosseous membranes* are broad flat membranous ligaments stretching between two bones. They provide stability by binding the two bones together, but allow limited movement. They also provide sites of muscle attachment. The interosseous membrane between the radius and ulna of the forearm, or that between the fibula and tibia in the leg, are good examples. A *cord* is a thick cordlike ligamentous band between two bones. For a more detailed discussion of ligaments, their functions and innervation, see Walker and Grod (2006).

Cartilaginous Joints

Cartilaginous joints are characterized by the presence of hyaline cartilage or fibrocartilage between the opposing joint surfaces.

Synchondroses

Synchondroses (singular: synchondrosis) are Characterized by the presence of hyaline cartilage. Epiphyseal plates in growing long bones are synchondroses, as are the costal cartilages. It is the ultimate fate of most synchondroses to undergo complete ossification. These are present early in development and are termed *primary cartilaginous joints*.

Symphyses

In *symphyses (singular: symphysis)*, Both joint surfaces are encased in hyaline cartilage and are joined by an intervening fibrocartilaginous disc. All are median. The *intervertebral discs* represent symphyses between the vertebral bodies. Other symphyses include the pubic symphysis between the two pubic bones, and that between the body and manubrium of the sternum. All symphyses are median, and all are associated with the axillary skeleton. Symphyses are more highly specialized than synchondroses and are referred to as *secondary cartilaginous joints*.

Articulations of The Vertebral Column

Joints between the vertebral bodies consist of the inferior surface of the body of the more cranial vertebra, the fibrocartilaginous *intervertebral disc* (refer to Figures 4-17 through 4-19 throughout the discussion of the intervertebral disc), and the superior surface of the more caudal vertebra. These joints are stabilized by the *anterior and posterior longitudinal ligaments*.

The *joints between vertebral arches* are the articulations between the inferior articular facets of the more cranial vertebra with the superior articular facet of the more caudal vertebra, as well as the ligaments running between laminae, transverse processes and spinous processes of adjacent vertebrae.

The Intervertebral Disc

The *intervertebral discs* are *symphyses*. They adhere to the superior and inferior surfaces of the adjacent vertebrae by thin layers of hyaline cartilage called *vertebral end plates*, which are considered to be an integral part of the disc (Bogduk, 1991). Each disc consists of *the vertebral end plates*, an outer laminated *annulus fibrosus* and an inner *nucleus pulposus*. The annulus fibrosus is fibrocartilage, while the nucleus pulposus (better developed in cervical and lumbar regions) is made up at birth of soft, gelatinous mucoid material and a few multinucleated notochordal cells. Notochordal cells disappear in the first decade of life, and the substance of the nucleus pulposus is replaced gradually by fibrocartilage. Thus with maturity the nucleus pulposus becomes less distinct from the rest of the disc (Walker and Grod, 2006).

Intervertebral discs are present between vertebral bodies from C2 to the sacrum, and a degenerate disc is present between S5 and the coccyx. They are generally thicker anteriorly than posteriorly in the cervical and lumbar regions, which creates the *spinal curvatures* in these regions. The Intervertebral discs are nearly flat in the thoracic spine. They are thinnest in the upper thoracic spine, and thickest in the lumbar spine. The outer a*nnulus fibrosus* has a narrow outer collagenous zone, and a wider, inner fibrocartilaginous zone. The annulus fibrosus consists of a number of laminae, with fibers of adjacent laminae set obliquely to one another, and at a $30°$ angle relative to the horizontal.

Recent studies suggest that this is not the case in the *cervical region*. Mercer and Bogduk (1999) report that the cervical annulus fibrosus has structure unlike that generally reported for the lumbar region. They note that cervical annuli are crescent-shaped, and deficient posteriorly, and do not consist of concentric lamellae of collagen fibers. Instead cervical intervertebral discs take the form of a

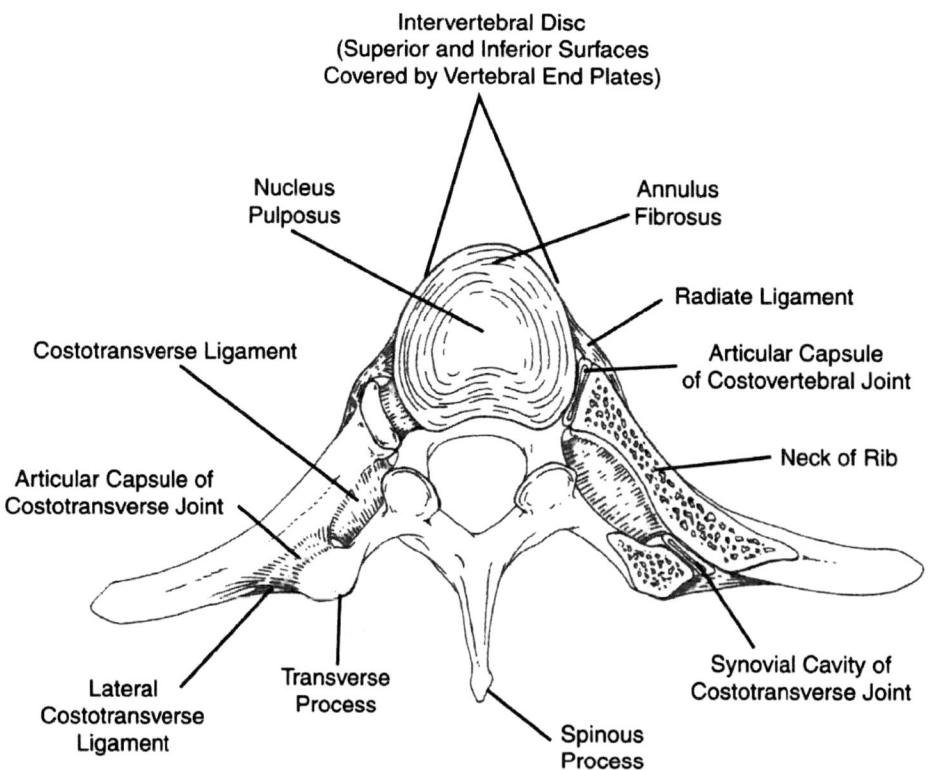

Figure 4-17 Ligaments of the ribs and the intervertebral disc, superior view.

crescentic mass of collagen anteriorly and laterally and are deficient posteriorly, where the deficiency is covered by the posterior longitudinal ligament. They conclude that "the three dimensional architecture of the cervical annulus fibrosus is more like a crescentic anterior interosseous ligament than a ring of fibers surrounding the nucleus pulposus" (Mercer and Bogduk, 1999: 619).

The oblique arrangement of fibers in adjacent laminae, as is found in the lumbar region, imparts great torsional strength to disc. Additionally, the outer laminae bulge externally, while the inner laminae bulge internally toward the nucleus pulposus. The annulus fibrosus is thickest anteriorly and thinnest posteriorly.

The functions of the annulus fibrosus are to enclose and retain the nucleus pulposus, absorb compressive shocks, form a structural unit between vertebral bodies, and allow restricted motion between adjacent vertebral segments.

The inner nucleus pulposus accounts for 40 percent of the bulk of the disc. The nucleus pulposus contains the remains of the notochord and has a high water content. Notochordal cells disappear during the first decade of life, and are replaced eventually by fibrocartilage. The nucleus pulposus is relatively larger at birth, but decreases in relative size as they dehydrate with age. The nucleus pulposus is better developed in the cervical and lumbar regions than in

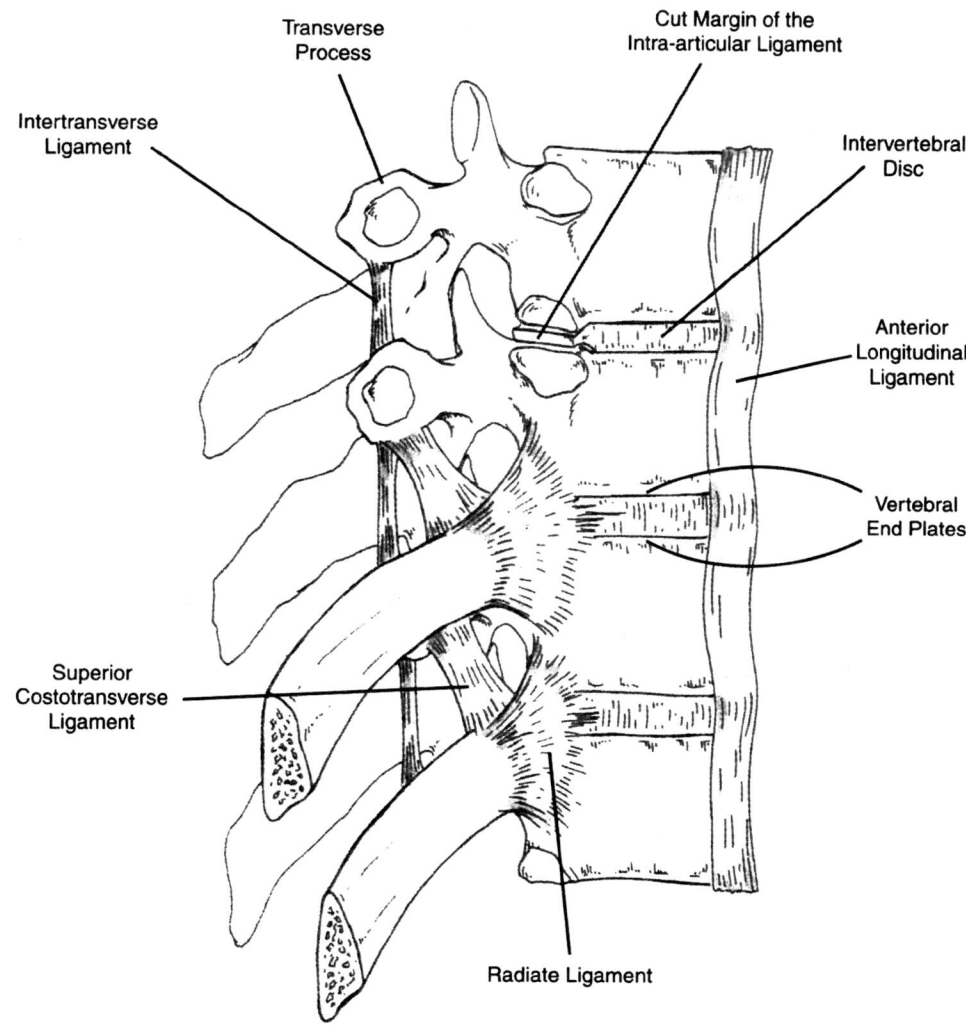

Figure 4-18 Articulations of the ribs and thoracic vertebrae, right lateral view.

the thoracic region, and lies nearer the posterior part of the disc than the anterior. Functions of the nucleus pulposus are to provide a resilient spacer allowing motion between vertebral segments, and to distributes compressive forces (but they are noncompressible, although deformable, themselves).

The vertebral end plates help prevent the vertebral body from suffering damage from pressure and contain the annulus fibrosus and nucleus pulposus within their normal borders. A rupture of the nucleus pulposus through the end plate is known as a *Schmorl's node*. When the nucleus herniates through the end plate, it allows the bodies of adjacent vertebrae to move closer together and can result in degenerative joint disease in the affected area of the vertebral column.

Intervertebral discs are avascular structures, but do have vasculature that surrounds their peripheries. They are attached to the anterior and posterior longitudinal ligaments, and to the heads of the ribs by the intraarticular ligaments. They contribute about 1/5 of the total height of the vertebral column, and are relatively thicker in the lumbar and cervical region, and thus contribute to greater mobility in these regions. They are thinnest, and essentially flat anteroposteriorly, in the thoracic region. Also, the discs are thicker anteriorly than posteriorly in the lumbar

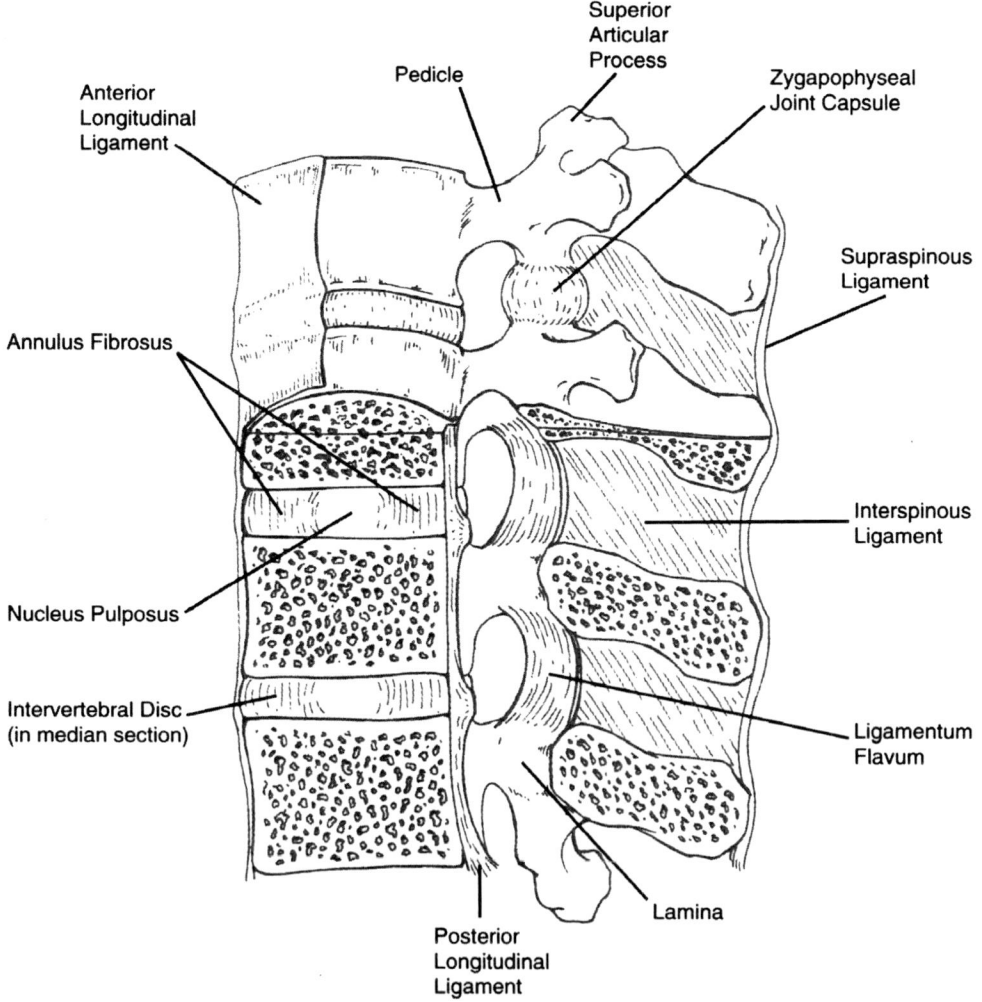

Figure 4-19 Joints between the vertebral arches, the anterior and posterior longitudinal ligaments, and the intervertebral disc, left lateral view.

and cervical regions, thus helping to create the lumbar and cervical curvatures.

The functions of the intervertebral discs are to bear weight (with bodies of vertebrae), distribute loads, and to dissipate shock (acting as "shock absorbers"). High water content aids this function. They become less capable of holding water with age as nucleus becomes progressively replaced by fibrocartilage. They also have a higher water content in the morning than later in the day. The intervertebral discs also act as a flexible buffer between rigid vertebrae, and permit adequate motion at low loads while providing stability at higher loads. When A disc is healthy, it is more likely that vertebrae will be damaged before the disc.

The intervertebral disc is innervated by the recurrent meningeal nerve (sinuvertebral nerve) posteriorly, the ventral rami and gray rami communicantes posterolaterally, and the gray rami communicantes and branches of the sympathetic trunk anterolaterally.

The Anterior Longitudinal Ligament

The *anterior longitudinal ligament* (Figure 4-20; see also Figures 4-18 and 4-19 throughout the discussion of the anterior longitudinal ligament) is a strong band of fibers forming a syndesmosis which extends along the anterior surface of the bodies of the vertebrae from the occipital to the sacrum. This ligament is specialized in the region of the atlas and axis. From its lateral aspects, fibers spread out to form the anterior atlantoaxial ligament between the atlas and axis and anterior atlantooccipital membrane between the atlas and the occipital bone.

Its morphology is more typical in lower cervical vertebrae and the rest of the column. In the cervical area it is thin and narrow, and in general it is *narrow cranially and wider caudally*. It is thicker in the thoracic than in the lumbar or cervical regions. It is somewhat thicker as well over the vertebral bodies than over the intervertebral discs. It attaches to the basilar portion of the occipital bone, the anterior tubercle of the atlas, the

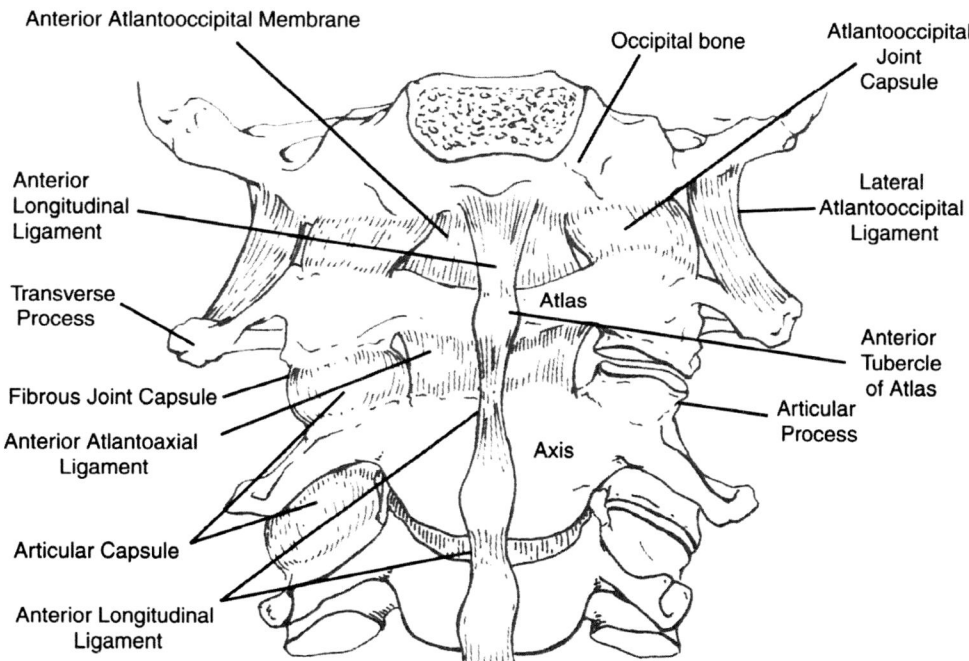

Figure 4-20 Anterior view of the superior portion of the anterior longitudinal ligament, anterior connections between the atlas, the axis and occiput, and the zygaphophyseal joint capsules of the cervical region.

body of the axis, the prominent margins of the vertebral bodies and intervertebral discs of the vertebrae down to the superior part of the anterior surface of the sacrum. The anterior longitudinal ligament is composed of superficial, intermediate, and deep fibers. The superficial fibers are the longest and extend over four to five vertebrae. Intermediate fibers extend two to three vertebral levels, while the deep fibers are shorter, crossing only between adjacent vertebrae. It functions to *limit extension* of the vertebral column.

The Posterior Longitudinal Ligament

The *posterior longitudinal ligament* (Figure 4-21; see also Figure 4-19 throughout the discussion of the anterior longitudinal ligament) forms another vertebral syndesmosis and attaches to the dorsal surfaces of the bodies of the vertebrae within the vertebral canal. The portion of the posterior longitudinal ligament superior to the axis is specialized as the tectorial membrane. The posterior longitudinal ligament extends from the axis to the sacrum and lies within the sacral canal. It consists of superficial and deep layers of fibers. The superficial fibers cross 3 to 4 vertebrae, while the deep fibers run between adjacent vertebral bodies. This ligament is not firmly attached to vertebral bodies, because the basivertebral veins and the vertebral venous plexus are situated between it and the vertebral bodies. The posterior longitudinal ligament is *broad cranially* and *narrow caudally*. In the lower thoracic region and lumbar region it is denticulated, being broad over the intervertebral discs and narrow over the vertebral bodies. It functions to limit flexion of the vertebral column.

Zygapophyseal Joints and Joints of the Vertebral Arches

The synovial joints between articular processes of the vertebrae are of the plane synovial type, and are known as *zygapophyseal joints* (also referred to as the *posterior joints, facet joints,* or *apophyseal joints*). The *articular capsules* of the zygapophyseal joints (see Figures 4-19 and 4-20) surround the margins of

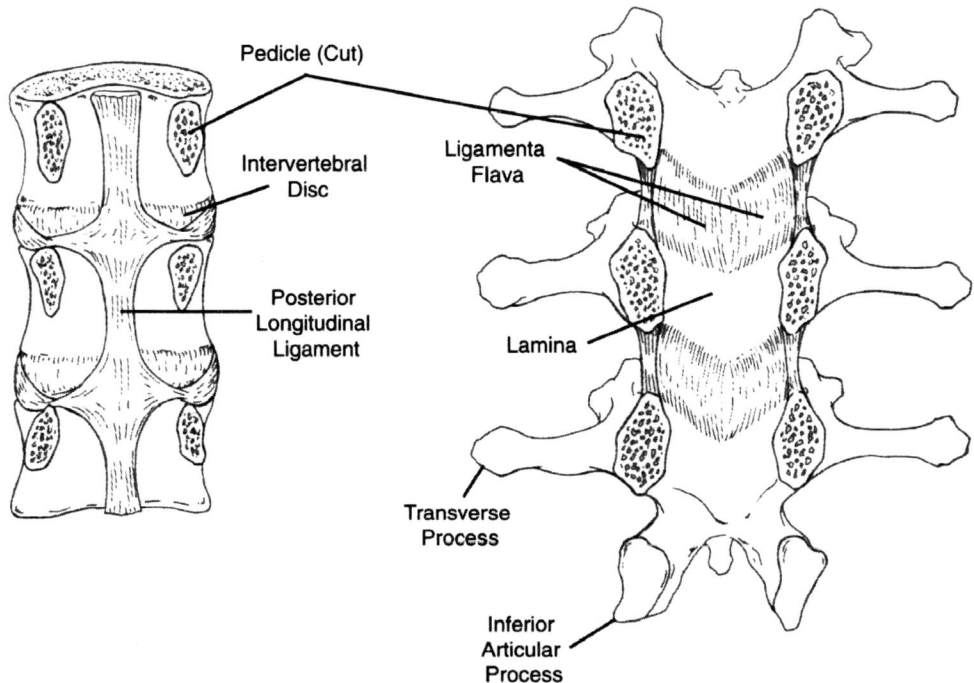

Figure 4-21 The posterior longitudinal ligament and the ligamenta flava in the lumbar region.

the articular processes. The fibrous articular capsules are thin and loose, and longer and looser in the cervical region than elsewhere. The size, shape, and movements of the zygapophyseal joints vary from vertebral level to vertebral level. These joints may contain fat pads or fibroadipose menisci (articular disks), especially in the lumbar region. They are innervated by medial branches of dorsal rami of spinal nerves; both the ascending branch of the dorsal ramus below and the descending branch of the dorsal ramus above the joint.

The vertebral arches are also connected to one another by the following ligaments:

- *The supraspinous ligament*
- The ligamentum nuchae
- *The ligamentum flavum* (plural: *ligamenta flava*)
- The interspinous ligament
- *The intertransverse ligament.*

The latter three ligaments (ligamentum flavum, interspinous ligament, and intertransverse ligament,) constitute the *intervertebral syndesmosis*.

Intervertebral, or Segmental, Syndesmoses

LIGAMENTA FLAVA: The *ligamenta flava* (*ligamentum flavum*, singular; see Figures 4-19 and 4-21) connect the laminae of adjacent vertebrae from the axis to the first segment of the sacrum. It is specialized as the *posterior atlantoaxial ligament or membrane* between the posterior arch of the atlas and the laminae of the axis, and the *posterior atlantooccipital membrane* between the posterior arch of the atlas and the posterior margin of the foramen magnum (Figure 34-22).

The ligamenta flava consist of two lateral portions which begin posterior to the articular processes and extend posteriorly to the point where the laminae meet to form the spinous process. They are composed primarily of yellow elastic tissue (Latin: *flavum* = "yellow). In the *cervical region* the ligamenta flava are thin, but broad and long. They are thickest in the *lumbar region*, and of intermediate thickness in the *thoracic region*. There is one pair between adjacent vertebrae, and they run from the anterior and inferior surface of the more superior lamina to the posterior and superior surface of the more inferior lamina. They function to brake the separation of the laminae in flexion of the vertebral column. Elastin fibers permit them to stretch and regain their shapes.

INTERSPINOUS LIGAMENTS: The interspinous ligaments (see Figure 4-19) are thin and membranous and connect adjoining spinous processes and extend from the root to the apex of each process. The interspinous ligaments function to limit flexion of the vertebral column. They meet the *ligamenta flava* ventrally and the supraspinous ligament dorsally. They are narrow and elongated in the thoracic region and thicker and quadrilateral in the lumbar region. In the cervical region, they are incorporated into the ligamentum nuchae.

INTERTRANSVERSE LIGAMENTS: The intertransverse ligaments (see Figure 4-18) are segmental syndesmoses connecting the transverse processes of adjacent vertebrae, and function to limit lateral flexion of the vertebral column. In the cervical region they are represented by only a few irregular fibers, and are largely replaced by the intertransverse muscles. In the thoracic region they are well developed cords that are blended intimately with adjacent muscles, while in the lumbar region they are thin and membranous.

Continuous, or Nonsegmental Syndesmoses

SUPRASPINOUS LIGAMENT: The *supraspinous ligament* (see Figure 4-19) is a strong fibrous cord which connects the apices of the spinous processes from the seventh cervical vertebra to the sacrum. It has superficial, intermediate, and deep fibers. The superficial fibers span 3-4 vertebrae, the intermediate span 2-3, and the deep run between adjacent vertebrae. It is thicker and broader in the lumbar region than in the thoracic, where it is relatively thin. In the cervical region above C7 the supraspinous and interspinous ligaments are specialized as the *ligamentum nuchae* (Figure 4-23).

LIGAMENTUM NUCHAE: Above C7 the interspinous ligaments along with the supraspinous ligament are replaced by the more specialized *ligamentum nuchae*. The ligamentum nuchae extends from the external occipital protuberance, external occipital crest, posterior tubercle of the atlas, and spines of the cervical vertebrae to the spinous process of C7. The fibers that comprise the ligamentum nuchae are developmentally homologous with the interspinous and supraspinous ligaments of lower vertebral levels. These fibers exist as a of a pair of dense, bilateral fibroelastic laminae separated by a layer of areolar tissue. The two laminae blend at free dorsal border. This structure functions to passively support the head. It is relatively much reduced in humans, being relatively much larger in quadrupedal mammals.

Craniovertebral Joints

The Atlantooccipital Joint

The *atlantooccipital joint* (Figure 4-24; see also Figure 4-23 throughout the discussion of the atlantooccipital joint), a *bicondylar synovial joint* is located between the superior articular facets of the atlas and the occipital condyles of the occipital bone of the cranium. It is a condyloid synovial joint. The atlas is attached to the occipital bone by the *articular capsules* around the synovial joint surfaces, and three ligaments:

- ᴄ The Anterior atlantooccipital membrane
- ᴄ The Posterior atlantooccipital membrane
- ᴄ The Two lateral atlantooccipital ligaments

The *fibrous joint capsules* (See Figure 4-20) are thickened posterolaterally, but are thin medially, and they sometimes medially communicate with a bursa

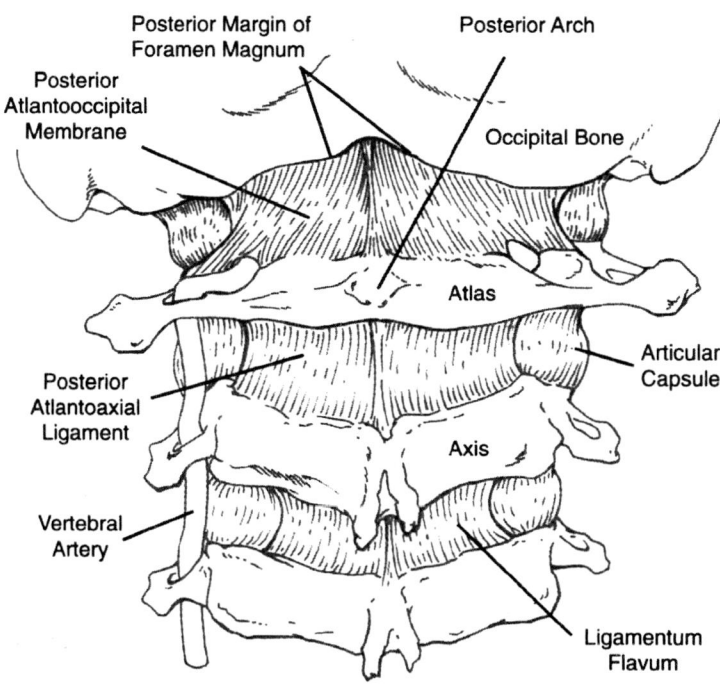

Figure 4-22 Posterior ligaments connecting the atlas, axis, and occiput.

between the dens and the transverse atlantal ligament (see below). The capsules surround the occipital condyles and the superior articular facets of the atlas. The *anterior atlantooccipital membrane* (Figure 4-20) forms a syndesmosis anteriorly, extending from the anterior margins of the foramen magnum to the cranial border of the anterior arch of the atlas. The *posterior atlantooccipital membrane* (figure 4-22) forms a syndesmosis posteriorly, and extends from the posterior margin of the posterior arch of the atlas to that of the foramen magnum. It is has an opening over the groove for the vertebral artery to allow the vertebral artery to enter the vertebral canal and the first cervical spinal nerve to exit. The *lateral atlantooccipital ligaments* are lateral thickenings of the articular capsules and run from lateral to the occipital condyles (the *jugular processes*) of the occipital bone to the transverse processes of the atlas (Figure 4-20).

Atlantoaxial Joints

There are two sets of articulations between the axis and atlas: the *lateral atlantoaxial joints* and the *median atlantoaxial joint*. The lateral atlantoaxial

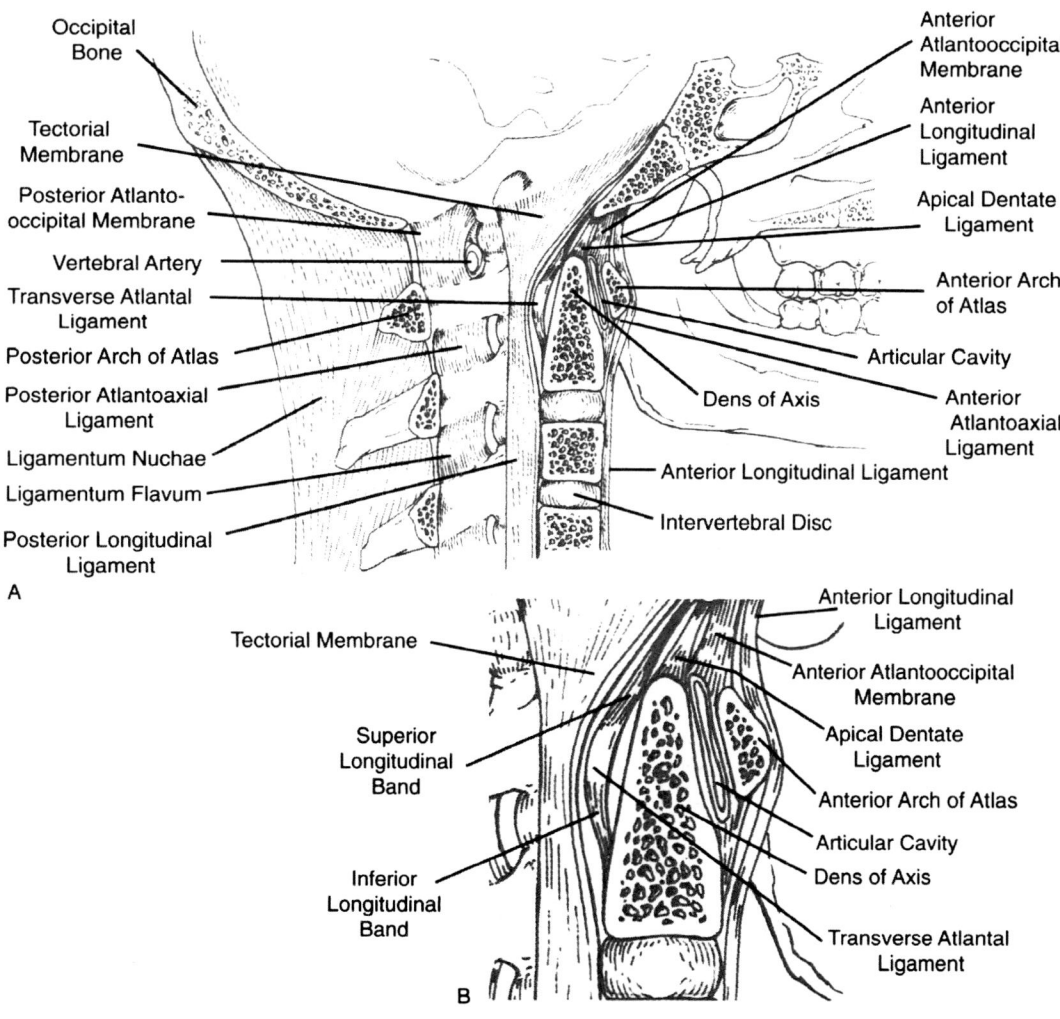

Figure 4-23 Median section through the cranium and cervical vertebral column, demonstrating the ligamentum nuchae, ligaments between the vertebral arches, anterior and posterior longitudinal ligaments, and ligaments of the atlantoaxial and atlanto-occipital joints. (*A*) Overview of median structures. (*B*) Close-up of the median atlantoaxial joint region.

joints are between the lateral masses of the atlas and the superior articular facets of the axis, while the median atlantoaxial joint is between the dens and the anterior arch of the atlas.

Ligaments associated with these joints are:

- *accessory atlantoaxial ligaments*
- anterior atlantoaxial ligament
- posterior atlantoaxial ligament
- *cruciform ligament,* consisting of:
  - transverse atlantal ligament
  - superior and inferior longitudinal bands.

The *lateral atlantoaxial joints* (Figure 4-25) are between the inferior articular facets of the atlas and the superior articular facets of the axis. This is a plane synovial joint capable of translational (gliding) movement. The *articular capsule* surrounds the margins of the inferior articular facets of the atlas and the superior articular facets of the axis. The fibrous part is thin and loose, but is strengthened at its posterior and medial aspects by an *accessory atlantoaxial ligament* (Figure 4-24), which is attached to the body of the axis near the base of the dens, and to the lateral mass of the atlas near the transverse ligament.

The *anterior atlantoaxial ligament* (see Figure 4-20) extends from the inferior border of the anterior arch of the atlas to the ventral surface of the body of the axis. It is an expansion of the cranial extension of the *anterior longitudinal*

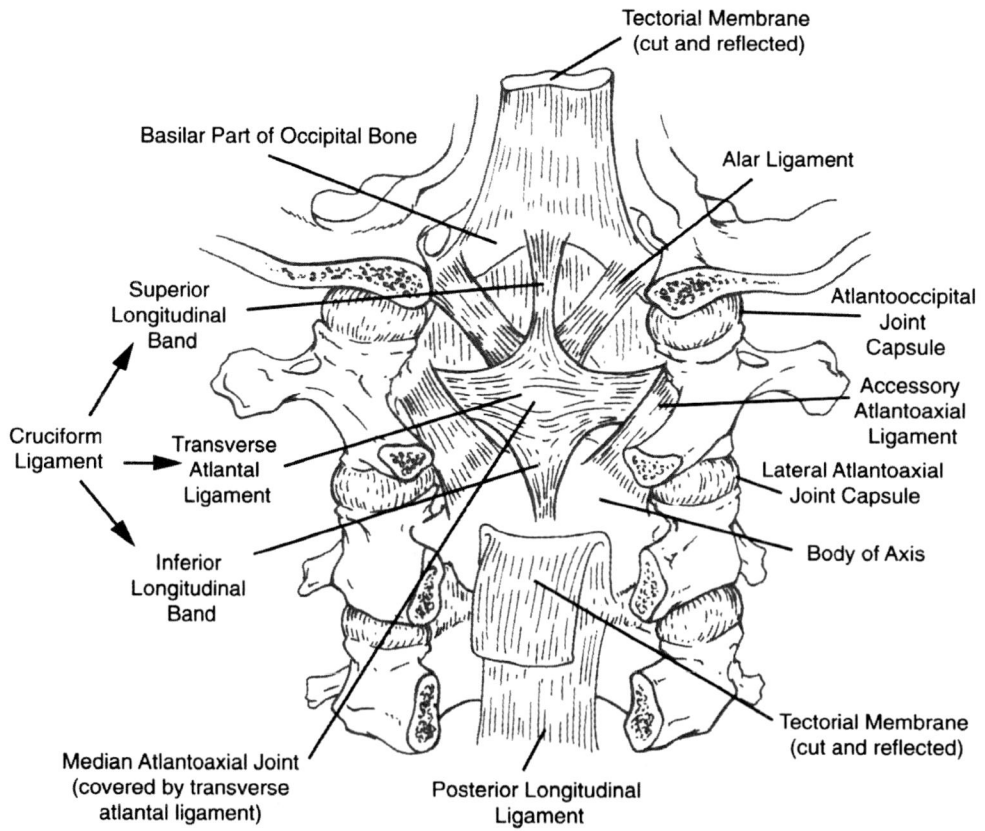

Figure 4-24 The cruciform ligament and associated structures, posterior view. A cervical laminectomy has been performed to expose the anterior view of the vertebral canal.

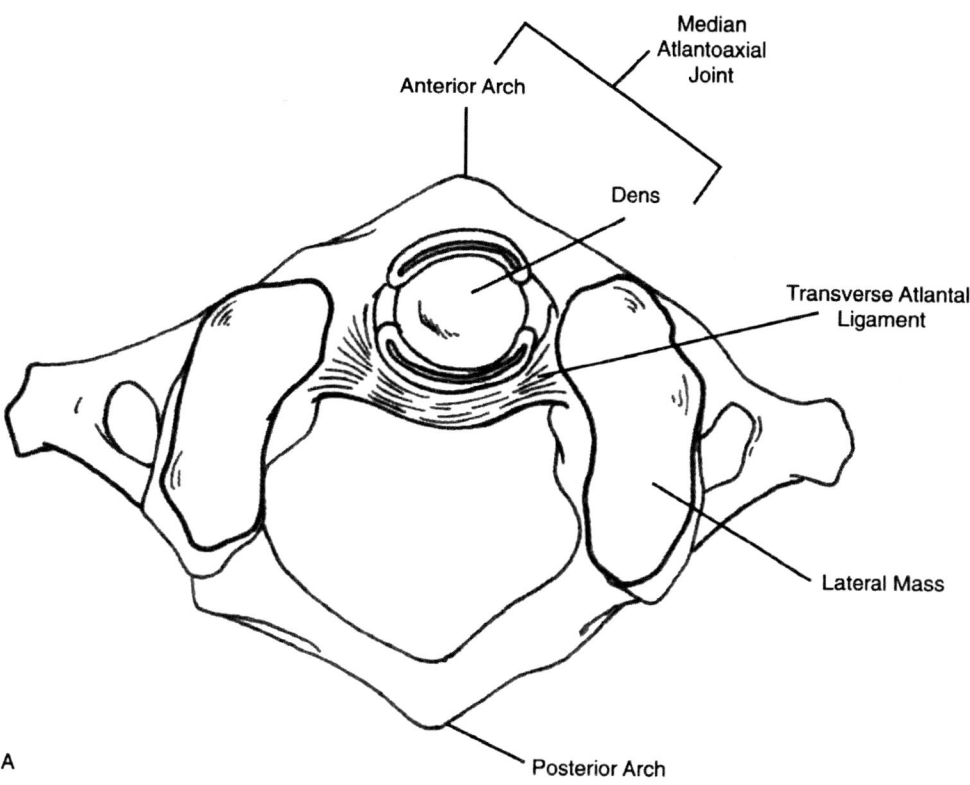

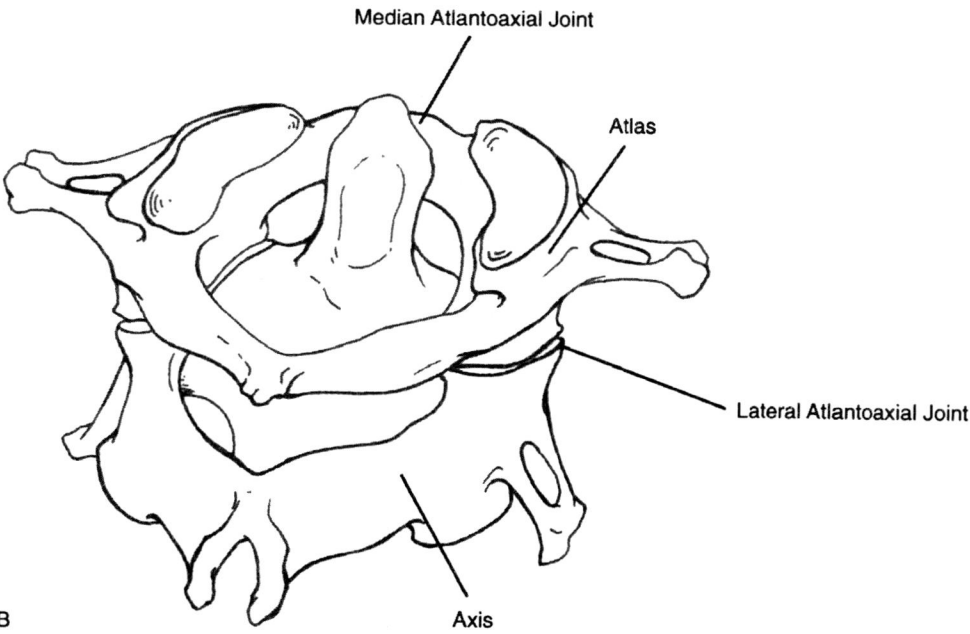

Figure 4-25 The median atlantoaxial joint: (*A*) superior view, (*B*) three-quarter view from lateral superoposterior angles, demonstrating the lateral atlantoaxial joints.

ligament (see Figure 4-20) which is a sheet of dense regular connective tissue running the length of the vertebral column and attached to the ventral aspects of the vertebral bodies and the intervening intervertebral discs.

The *posterior atlantoaxial ligament* (see Figure 4-22) extends from the inferior border of the posterior arch of the atlas to the superior edges of the laminae of the axis. It is homologous with the *ligamentum flavum* which runs between the adjacent laminae of all other vertebrae.

The *median atlantoaxial joint* (refer to Figures 4-23 through 4-25 throughout the discussion of the median atlantoaxial joint) is located between the anterior articular facet of the dens and the fovea dentis of the anterior arch of the atlas. This is a *trochoid* (or pivot) joint. The dens is held against the anterior arch of the atlas by the *transverse atlantal ligament*. There are two synovial cavities related to the dens, one located between the dens and the anterior arch of the atlas, and the other between the dens and the transverse atlantal ligament. This ligament attaches to small tubercles on the medial surfaces of the lateral masses of the atlas and the fibers pass transversely posterior to the dens to help hold the dens in place.

As the transverse atlantal ligament crosses the dens posteriorly, a small fasciculus extends superiorly and another inferiorly to form the *superior and inferior longitudinal bands* of the *cruciform ligament*. The superior band attaches to the internal aspect of the basilar part (or *clivus*) of the occipital bone deep to the tectorial membrane, while the inferior band attaches to the body of the axis deep to the posterior longitudinal ligament. The transverse atlantal ligament together with the longitudinal bands forms the *cruciform ligament*.

Ligaments attaching the axis to the occipital bone

The axis not only articulates with the atlas, but is also connected to the occipital bone by ligaments and membranes (refer to Figures 4-23 and 4-24 throughout the discussion of these structures). Movements occurring between the axis and the occipital bone are flexion, extension, and slight lateral tilting. There are three primary attachments between the axis and the occipital. They are the following:

- *The Tectorial Membrane*
- The Two Alar Ligaments
- *The Apical Ligament of the Dens.*

The *tectorial membrane* extends from the dorsal surface of the body of the axis to the *clivus* (internal aspect of the basilar portion of the occipital bone). It attaches along the anterior border of the foramen magnum and blends with the periosteal layer of the cranial dura mater that surrounds the brain. It expands to cover the dens and its ligaments posteriorly within the vertebral canal. The tectorial membrane is the continuation of the posterior longitudinal ligament above the second cervical vertebra. The two *alar ligaments* extend from each side of the cranial part of the dens to rough depressions on the medial sides of the occipital condyles.

The alar ligaments are also known as the *check ligaments* because they limit or "check" the rotation of the cranium to the contralateral side on the vertebral column. They also tighten with head flexion, but are relaxed during head extension. The *apical ligament of the dens* extends from the tip of the dens to the anterior margin of the foramen magnum between the alar ligaments. It is intimately blended with the deep portion of the anterior atlantooccipital membrane anteriorly and *superior longitudinal band* of the *cruciform ligament* posteriorly. It is believed to developmentally be the core of the *proatlas*, and contains some notochordal cells. The proatlas is a transitional structure that normally does not form a bone in humans, but has formed a separate bone between the atlas and occipital bone in experimental mice (Carlsen, 1994).

Innervation of Intervertebral Joints

All intervertebral joints are segmentally innervated by the adjoining spinal nerves.

Lumbosacral Articulation

Lumbar vertebrae are joined with the sacrum and pelvis via the anterior and posterior longitudinal ligaments, the intervertebral disc between L5 and S1, the ligamenta flava, the interspinous and supraspinous ligaments, and articular capsules of the zygapophyseal joints. Additionally, the fifth lumbar vertebra, and occasionally the 4th, is joined with the pelvis via one additional ligament on each side, the iliolumbar ligaments (Figures 4-26 and 4-27). The *iliolumbar ligament* consists of a *superior band* and an *inferior band,* which is also known as the *lumbosacral ligament*. Both bands attach medially to the transverse process of L5. The *superior band* runs from the transverse process of L5 to crest of the ilium immediately ventral to the sacroiliac articulation. The *inferior band* runs from the transverse process of L5 to the base of the sacrum, where it blends with the *anterior sacroiliac ligament*. These two bands limit lateral flexion of the lumbar spine.

Sacrococcygeal Joint

The *sacrococcygeal joint* (refer to Figures 4-26 and 4-27 throughout the discussion of the sacrococcygeal joint) is a slightly moveable joint between the first coccygeal vertebra and the fifth sacral vertebra. A small intervertebral disc is present between these two vertebral bodies. The following ligaments strengthen this joint:

- *The Ventral Sacrococcygeal Ligament*
- *The Superficial Dorsal Sacrococcygeal Ligament*
- *The Deep Dorsal Sacrococcygeal Ligament*
- *The Lateral Sacrococcygeal Ligaments*
- *The Intercornual (Interarticular) Ligaments.*

The *ventral sacrococcygeal ligament* consists of a few irregular fibers running from the ventral surface of the sacrum to the ventral surface of the coccyx which blend with the periosteum. The *superficial dorsal sacrococcygeal ligament* runs from the free margin of the sacral hiatus and attaches to the dorsal surface of the coccyx. It closes the posterior aspect of the most distal part of the sacral canal and corresponds to the ligamenta flava. The *deep dorsal sacrococcygeal ligament* corresponds to the posterior longitudinal ligament and is a flat band which arises from the posterior aspect of the fifth sacral segment inside the sacral canal and attaches to the dorsal surface of the coccyx deep to the superficial dorsal sacrococcygeal ligament.

The *lateral sacrococcygeal ligaments* are present on either side of the coccyx. They run from the rudimentary transverse processes of the first coccygeal vertebra to the lower lateral angle of the sacrum. They complete the foramina for the fifth sacral nerve. The *intercornual ligaments* connect the cornua of the sacrum and coccyx.

Motions of The Joints of The Human Skeleton

Movements of the vertebral column are in large part allowed by, and likewise limited by, the forms of the articular surfaces of the zygapophyses. Other factors, including the ligaments, the intervertebral discs, and the surrounding bones and muscles, also affect the degree of motion possible in a particular region of the vertebral column. This discussion of motions of the skeleton will serve as an introduction to the terminology of joint motion that will be useful in later discussions of motions of joints in other regions of the skeleton.

Before we can discuss the motions of the vertebral column, we must review some of the terminology used in the description of the movements of joints. In general, we can describe three types of joint motion: flexion/extension; abduction/adduction; and rotation (Figure 4-28). Other motions can be described as variations or combinations of these basic motions.

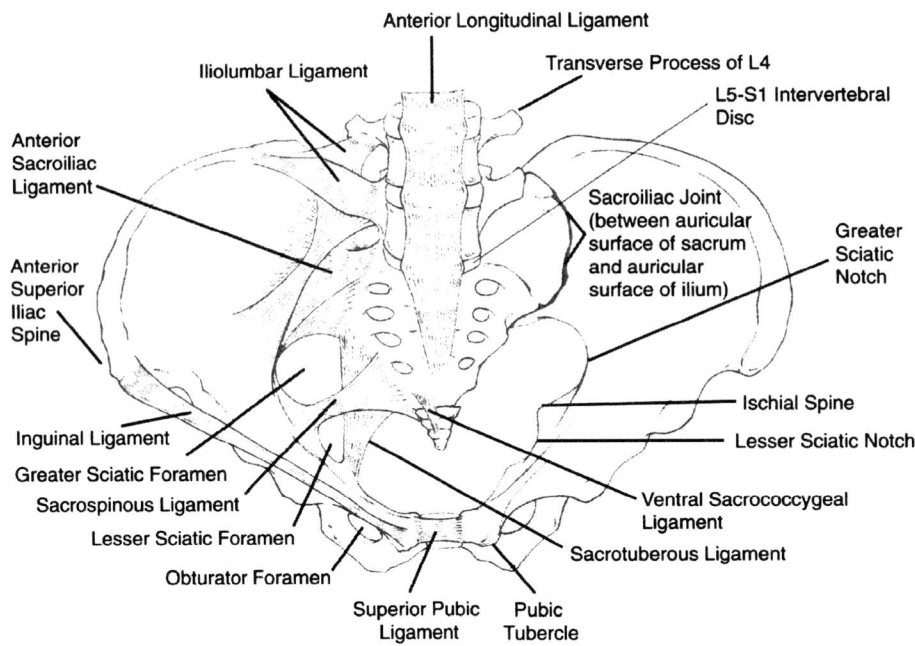

Figure 4-26   Ligaments of the pelvis, anterior view.

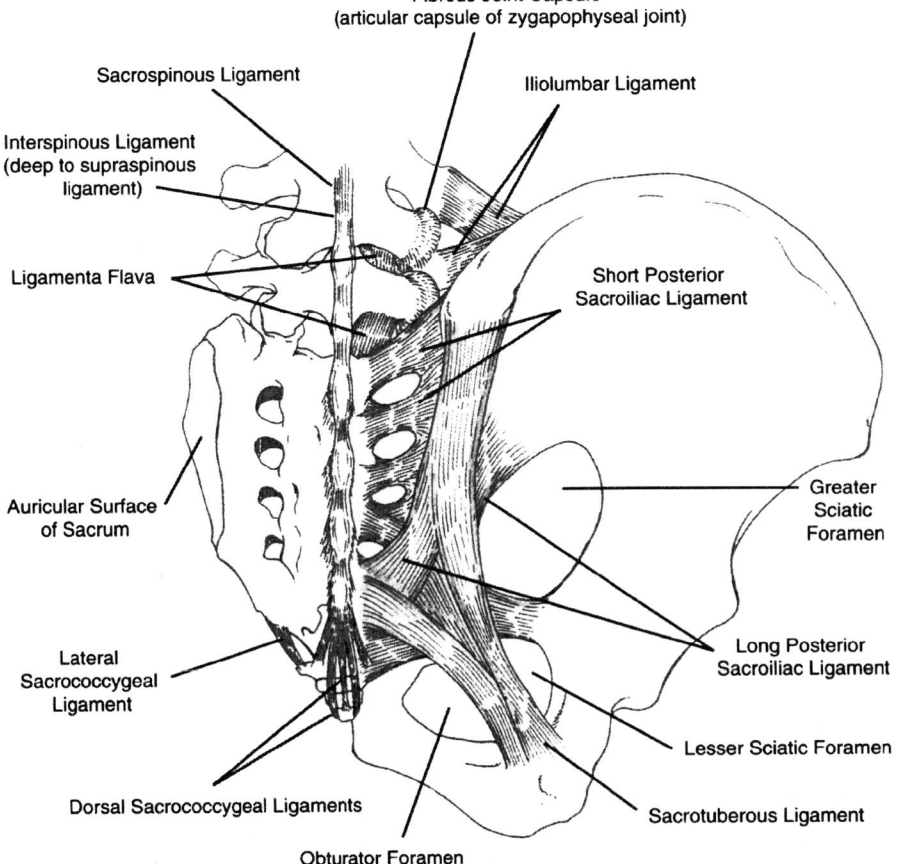

Figure 4-27   Ligaments of the pelvis, posterior view.

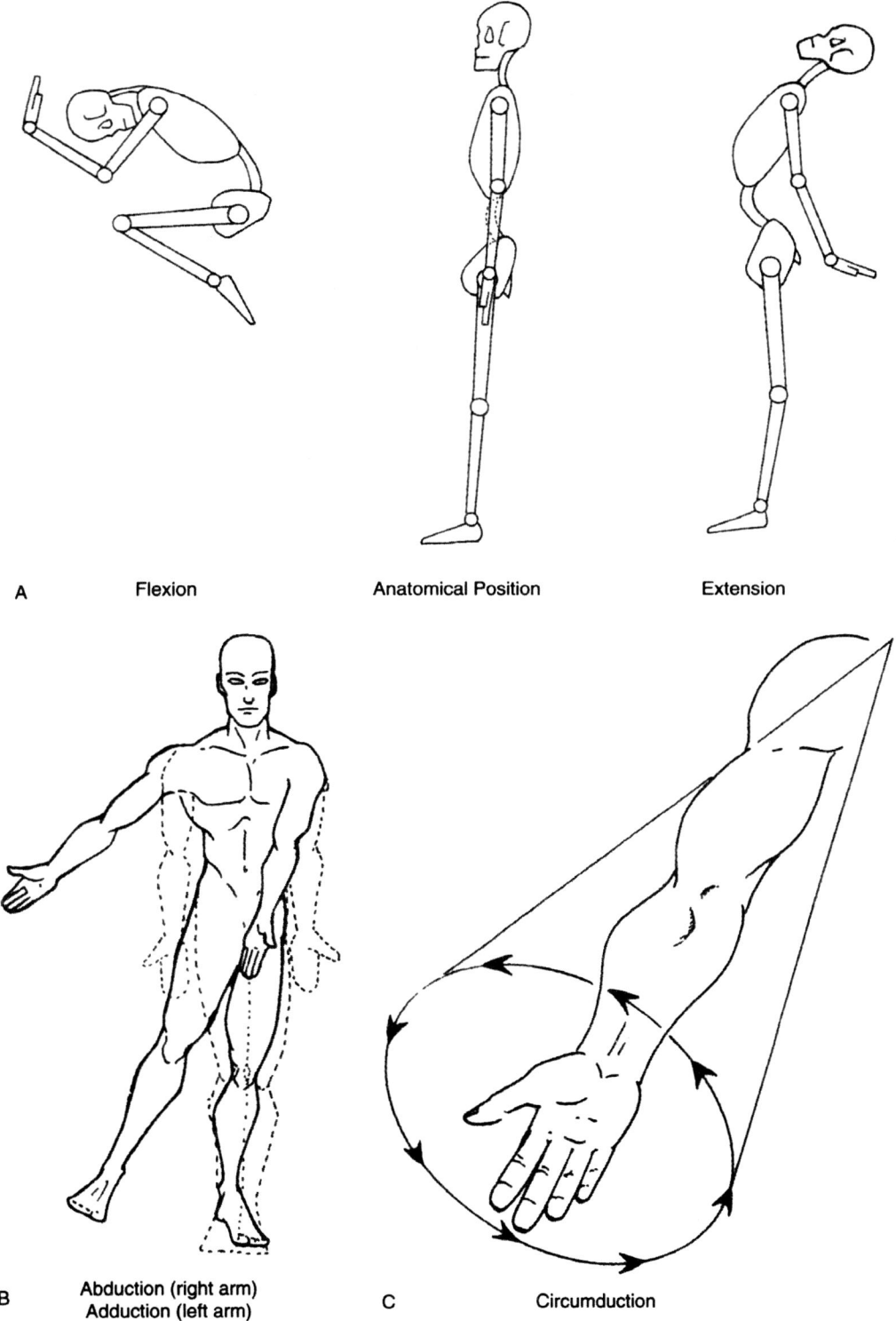

Figure 4-28 (*A*) Flexion/extension of the body. (*B*) Abduction/adduction of the arm and leg, with the broken lines indicating the anatomical position. The solid lines on the right side of the subject indicate abduction of the arm and leg; the solid lines on the left side demonstrate adduction of the arm and leg. (*C*) Circumduction of the arm. (*Continues on next page*)

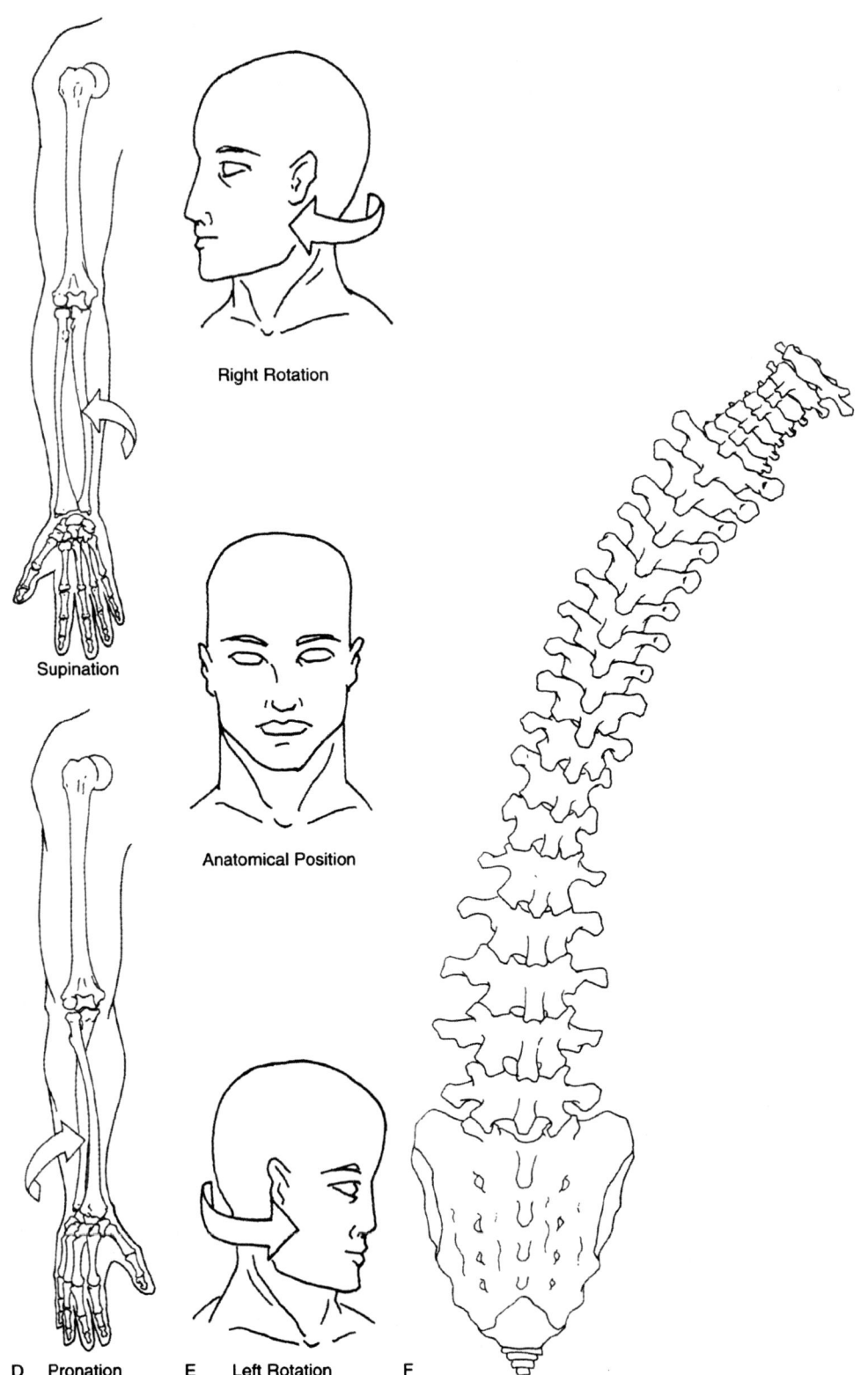

Figure 4-28 (*Continued*) (*D*) Supination/pronation of the forearm. (*E*) Rotation of the head. (*F*) Right lateral flexion of the vertebral column, posterior view.

Flexion and extension are motions along a parasagittal plane. *Flexion* refers to a motion along a parasagittal plane in which the angle formed between two designated body surfaces is *decreased*. In general, it is generally implied that we are decreasing the angle between anterior body surfaces in flexion, though this is not always the case. For example, in flexion of the elbow joint, we decrease the angle between the anterior surfaces of the arm and forearm, or in osteological terms, between the anterior surface of the humerus on the one hand and the anterior surfaces of the radius and ulna on the other (the upper limb is bent at the elbow as in taking a drink). In the case of the knee, however, flexion brings the posterior surfaces of the tibia and femur closer together.

Extension refers to a motion along a parasagittal plane in which the angle formed between two designated body surfaces is *increased*. To return to our example of the elbow joint, in extension the angle between the anterior surfaces of the humerus superiorly and the radius and ulna inferiorly is *increased* (the upper limb is straightened out at the elbow).

Abduction and adduction are motions along a coronal plane. Abduction and adduction are motions carried out at a 90 degree angle to flexion and extension. In *abduction*, a body part is moved away from the midline of the body. Conversely, in *adduction*, a body part is moved back toward the midline of the body. An easy way to remember the difference: In abduction, a body part is taken away from the midline just as an *abducted* child is taken away from its parents. In adduction, a body part is *added* to the midline of the body. Let us again turn to an example from the upper limb. In the anatomical position, the upper limb is in the adducted position. When the upper limb is raised away from the side of the body in a coronal plane, it is abducted. When the fingers are spread apart, they are abducted. When they are brought back together, they are adducted.

Rotation, as the name implies, is a turning of a body part about a longitudinal axis. *Supination* and *pronation* are the turning of the forearm such that the palm faces forward (anatomical position) and backward, respectively. In supination and pronation, the radius rotates about an axis that passes through the radial head and the ulnar styloid.

Circumduction is a motion that combines elements of flexion, abduction, extension, and adduction. In circumduction, a body part describes a cone of motion. One end of the concerned body part is more fixed, whereas the other end is more free to move. Using another example from the upper limb, think of this: if the entire upper limb is circumducted, the shoulder joint and head of the humerus constitute the more fixed end and the tips of the fingers constitute the freer end.

In circumduction of the upper limb, the shoulder joint is flexed, abducted, extended, then adducted (or the order is reversed). During this movement, the tips of the fingers roughly describe a circle in space, while the proximal humerus remains relatively fixed at the shoulder joint. The motion of the upper limb in circumduction thus describes a cone in which the shoulder represents the apex and the motion of the tips of the fingers describes its base. It is perhaps easier to explain these motions with respect to the limb bones than to those of the vertebral column; however, now that we have defined the basic terms of motion as they apply to the limbs, we turn to the motions of the vertebral column.

Movements of The Vertebral Column

Movements of the vertebral column may involve the entire column or only a part of the column. In bending over to touch your toes, for example, the entire vertebral column is flexed. Bowing from the waist involves flexion of the lumbar column only. If you bend your head forward so that your chin rests on your chest, you have flexed only the cervical vertebral column. As we shall see, the degree of flexion and extension allowed varies throughout the vertebral column.

Also, the range of movements of the vertebral column is restricted by the intervertebral disc. These joints at the discs are slightly movable, allowing flexion, extension, and lateral flexion, a movement of the vertebral column in the coronal plane. In general, only small movements occur between adjacent vertebrae, but the cumulative range is great. When discussing motions of the vertebral column, it is also important to consider the occipital bone of the cranium as well (discussed in Chap. 8). In a sense, we can think of the occipital bone and its condyles as cervical vertebra 0.

Now we can consider the individual motions of the vertebral column.

Flexion of the Vertebral Column

In flexion of the vertebral column, the angle between the anterior surfaces of the abdomen and thorax and the anterior surfaces of the thighs is decreased, as in bending forward. During flexion, the inferior articular processes of the more superior vertebrae glide superiorly relative to the superior articular processes of the lower vertebrae. The fibrous joint capsules of the zygapophyseal joint capsules become taut. When the vertebral column is flexed, the anterior longitudinal ligament becomes relaxed; the intervertebral discs are compressed anteriorly; the laminae and spinous processes spread apart; and the posterior longitudinal ligament, ligamentum flavum, interspinous and supraspinous ligaments, and posterior fibers of the intervertebral discs are all tensed. The vertebral extensor muscles resist and help control flexion of the vertebral column.

Extension of the Vertebral Column

Extension of the vertebral column increases the angle between the anterior surfaces of the thighs and the anterior surfaces of the thorax and abdomen, as in straightening up or even bending backward. Events that occur during extension are opposite those that occur during flexion. The anterior longitudinal ligament and anterior fibers of the annulus fibrosus tense, the intervertebral discs are compressed posteriorly, and the spinous processes and zygapophyses are approximated.

Rotation of the Vertebral Column

Rotation of a vertebra involves the turning of the vertebra about its superoinferior axis relative to adjacent vertebrae. Rotation of a vertebra upon another results in torsional deformation of the intervertebral discs. In any given vertebra, the axis of rotation lies somewhere between the anterior aspect of the body and the posterior tip of the spinous process. Rotation is defined according to the motion of the anterior aspect of the vertebral body. In *left rotation* of a vertebra, any given point on the anterior surface of the body of that vertebra moves toward the left, whereas the tip of the spinous process of that vertebra moves toward the right. In *right rotation*, the opposite events occur: the anterior surface of the vertebral body turns to the right whereas the tip of the spinous process moves to the left.

Lateral Flexion of the Vertebral Column

A special motion of the vertebral column is *lateral flexion*. Lateral flexion of the vertebral column corresponds to abduction and adduction of the limbs. It is a motion primarily in the coronal plane, as in leaning to one side from the waist. As with flexion and extension, we can laterally flex a portion of the vertebral column or the entire column. If we keep the trunk still but bend the head in a coronal plane toward the left shoulder, we are laterally flexing the cervical vertebral column to the left.

The opposite motion is lateral flexion to the right. Because of the forms of the zygapophyses, lateral flexion always involves some degree of rotation. This rotation can be either *ipsilateral* (to the same side) or *contralateral* (to the opposite side), depending on the region of the column; for example, lateral flexion of the cervical vertebral column is accompanied by ipsilateral rotation, meaning that when we flex the cervical vertebral column to the left, the cervical vertebrae undergo left rotation. During lateral flexion, the intervertebral discs are compressed on the side flexed to and they are tensed on the contralateral side.

Circumduction of the Vertebral Column

Circumduction of the vertebral column combines elements of vertebral flexion, lateral flexion, extension, and rotation. Circumduction may involve only a part of the column, such as the cervical region, or it may involve the entire vertebral column above the sacrum. In the case of circumduction of the cervical vertebral column, for example, the top of the head is the free point describing the base of our cone of motion, and the top of the first thoracic vertebral body represents the apex of the cone.

Regional Differences in Motion of the Vertebral Column

The extent of all motions of the vertebral column is largely a result of the shape of the zygapophyseal joints, which vary from region to region. In the cervical region, the orientation of superior articular facets from anterosuperior to posteroinferior (facing dorsally and superiorly) facilitates flexion and extension. The relatively thick intervertebral discs in the cervical region also contribute to greater flexion and extension in this region. Between adjacent cervical vertebrae, on average 15 degrees of flexion/extension takes place. There is moderate rotation in the cervical region, except between C1 and C2, where rotation is considerable.

The cervical vertebral column shows the greatest degree of lateral flexion of the vertebral regions. Between adjacent cervical vertebrae, approximately 10 degrees of lateral flexion is possible. Lateral flexion in this region is coupled with ipsilateral rotation of vertebral bodies. The articular facets slope downward laterally and posteriorly, guiding the superior vertebra into ipsilateral rotation during lateral flexion.

In the thoracic region, the vertical orientation of the articular facets limits flexion, and overlapping laminae and spinous processes limit extension. The thin intervertebral discs in the thoracic region also limit flexion and extension. An average of 6 degrees of flexion/extension occurs in the thoracic region, varying from the top to the bottom of the thoracic column, with 4 degrees allowed in the upper thorax and 12 degrees possible in the lower thorax. The articular facets lie along a horizontal arc, permitting considerable rotation. The greatest rotation is in the upper thoracic region. Lateral flexion of thoracic vertebrae is about 6 degrees between adjacent vertebrae; this motion is restricted by the ribs. In the upper thoracic region, lateral flexion is coupled with ipsilateral rotation of vertebral bodies. In the lower thoracic region, it is coupled with contralateral rotation of vertebral bodies.

In the lumbar region, flexion, extension, and lateral flexion occur freely. Approximately 15 degrees of flexion/extension takes place between adjacent vertebrae in the lumbar region. The greater thickness of the intervertebral disc in the lumbar region increases the allowable range of movement here. When the vertebral column as a whole is flexed (as in touching the toes) and extended, about 75 percent of this motion is movements of the lumbar region. Rotation is most restricted in the lumbar region because the articular facets are oriented along parasagittal planes. Lateral flexion is about 6 degrees between adjacent lumbar vertebrae, as in the thoracic region, and it is coupled with contralateral rotation of the vertebral bodies.

It is especially interesting, considering the effects of the zygapophyseal joints in restricting flexion and extension, that these joints in the lumbar region of humans are unlike those of other primates. To obtain and maintain an erect trunk position, humans, as well as other primates, must ensure that their center of mass falls on a vertical line that intersects some portion of their foot. If that line falls forward of the foot, they fall forward; if the line falls behind the foot, they fall backward (this assumes that they are standing still-in motion this is not true because of the effects of inertia).

But positioning the center of mass over the foot in the erect stance is virtually impossible for other primates because the inferior zygapophyseal joints of each lumbar vertebra are in the same position as the superior ones in the subjacent vertebra. This prevents any significant lordosis, forcing primates such as chimpanzees to adopt a bent-hipped, bent-kneed posture to position their center of mass over the ground contact points of their feet. In humans, the progressive widening of each pair of inferior zygapophyseal joints in the lumbar column allows considerably more imbrication (sliding). This in turn permits significant lordosis and a fully erect stance without flexion at either the hip or the knee.

Variations and Disorders of The Vertebral Column

There are many anomalies of the vertebral column. Only a few will be discussed here, most of which can be attributed to abnormal development of the vertebral column.

Variation in numbers of vertebral segments

One of the most common anomalies of the vertebral column is a variation in the numbers of vertebral segments. Variation occurs either in the total number of vertebral segments in the column, — that is, more or fewer than 33 — or in how many vertebrae are found in the different regions while the total number of segments remains 33. In the latter case, one vertebra is usually "shifted" from one region to an adjacent region, e.g., there may be 4 lumbar vertebrae and 6 segments in the sacrum. Variation in the number of cervical vertebrae is rare, but may occur in pathological conditions such as *Klippel-Feil syndrome*, in which there is often fusion and a reduction in number of cervical vertebrae, among other anomalies.

Abnormal Curvatures of the Vertebral Column

As described earlier, there are four curvatures to the vertebral column that are normal: cervical, thoracic, lumbar, and sacral (which includes the coccyx). The thoracic and sacral curvatures develop first (primary curvatures) and are convex posteriorly. The cervical and lumbar curvatures develop somewhat later and are convex anteriorly.

Any of these curvatures can become pathologically exaggerated. When there is a pathological curvature that is abnormally convex posteriorly, it is called a *kyphosis* (or "hyperkyphosis") (Figure 4-29). This is most frequently seen in the thorax, and is often associated with osteoporosis in the elderly ("Dowager's Hump"). It also develops in adolescents with *Scheuermann's Disease,* in which the anterior parts of the annular epiphyses do not develop properly.

If there is a curvature that is abnormally convex anteriorly, this is a *lordosis* (or "hyperlordosis"). This condition is most often seen in the lumbar region. It often occurs with obesity or temporarily with pregnancy, both of which place the body's center of

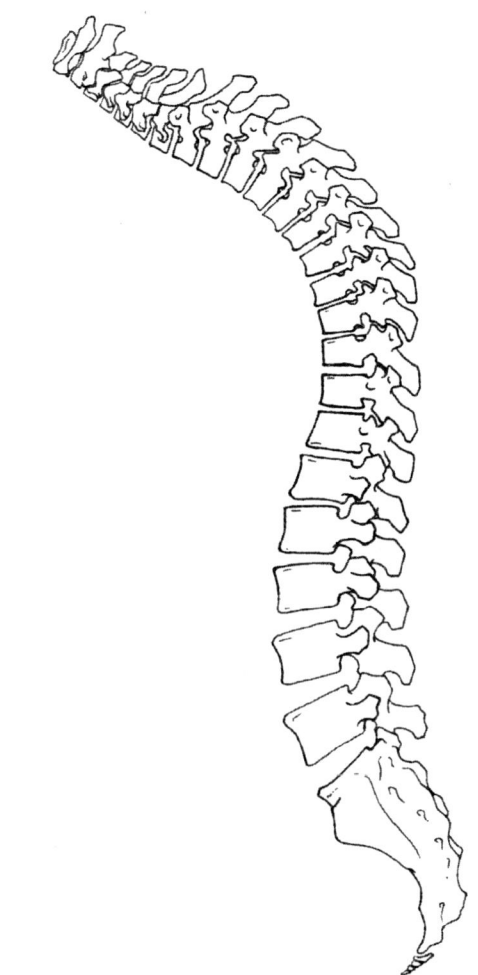

Figure 4-29 Kyphosis of the thoracic region.

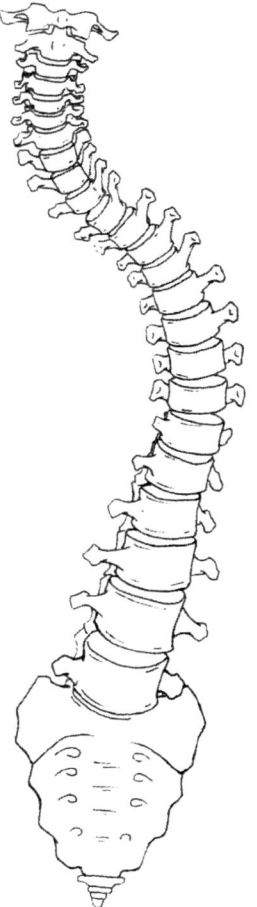

gravity farther forward than normally seen, displacing the lumbar vertebrae.

NOTE: anatomists usually refer to normal curvatures, and reserve the terms *kyphosis* and *lordosis* for pathological curvatures. Some clinicians use *kyphosis* and *lordosis* to refer to normal curvatures, and qualify them as *hyperkyphosis* or *hyperlordosis* to indicate a pathological condition.

Another pathological curvature is *scoliosis* (Figure 4-30), which is a mediolateral curvature of the vertebral column. This condition has a frequency of about 1 in 200, and is more often seen in females than in males. It may have different causes, such as a muscular weakness on one side ("myopathic"), or the presence of a hemivertebra (Figure 4-31). In the latter case, only one side of a vertebral body develops, causing the column above it to tilt, or curve, to the side. In other cases, the cause of scoliosis is unknown ("idiopathic").

Figure 4-30 Scoliosis.

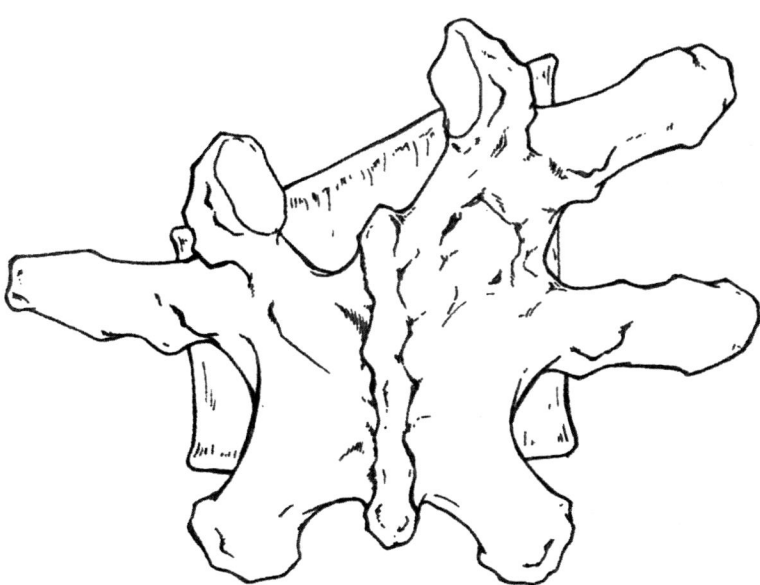

Figure 4-31 Hemivertebra on the right side, with ankylosis to the inferior vertebra, posterior view.

Herniated Discs

A herniated disc involves a defect in the annulus fibrosus of an intervertebral disc, either through degeneration (which occurs after the second decade of life) or trauma, and protrusion of the nucleus pulposus through this defect. This most commonly occurs in the lumbar region, with 95% of herniations occurring at the level of L4/L5 or L5/S1. The herniation is usually to the posterolateral aspect of the disc, where the annulus fibrosus is thin and unprotected by the posterior longitudinal ligament. The nucleus pulposus thus tends to herniate into the intervertebral foramen, compressing the spinal nerve roots. At the lower lumbar levels, this often results in *sciatica,* or severe pain in the leg felt along the course of the sciatic nerve. This nerve is a branch of the lumbosacral plexus and gets contributions from spinal nerves L4-S3.

Spondylolysis

In *spondylolysis*, the spinous processes, laminae and inferior articular processes are not attached to the body of the vertebra. This can be unilateral or bilateral, and most often occurs at L5. It is found in about 5-10% of people, and although the cause is unknown, it is thought that it may be congenital and/or caused by trauma or stress fractures. It is not seen in newborns, but the frequency increases with age up to age 20. If congenital, it may result from a failure of fusion of the vertebral arch(es) with the body during development.

Spondylolisthesis

Spondylolisthesis is an anterior displacement of the vertebral column, and usually occurs at the lumbosacral joint. It may have several causes, including bilateral spondylolysis at L5, fracture of the pedicles or inferior articular processes of L5, fracture of the superior articular processes of the sacrum, or congenital abnormalities of the zygapophyseal joints or vertebral arches. If both vertebral arches aren't attached to the vertebral body, the entire body's center of gravity (located just anterior to the lumbosacral joint) will cause L5 to slip forward on the sacrum. Spondylolisthesis at the lumbosacral joint usually results in compression of spinal nerves S1 and S2, producing symptoms of sciatica.

Spina Bifida

Spina bifida results from the failure of the two halves of the neural arch to fuse during development, leaving a defect in the arch. Next to variation in segment numbers, this is the most frequent congenital anomaly of the vertebral column. Spina bifida can be classified into two types: *Spina bifida occulta* and *Spina bifida cystica.* In S*pina bifida occulta* (Figure 4-32), the laminae fail to fully develop and fuse. It may occur in only one or two vertebrae, and does not involve herniation of the spinal cord or meninges. It is usually asymptomatic, but may occasionally be associated with some neurological disturbances. About 2% of people have spina bifida occulta of L5. *Spina bifida cystica (Spina bifida*

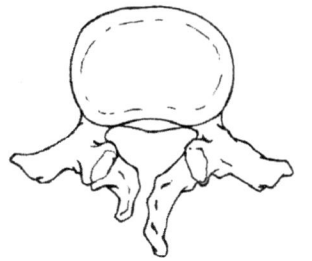 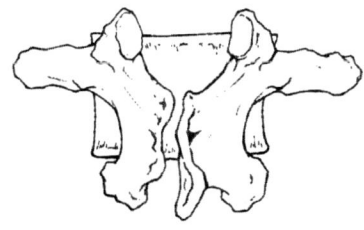

Figure 4-32 Spina bifida occulta of a lumbar vertebra.

vara) is a much more severe condition. It involves a cyst-like herniation of the spinal cord and/or meninges through the defect in the vertebral arch. There are three types of Spina bifida cystica. *Spina bifida cystica with meningocoele* occurs when there is herniation of the meninges, but not the spinal cord itself. When the spinal cord as well as the meninges is included in the herniation, it is called *Spina bifida cystica with meningomyelocele*. The most severe form is *Spina bifida cystica with myeloschisis*. In this condition, there is not only herniation of the spinal cord and meninges, but the spinal cord is also malformed and flattened due to a failure in closure of the neural tube during early development of the central nervous system. Spina bifida cystica occurs about once in every 1000 births, and usually involves meningomyelocele. It is often associated with other defects, such as clubfoot and hydrocephaly, and can be fatal to the infant. Infants that survive often show evidence of damage to the nervous system, such as paralysis of the limbs and disturbances in control of the bladder and bowel.

Whiplash

Whiplash is a soft tissue injury of the cervical spine. It is commonly caused by automobile accidents. The term *whiplash* is descriptive of the motions of the cervical spine which result in the injury. It is generally induced when a stationary automobile is struck in the rear by a moving vehicle. The patient's body and automobile seat are driven forward by the impact, but the head does not move with the body and as a result, the cervical spine is thrown into hyperextension. The anterior cervical muscles are severely stretched as a result, and as a consequence, they contract and throw the head forward by flexion of the cervical vertebral column. As the car stops, the body stops as well, but the head continues moving forward. Motion is stopped posteriorly by the occiput hitting the trunk, and motion is stopped anteriorly by the chin striking the sternum. This is referred to as an "acceleration injury."

A "deceleration injury" can result when the driver of a moving vehicle strikes a stationary vehicle. The opposite motions occur in this instance with flexion followed by hyperextension. It would be more appropriate to describe whiplash as musculoligamentous sprain or strain of the cervical spine (Hirsch, et al., 1988; Edwardson, 1995).

In the *extension phase* of whiplash, damage can occur to the prevertebral muscles, the anterior longitudinal ligament, the intervertebral discs, odontoid process, zygapophyseal joints, spinous processes, the temporomandibular joints, frontal and temporal lobes of the brain, and the esophagus. In the *flexion phase*, vertebral bodies, intervertebral discs, the interspinous ligaments, zygapophyseal joints, ligamentum nuchae and posterior neck muscles can all be damaged. (Bogduk, 1986; Edwardson, 1995). Additionally, nerve root compression has also been identified as a result of whiplash (Downs and Twomey, 1979; Edwardson, 1995). See Edwardson (1995) for a fuller discussion of the clinical manifestations of whiplash.

Degenerative Joint Disease

The vertebral column in whole or in part may be affected by degenerative joint diseases. Osteoarthritis is a common affliction of the vertebral column. Substantial lipping of the vertebral bodies can occur. Likewise, the zygapophyseal joints may be affected as well. *Cervical spondylosis* (*spondylosis* = breakdown of vertebral structures) affects up to 40 percent of the population. It can be asymptomatic, or may result in compression of cervical spinal nerves. It can affect intervertebral discs, vertebral bodies, zygapophyseal joints, the uncovertebral joints (of Luschka), and the ligaments. It is most frequent at C5-C6, and then C6-C7 (Edwardson, 1995 and references therein).

References Cited

Adams, T; Heisy, RS; Smith, MC; and Briner, BJ (1992) Parietal bone mobility in the anesthetized cat. *J. Am. Osteopathic Assoc.* 92: 599-600, 603-610, 615-622.

Bhaskar, SN (1990) *Orban's Oral Histology and Embryology,* Eleventh Edition. Mosby-Yearbook Publishers, Chicago.

Bogduk, N (1986) The anatomy and physiology of whiplash. *Clinical Biomechanics* 1: 92-101.

Breathnach, AS, ed. (1965) *Frazer's Anatomy of the Human Skeleton,* Sixth Edition. J and A Churchill, London.

Carlsen, BM (1994) *Human Embryology and Developmental Biology.* Mosby, St. Louis.

Downs, J and Twomey, L (1979) The whiplash syndrome. *Australian Journal of Physiotherapy* 25: 23 -241.

Duray, SM; Morter, HB; and Smith, FJ (1999) Morphological variation in cervical spinous processes: Potential applications in forensic identification of race from the skeleton. *Journal of Forensic Sciences,* September, 1999: 937-944.

Edwardson, BM (1995) *Musculoskeletal Disorders: Common Problems.* Singular Publishing Group, Inc., San Diego.

Gray, Henry (1985) *Gray's Anatomy,* 30th American Edition. C.D. Clemente, ed. Lea and Febiger, Philadelphia.

Gray, Henry (1995) *Gray's Anatomy,* 38th British Edition. P.L. Williams, L.H. Bannister, M.M. Berry, P. Collins, M. Dyson, J.E. Dussek and M.W.J. Ferguson, eds. Churchill Livingstone, New York.

Heisey, SR and Adams, T (1993) Role of cranial bone mobility in cranial compliance. *Neurosurgery* 33: 869-877.

Hirsch, SA, Hirsch, PJ, Hiramoto, H, and Weiss, A (1988) Whiplash syndrome: Fact or fiction? *The Orthopedic Clinics of North America* 19: 791-795.

Krogman, W.M. (1962). *The Human Skeleton in Forensic Medicine.* Charles C. Thomas, Springfield, Illinois.

McKern, TW and Stewart, TD (1957) *Skeletal Age Changes in Young American Males.* Technical Report EP-45, U.S. Army Quartermaster Research and Development Center, Natick, Massachusetts.

Mercer, S and Bogduk, N (1999) The ligaments and anulus fibrosus of human adult cervical intervertebral discs. *Spine* 24: 619-628.

Rogers, JS and Witt, PL (1997) The controversy of cranial bone motion. *J. Orthop. Sports Phys. Ther.* 26: 95-103.

Walker, R.A. and J.P. Grod, (2006) Functional Anatomy of the Lumbar Spine. In: Morris, CE (ed.) *Low Back Syndromes: Integrated Clinical Management*, pp. 19-62. McGraw-Hill Medical Publishing Division.

CHAPTER
| Five |

Bones of The Postcranial Skeleton: Ribs and Sternum

We next turn our attention to the ribs and sternum. The ribs and sternum are closely associated with the vertebral column. They add substantial stability to the vertebral column. Additionally they serve to protect the delicate organs housed in the thorax, specifically the heart and lungs. They are also the site of attachment of many of the muscles of the trunk. These muscles, like the flat muscles of the abdominal wall, are responsible for movements that help move the vertebral column. The sternum additionally is an important site of muscle attachments for muscles of the upper limb, and for those of the anterior part of the neck, such as the sternocleidomastoid.

The Ribs

The *ribs* are elastic arches of bone which, together with the thoracic vertebrae, the sternum, and the costal cartilages, form the skeleton of the thorax (Figure 5-1). There are twelve pairs of ribs. The first seven pairs articulate posteriorly with the bodies and transverse processes of the thoracic vertebrae, and anteriorly, via their costal cartilages, with the sternum. The costal cartilages are composed of hyaline cartilage. Their dorsal ends articulate with the ventral extremities of ribs 1 to 10. Their ventral ends articulate either with the sternum (ribs 1-7), or with the costal cartilage of the rib above them (ribs 8-10).

Ribs 1 through 7 are referred to as the *true* or *vertebrosternal ribs*. The remaining five pair are referred to as the *false rib*s. Pairs 8, 9, and 10 have their costal cartilages attaching to the costal cartilage of the rib above, and are also known as the *vertebrochondral ribs*. Pairs 11 and 12 are also known as the *floating ribs*. Their extremities are embedded in the musculature of the posterior thoracic wall. They are tipped with hyaline cartilage, but do not articulate with the sternum or other costal cartilages. Furthermore, unlike the cranial ten pairs of ribs, the 11th and 12th pairs do not articulate with the transverse processes of their corresponding vertebrae.

The description that follows applies to the more typical 3rd through 9th ribs (Figure 5-2). The 1st and 2nd ribs and 10th, 11th, and 12th ribs present features unique to them which diverge from this common pattern.

Each rib has two extremities, a *dorsal* or *vertebral, extremity* and a *ventral,* or *sternal, extremity*. The intervening portion of the rib is the *body,* or *shaft*. The dorsal extremity bears a *head*, a *neck*, and a *tubercle* (Refer to Figure 5-2 throughout the discussion of a typical rib).

The *head* bears a kidney-shaped *articular facet* divided into superior and inferior *demifacets* by a horizontal *interarticular crest* (crest of the head). The superior portion of this articular facet is for articulation with the inferior demifacet of a thoracic vertebra, while the inferior portion is for articulation with the superior demifacet of the vertebra below. These *demifacets* on the heads of the ribs vary in their configurations in the same manner as the costal facets and demifacets of the thoracic vertebrae.

The *neck* is a flattened portion of the rib extending laterally from the head. The ventral surface of the neck is flat and smooth, while the dorsal surface is roughened for attachments of various ligaments. It has a cranial and a caudal border. The cranial border bears a rough *crest of the neck* for the attachment of the anterior layer of the superior costotransverse ligament. The caudal border is rounded.

The *tubercle* of the rib is a raised area on the posterior surface of the dorsal extremity at the junction of the neck and the body. It consists of an *articular portion* and a *nonarticular portion*. The articular portion is more caudal and medial, and bears a small oval facet for articulation with the transverse process of the more caudal of the two vertebrae with which the head articulates. The lateral nonarticular portion is roughened and receives the attachment of the lateral costotransverse ligament.

The body of the rib is thin and flat, and has two surfaces, an *external surface* and an *internal surface*. The body has a rounded *superior border* and a sharp *inferior border*. A little beyond the tubercle, the body is marked by a prominent line which gives attachment to a tendon of the iliocostalis muscle. This is the *angle* of the rib. At the angle of the rib, the shaft turns anteromedially. The distance between the tubercle and the angle progressively increases from the second to the tenth ribs. Along the inferior margin of the internal surface of the rib is the *costal groove*. This groove is on the inferior border of the rib dorsally, but just ventral to the angle it comes to lie on the internal surface of the rib. The superior edge of the groove is rounded and gives attachment to the internal intercostal muscle. The inferior border of the groove, which is also the inferior border of the body of the rib, is sharp. The groove houses the intercostal nerve, artery and vein. The *sternal extremity* of the body is

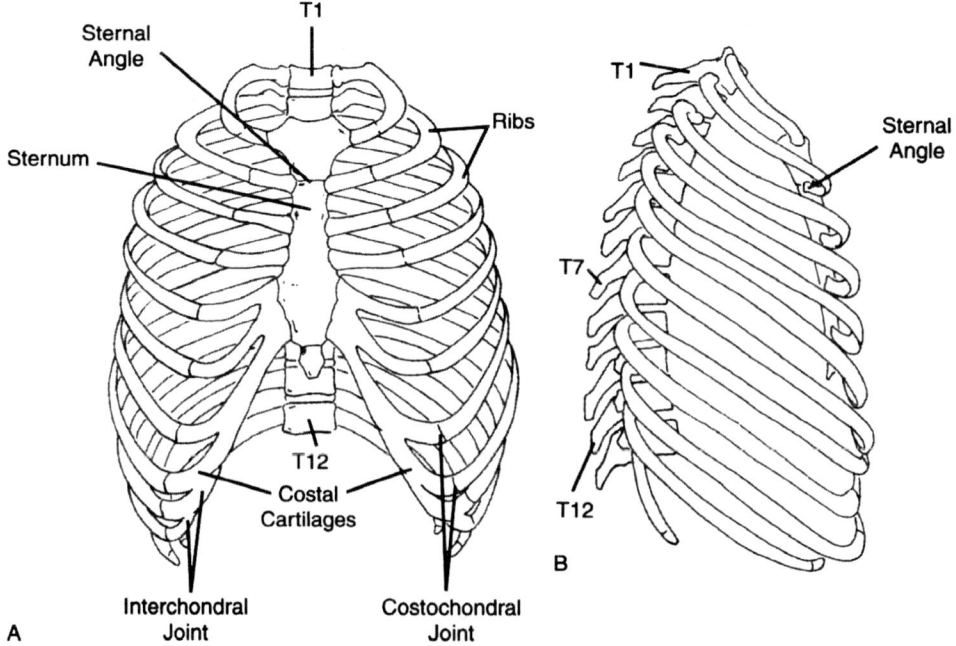

Figure 5-1 The bony thorax: (*A*) anterior view and (*B*) lateral view.

flattened and presents a hollow oval which receives the rounded end of the costal cartilage.

The *first rib* (Figure 5-3) is the most curved and the shortest. It has no angle and the body is flattened horizontally. Its cranial surface is marked by two grooves and the *scalene tubercle*. The *scalene tubercle* is toward the sternal extremity near the internal edge of the rib and is the site of attachment of the scalenus anterior muscle. It separates two grooves. The ventral groove is the *groove for the subclavian vein*, while the more dorsal groove is the *groove for the subclavian artery*. The head bears no interarticular crest, and so has only a single articular facet.

Near the middle of the external surface of the *second rib* (see Figure 5-3) is a rough *tuberosity for the serratus anterior*. While the serratus anterior attaches along the external surface of the first eight or nine ribs, its attachment to the second rib is particularly well-marked.

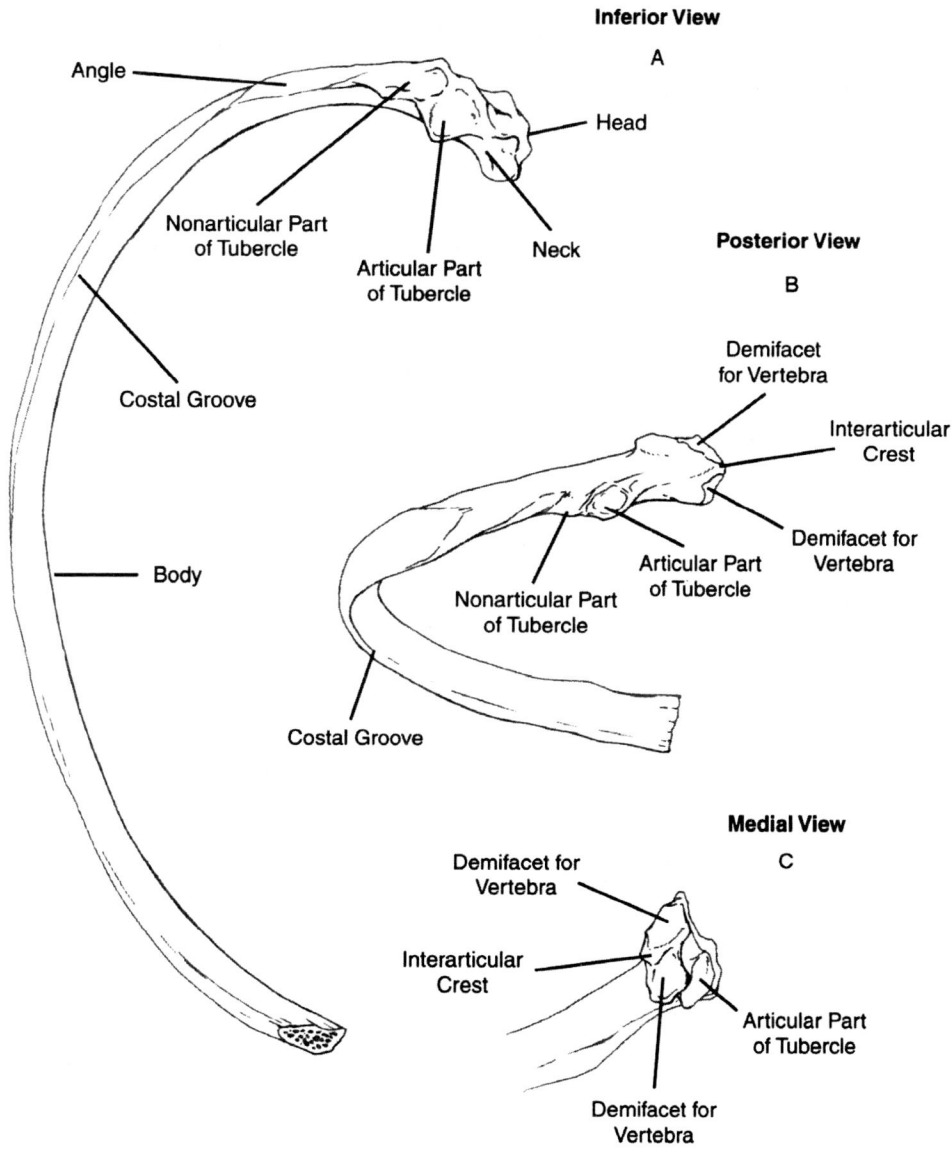

Figure 5-2 Features of typical ribs: (*A*) inferior view, (*B*) posterior view, (*C*) medial view.

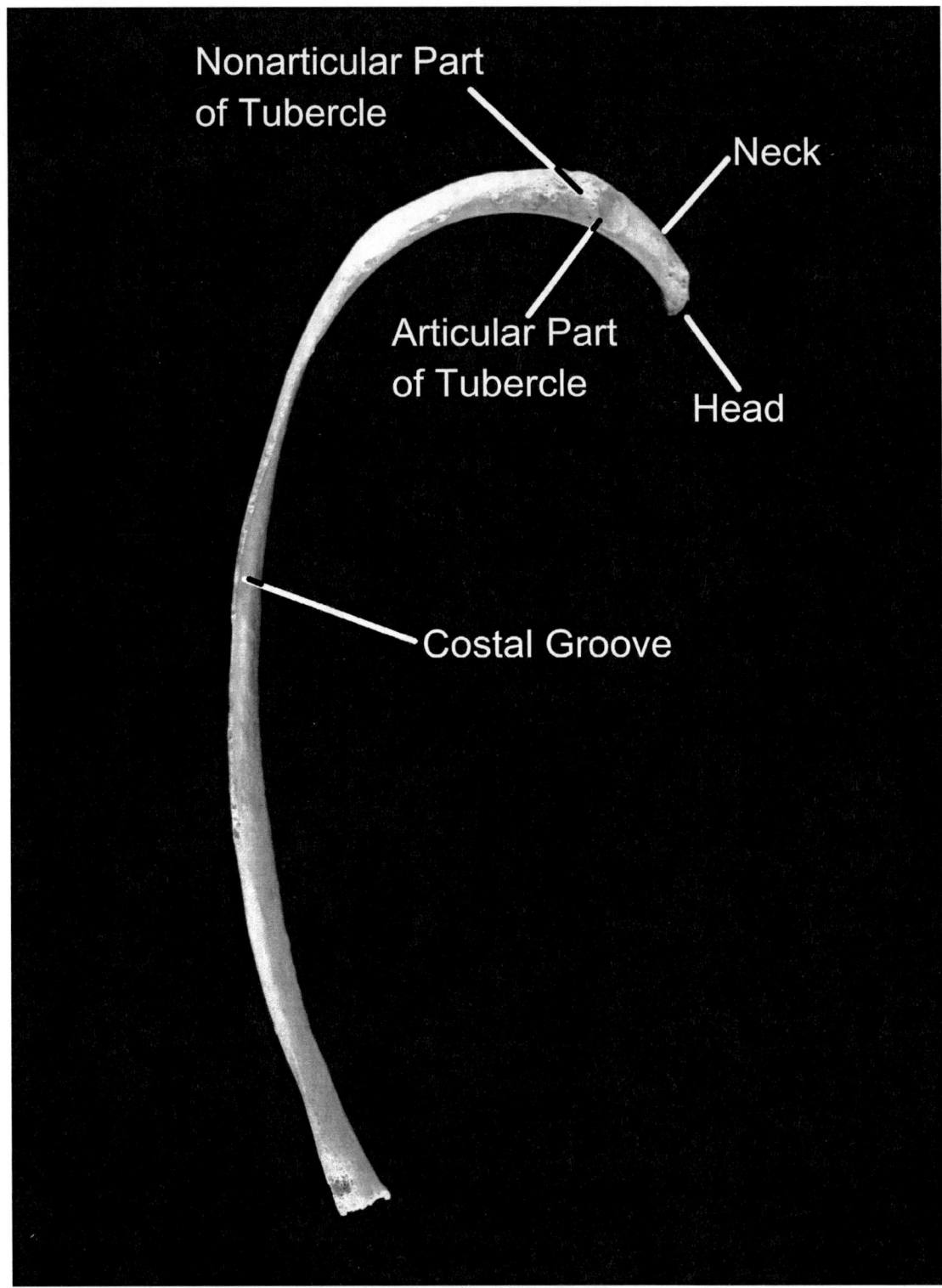

Figure 5-2 *(Continued)* Photograph of the inferior view of a typical rib. See accompanying line drawing for identification of additional landmarks.

The *10th rib* is similar to typical ribs except that there is usually no interarticular crest at the head, so that it has a single articular facet. The *eleventh and twelfth rib*s (Figure 5-4) are the floating ribs. Each has a single, large articular facet on the head, and no interarticular crest. They also have no neck or tubercle. The eleventh has a slight angle and a shallow costal groove, while the twelfth demonstrates neither.

The Sternum

The *sternum* (Figure 5-5; see also Figure 5-1) is composed of three parts: the manubrium, the body, and the xiphoid process.

The *manubrium* is the most superior part of the sternum. At the center of its superior border is the *jugular (suprasternal) notch.* On either side of the jugular notches are the *clavicular notches,* which bear articular facets for articulations with the sternal ends of the clavicles.

The *body* (or *gladiolus*) is the largest portion of the sternum, and is composed of four sternebrae or segments which become fused in adulthood. The manubrium joins the body of the sternum at the *sternal angle* (Angle of Louis) which is an important anatomical landmark. At the sternal angle, the *second costal cartilages* articulate with both the manubrium and body. The costal cartilages are composed of hyaline cartilage and run between the sternum and the ventral extremities of the ribs.

Because the second ribs lie at the level of the easily palpable sternal angle, the ribs can be counted with accuracy in the living subject. On either side of the sternum are seven *costal notches* for articulations with the costal cartilages. There is one complete notch and one half costal notch on either side of the manubrium. There are four complete notches and two half-notches on each side of the body. The xiphoid process also possesses a half-notch for the costal cartilage of the seventh rib on each side.

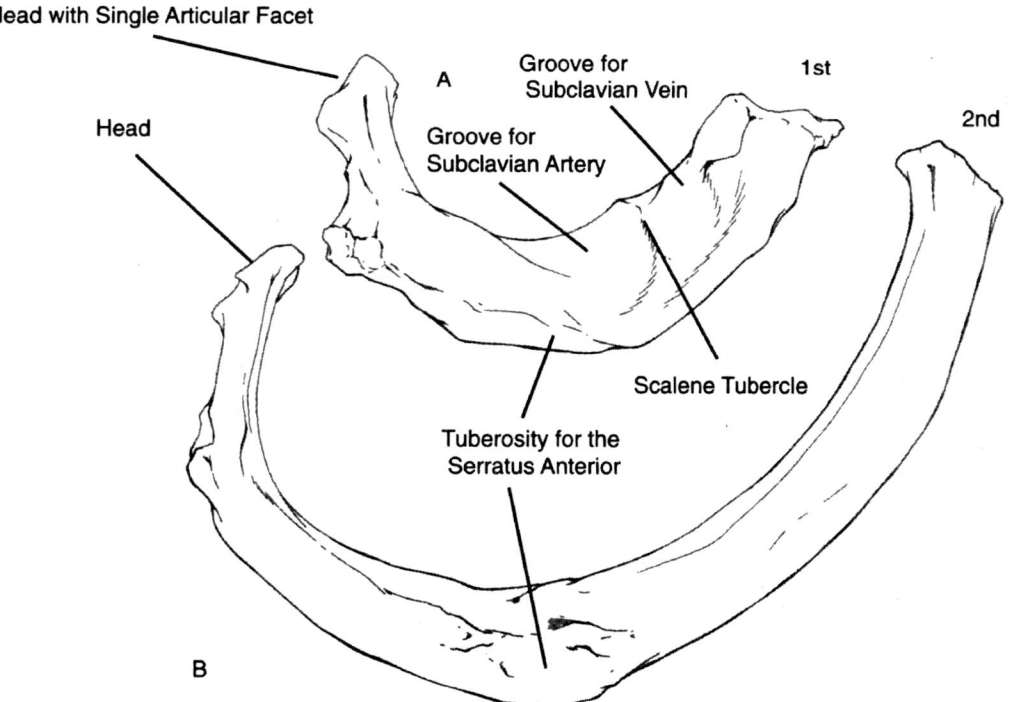

Figure 5-3 (A) The first rib and (B) the second rib, superior view.

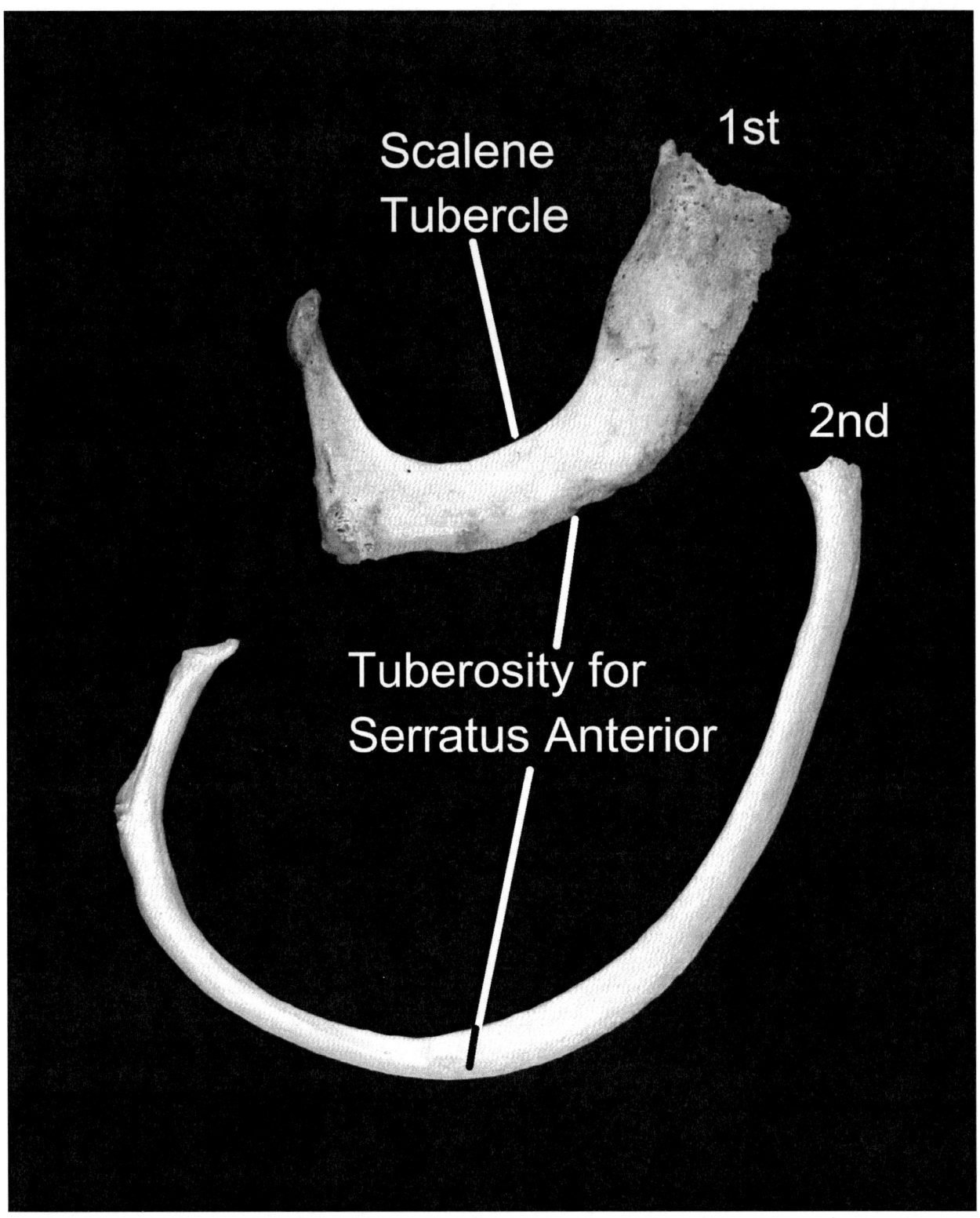

Figure 5-3 *(Continued)* Photograph of the superior view of the first (top) and second ribs. See accompanying line drawing for identification of additional landmarks.

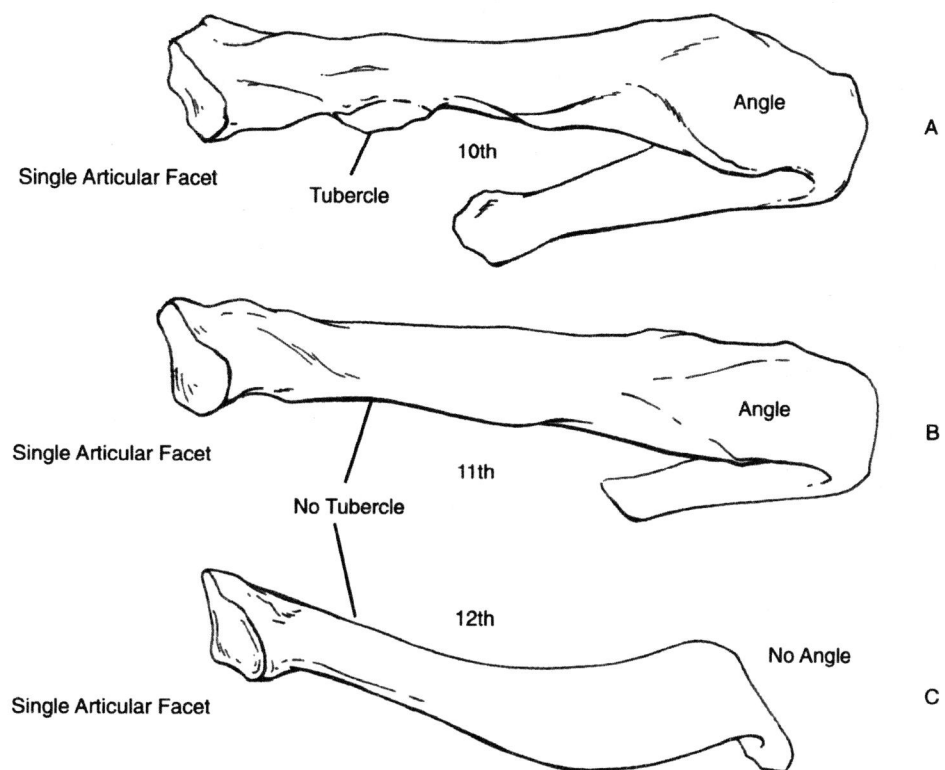

Figure 5-4 (A) The 10th rib, (B) the 11th rib, and (C) the 12th rib.

The most inferior and smallest segment of the sternum is the *xiphoid process*. It is thin and elongated. It is cartilaginous throughout youth, and progressively ossifies with age. It may remain partly cartilaginous, however, throughout life. It articulates with the body of the sternum at the *xiphisternal junction*. The costal cartilages of the seventh ribs articulate with the xiphoid process.

Articulations of the Ribs

The Costovertebral Joints

The *costovertebral joints* (Figure 5-6) are usually described as plane synovial joints and allow gliding movement. The fibrous *articular capsule* surrounds the head of the rib and the socket formed by the demifacets on the cranial and caudal vertebrae and their intervertebral disc. The *synovial cavity* is divided in two by the *intra-articular ligament* for ribs 2-9 (which articulate with two vertebral bodies), and is single for ribs 1, 10, 11, and 12, which articulate with a single large costal facet on a single vertebral body. The *intraarticular ligament* consists of short fibers extending from the interarticular crest of the head of the rib to the intervertebral disc. Joints of ribs 1, 10, 11, and 12 do not have intraarticular ligaments.

The *radiate ligaments* attach the anterior parts of each costal head to the bodies of two vertebrae and their intervertebral disc. The radiate ligament always attaches to two vertebrae even if the rib only articulates with a single vertebrae. For example, the first rib articulates only with T1, but the radiate ligament of the first rib attaches both to C7 and T1.

The Costotransverse Joints

The *costotransverse joints* (see Figure 5-6) are formed by the tubercle of the rib articulating with the *transverse costal facet* of the transverse process of a thoracic vertebra. These are also plane synovial joints, which allow sliding motion. A rib articulates with the transverse process of its corresponding vertebrae, i.e.: the 10th rib articulates with the transverse process of the 10th thoracic vertebra. There are no costotransverse joints associated with T11 and T12.

Each costotransverse joint is surrounded by a fibrous *articular capsule*. The capsule surrounds the tubercle of the rib and the transverse costal facet of

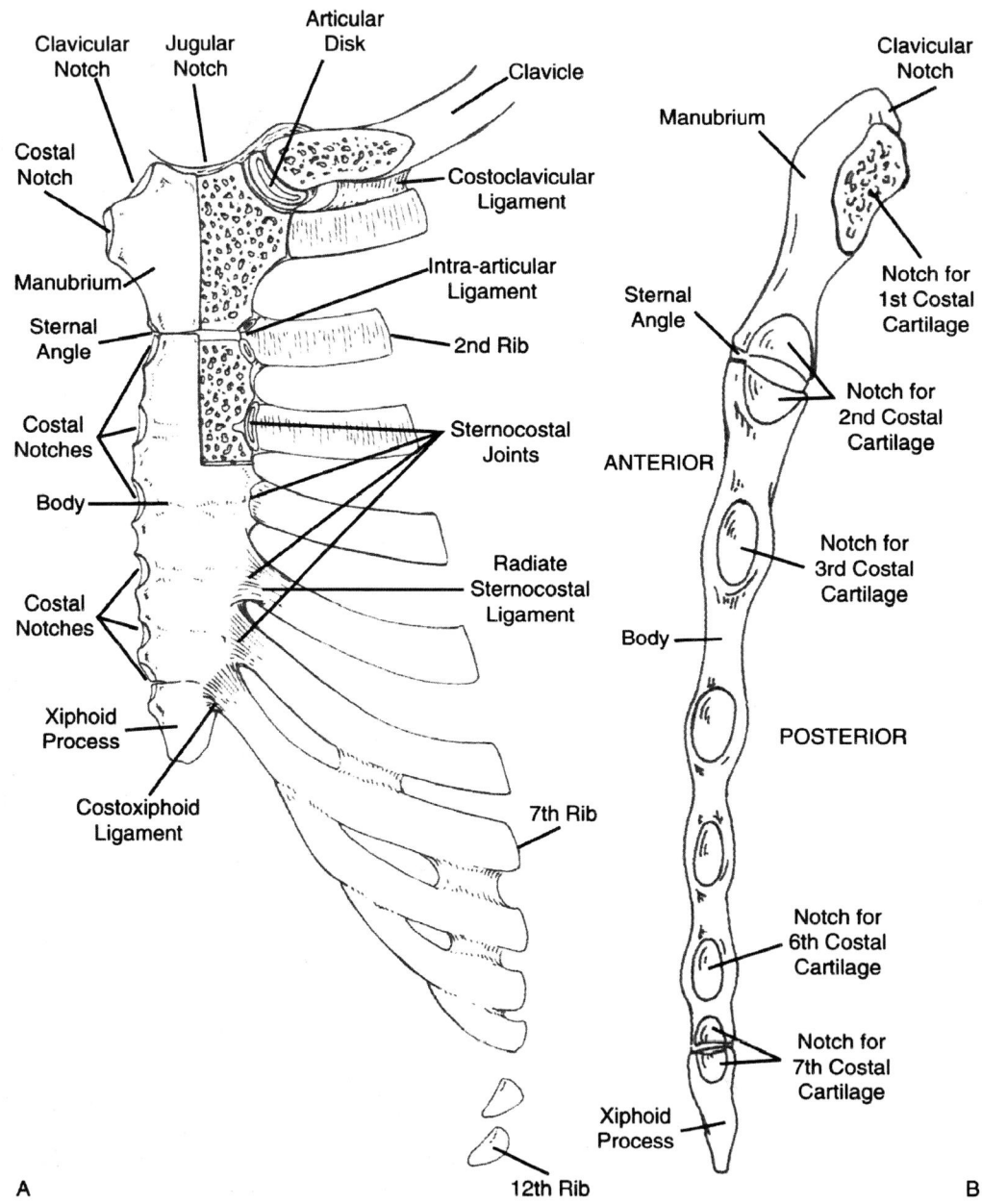

Figure 5-5 (A) The sternum and costal cartilages, anterior view, and (B) the sternum, lateral view.

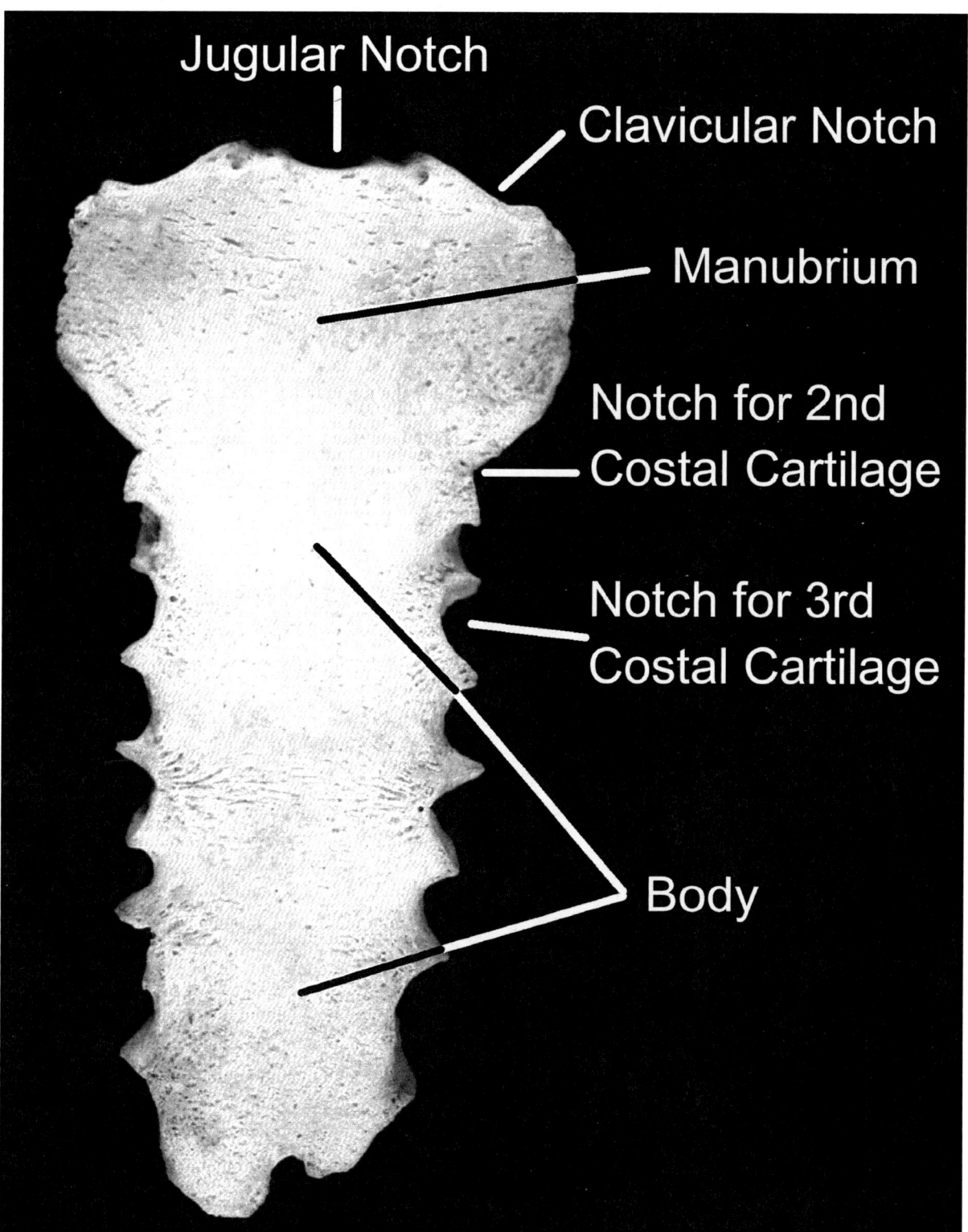

Figure 5-5 *(Continued)* Photograph of the anterior aspect of the sternum. See accompanying line drawing for identification of additional landmarks.

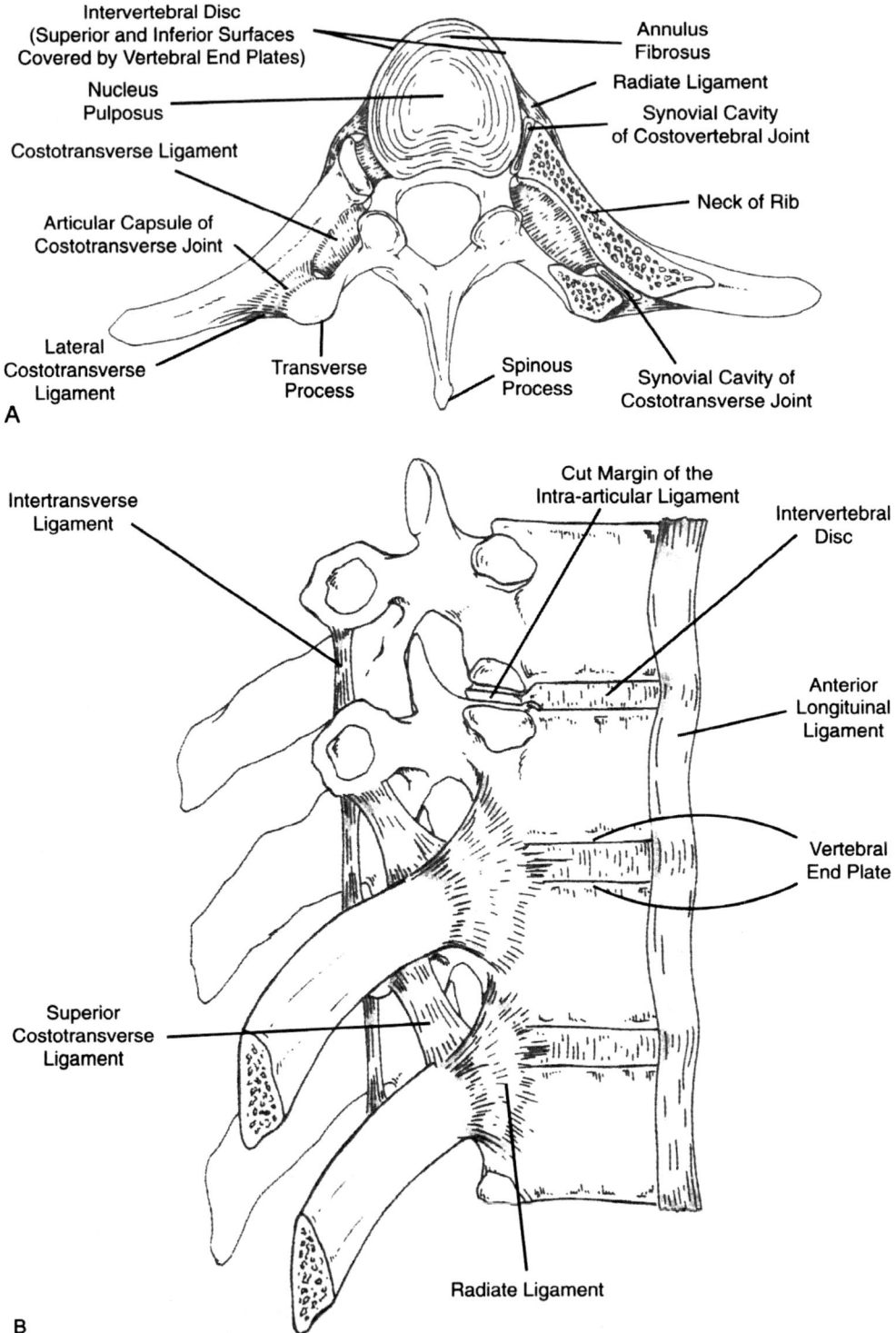

Figure 5-6 (A) Ligaments of the ribs and the intervertebral disc and (B) articulations of the ribs and thoracic vertebrae.

the vertebra. In addition to the articular capsule, these joints are strengthened by three other ligaments:

- *Superior Costotransverse Ligament*
- Costotransverse Ligament
- *Lateral Costotransverse Ligament.*

The *superior costotransverse ligament* runs from the neck of the rib to the transverse process above. Each ligament is composed of an anterior and a posterior layer. The *anterior layer* extends from the crest of the neck of the rib to the lower border of the transverse process above it, while the *posterior layer* extends from the dorsal aspect of the rib's neck. The first rib possesses no superior costotransverse ligament. The 12th rib possesses an additional *lumbocostal ligament* that extends from the neck of the 12th rib to the base of the transverse process of the first lumbar vertebra.

The *costotransverse ligament*, or ligament of the neck of the rib (see Figure 5-6), extends from the dorsal aspect of the neck posteriorly to the anterior surface of the adjacent transverse process. This ligament may be absent or rudimentary on the 11th and 12th ribs.

The *lateral costotransverse ligament*, or ligament of the tubercle of the rib (see Figure 5-6), extends from the apex of the transverse process to the tubercle of the corresponding rib. This ligament is not present for ribs 11 and 12.

Joints of the Costal Cartilages:

The joints of the costal cartilages are:

- *The Sternocostal Joints*
- The Interchondral Joints
- The Costochondral Joints.

The *sternocostal joints* (see Figures 5-1 and 5-5) occur between the costal notches of the sternum and the ventral ends of the costal cartilages. The *first sternocostal joint* is a synchondrosis. The *second through seventh* are plane synovial joints. They possess a fibrous articular capsule and three ligaments:

- *The Radiate Sternocostal Ligament*
- The Intraarticular Ligament
- The Costoxiphoid Ligament.

The *radiate sternocostal ligament* extends from the sternal extremity of the costal cartilages to the anterior and posterior borders of the sternum. The *intraarticular ligaments* are constant only between the second costal cartilages and the sternum. It extends from the costal cartilage to the symphysis between the manubrium and the body of the sternum. Thus it is intraarticular. Occasionally there may be intraarticular ligaments associated with the third costal cartilages. The *costoxiphoid ligaments* connect the anterior and posterior surfaces of the seventh costal cartilage with the same surfaces of the xiphoid process. Slight movements occur at the sternocostal joints with respiration.

The Interchondral Joints

The *interchondral joints* (see Figure 5-1 A) are found between the 6th, 7th, 8th, 9th, and 10th costal cartilages as they articulate with each other. Those between the 6th to 9th cartilages are plane synovial joints, and are enclosed by thin fibrous articular capsules, and strengthened by medial and lateral interchondral ligaments. The joint between the 9th and the 10th costal cartilages is never synovial and is sometimes absent.

The Costochondral Joints

The *costochondral joints* (see Figure 5-1 A) are synchondroses occurring between the ventral extremities of the ribs and their costal cartilages. The costal cartilages are persistent, unossified anterior parts of cartilaginous models preceding fully developed ribs. When separated at their junctions, the end of the costal cartilage is rounded, while the ventral end of the rib is depressed. The periosteum and the perichondrium are continuous across this junction. No movement occurs at these joints

Articulations of the Sternum:

The articulations of the sternum are:

- *The Manubriosternal Joint*
- The Xiphisternal Joint
- The Sternoclavicular Joints.
- *The Sternocostal Joints* (discussed with joints of the costal cartilages).

The *manubriosternal joint* is a symphysis joint between the manubrium and the body of the sternum. It is located at the sternal angle. It may occasionally ossify in later life. The *xiphisternal joint* is also a symphysis. It is located at the junction of the body of the sternum with the xiphoid process. It is usually ossified by age 40. The *sternoclavicular joints* are located between the clavicles, the clavicular notches on the superior border of the manubrium, and the costal cartilages of the 1st ribs. At each joint there is an *articular disc* which divides it into two synovial cavities. These are *sellar synovial joints*. Each joint is surrounded by a fibrous *articular capsule*. In addition, the joints are strengthened by several ligaments, which include:

- *The Anterior Sternoclavicular Ligament*
- The Posterior Sternoclavicular Ligament
- The Interclavicular Ligament
- The Costoclavicular Ligament

The *anterior sternoclavicular ligament* attaches to the anterior surface of the clavicle, manubrium, and first costal cartilage. The *posterior sternoclavicular ligament* covers the posterior surface of the joint. It runs from the posterior surface of the clavicle's sternal end to the posterior aspect of the upper manubrium. The *interclavicular ligament* unites the superior aspects of the sternal ends of both clavicles with one another. Some fibers attach to the superior margin of the manubrium. The *costoclavicular ligament* unites the upper part of the first rib and costal cartilage with the inferior medial surface of the clavicle.

Movement at the sternoclavicular joint occurs in conjunction with movements of the pectoral girdle and upper limbs. The joint moves anteriorly, posteriorly, and vertically and allows the lateral end of the clavicle to be raised for full elevation of the upper limb (Gray, 1995).

Ossification of the Ribs and Sternum

The Ribs

The cartilage models for the ribs are present by the eighth week *in utero*. At this same time, a primary center of ossification appears near the angle of the rib, usually appearing first in the sixth and seventh ribs. Three secondary centers also develop: one for the head of the rib, and one for the articular portion of the tubercle of the rib and one for the nonarticular portion of the tubercle of the rib. These secondary centers develop postnatally. They appear between 16 and 20 years, and synostose with the rib by age 25. The first rib and ribs 7 to 10 have only one epiphysis for the tubercle, and the eleventh and twelfth lack centers for the tubercle (Breathnach, 1965; Gray, 1985; 1995).

The Sternum

The sternum (Figure 5-7) is initially represented by two parallel mesenchymal bars, the *sternal bars*, that are widely separated from one another. At about the sixth week in utero the superior ends of the bars approach one another and unite, and then fuse increasingly with one another progressing inferiorly.

Ossification of the sternum begins in the fifth or sixth week in utero with the manubrium. The manubrium frequently has two centers of ossification. The body has four centers of ossification, the first appearing at 6 months in utero, the second at 7 months in utero, the third in the eighth or ninth month in utero, and the fourth at birth or within the first year.

The xiphoid process has a single center of ossification that appears between ages 5 and 18. Two small *episternal centers of ossification* may appear on either side of the jugular notch, and are probably vestiges of the *episternal bone* of lower vertebrates.

The separate ossification centers of the sternum before their synostosis with one another are known as *sternebrae*. Synostosis of the sternebrae usually begins inferiorly and works its way superiorly. By age 25 the four segments of the sternal body will have synostosed. The xiphoid process synostoses with the body of the sternum most frequently after age 40, though it may occur well earlier than that. In older ages, the symphysis between the manubrium and the body of the sternum may undergo synostosis. Generally, the synostosis is only superficial, and remnants of the cartilaginous symphysis remain internally (Breathnach, 1965; Gray, 1985; 1995).

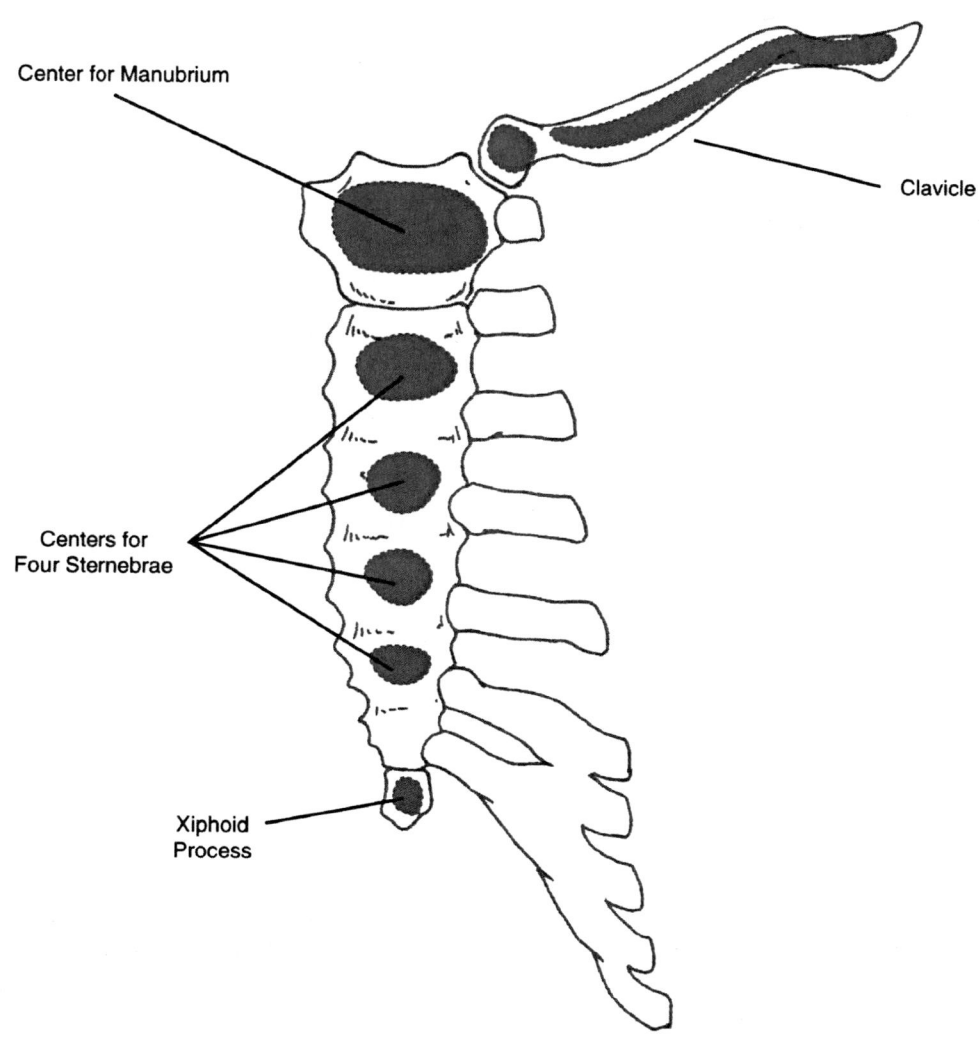

Figure 5-7 Centers of ossification of the sternum and the clavicle.

Disorders of the Ribs and Sternum

The ribs are noticeably affected by *rickets*. In rickets, the ends of the ribs that adjoin the costal cartilage become enlarged. This condition is referred to as *"rachitic rosary."* In mild cases, the enlargement may be restricted to the internal aspect of the thorax. Lateral to the costochondral junction, the ribs may become bowed and depressed, leaving a groove to either side of the sternum. The sternum is often forced outward by the bowing of the ribs, increasing the anteroposterior diameter of the thorax. This forward projection of the sternum is known as "pigeon breast."

"Funnel breast" is the opposite condition from "pigeon breast." It occurs in sufferers of rickets, and also in others with no evidence of rickets. In "funnel breast", the inferior end of the sternum is deeply depressed posteriorly.

Phthisical chest, is sometimes associated with pulmonary tuberculosis. In this condition, the ribs show great obliquity and the scapulae are projecting. *Barrel chest* is another condition that involves an enlargement of the chest in all dimensions and is associated with pulmonary emphysema.

Scoliosis or other *abnormalities of the thoracic spinal curvature* may distort the shapes of the ribs and thoracic cavity. In severe cases they may compromise heart and lung functions.

References Cited

Breathnach, AS, ed. (1965) *Frazer's Anatomy of the Human Skeleton,* Sixth Edition. J and A Churchill, London.

Gray, Henry (1985) *Gray's Anatomy,* 30th American Edition. C.D. Clemente, ed. Lea and Febiger, Philadelphia.

Gray, Henry (1995) *Gray's Anatomy,* 38th British Edition. P.L. Williams, L.H. Bannister, M.M. Berry, P. Collins, M. Dyson, J.E. Dussek and M.W.J. Ferguson, eds. Churchill Livingstone, New York.

CHAPTER
| Six |

Bones of the Postcranial Skeleton: Upper Limb

The *upper limb* is attached to the trunk through the *pectoral girdle* (Figure 6-1). The pectoral girdle consists of the *scapula* and *clavicle*, while the upper limb skeleton consists of the *humerus* of the *arm* *(brachium)*, the *radius* and *ulna* of the *forearm (antebrachium)*, the *carpal bones* of the *wrist (carpus)*, and the *metacarpals* and *phalanges* of the *hand (manus)*.

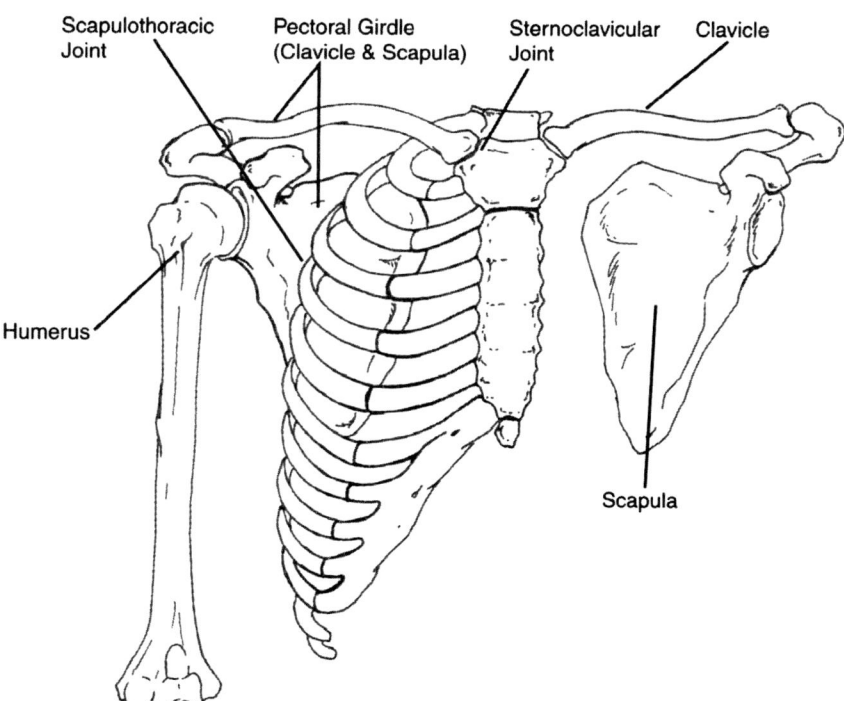

Figure 6-1 The pectoral girdle and thorax. The clavicle and scapula are visible on the right; the pectoral girdle is seen on the left in relationship to the bony thorax and the upper limb.

The Pectoral Girdle

The pectoral girdle consists of two bones, the *scapula* and the *clavicle*. The upper limb attaches to the trunk via the glenohumeral joint between the glenoid fossa of the scapula and the head of the humerus. The upper limb is attached to the axial skeleton via the clavicle, which articulates distally with the acromion process of the scapula and proximally with the manubrium of the sternum. This sternoclavicular joint is the sole bony attachment of the upper limb to the axial skeleton.

The Clavicle

The clavicle (Figure 6-2; see also Figure 6-1) forms the ventral portion of the pectoral girdle. It is a long bone, and curved somewhat like the letter "S". It is situated nearly horizontally above the first rib. It has a *sternal extremity* and an *acromial extremity*. The double curve of the bone lies in the horizontal plane. The medial curve occupies about the medial 2/3 of the bone and has a posterior concavity. The lateral curve has an anterior concavity which is shorter and sharper than its medial counterpart. It occupies the lateral 1/3 of the bone. The medial 2/3 of the shaft of the clavicle is rounded, while the lateral 1/3 is flattened. The *sternal extremity* of the clavicle presents a small articular facet for articulation with the clavicular notch of the manubrium. The flattened *acromial extremity* bears an articular facet for articulation with the acromion process of the scapula. The inferior aspect of the clavicle is readily identifiable by the markings for ligamentous attachments. Chief among these are the *impression for the costoclavicular ligament* on the inferior aspect of the sternal extremity, the *groove for the subclavius muscle* along the inferior surface of the midshaft, the *conoid tubercle*, and the *trapezoid line*. The *conoid tubercle* is located on the inferior surface at the dorsal border near the acromial extremity. The *trapezoid line* is also on the inferior surface of the acromial extremity and runs from the conoid tubercle ventrally and laterally. The conoid

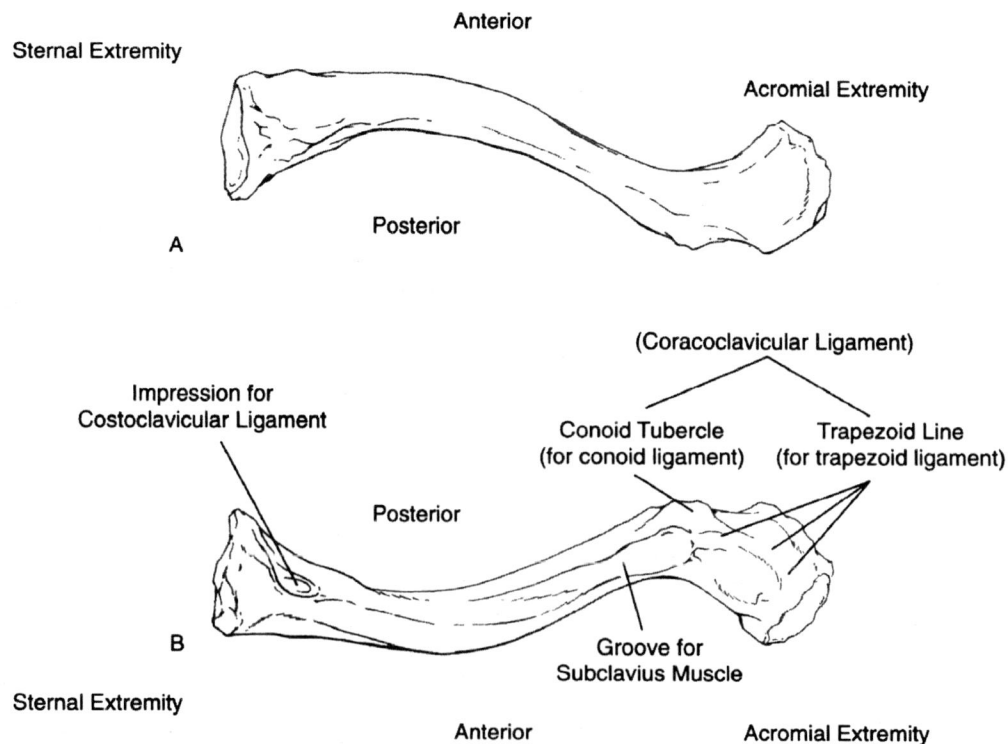

Figure 6-2 The right clavicle: (*A*) superior and (*B*) inferior views.

tubercle and trapezoid line provide attachment for the conoid and trapezoid ligaments, respectively. These two ligaments constitute the coracoclavicular ligament, which binds the clavicle to the scapula.

Scapula

The scapula (refer to Figures 6-3 through 6-5 throughout the discussion of the scapula) is a flat bone lying on the posterolateral aspect of the superior part of the thorax. Superiorly it articulates with the acromial extremity of the clavicle, and laterally with the head of the humerus. It has *three borders, three angles, and two surfaces*. The borders are the *superior, vertebral* or medial, and *axillary* or lateral. The angles are the *superior, inferior*, and *lateral*. The surfaces are the *dorsal* and *ventral* (or *costal*).

The *ventral surface* is concave and forms the *subscapular fossa* which gives origin to the subscapularis muscle. The ventral surface has a raised rim along its *vertebral border* which gives origin to the serratus anterior muscle. The most prominent feature of the *dorsal surface* is the *scapular spine*. The spine divides the dorsal surface into a smaller *supraspinatus fossa* for origin of the supraspinatus muscle, and a much larger *infraspinatus fossa* for the origin of the infraspinatus muscle. The lateral end of the spine is free as the large *acromion process*, which projects posterior to the *glenoid fossa* of the lateral angle. Deep to the acromion and posterior to

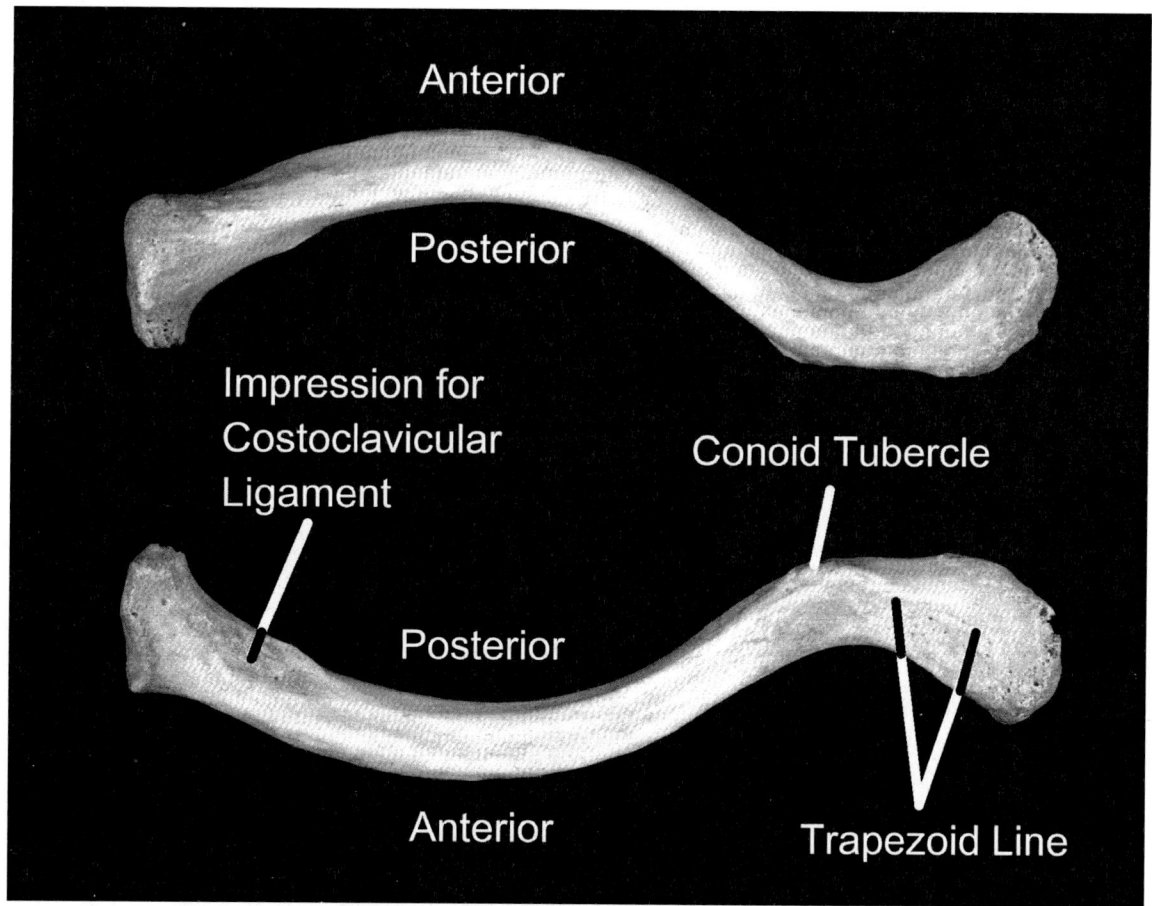

Figure 6-2 *(Continued)* Photographs of superior (top) and inferior views of the right clavicle. See accompanying line drawing for identification of additional landmarks.

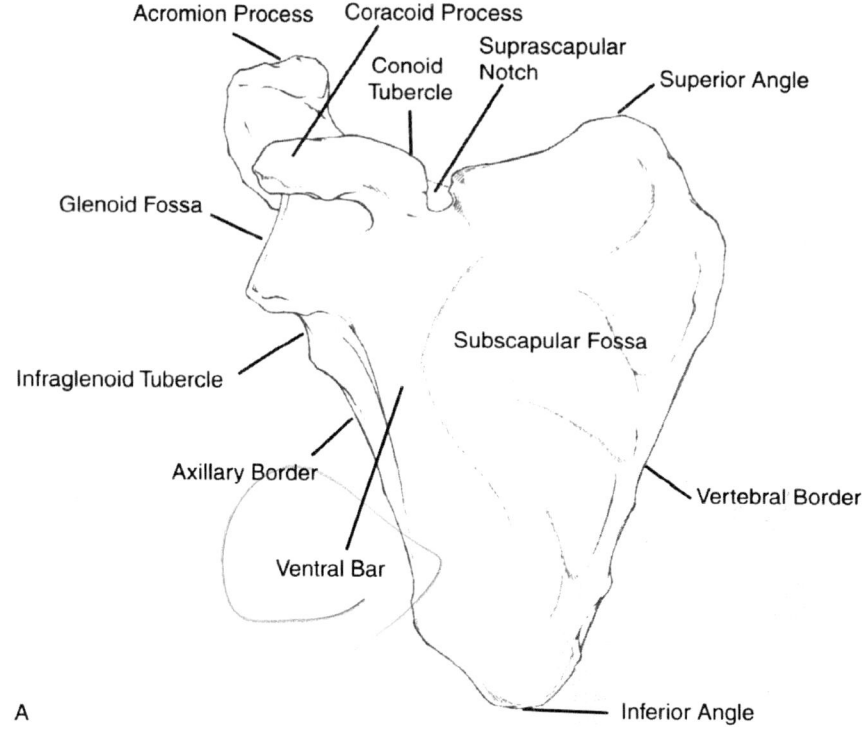

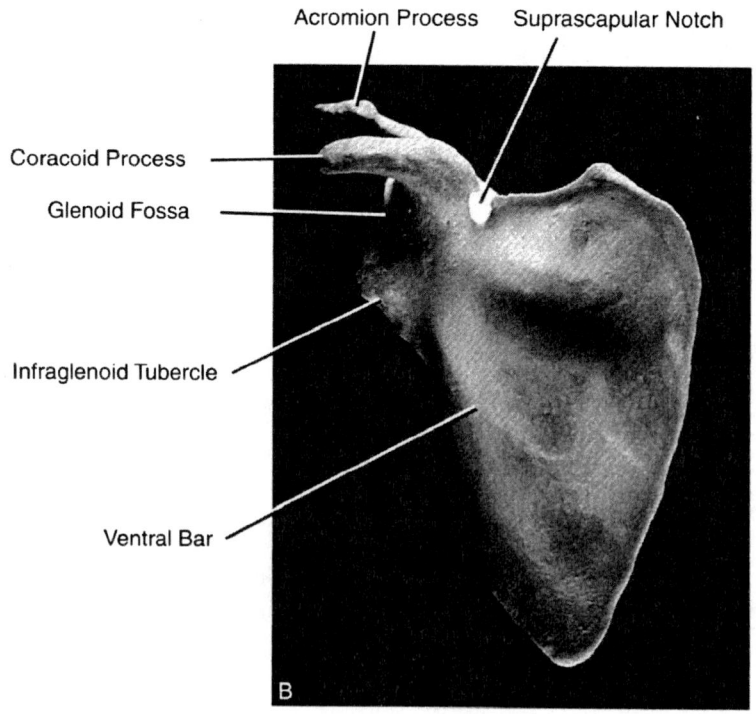

Figure 6-3 (*A*) Line drawing of the right scapula, ventral (costal) surface. (*B*) Photograph of the costal surface of the scapula.

the glenoid fossa is the *greater scapular notch*. The acromion process projects ventrally, cranially, and laterally. It bears a small articular facet for articulation with the lateral end of the clavicle.

The spine begins on the dorsal surface of the *vertebral border*. The *base* of the spine is located here at approximately the junction of the superior 1/3 and inferior 2/3 of the vertebral border of the scapula. The levator scapulae muscle attaches along the vertebral border superior to the base of the spine. The rhomboideus minor muscle attaches to the vertebral border at the base of the spine, while the rhomboideus major attaches along the remaining inferior portion of the vertebral border inferior to the base of the spine.

Along the dorsal surface of the *axillary border* are markings for the origins of the teres major and teres minor muscles. They run from the inferior angle to the *infraglenoid tubercle* at the lateral angle. On the costal surface of the lateral border is a structural thickening, the *ventral bar* of the scapula. It runs from the glenoid fossa to the inferior angle.

The *superior border* is sharp in the medial portion, and then presents a marked *suprascapular notch* for the suprascapular nerve. This is bridged by the suprascapular ligament, which is occasionally ossified, transforming the notch into a foramen. Arising lateral to the suprascapular notch is the finger-like *coracoid process*, which projects superiorly and then bends sharply laterally to project above the glenoid fossa. Attaching to the coracoid process are the trapezoid and conoid ligaments. On the superior surface of the bend or "knuckle" of the coracoid process is a small, roughened tubercle for attachment of the conoid ligament. This is the *conoid tubercle* of the scapula. The coracoacromial ligament runs between the coracoid process and the acromion process. The coracobrachialis, short head of the biceps brachii and the pectoralis minor all attach to the coracoid process.

The *lateral angle* of the scapula is not "pointed" like the superior and inferior angles, and is dominated by the *glenoid fossa*. The glenoid fossa is oval in outline and concave. It faces laterally and articulates with the head of the humerus. The glenoid fossa in life is covered by hyaline articular cartilage, and is connected with the body of the scapula by a somewhat constricted *neck*. At its apex, near the base of the coracoid process is a slight elevation, the *supraglenoid tubercle*, to which the long head of the biceps brachii is attached. The *infraglenoid tubercle* is located immediately inferior to the glenoid fossa. It is a rough triangular impression about an inch in length which commences at the inferior border of the glenoid fossa and proceeds inferiorly along the lateral border. It is the site of origin of the long head of the triceps brachii.

Bones of the Upper Limb

The upper limb is divisible into the *arm (brachium)*, the *forearm (antebrachium)*, the *wrist (carpus)* and *hand (manus)*. The arm is the more proximal portion of the limb, between the shoulder and the elbow. The antebrachium is the portion of the upper limb from the elbow to the wrist. The *humerus* is the bone of the *brachium*, while the *radius and ulna* are the two bones of the *antebrachium*.

The Humerus

The *humerus* (Figure 6-6) is the large bone forming the skeleton of the upper arm or brachium. It consists of a proximal *head*, a *shaft*, and a *distal extremity*.

Humerus: The Proximal End

When held in anatomical position, the *head* faces medially, somewhat posteriorly and superiorly. The head is joined to the rest of the shaft by a slightly restricted *anatomical neck*. The head of the humerus articulates with the glenoid fossa of the scapula. Just distal to the head of the humerus, and separated from it by the anatomical neck, are two raised processes that serve as the site of muscular attachments, the *greater and lesser tubercles* (sometimes referred to as *tuberosities*).

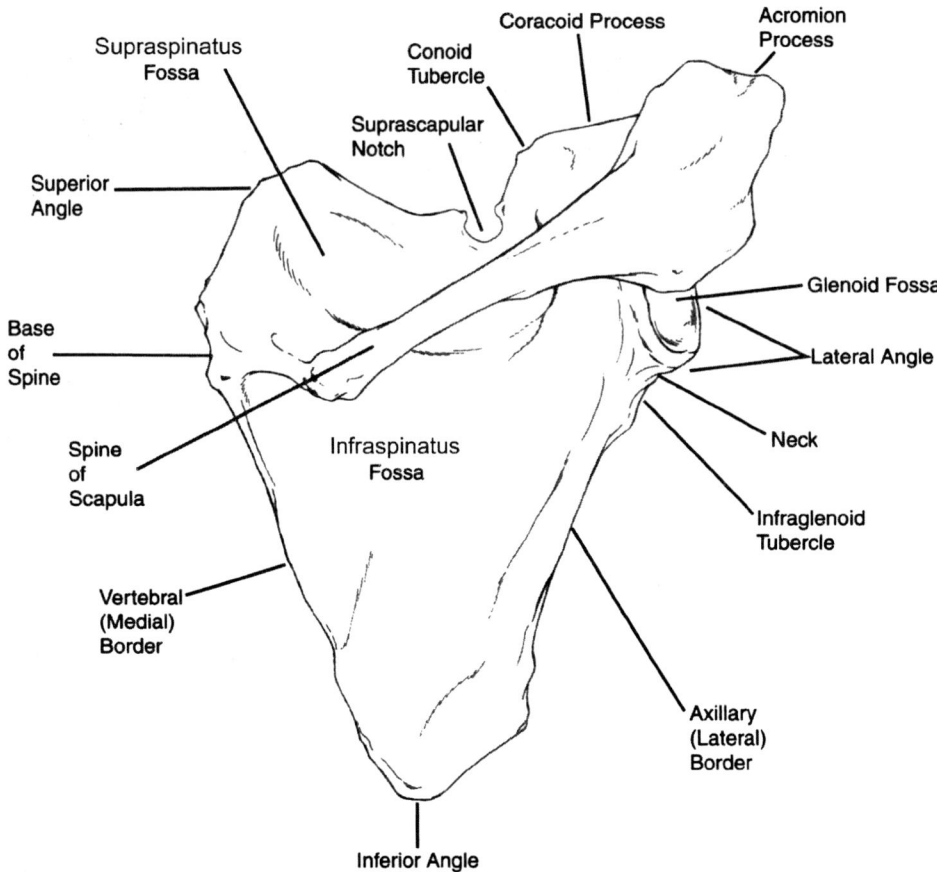

Figure 6-4 The right scapula, dorsal surface.

On the anterior surface of the shaft of the humerus is the *lesser tubercle* of the humerus. On the lateral aspect of the proximal shaft, and separated from the lesser tubercle by the *intertubercular sulcus* (or *bicipital groove*) is the *greater tubercle*. The lesser tubercle is the site of insertion of the subscapularis muscle, while the greater tubercle receives the supraspinatus, infraspinatus and teres minor muscles. The intertubercular sulcus serves as a passageway for the tendon of the long head of the biceps brachii muscle as it passes through the capsule of the glenohumeral joint. Just distal to the greater and lesser tubercles the shaft of the humerus narrows. This is referred to as the *surgical neck* of the humerus. The surgical neck is a common site of fractures, and fractures here can damage the anterior and posterior humeral circumflex arteries and the axillary nerve (supplying part of the shoulder region) which are adjacent structures.

Humerus: The Shaft

The *shaft* of the humerus runs from the surgical neck to the medial and lateral epicondyles at the distal end of the bone. On the anterior aspect of the superior portion of the shaft is the continuation of the intertubercular sulcus. It is demarcated by *medial* and *lateral lips* that are continuous with the lesser and greater tubercles respectively. The medial lip of the groove has rough markings for the insertion of the teres major muscle, while the lateral lip serves as a site of attachment for the pectoralis major muscle. The latissimus dorsi muscle inserts into the floor of this shallow groove between the sites of attachment of the teres major and pectoralis major. The lateral lip of the intertubercular sulcus continues inferiorly to the anterior part of the *deltoid tuberosity,* which is located on the lateral aspect of the shaft.

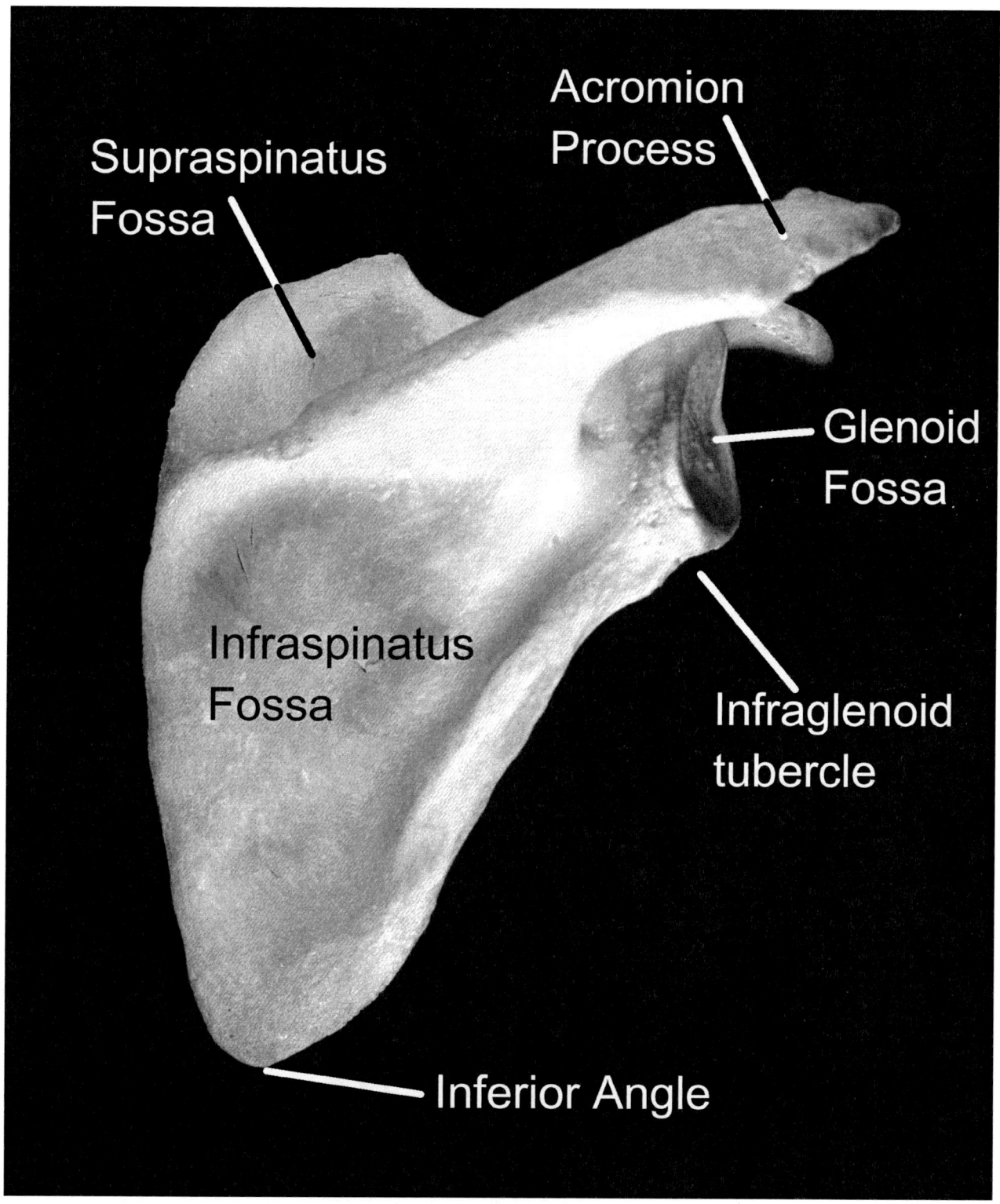

Figure 6-4 *(Continued)* Photograph of the dorsal surface of the right scapula. See accompanying line drawing for identification of additional landmarks.

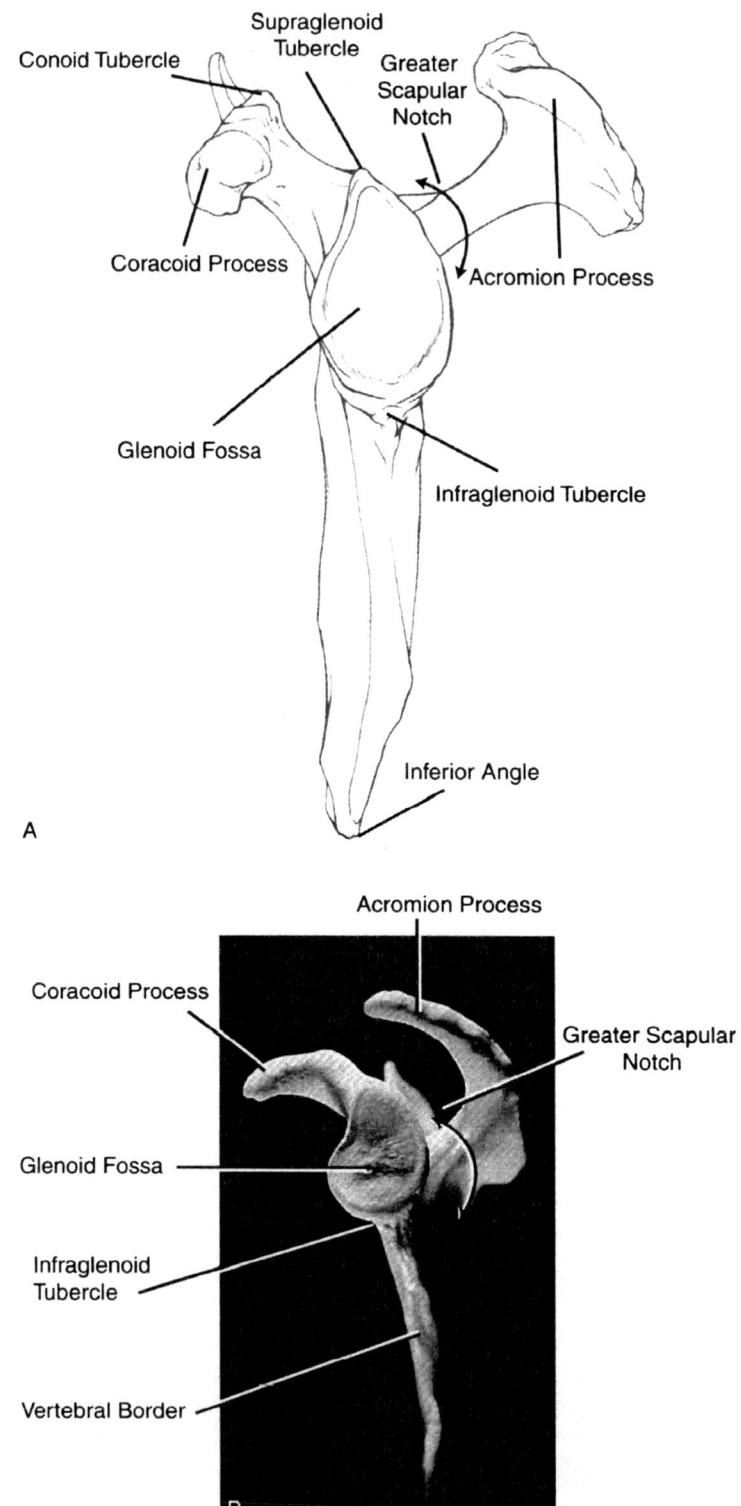

Figure 6-5 (A) The left scapula: axillary border demonstrating features of the lateral angle. (B) Photograph of the left scapula demonstrating the same features.

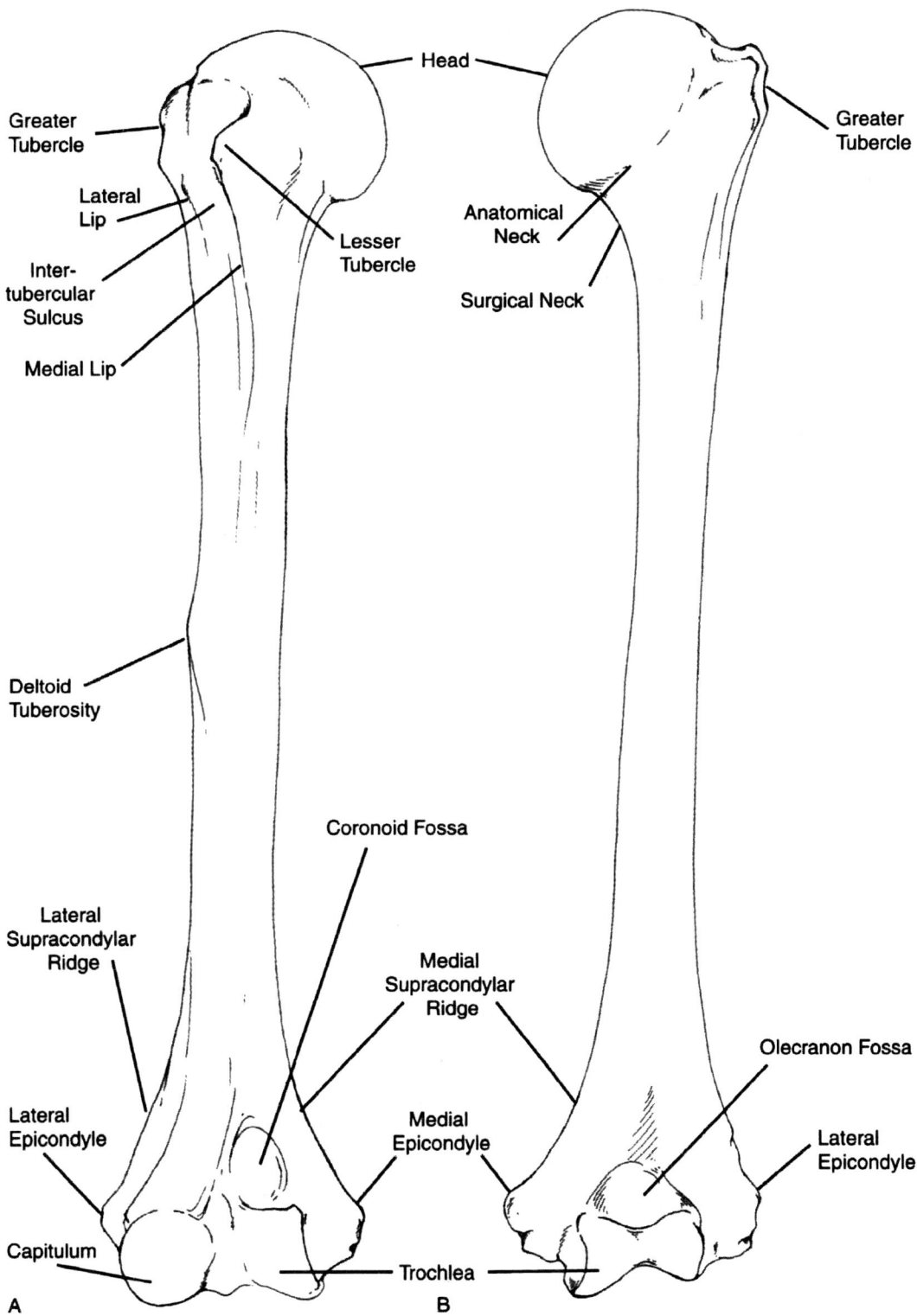

Figure 6-6 The right humerus: (A) anterior and (B) posterior views.

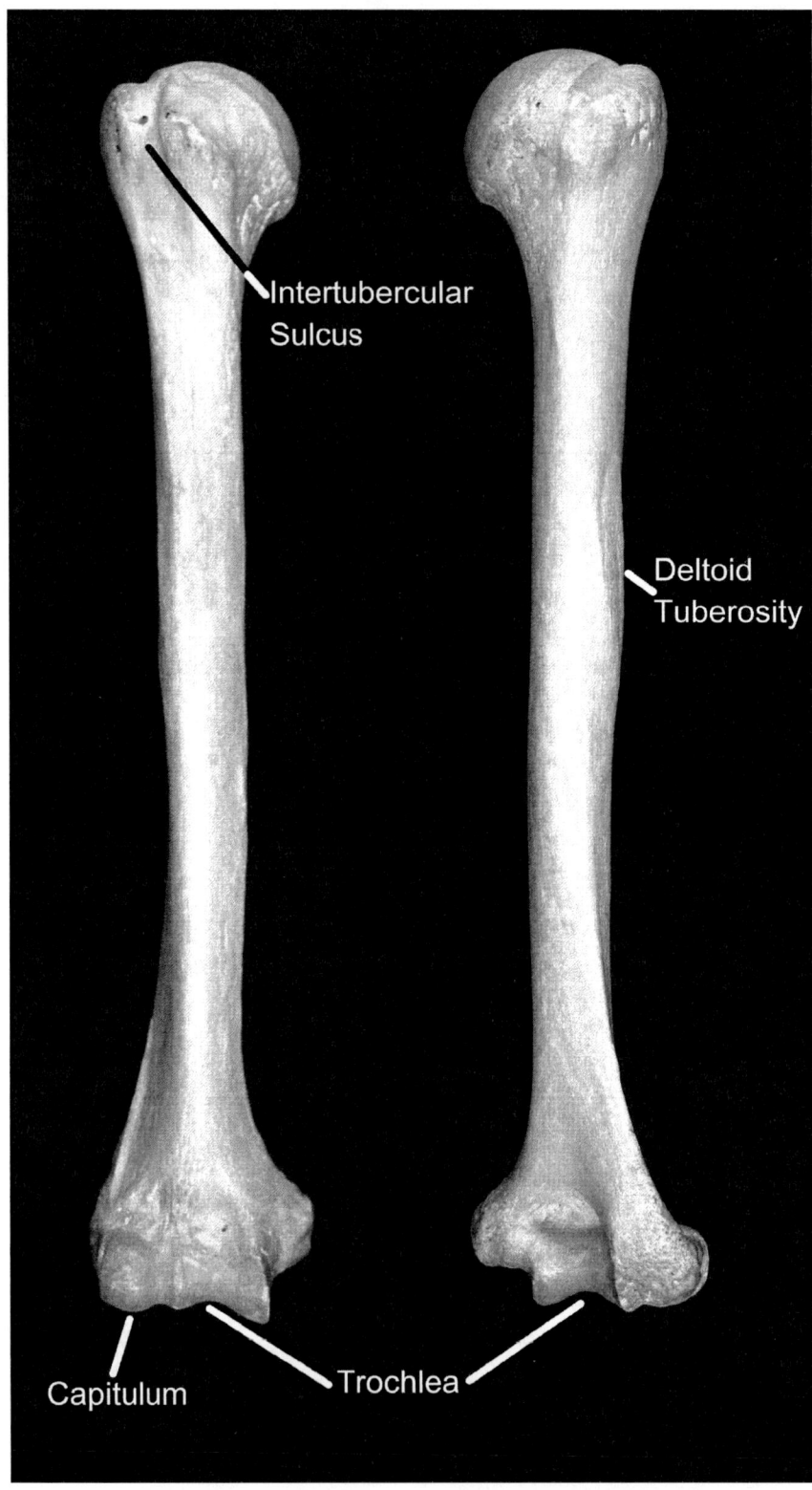

Figure 6-6 *(Continued)* Photographs of anterior and posterior views of the right humerus. See accompanying line drawing for identification of additional landmarks.

The deltoid muscle inserts here. Nearly opposite the deltoid tuberosity, on the medial aspect of the shaft, is the *impression for the insertion of the coracobrachialis muscle*. Generally, this is located slightly more distal on the shaft than is the deltoid tuberosity. The *radial sulcus* is a shallow, ill-defined impression on the posterior aspect of the shaft. It passes from superomedial to inferolateral across the posterosuperior part of the humeral shaft between the sites of origin of the lateral head of the triceps superolaterally and the medial head of the triceps inferomedially. The radial nerve and profunda brachii artery run in this groove under cover of the triceps muscle.

Humerus: The Distal End

The distal end of the humerus is broad mediolaterally and consists of, more superiorly, *medial* and *lateral epicondyles*, and inferiorly articular surfaces for the ulna and radius: the *trochlea* and *capitulum*.

The *medial* and *lateral epicondyles* are raised protuberances located at either side of the distal end of the humerus superior to the distal articular surfaces. Proceeding superiorly from the medial and lateral epicondyles are *medial* and *lateral supracondylar (or epicondylar) ridges*, which are located at the medial and lateral borders of the distal shaft, respectively. The medial epicondyle and medial epicondylar ridge serve as a common origin for the superficial flexor muscles in the forearm, while the lateral epicondyle and epicondylar ridge serve a similar function for the superficial extensor muscles in the forearm. The posterior surface of the medial epicondyle is marked by a groove for the ulnar nerve.

The distal articular surface of the humerus is divided into two portions, the *trochlea* and the *capitulum*. The trochlea is larger, and is located more medially. As its name implies, it is somewhat spool-shaped, and consists of a deep groove defined by two well-marked borders. It receives the trochlear notch of the ulna and forms a hinge joint with it. The trochlea protrudes somewhat anteriorly, and most of its articular surface is located anteriorly and inferiorly.

Superior to the trochlea on the anterior humeral surface is the *coronoid fossa*, which receives the coronoid process of the ulna during flexion. Posterior to the trochlea is the much larger *olecranon fossa*, which receives the end of the olecranon process of the ulna during extension. The bone in the olecranon process is very thin, and is sometimes perforated by a supratrochlear foramen. The *capitulum* is the smaller of the two distal articular surfaces, and is located laterally. The capitulum is smooth and rounded and is only about half the width of the trochlea. It articulates distally with the central fossa of the head of the radius. The elbow joint is formed by articulations between the trochlea and ulna, capitulum and radius, and radial head and ulna.

The Radius

The *radius* (Figure 6-7) consists of a *head, shaft*, and *distal extremity* ending in a *styloid process*. It has four articular surfaces. Its proximal extremity has a surface for articulation with the capitulum of the humerus, and a surface for its proximal articulation with the ulna. Distally the radius has one articular surface for articulation with the carpus, or wrist bones, and one for its distal articulation with the ulna.

Radius: The Proximal End

The proximal end of the radius can be divided into a head, radial neck, and articular circumference. The *radial head* is an expanded disc with a central fossa for articulation with the capitulum of the humerus. Surrounding the head of the radius is a ring of subchondral bone that rests in the radial notch of the ulna, forming the *proximal radioulnar joint*. Distal to the head is a short constricted region of the proximal shaft termed the *radial neck*. Distal to the neck, on the anteromedial part of the shaft, is the *radial* (or *bicipital*) *tuberosity*, which is the site of the insertion of the tendon of the biceps brachii muscle. The bicipital tuberosity has a rough portion for the actual muscle insertion, and a smooth portion for a bursa underlying the biceps tendon.

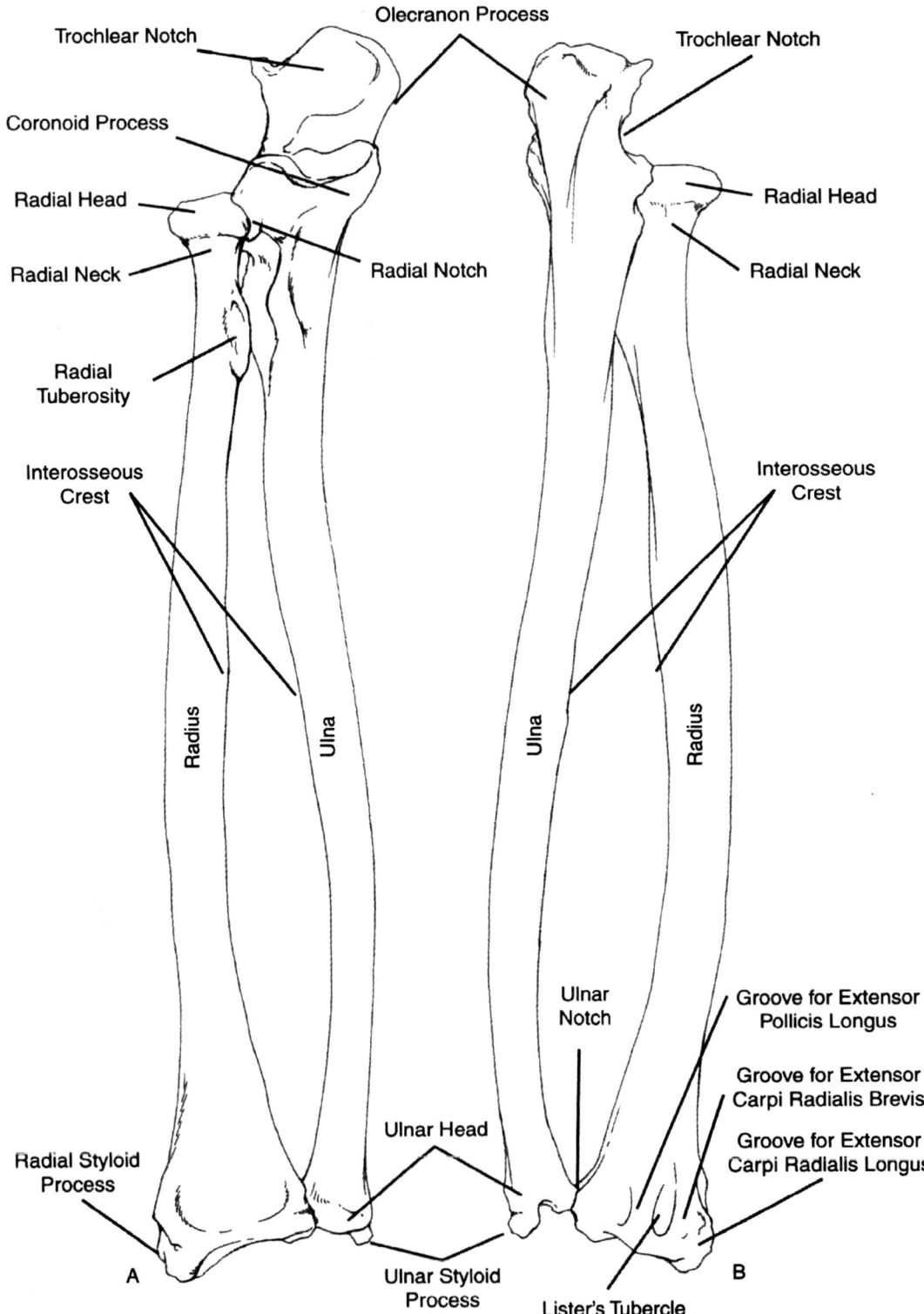

Figure 6-7 The right radius and ulna: (*A*) anterior and (*B*) posterior views.

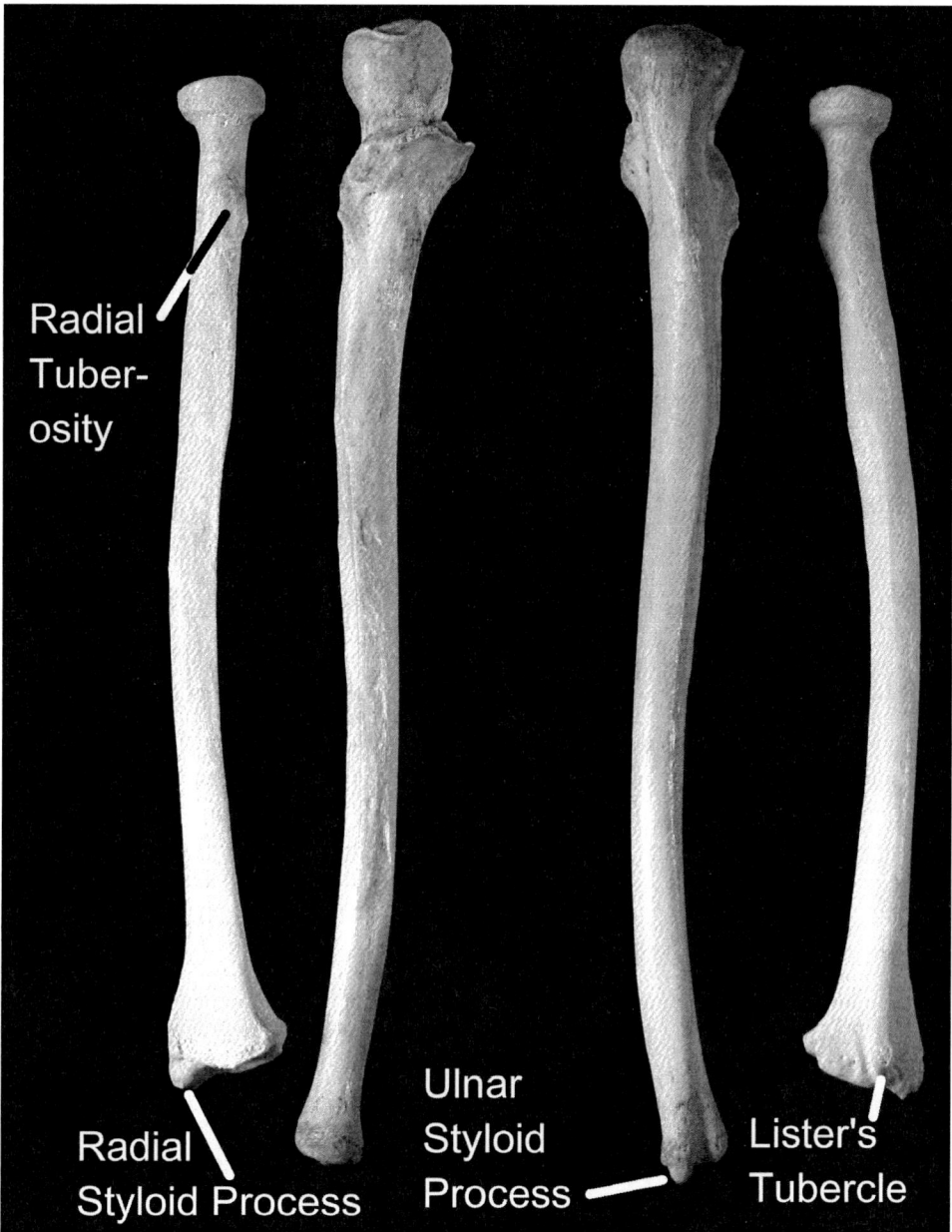

Figure 6-7 *(Continued)* Photographs of anterior and posterior views of the right radius and ulna in their relative positions in the forearm. See accompanying line drawing for identification of additional landmarks.

Radius: The Shaft

The proximal end of the radius attaches by its neck to the *radial shaft*. The shaft is more cylindrical superiorly, but becomes broadened and flattened distally. The shaft has a smooth rounded *lateral border* and a sharp, medial, *interosseous crest*. To this interosseous crest attaches an interosseous membrane, which also attaches to a similar interosseous crest on the lateral aspect of the ulna. The interosseous crest continues for most of the length of the bone, and ends in a small triangular area just superior to the *ulnar notch*. The ulnar notch is located just superior to the distal articular surface of the radius. The ulnar notch articulates with the head of the ulna and together they form the *distal radioulnar joint*. Approximately one-third of the way from the proximal end of the bone on its anterior surface is the *nutrient foramen*, which is directed proximally. Running inferolaterally from the bicipital tuberosity across the anterior aspect of the shaft is the *anterior oblique line*. In a similar position on the posterior aspect of the shaft is the *posterior oblique line*. The flexor digitorum superficialis muscle arises from the anterior oblique line. Distal to the anterior oblique line is a hollow for the origin of the flexor pollicis longus muscle. The posterior oblique line divides the area of attachment of the supinator muscle superiorly from the area of attachment of the abductor pollicis longus muscle inferiorly.

Radius: Distal End

At the *distal end* of the radius is a large articular surface for the *carpus*, or wrist bones, a lateral, distally projecting *styloid process*, and on its posterior surface a series of ridges and grooves for the passage of tendons of the extensor muscles of the hand from the posterior compartment of the forearm to their distal sites of attachment.

The *carpal articular surface* of the radius is located at the distal end of the bone. It may be divided by a slight ridge or other surface irregularities into two portions. The medial portion is for articulation with the *lunate bone* while the lateral portion is for the *scaphoid bone*. The articular surface of the ulnar notch is continuous with the carpal articular surface, the two surfaces meeting at an angle of about 90 degrees. The lateral aspect of the distal end comes to a projecting distal point, the *styloid process*. The radial collateral ligament of the wrist joint attaches to this process. The styloid process is also the distal attachment site of the brachioradialis muscle. The posterior aspect of the styloid process is marked by two slight grooves. The more lateral is the *groove for the extensor carpi radialis longus*. The more medial is the *groove for the extensor carpi radialis brevis*.

The central region of the posterior aspect of the distal end of the radius is characterized by a raised area, known as the *extensor retinaculum tubercle* or *Lister's Tubercle*. Medial to this is the prominent *groove for the extensor pollicis longus*. Medial to this groove is a second broad *groove for the tendons of the extensor digitorum and the extensor indicis*. Most medial is a third *groove for the tendon of the extensor digiti minimi*. This groove is completed by articulation with the head of the ulna. Lateral to Lister's Tubercle are the posterior surface of the styloid process and the *grooves for the extensor carpi radialis longus and brevis tendons*.

The Ulna

The principal parts of the *ulna* (see Figure 6-7) are the proximal *olecranon process and coronoid process*, which contain the trochlear notch for articulation with the humerus, the *shaft*, and the distal *ulnar head* and *ulnar styloid process*.

Ulna: The Proximal End

The proximal end of the ulna is composed of two prominent projections, the *olecranon process* and the *coronoid process*, and two concave articular facets, the *trochlear notch* and the *radial notch*. The *olecranon process* is the proximal extension of the shaft of the ulna and it forms the point of the elbow. Its posterior surface receives the large triceps tendon. It has a rough, more posterior, portion for the actual tendon insertion, and a smoother, more anterior portion for a subtendonous bursa. The *coronoid process* projects from the proximal and anterior surface of the shaft of the ulna. Its proximal surface is smooth and concave and forms the distal part of the trochlear notch of the ulna. At the junction of the coronoid process and the shaft of the ulna is the *ulnar tuberosity*, to which attaches the tendon of the brachialis muscle. On the lateral surface of the coronoid process is the *radial notch* of the ulna. This notch receives the circumferential articular surface of the radial head, and together they form the *proximal radioulnar joint*. The *trochlear notch* (or semilunar notch) is a large depression formed by the coronoid and olecranon processes. It articulates with the trochlea of the humerus.

Ulna: The Shaft

The *shaft* or body of the ulna is three-sided. The most prominent feature is its *interosseous crest* along its lateral margin. An interosseous membrane attaches here and to the interosseous crest of the radius. The nutrient foramen of the ulna is usually located on the anterior surface of the proximal portion of the shaft. The *oblique line* separates the origin of the flexor digitorum profundus muscle from the origin of the pronator quadratus muscle. It is usually located in the distal 1/5 of the anterior surface of the shaft and runs distomedially. The posterior surface of the ulna is divided along most of its length by a low ridge. The more medial area is occupied by the origin of the extensor carpi ulnaris. From proximal to distal, the more lateral area is occupied by the origins of three muscles: abductor pollicis longus, extensor pollicis longus, and extensor indicis. Each occupies about 1/3 of this area, and you may be able to detect slight separations of their origins.

Ulna: The Distal End

The distal end of the ulna is its *head* (just the opposite of the radius). The head is composed of subchondral bone and is separated from the *ulnar styloid process* by a deep groove. To this groove attaches an

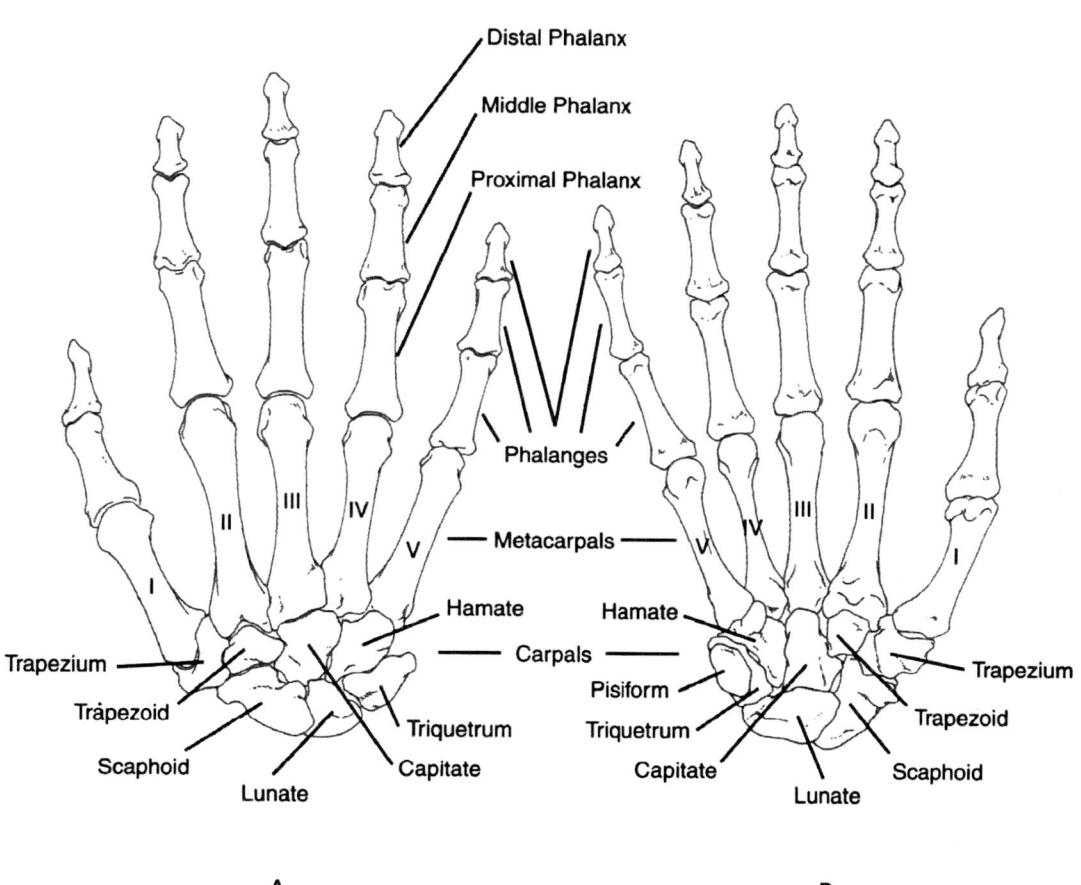

Figure 6-8 The skeleton of the right hand: (*A*) posterior or dorsal and (*B*) anterior or palmar views.

intraarticular disc that separates the ulnar head from the carpal bones. The distal surface of the ulnar head is in contact with the intraarticular disc, while its *articular circumference* articulates with the ulnar notch of the radius, at the *distal radioulnar joint*. On the posterior surface of the distal end of the ulna is a groove that becomes continuous with the groove between the head and the ulnar styloid process. This is the *groove for the tendon of the extensor carpi ulnaris muscle*.

The Hand

The skeleton of the *hand* (Figure 6-8) is composed of 27 bones. These bones consist of three groups: the *carpal bones*; the *metacarpal bones*; and the *phalanges*. The *carpal bones* are most proximal, form the wrist, and articulate proximally with the radius. The *metacarpals* are a set of five small long bones. They are numbered one through five from the radial side. Their bases articulate with the distal row of carpal bones, and their heads articulate with the proximal phalanges of each of the digits. Each digit has three *phalanges*, a proximal, a middle, and a distal, with the exception of the thumb or pollex. The thumb has only two phalanges, termed the proximal and the distal.

The Hand: The Carpus

The *carpus* (*wrist*) consists of eight small, irregularly shaped bones (Figure 6-9; refer also to figure 6-8 throughout the discussion of the carpal bones) arranged in two rows, a proximal and a distal. They are closely applied to one another and are held firmly in their rows by strong ligaments. As a whole they form a strong fibrous and bony mass that is concave

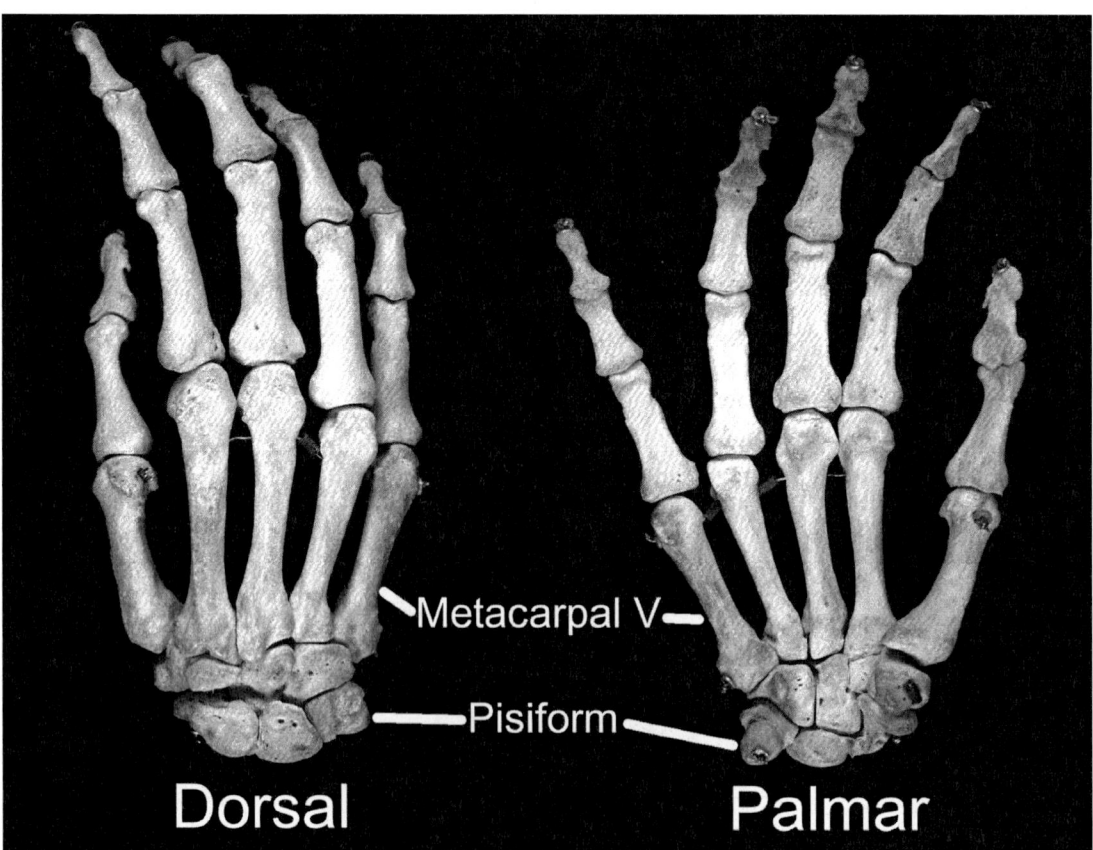

Figure 6-8 *(Continued)* Photographs of dorsal (left) and palmar views of the right hand. See accompanying line drawing for identification of additional landmarks.

on the palmar side and convex dorsally. Each row consists of four bones. The *proximal row* consists of, from lateral to medial, (1) the *scaphoid*, (2) the *lunate*, (3) the *triquetrum* (or *triangular*), and (4) the *pisiform*. The *distal row* consists, from lateral to medial, of (5) the *trapezium*, (6) the *trapezoid*, (7) the *capitate*, and (8) the *hamate*.

The *scaphoid* articulates medially with the lunate, proximally with the radius, and distally with the trapezium, trapezoid, and capitate. It is the largest bone of the proximal row. On the lateral part of the palmar surface of the scaphoid is its *tubercle*, to which attaches the flexor retinaculum. Some older texts may refer to the scaphoid as the *navicular*.

The *lunate* articulates with the scaphoid laterally and the triquetral medially, superiorly with the radius and the intraarticular disc, and inferiorly with the capitate and a small area of the hamate. The lunate may be distinguished by its deep concavity and crescentic outline. Its *proximal surface* is convex and smooth and articulates with the radius. Its *distal surface* is deeply concave and articulates with the head of the capitate.

The *triquetral* articulates with the lunate laterally and the pisiform anteriorly. Inferiorly it articulates with the hamate, and superiorly with the intraarticular disc at the distal ulna. It is pyramidal in shape, and may be distinguished by an isolated oval facet for articulation with the pisiform bone.

The *pisiform* is a very small, more-or-less "pea" shaped bone and articulates posteriorly with the triquetral. It "sits" on the palmar surface of the triquetral. It is easily distinguished by its small size, oval shape, and its single articular facet. The flexor carpi ulnaris, abductor digiti minimi, and flexor retinaculum all attach to this small bone.

The *trapezium* articulates medially with the trapezoid, superiorly with the scaphoid, inferiorly with the first metacarpal, and with a small area of the base of the second metacarpal. It may be distinguished by a deep groove on its palmar surface, which transmits the tendon of the flexor carpi radialis muscle, and its saddle-shaped articular surface for the base of the first metacarpal.

The *trapezoid* articulates laterally with the trapezium and medially with the capitate. Superiorly it articulates with the scaphoid and inferiorly with the second metacarpal. The trapezoid is the smallest bone in the distal row. It can be identified by its wedge shape. The broad end of the wedge is its dorsal surface, while the narrow end of the wedge is its palmar surface.

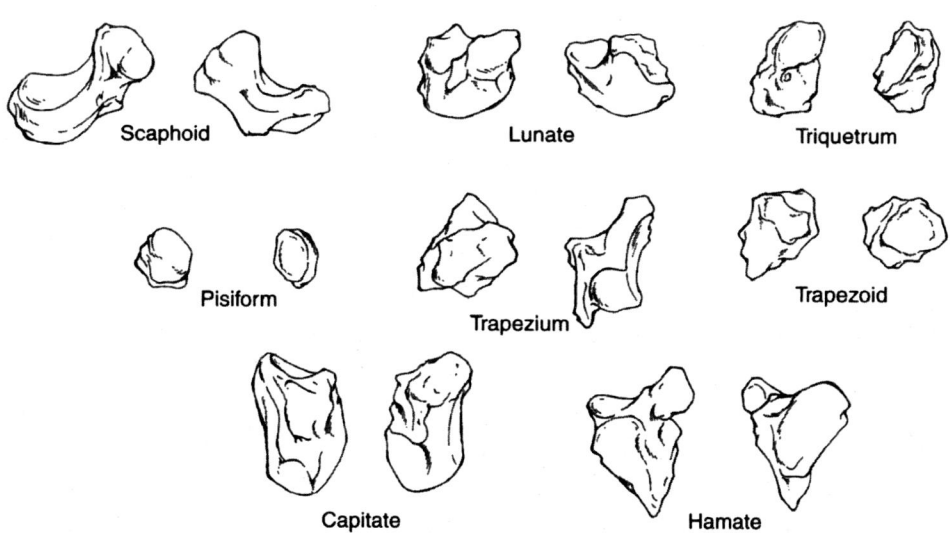

Figure 6-9 The individual carpal bones of the right hand.

The *capitate* articulates laterally with the trapezoid and medially with the hamate. Superiorly it articulates with the scaphoid and lunate. Inferiorly it articulates primarily with the third metacarpal, but also with the second and fourth. It is easily identified as the largest of the carpal bones and occupies the center of the wrist. It has a rounded portion or *head* that is received into the concavity formed by the scaphoid and lunate bones. Distal to the head is a constricted *neck* that connects the head to the capitate *body*.

The *hamate* articulates laterally with the capitate, superiorly with the triquetral and lunate, and inferiorly with the fourth and fifth metacarpals. It is readily identified by the hook-like process which projects from its palmar surface, the *hamulus*. To the apex of the hamulus attach the flexor retinaculum and the flexor carpi ulnaris. It is one of four eminences on the palmar aspect of the carpus to which attaches the flexor retinaculum. The other three attachments of the flexor retinaculum are the pisiform medially and the oblique ridge of the trapezium and the tubercle of the scaphoid laterally.

The Hand: The Metacarpus

The *metacarpus* is formed by the five *metacarpal bones* (see Figure 6-8). Each of the five metacarpal bones has a *shaft*, a proximal *base* and a distal *head*. With the exception of the first metacarpal (for the thumb) each base has articular facets on its sides for articulation with the bases of adjacent metacarpals. On the proximal surface of the base of each is an articular facet for the particular carpal bone(s) with which that metacarpal articulates.

The Hand: The Phalanges

Each *proximal phalanx* of the phalanges (see Figure 6-8) possesses an oval, concave facet for articulation with the rounded articular surface of the head of the metacarpals. *Middle phalanges* all possess double facets for articulation with the double condyles at the distal ends of the proximal phalanges. The distal ends of the middle phalanges similarly possess these double condyles and articulate with the bases of the distal phalanges. The medial and lateral sides of the palmar (volar) surface of each proximal and middle phalanx is marked by a longitudinal ridge to which the flexor retinacula attach. The *distal phalanges* possess double concave facets at their proximal ends for articulation with the condyles of the middle phalanges. The distal tips of the distal phalanges possess flattened expansions termed the *ungual tuberosities* that support the fibro-fatty pulp of the finger tip. The *pollex* (thumb) does not possess a middle phalanx, but its distal and proximal phalanges resemble those of the other four digits. The palmar surfaces of the shafts of the phalanges of the hand are flattened. The palmar (volar) surface of the terminal phalanx of the thumb is normally marked by a deep pit or insertion of the flexor pollicis longus.

Articulations of The Upper Limb

The Sternoclavicular Joint

The *sternoclavicular joints* (see Figure 6-1) are located between the clavicles, the clavicular notches on the superior border of the manubrium, and the costal cartilages of the 1st ribs. At each joint there is an *articular disc* which divides it into two synovial cavities. These are *sellar synovial joints*. Each joint is surrounded by a fibrous *articular capsule*. In addition, the joints are strengthened by several ligaments, which include:

- *The Anterior Sternoclavicular Ligament*
- *The Posterior Sternoclavicular Ligament*
- *The Interclavicular Ligament*
- *The Costoclavicular Ligament*

The *anterior sternoclavicular ligament* attaches to the anterior surface of the clavicle, manubrium, and first costal cartilage. The *posterior sternoclavicular ligament* covers the posterior surface of the joint. It runs from the posterior surface of the clavicle's sternal end to the posterior aspect of the upper manubrium. The *interclavicular ligament* unites the superior aspects of the sternal ends of both clavicles with one another. Some fibers attach to the superior margin of the manubrium. The *costoclavicular ligament* unites the upper part of the first rib and costal cartilage with the inferior medial surface of the clavicle.

Movement at the sternoclavicular joint occurs in conjunction with movements of the pectoral girdle and upper limbs. The joint permits motion anteriorly, posteriorly, and vertically and allows the lateral end of the clavicle to be raised for full elevation of the upper limb (Gray, 1995).

The Acromioclavicular Joint

The acromioclavicular joint (Figure 6-10) is a plane synovial joint. There may be, however, a slight concave curvature on either surface with a matching slightly convex curvature on the opposite surface. The clavicle has a narrow, oval facet that faces inferolaterally and articulates with the corresponding facet on the medial border of the acromion process.

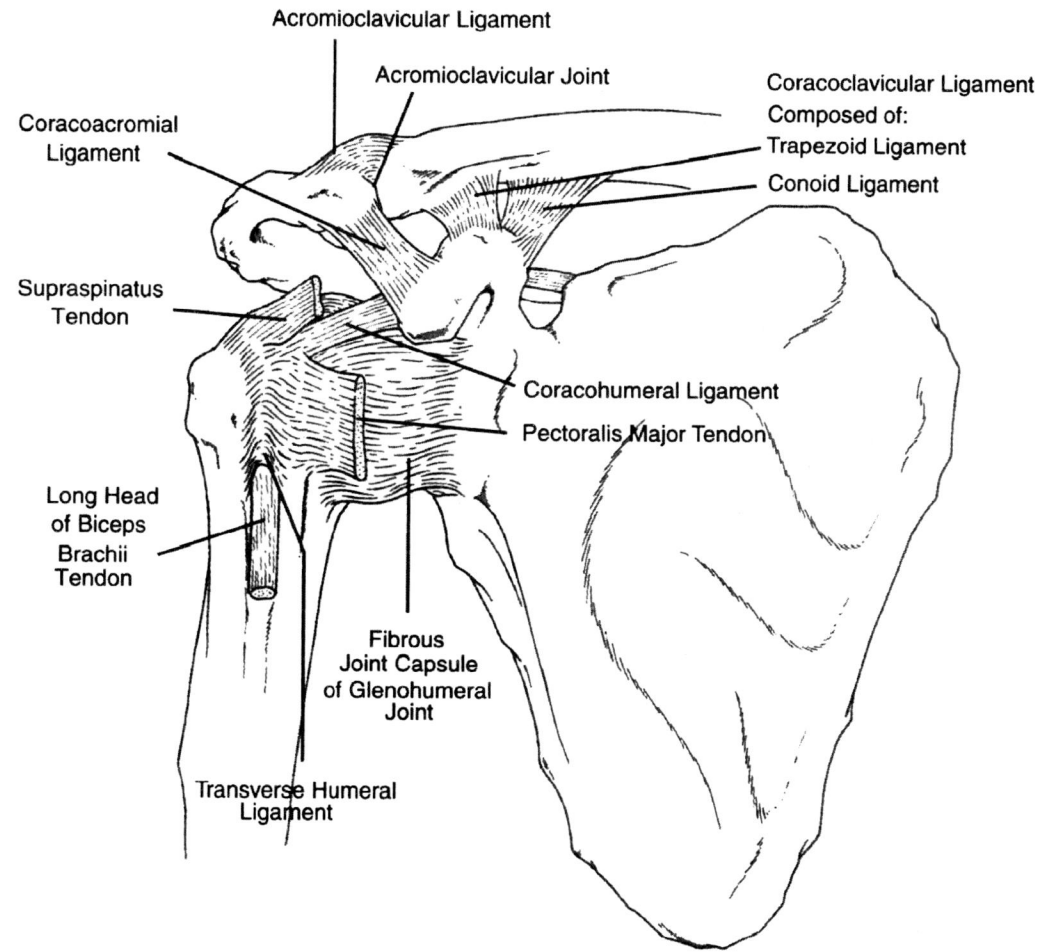

Figure 6-10 The acromioclavicular and glenohumeral joints, right side, anterior view.

The joint is reinforced by the *fibrous joint capsule* as well as two ligaments:

- acromioclavicular ligament.

- *coracoclavicular ligament*, composed of :

    - conoid ligament

    - *trapezoid ligament.*

The fibrous joint capsule completely surrounds the joint and is reinforced superiorly by the quadrilaterally shaped *acromioclavicular ligament*. It runs from the lateral end of the superior aspect of the clavicle to the adjoining surface of the acromion process. Its fibers blend with the aponeuroses of the trapezius and deltoid muscles. The inferior aspect of the clavicle is marked by the sites of attachment for the *coracoclavicular ligament*, which include the *conoid tubercle* and the *trapezoid line*.

The conoid tubercle and trapezoid line provide attachment for the *conoid and trapezoid ligaments*, respectively. These two ligaments constitute the *coracoclavicular ligament*, which binds the clavicle to the scapula. It is separate from the acromioclavicular joint, but is an important extrinsic ligament of the joint. There is frequently an incomplete articular disc in the superior part of the acromioclavicular joint cavity separating the two articular surfaces from one another. On rare occasions the disc is extensive enough to completely separate the joint cavity into two distinct parts.

Movements of the acromioclavicular joint are similar to those of the sternoclavicular joint. Each allows about 30 degrees of axial rotation of the clavicle, and therefore between them permit about 60 degrees of rotation. Angulation between the scapula and clavicle can occur in any direction.

Glenohumeral Joint

The glenohumeral joint (see Figure 6-10) is a spheroidal or ball-and-socket joint between the oval-shaped humeral head and the glenoid fossa of the scapula. The humeral head has a much larger articular surface than the glenoid fossa of the scapula. Because of the relatively small bony contact between the head of the humerus and the glenoid fossa, the head of the humerus is held in contact with the glenoid fossa largely by the activities of the muscles surrounding the joint.

The fibrous joint capsule that surrounds the glenohumeral joint is very loose. It is weakest posteriorly, and allows considerable displacement of the head of the humerus from the glenoid in the absence of the surrounding muscles and ligaments. The advantage of this arrangement is in the tremendous range of motion permitted by the glenohumeral joint. A disadvantage is the head of the humerus is relatively easily displaced (Matsen, 1980).

The articular surfaces of the glenoid and humeral head, as in all synovial joints, are covered with hyaline articular cartilage. The humeral articular surface is an oblate spheroid with its long axis located superoinferiorly. The glenoid fossa of the scapula is likewise covered with hyaline cartilage, and is oval and concave to receive the head of the humerus. The area of contact between the two is made somewhat greater by the *glenoid labrum* (see below).

In addition to the head of the humerus and glenoid fossa, the glenohumeral joint includes a number of other structures.

The Coracoacromial Ligament

The *coracoacromial ligament* extends superiorly over the top of the humeral head and reinforces the joint superiorly. The coracoacromial ligament attaches to the coracoid process more anteriorly and medially and to the acromion process more posteriorly and laterally.

The Coracoacromial Arch

The *coracoacromial arch* is an osteoligamentous arch formed by the acromion process, the coracoacromial ligament, and the coracoid process. Beneath the arch passes the tendon of the supraspinatus muscle which arises from the supraspinatus fossa of the scapula and inserts into the greater tubercle of the humerus. A bursa lies between the coracoacromial ligament and this tendon. The coracoacromial arch protects the glenohumeral joint from superior displacement, and the muscle adds stability to the joint.

The Joint Capsule.

The *joint capsule* of the glenohumeral joint completely surrounds the joint. It attaches to the circumference of the glenoid fossa medial to, and to the margins of, the *glenoid labrum* at the neck of the scapula. Laterally it attaches to the margins of the humeral head following the course of the anatomical neck of the humerus, and inferiorly it encroaches on the surgical neck. In the absence of surrounding musculature, the head of the humerus can be separated from the glenoid fossa by up to 2.5 cm. The looseness of the articular capsule allows great freedom of motion of the humeral head upon the glenoid fossa. The surrounding musculature reinforces the joint and keeps the humeral head in contact with the glenoid fossa.

The Glenohumeral ligaments

In places, the fibrous joint capsule is thickened to form intrinsic ligaments of the joint. These are the *glenohumeral ligaments*. These ligaments are three thickenings of the anterior aspect of the joint capsule, and are termed the *superior, middle,* and *inferior glenohumeral ligaments*. These ligaments are only visible from the interior of the joint cavity. They are best seen when the joint capsule is opened posteriorly and the head of the humerus is removed with a saw or chisel. The *superior glenohumeral ligament* runs from the top of the glenoid cavity to the greater tubercle of the humerus. The *middle glenohumeral ligament* goes from the anterior margin of the glenoid fossa to the middle of the lesser tubercle of the humerus. The *inferior glenohumeral ligament* runs from the inferior margin of the glenoid cavity to the anatomical neck of the humerus. Often, it is difficult to separate the three individual ligaments.

The Coracohumeral ligament.

The *coracohumeral ligament* is a broad band of tissue strengthening the upper part of the joint capsule. It runs from the root of the coracoid process to the anterior margin of the greater tubercle of the humerus. The tendon of the long head of the biceps brachii separates this ligament from the superior glenohumeral ligament.

The Glenoid Labrum

The *glenoid labrum* is a fibrocartilagenous rim that surrounds the margins of the glenoid fossa and serves to enlarge and deepen the articular surface for the reception of the head of the humerus. It is triangular in cross section with a thin, sharp free border. The fibers of the tendon of the long head of the biceps blend into the labrum as that tendon inserts into the supraglenoid tubercle above the glenoid fossa.

The Transverse Humeral Ligament.

The *transverse humeral ligament* is a bundle of short fibers that run from the greater to the lesser tubercles of the humerus running from the lateral to the medial border of the bicipital groove. In doing so, they close the bicipital groove and help form a tunnel for the passage of the tendon of the long head of the biceps brachii as it enters the glenohumeral joint to attach to the supraglenoid tubercle.

Tendon of Long Head of The Biceps Brachii Muscle

The *tendon of the long head of the biceps brachii* is structurally important to the glenohumeral joint. This tendon of muscle origin is unusual in that it passes *through* the joint capsule of the glenohumeral joint. (Tendons of muscles are generally located outside joint capsules rather than within them.) The tendon originates from the supraglenoid tubercle of the scapula, passes through the top of the glenohumeral joint and over the head of the humerus and passes out of the joint capsule by passing under the transverse humeral ligament and into the bicipital groove. In addition to providing attachment to the muscle, this tendon helps maintain the humeral head in the glenoid fossa and limits the upward displacement of the humeral head.

The Scapulothoracic Joint

The *scapulothoracic joint* (see Fig. 6-1) is unusual in that it is neither a synarthrosis (solid connective tissue joint) nor a diarthrosis (cavitated, or synovial, joint) but rather a joint between the muscle-covered scapula and the muscle-covered thoracic wall. It is therefore sometimes referred to as a synsarcosis. The only bony articulations between the scapula and the rest of the skeleton are at the acromioclavicular and sternoclavicular joints. Between the subscapularis muscle, which covers the ventral or costal surface of the scapula, and the ribs and intercostal muscles lies the serratus anterior muscle, which attaches the vertebral border of the scapula to the first eight or nine ribs.

Ranges of Motion of the Shoulder Joint

As with the hip joint (the other spheroidal or ball-and-socket joint in the human body), the shoulder joint has a wide range of motion. Because there is very little bony constraint, the upper limb can move through a very wide range of motion at the shoulder. Abduction and adduction of the upper limb through the full range of 180 degrees of motion are due to movement at both the glenohumeral joint and the scapulothoracic joint. Approximately 120 degrees of this motion are due to motion at the glenohumeral joint, while the remaining 60 degrees or so of this motion is due to motion at the scapulothoracic joint. However, the great freedom of flexion, extension, abduction, and adduction permit the upper limb to be fully circumducted.

The scapula is acted on by a number of muscles that serve to direct it along the thoracic wall. The scapula may be pulled forward *(protraction)* or pulled backward *(retraction)*. The scapula can also be rotated upward so that the glenoid fossa faces more superiorly or rotated downward so that the glenoid fossa faces more inferiorly. In movements of the upper limb, motion at the scapulothoracic joint is almost always involved in addition to motion at the glenohumeral joint. As noted above, approximately one third of abduction and adduction of the upper limb occurs at the scapulothoracic joint. As the scapula moves, there is also motion at the sternoclavicular and acromioclavicular joints.

Disorders of the Shoulder Joint

Because the shoulder joint is relatively unconstrained, it is easily subject to dislocation. Since the shoulder joint capsule is weakest inferiorly, the most frequent pattern of dislocation is the *subglenoid position*, in which the head of the humerus is displaced inferiorly below the glenoid fossa. This is the region in which the joint capsule has the least reinforcement. The long head of the triceps brachii muscle attaches to the infraglenoid tubercle and usually prevents the head of the humerus from being pushed posteriorly. As a consequence, the head of the humerus is frequently secondarily pushed into a *subcoracoid* or *subclavicular position*. (For further discussion of clinical problems in the shoulder joint See Edwardson (1995), Frankel and Nordin (1980) and Nordin and Frankel (2001) for a fuller discussion of joint mechanics.)

In *ankylosis*, the bony union of the glenohumeral joint or the acromioclavicular joint, some motion of the upper limb is still possible because of motion at the scapulothoracic joint.

The Elbow Joint

The elbow is comprised of a complex of joints (Figure 6-11), including the *humeroulnar, humeroradial,* and *proximal radioulnar* joints.

The Humeroulnar Joint

The *humeroulnar joint* is a ginglymus or hinge joint. It is located between the trochlea of the humerus and the

Bones of the Postcranial Skeleton: Upper Limb

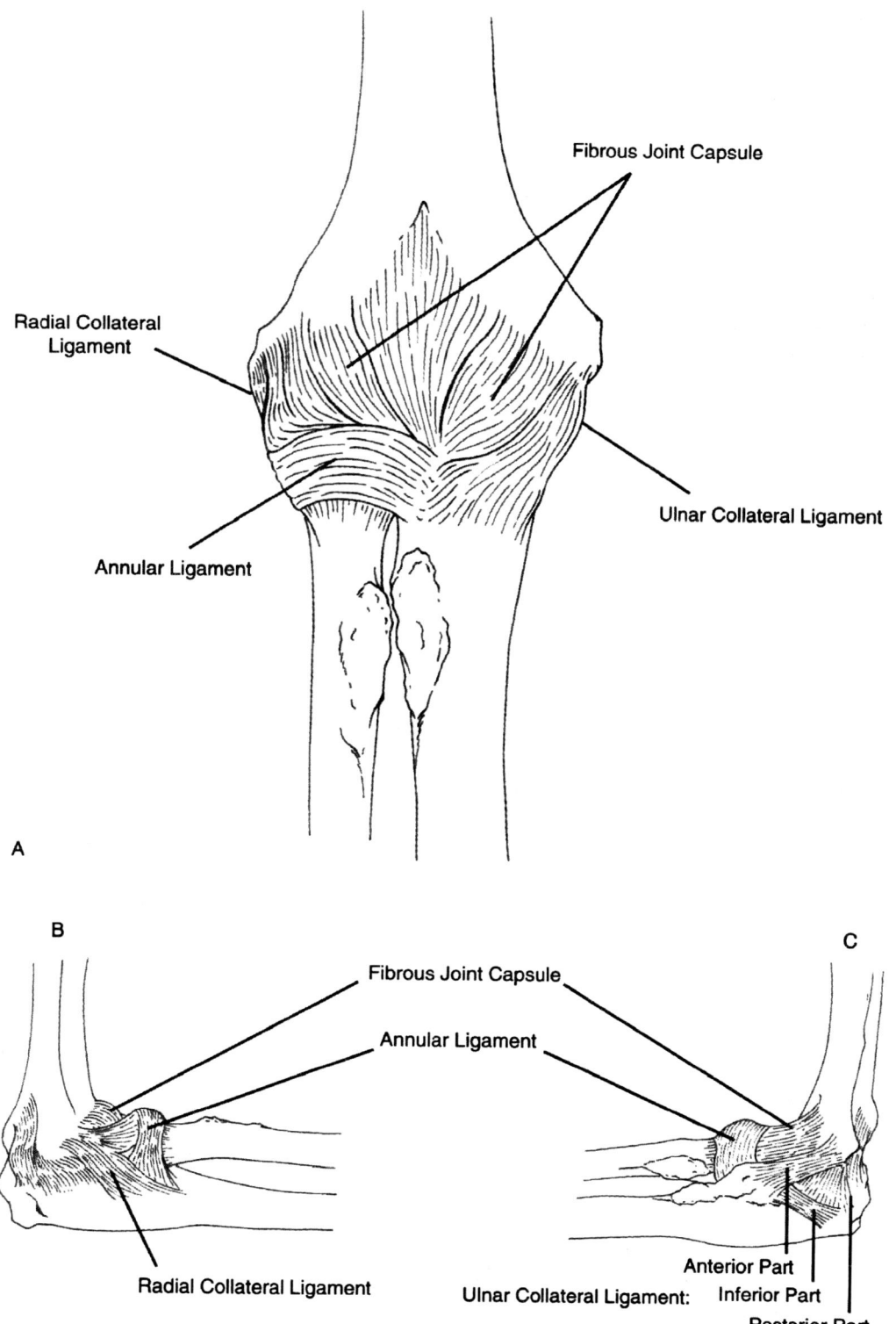

Figure 6-11 The right elbow joint: (A) anterior, (B) lateral, and (C) medial views.

trochlear notch, olecranon process, and coronoid process of the ulna. The trochlear notch of the ulna receives the trochlea of the humerus. The form of these joint surfaces and their investing musculature (see above) essentially permit only flexion and extension. Hyperextension is prevented by the biceps brachii and brachialis muscles and is limited by the olecranon process.

The Humeroradial Joint

The *humeroradial joint* is located between the radial head and the capitulum of the distal end of the humerus. The joint combines elements of a ginglymus, a trochoid or pivot joint, and a gliding joint. In fact, the humeroradial articulation resembles a ball-and-socket joint, but other associated structures (*viz.*, the humeroulnar joint, annular ligament and associated ligaments) effectively restrict its motion in abduction and adduction. However, the architecture of the joint permits a broad range of flexion/extension and rotation. The capitulum of the humerus is rounded and convex, while the superior surface of the radial head is round and concave. As the elbow flexes and extends, the radial head slides alternately anteriorly and posteriorly over the capitulum. In pronation and supination of the forearm, the shaft and head of the radius rotate. Thus the radial head is rotating about its center relative to the surface of the capitulum.

The Proximal Radioulnar Joint

The *proximal radioulnar joint* is a trochoid (pivot) synovial joint. Around the head of the radius is a flat, circular surface that articulates with the radial notch of the lateral aspect of the proximal ulna. The *annular ligament* holds the head of the radius in contact with the radial notch of the ulna and guides its motion. This ligament attaches to the anterior and posterior borders of the radial notch, and with the notch it completes a circular collar in which the head of the radius can rotate.

The Ulnar Collateral Ligament

The *ulnar (medial) collateral ligament* is triangular in shape, with thickened anterior, posterior and inferior borders. The anterior part arises off the medial epicondyle of the humerus and attaches to the coronoid process of the ulna. The posterior part runs from the back of the medial epicondyle to the medial border of the olecranon process. The inferior part, which is much more weakly developed than the anterior and posterior parts, attaches the distal portions of the anterior and posterior parts. This ligament reinforces the medial aspect of the elbow joint. A large portion of it becomes taut in flexion.

The Radial Collateral Ligament

The *radial (lateral) collateral ligament* arises low on the lateral epicondyle of the humerus and attaches inferiorly to the annular ligament and to the supinator crest of the ulna. It blends with the attachments of the supinator muscle and the extensor carpi radialis brevis muscle. It reinforces the lateral aspect of the elbow joint and tightens in flexion.

The Joint Capsule

The *fibrous joint capsule* attaches on the humerus anteriorly above the coronoid fossa and blends with the radial and ulnar collateral ligaments. Anteriorly it shares some fibers with the brachialis muscle. Posteriorly it attaches behind the trochlea and capitulum and extends down to attach to the medial and lateral margins of the olecranon process. The synovial membrane lines the fibrous capsule and is thrown into a fold that projects from the posterior part of the capsule to partly separate the humeroradial from the humeroulnar joint.

Ranges of Motion of the Elbow Joint

Because of the shapes of the olecranon fossa and the humeral trochlea, flexion and extension of the humeroulnar joint are accompanied by a small degree of torsion or rotation of the ulna. The ulna is *very slightly* pronated in extension and *very slightly* supinated in flexion. In the anatomical position, when the forearm is extended and supinated, the long axis of the ulnar shaft and the long axis of the humeral shaft meet at an obtuse angle due to the obliquity of the trochlea with relation to the humeral shaft. This is called the *carrying angle*. This angle deflects the hand and distal ends of the radius and ulna laterally away from the body. From extreme flexion (hyperflexion) to maximum extension (hyperextension), the elbow joint is capable of motion through about 140 degrees.

Disorders of the Elbow Joint

Lateral epicondylitis ("tennis elbow") and medial epicondylitis ("golfer's elbow") are two of the most common problems encountered at the elbow joint. *Lateral epicondylitis* is an inflammation of the common extensor tendon, which attaches to the lateral epicondyle. This is results from repetitive motion injuries of the wrist extensors that attach here, particularly the extensor carpi radialis brevis muscle. *Medial epicondylitis* is an inflammatory response to repetitive motion injuries of the wrist flexors.

The elbow may be displaced posteriorly by a dislocation of the ulna, and is complicated by fractures of the coronoid process of the ulna. The medial epicondyle of the humerus is frequently separated from the remainder of the bone in lateral dislocations, due to the presence of the very strong ulnar collateral ligament.

The radial head is frequently dislocated anteriorly with rupture of the annular ligament, particularly in falls when the hand is used to break a fall with the forearm extended. This is most common in children. In children, the head of the radius may be displaced inferiorly through the annular ligament. In this situation, the annular ligament becomes entrapped between the capitulum and the radial head, and the forearm becomes fixed in semiflexion, midway between supination and pronation. The small radial head of children predisposes them to this disorder.

Tears and dislocations of bony articulations in children are also attributable to inherent weakness of the epiphyses, particularly at the medial epicondyle of the humerus and the head of the radius. (See Edwardson (1995) for a further discussion of clinical problems in the elbow joint and Frankel and Nordin (1980) and Nordin and Frankel (2001) for a fuller discussion of joint mechanics.)

Radiounlar Articulations

There are three joints between the radius and ulna: the *proximal radioulnar joint*, the *middle radioulnar joint (radioulnar syndesmosis)* and the *distal radioulnar joint* (Figure 6-12). The proximal radioulnar joint has been discussed with the elbow joint.

Radioulnar syndesmosis

The shafts of the radius and ulna are joined to one another throughout their lengths by two interosseous ligaments: the interosseous membrane and the oblique cord, creating the *radioulnar syndesmosis*, or *middle radiouln.ar joint*.

Interosseous membrane

The *interosseous membrane* is a broad, flat membrane composed primarily of dense connective tissue whose fibers run inferiorly and medially to attach the interosseous crest of the radius to the interosseous crest of the ulna. The membrane helps bind the two bones together and creates a large surface for origins of the flexor and extensor muscles in the forearm. In fractures of the upper limb that require immobilization of this membrane, the forearm is always cast (or immobilized) in the supine position (figure 6-12). This keeps the membrane taut, preventing atrophy.

Oblique cord

The *oblique cord* is a ligament that runs from the coronoid process of the ulna to the shaft of the radius a little distal to the radial tuberosity. Its fibers run inferiorly and laterally, and is sometimes absent.

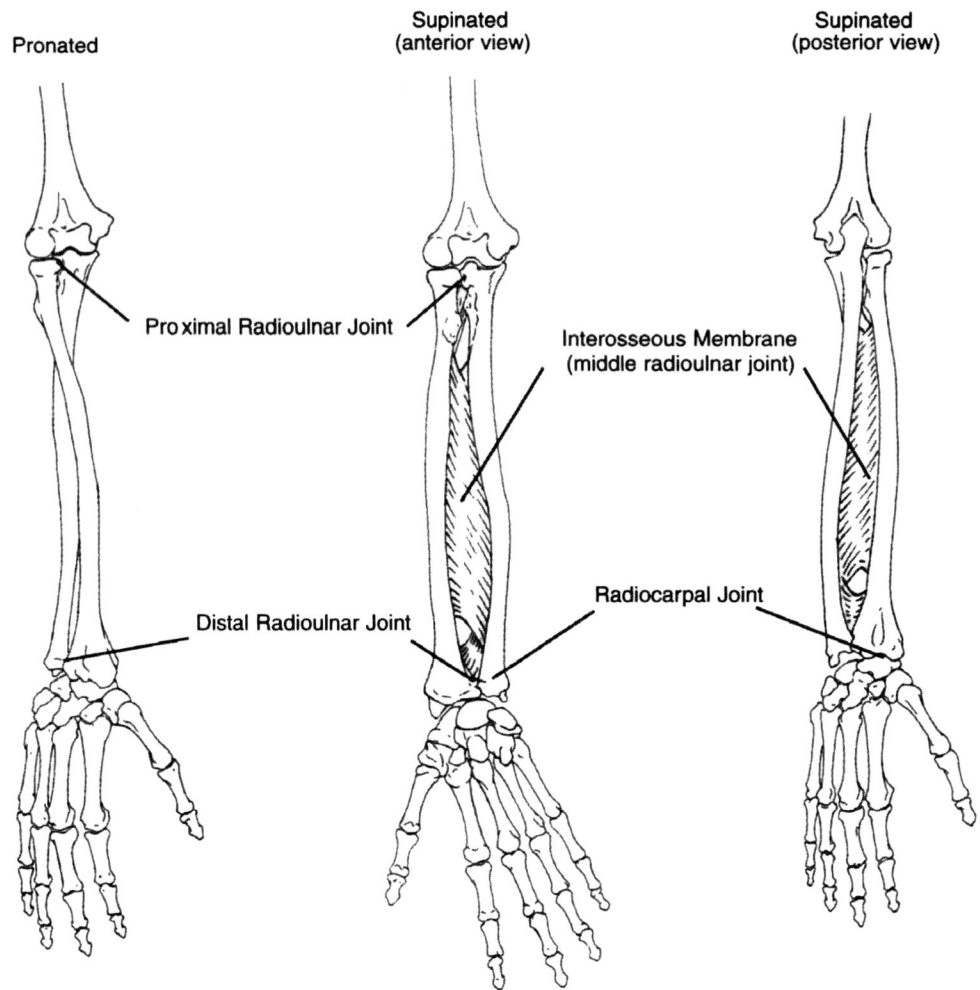

Figure 6-12 Pronation and supination of the forearm skeleton, right side.

Distal Radioulnar Joint

The *distal radioulnar joint* is nearly the inverse of the proximal joint. The head of the ulna is located distally, and presents a circular margin which articulates with the ulnar notch on the distal radius. This trochoid joint is surrounded by a loose articular capsule lined with a synovial membrane. The joint is reinforced by the presence of an *articular disc*. The fibrous joint capsule is thickened posteriorly and anteriorly. The synovial lining of the joint extends a short distance proximally between the radius and ulna to form a *saciform recess*.

The *articular disc* is triangular in shape and extends transversely across the distal end of the ulnar head. Its apex attaches to the depression between the head and styloid process of the ulna, and its base attaches to the medial edge of the distal articular surface of the radius. Its center is thinner than its periphery, and may occasionally be perforated. It serves to help bind the distal ends of the radius and ulna together and separates the ulnar head from articulation with the carpals. It also separates the ulnar notch from the wrist joint. The disc's inferior surface is concave to receive the medial part of the lunate bone. Proximally, the disc itself rotates relative to the head of the ulna, while distally the carpal bones move relative to the disc in flexion/extension or abduction/adduction of the wrist.

The Wrist

The *wrist* (Figures 6-13; Refer also to figure 6-12 throughout the discussion of the wrist) can be loosely defined as the region between the hand and the forearm. Two separate joints lie within the wrist: the *distal radioulnar joint* and the *radiocarpal joint.* The distal radioulnar joint is described in the preceding section.

Radiocarpal Joint

The *radiocarpal joint* (wrist joint proper) is an ellipsoid synovial joint. It is formed by the articulation of the distal end of the radius and the articular disc superiorly with the scaphoid, lunate and triquetral of the proximal row of carpal bones inferiorly.

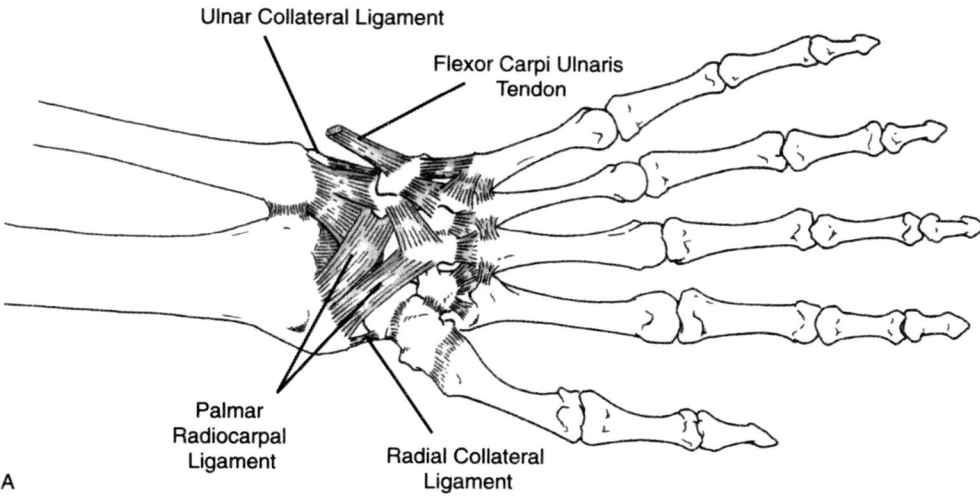

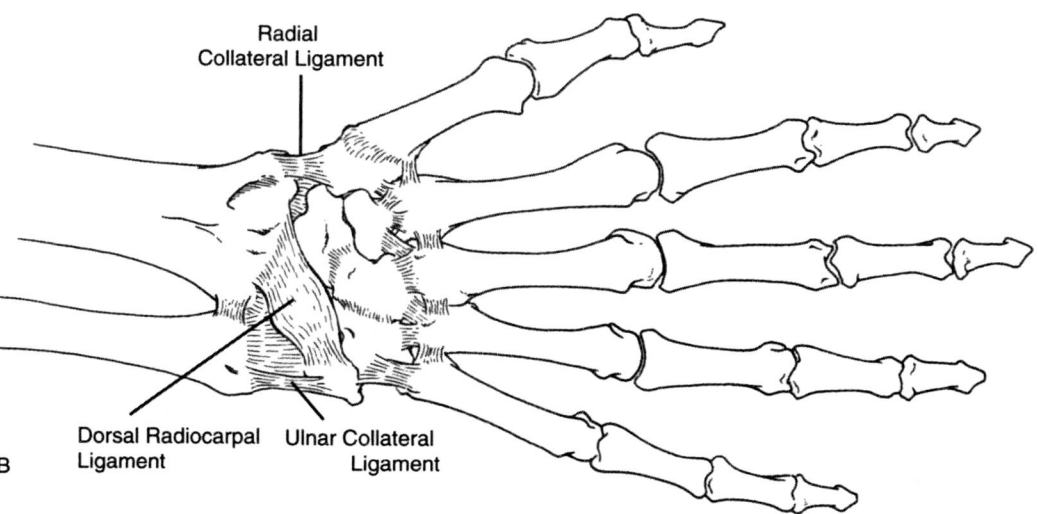

Figure 6-13 Ligaments of the wrist: (A) anterior or palmar and (B) posterior or dorsal views, right side.

The Ligaments

A fibrous articular capsule surrounds the joint. The joint is reinforced by a number of ligaments including the following:

PALMAR RADIOCARPAL LIGAMENT: The dorsal radiocarpal ligament runs from the anterior distal radius and ulna to the palmar surfaces of the scaphoid, lunate and triquetral bones. A few fibers may attach to the capitate. It is essentially a reinforcement of the anterior portion of the joint capsule.

DORSAL RADIOCARPAL LIGAMENT: The dorsal radiocarpal ligament is weaker and less well developed than the palmar. It attaches to the posterior distal radius and runs inferiorly to attach to the dorsal aspects of the scaphoid, lunate and triquetral. Like its palmar counterpart it is again a strong posterior reinforcement of the radiocarpal joint capsule.

ULNAR COLLATERAL LIGAMENT: The ulnar collateral ligament attaches superiorly to the styloid process of the ulna and runs inferiorly along the medial aspect of the wrist to attach below to the triquetral and to the pisiform and transverse carpal ligament. The ulnar collateral ligament, along with the radial styloid process, serves to guide adduction or radial deviation of the hand at the wrist joint.

RADIAL COLLATERAL LIGAMENT: The radial collateral ligament runs from the styloid process of the radius to the lateral aspect of the scaphoid with a few additional fibers running to the trapezium and flexor retinaculum. The radial collateral ligament, along with the ulnar styloid process, serves to guide adduction or ulnar deviation of the hand at the wrist joint. The aforementioned four ligaments reinforce the fibrous joint capsule and essentially surround its entire external aspect.

FLEXOR RETINACULUM (TRANSVERSE CARPAL LIGAMENT): The flexor retinaculum is a strong ligamentous band that runs from the pisiform and hook of the hamate laterally to the scaphoid tuberosity and palmar side of the trapezium. The space beneath the retinaculum and anterior to the carpal bones is the *carpal tunnel*. The carpal tunnel is completely lined with synovial membranes that provide lubrications for the tendons of the long flexors of the hand. The tunnel provides a passage for these tendons to reach their distal attachments in the hand. Accompanying these tendons is the median nerve.

Ranges of Motion

As noted, the wrist joint is an ellipsoid synovial joint. Motions permitted include abduction, adduction, flexion and extension. The long axis of the articular surfaces is oriented in a coronal plane, while the short axis is along a parasagittal plane. As a result of the differences in lengths of these axes, abduction and adduction along the coronal plane are permitted, and flexion and extension along a parasagittal plane are permitted, but rotation is much more limited. Movement also occurs between the carpal bones, as noted below. *Circumduction* of the hand at the wrist is carried out by sequential movement of flexion, abduction, extension and abduction, as little rotation can occur. The degree of abduction of the hand is considerably less than the amount of adduction, perhaps due to a smaller styloid process on the radius relative to that of the ulna.

Disorders of the Wrist Joint

Of the bones of the wrist, the scaphoid is the most frequently fractured, usually fracturing transversely across its long axis. These fractures occasionally fail to unite, generally because the proximal fragment frequently has no nutrient artery and so loses its blood supply. As a result, avascular necrosis is a common complication in fractures of the scaphoid. *Palmar dislocation of the lunate* is also a frequent complication of fractures of the scaphoid. When this occurs, the displaced lunate may compress the median nerve against the flexor retinaculum.

Carpal tunnel syndrome can be considered a wrist disorder. It is caused by inflammation of the synovium surrounding the tendons as a result of repetitive motion injuries, fractures, arthritis, or other conditions. The inflammation results in compression or impingement of the median nerve, causing weakness, numbness, and pain of the hand. (For more details, see Edwardson, 1995.)

Joints of the Hand

The hand is composed of a number of small bones, separated by a multitude of synovial joints. We will touch on the basics of their anatomy here.

The Intercarpal Joints

The joints between the carpal bones may be defined as plane synovial joints. The basic type of motion that occurs between carpal bones is gliding, or translatory, motion.. The *midcarpal joint* is the joint between the *proximal* (scaphoid, lunate, and triquetral) and *distal* (trapezium, trapezoid, capitate, and hamate) rows of carpal bones. Motion at the midcarpal joint is largely analogous to that in a ball and socket joint, with the proximal row approximating the sides of the "socket", and the distal row forming a proximally projecting "ball", centered around the head of the capitate. This is only a partial analogy, however, as motion among these various elements is complex. Bones in the individual rows are bound to one another by *dorsal* and *palmar ligaments* and by *interosseous ligaments*.

Range of Motion of the Hand

Flexion/extension and abduction/adduction occur at the *midcarpal joint* as well as at the *radiocarpal joint*. Movements of the hand and wrist involve movements at both of these joints. The combined range of typical motion is *flexion*: 85 degrees, *extension*: 85 degrees, *adduction (ulnar deviation)*: 45 degrees and *abduction (radial deviation)*: 45 degrees. Thus, these joints can be circumducted.

The Carpometacarpal Joints

The first carpometacarpal joint is between the base of the proximal phalanx of the thumb (pollex) and the trapezium. This is a *sellar synovial joint*. This joint is capable of flexion/extension and abduction/adduction as well as circumduction. This is also true of the other four carpometacarpal joints, but is much more limited in them.

There first carpometacarpal joint has unique features that greatly enhance its motion. First, as noted, it is a sellar synovial joint. Secondly, the thumb (pollex) is rotated 90 degrees relative to the other fingers. For the pollex, the plane of flexion/extension is parallel to the plane of the palm of the hand, rather than perpendicular to it as for the other fingers. The plane of abduction/adduction for the pollex is perpendicular to the plane of the palm, whereas for the other fingers it is parallel to the palm. Also, the pollex may be *opposed* to the other fingers (bringing the pad of the tip of the thumb into contact with the pad, or palmar surfaces, of the other fingers). *Opposition* of the pollex is carried out by simultaneously flexing and medially rotating the abducted thumb. The sellar-shaped joint surfaces of the trapezium and base of the first metacarpal allow this action to be brought about easily.

Opposition of the pollex is a hallmark of primates in general, and humans in particular. In humans opposition is particularly strong because of the relative shortness of the other digits, which permits "pulp to pulp" contact of their tips with that of the thumb. Opposition allows the manipulation of objects and the environment and enables human beings to carry out complex tasks, such as writing, with great ease.

The Second through fifth carpometacarpal joints

The remaining carpometacarpal joints do not permit the freedom of motion of the first. Their joint surfaces are usually described as plane, though they possess some curvature and are somewhat more complex than the term *plane* implies. The carpals and metacarpals are joined together by synovial articular capsules, by *dorsal* and *palmar* ligaments, and by *interosseous ligaments*. These joints allow flexion/extension and abduction/adduction. They also permit a degree of rotation. This can be demonstrated by grasping a small, round, object, such as a tennis ball or an apple.

When the hand cups, the metacarpals rotate to a degree, resulting in a cupping of the palm.

The fifth metacarpal rotates to a greater degree than the second through fourth. Thus the fifth metacarpophalangeal joint may be easily circumducted. During *opposition*, the second through fifth metacarpals will rotate to a degree to allow the second through fifth digits to come into contact with the first, though most of the movement takes place in the pollex.

The Metacapophalangeal Joints

The *metacarpophalangeal joints* are usually considered *ellipsoid joints*, but the metacarpal heads are divided on their palmar aspects and thus these joints are almost bicondylar. The joints are surrounded by a fibrous joint capsule. Each also has a *palmar ligament* reinforcing the capsule anteriorly, and two *collateral ligaments*, one on each side of the joint.

The second through the fifth metacarpophalangeal joints are joined to one another on their palmar surfaces by the *deep transverse metacarpal ligaments*. These joints permit flexion/extension, abduction/adduction, and a limited amount of rotation that may accompany flexion/extension. These joints may be flexed nearly 90 degrees, but extension is limited to just a few degrees.

The Interphalangeal Joints

Because each digit except the thumb has three phalanges, digits two to five have both a proximal and a distal *interphalangeal joint*. The interphalangeal joints are *ginglymi*, or uniaxial hinge joints. Each has a *medial* and *lateral collateral ligament* and a *palmar ligament* to reinforce the fibrous joint capsule. These joints allow a great deal of flexion, but extension is considerably limited by the tension of the tendons of the flexor muscles. A *slight* amount of rotation occurs during flexion and extension. In flexion, the phalanges rotate slightly laterally such that the pads of the fingers meet the pad of the terminal pollex. In extension, slight medial rotation occurs.

Ossification of the Upper Limb Bones

The Scapula

There are normally at least seven centers of ossification for the scapula (Figure 6-14). The primary center of ossification is for the scapular body, which appears about the second month *in utero*. Ossification begins near the glenoid fossa and spreads throughout the body. The spine of the scapula arises from the dorsal aspect of the body around three fetal months as an extension of the center for the body. At birth, the glenoid, vertebral border, inferior angle, coracoid process and acromion process are still cartilaginous. The secondary center of ossification in the middle of the coracoid appears at about one year, and unites with the rest of the bone at about 15 years. The upper third of the glenoid ossifies from a subcoracoid center of ossification. It appears between ages 10 and 11 and fuses with the remainder of the bone between ages 16 and 18. Between the fourteenth and twentieth years, ossification of the remainder of the scapula takes place rapidly. In order, secondary centers for the root of the coracoid process, base of acromion, inferior angle, tip of the acromion and the remainder of the vertebral border all ossify between 14 and 20 (Breathnach, 1965; Gray, 1985; Krogman, 1962).

The Clavicle

The clavicle (Figure 6-15) ossifies from three centers: a medial and a lateral for the shaft (both primary centers), and a secondary center for the sternal end of the clavicle. The clavicle demonstrates both intramembranous and endochondral ossification. The shaft forms by intramembranous ossification, while the sternal end forms by endochondral ossification. The primary centers for the shaft appear between 5 and 6 weeks *in utero*. As a consequence, the clavicle is the first bone in the body to ossify. The secondary center at the sternal end appears at about ages 18 to 20, and unites with the shaft of the bone at about age 25. As a result, it is one of the last epiphyses in the body to undergo synostosis (Breathnach, 1965; Gray, 1985; Krogman, 1962).

The Humerus

The humerus (Figure 6-16) normally has eight centers of ossification: one primary center for the diaphysis, and seven secondary centers. The proximal end of the humerus forms from a compound epiphysis formed by the union of three secondary centers of ossification. The distal epiphysis is also a compound epiphysis formed from the union of three secondary centers of ossification. The medial epicondyle has a single center of ossification and is isolated from the others. The primary center for the shaft of the humerus appears at about 7 to 8 weeks in utero.

PROXIMAL END: The center for the head of the humerus appears either at birth or in the first few months of life, that for the greater tubercle at three years, and that for the lesser tubercle at five years. The three unite with one another between ages 5 and 6. This compound epiphysis finally synostoses with the shaft at about 18 to 20 years. As with most other epiphyses, this occurs at a younger chronological age in girls than in boys.

DISTAL END: The ossification center for the capitulum appears at approximately two years, that for the trochlea at nine to ten years, and for the lateral epicondyle at 11 to 13 years. These three fuse with one another at about puberty, and the resulting compound epiphysis fuses with the distal end of the diaphysis at 14 to 17 years.

MEDIAL EPICONDYLE: The secondary center of ossification for the medial epicondyle appears at 5 to 7 years, and fuses with the rest of the bone at about 15 to 18 years (Breathnach, 1965; Gray, 1985; Krogman, 1962).

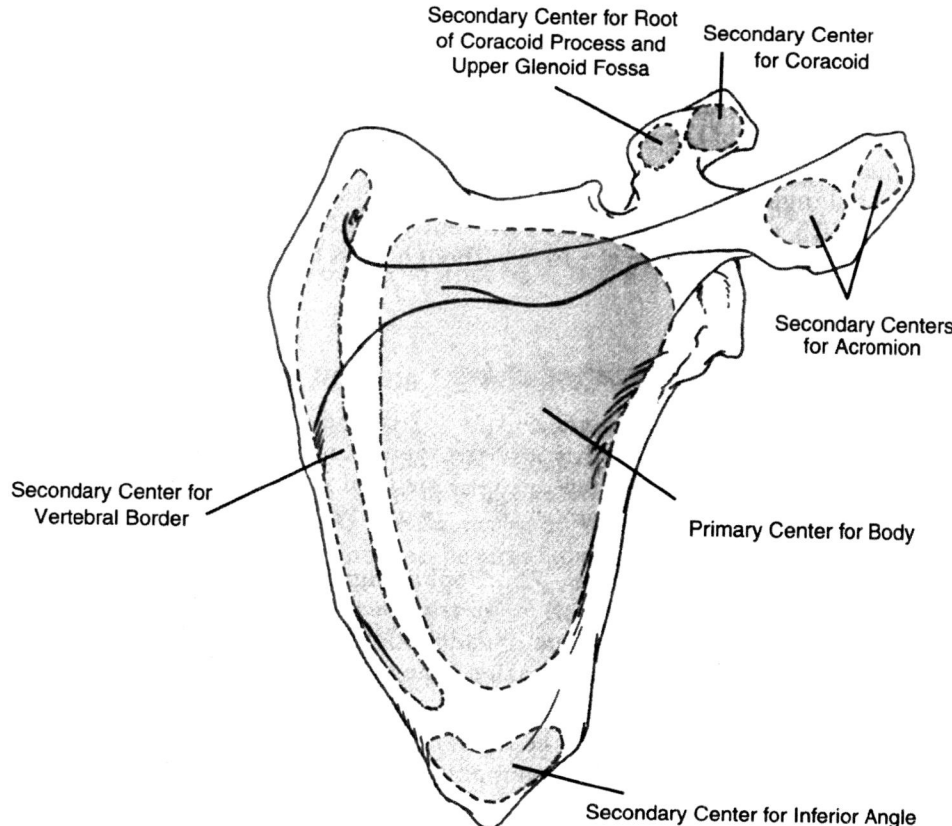

Figure 6-14 Centers of ossification of the scapula.

The Radius

The radius (see Figure 6-16) ossifies from three centers, one for the head, one for the shaft, and one for the distal end. The primary center of ossification is in the diaphysis, which begins ossification at about eight weeks in utero. The secondary center of ossification in the radial head appears at five years and fuses with the shaft between 15 and 18 years. The secondary center for the distal end appears at 2 years of age, and fuses with the shaft between 17 and 20 years of age. As with other epiphyses, ossification and fusion occur somewhat earlier in girls than in boys (Breathnach, 1965; Gray, 1985; Krogman, 1962).

The Ulna

Like the radius, the ulna (Figure 6-16) ossifies from three centers, one at the proximal end of the olecranon process, one for the shaft, and one for the distal end (ulnar head). The center of ossification for the shaft

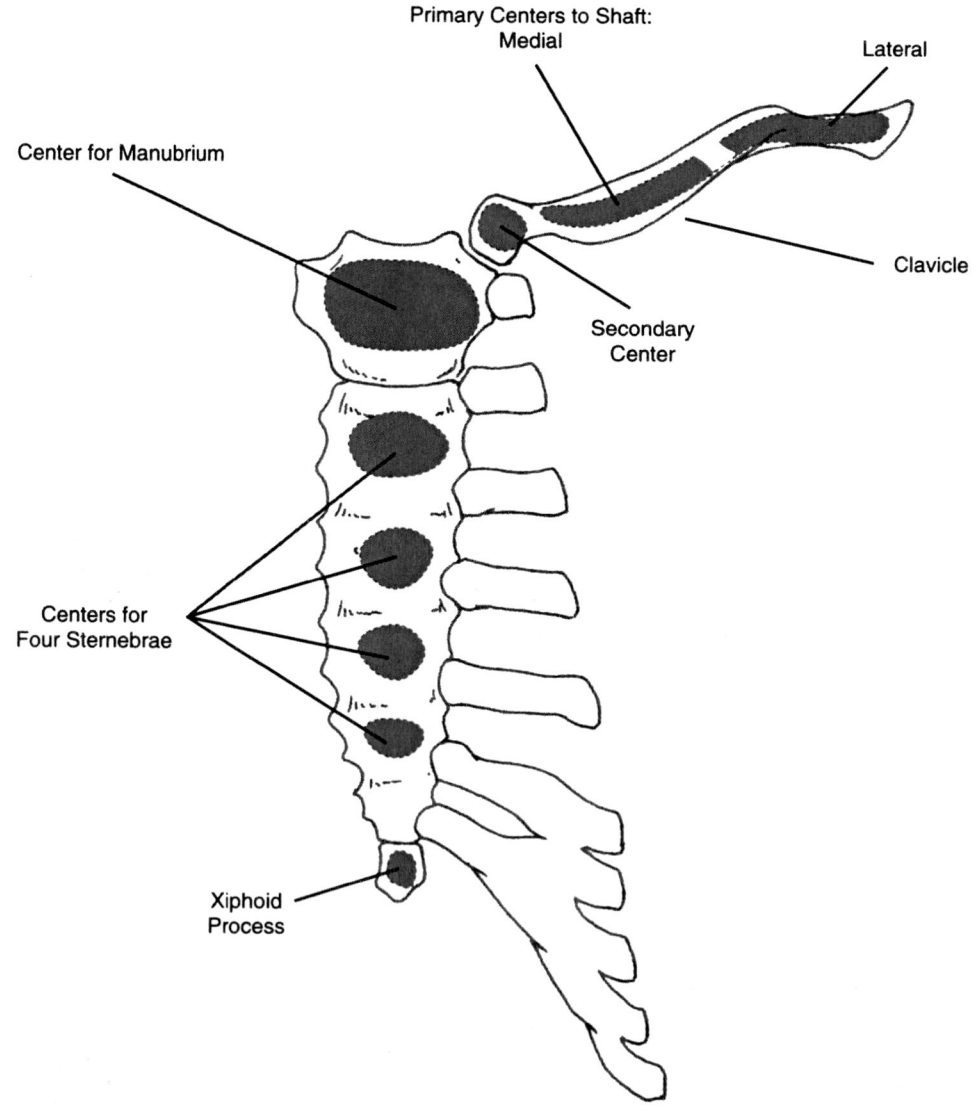

Figure 6-15 Centers of ossification of the sternum and the clavicle.

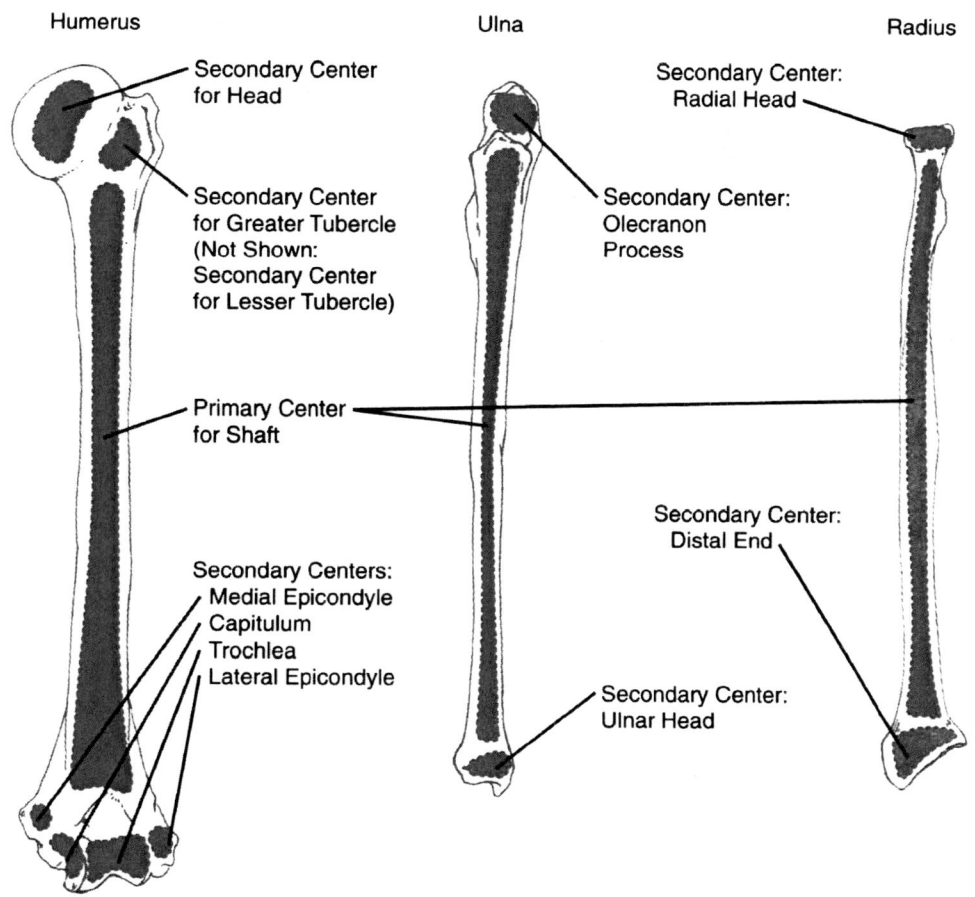

Figure 6-16 Centers of ossification of the humerus, the ulna, and the radius.

appears at about eight weeks *in utero*. The proximal epiphysis appears at about 10 years, and fuses with the rest of the olecranon at about age 16. Most of the olecranon ossifies by a continuation of the ossification of the shaft. The center for the ulnar head appears at about 5 to 6 years, and fuses with the shaft by age 20. As with all the long bones, union of epiphyses occurs earlier in girls than in boys (Breathnach, 1965; Gray, 1985; Krogman, 1962).

The Carpals

Each carpal has only a single center of ossification. The *capitate* and *hamate* ossify at one year, the *triquetral* at three years, the *lunate* at four years in girls and five years in boys, the *trapezium, trapezoid* and *scaphoid* in the fifth year in girls and in the sixth and seventh in boys, and the *pisiform* at age 9 in girls and age 11 in boys. Occasionally, a small bone, the *os centrale*, may appear in the wrist. When present, it is found on the dorsum of the wrist between the scaphoid, trapezoid and capitate. This bone is constant in some other primates, but not in human beings. Ordinarily the cartilage anlage for the os centrale fuses with the anlage of the scaphoid at about two months *in utero* (Breathnach, 1965; Gray, 1985; Krogman, 1962).

Metacarpals

Metacarpals II through V all ossify from two centers, one for the shaft and one for the head (distal end). Metacarpal I (for the pollex) ossifies from two centers as well, but its epiphysis is located at its proximal rather than its distal end. In this regard it follows the pattern of the phalanges. (It has been suggested that the first metacarpal is phylogenetically the proximal phalanx of the first digit, the bone homologous to the other four metacarpals having been lost in the course of evolution (Gray, 1985).) The shafts of the metacarpals begin ossification at the eighth or ninth week in utero. The centers for the second and third metacarpals appear first, and that for the first appears last. At about age three the epiphyses of all five metacarpals begin to ossify, and they unite with the shafts of the metacarpals between ages 18 and 20. There can be a secondary center at the base of the third metacarpal (forming its "styloid process"). It is occasionally present, and may fail to fuse with the rest of the bone, forming a separate bone (Breathnach, 1965; Gray, 1985; Krogman, 1962).

The Phalanges

The phalanges, like the first metacarpal, all ossify from two centers: one for a proximal epiphysis and one for the shaft. Like the metacarpals, the centers for the shafts of the phalanges appear at the eighth or ninth week *in utero*. The distal phalanges ossify from their distal ends first, instead of from the centers of the shafts as with the proximal and middle phalanges. Along with the clavicle, the distal phalanges are among the first bones in the body to ossify. The proximal epiphysis, or phalangeal base, begins to ossify at age two in the proximal phalanges, and at age three in the middle and distal phalanges. The epiphyses of all three phalanges of each digit fuse with the shaft by age 20 (Breathnach, 1965; Gray, 1985; Krogman, 1962).

References Cited

Breathnach, AS, ed. (1965) *Frazer's Anatomy of the Human Skeleton,* Sixth Edition. J and A Churchill, London.

Edwardson, BM (1995) *Musculoskeletal Disorders: Common Problems.* Singular Publishing Group, Inc., San Diego.

Gray, Henry (1985) *Gray's Anatomy,* 30th American Edition. C.D. Clemente, ed. Lea and Febiger, Philadelphia.

Krogman, W.M. (1962). *The Human Skeleton in Forensic Medicine.* Charles C. Thomas, Springfield, Illinois.

Matsen, FA (1980) Biomechanics of the shoulder. In: Frankel, VH, and Nordin, M (1980) *Basic Biomechanics of the Skeletal System.* Lea and Febiger, Philadelphia.

Nordin, M, and Frankel, (2001) *Basic Biomechanics of the Skeletal System,* 3rd Edition. Lippincott, Williams and Wilkins, Baltimore.

CHAPTER
|Seven|

Bones of the Postcranial Skeleton: Lower Limb

The Pelvic Girdle: The Os Coxae

The *pelvic girdle* attaches the lower limb to the axial skeleton. The pelvic girdle consists of the two *ossa* (singular *os*) *coxae*, better known as the *hip bones* or *innominate bones* (refer to figures 7-1 to 7-3 throughout the discussion of the innominate or hip bones). The innominates articulate with the sacrum at the sacroiliac joint, and together they form the bony *pelvis*. Each innominate articulates with the lower limb at the hip joint.

In the adult, the *innominate* is present as a single, irregularly shaped bone In the child, however, it consists of three separate bones, the *ischium, the ilium and the pubis*, which are united by a Y-shaped hyaline growth cartilage in a central depression called the *acetabulum* (the hemispherical socket for the femur). These three bones fuse to one another at adolescence. The upper blade-like bone is the *ilium*. At the *acetabulum* (Latin meaning "vinegar cup") it is fused to the *pubis* anteriorly and the *ischium* inferiorly. The two innominates are joined to the sacrum at the bilateral *sacroiliac joints*, and to each other at the anterior *pubic symphysis*.

Much of the internal surface of the ilium is taken up by the *iliac fossa*, from which the iliacus muscle originates. The blade itself is surmounted by the *iliac crest*, which is formed by a separate *apophysis* during development. (An *apophysis* is a secondary center of ossification forming under a muscle attachment, also known as a *traction epiphysis*.) The iliac crest serves as attachment for the internal and external oblique and transversus abdominus muscles of the abdominal wall. Its anteriormost projection is the *anterior superior iliac spine*, which provides attachment for the sartorius muscle. Inferior to this is the *anterior inferior iliac spine* which is also formed from a separate apophysis. The upper portion of the anterior inferior iliac spine gives origin to the straight head of the rectus femoris muscle. A teardrop shaped lower portion (with the point of the tear drop pointed posteriorward) gives origin to the iliofemoral ligament and directly abuts the rim of the acetabulum.

The second head of the rectus femoris muscle arises from a small depression a few millimeters above the acetabular rim more posteriorly. Just anteromedial and inferior to the anterior inferior iliac spine is the *iliopsoas groove*, which provides passage for the iliopsoas muscle as it passes over the femoral head to its insertion on the lesser trochanter of the femur. The medial wall of this groove is sometimes formed by the bossing (swelling) which results from the fusion point of the ilium and the pubis. This is called the *iliopectineal (iliopubic) eminence*. There is a sharp border between the iliac blade bearing the origin of the iliacus and its more inferior portion which forms the true pelvis. This sharp border is called the *iliac arcuate line*. It junctures the *auricular surface* at its apex. Just inferior to the auricular surface is the *preauricular sulcus* (also known as the *paraglenoid sulcus*). This groove is more common and better developed in females than in males. Its origin and function are controversial, but may be related to changes in adjacent ligaments induced by hormones associated with pregnancy and/or childbirth. Between the auricular surface and the iliac crest is a very rugose (roughened) region called the *iliac tuberosity*. This tuberosity is for the attachment of the *interosseous sacroiliac ligament*.

The *exterior surface of the ilium* (Figure 7-4; refer also to Figures 7-2 and 7-3 throughout the remainder of the discussion of the ilium) gives rise two important muscles of gait, the gluteus medius and minimus. The minimus occupies the

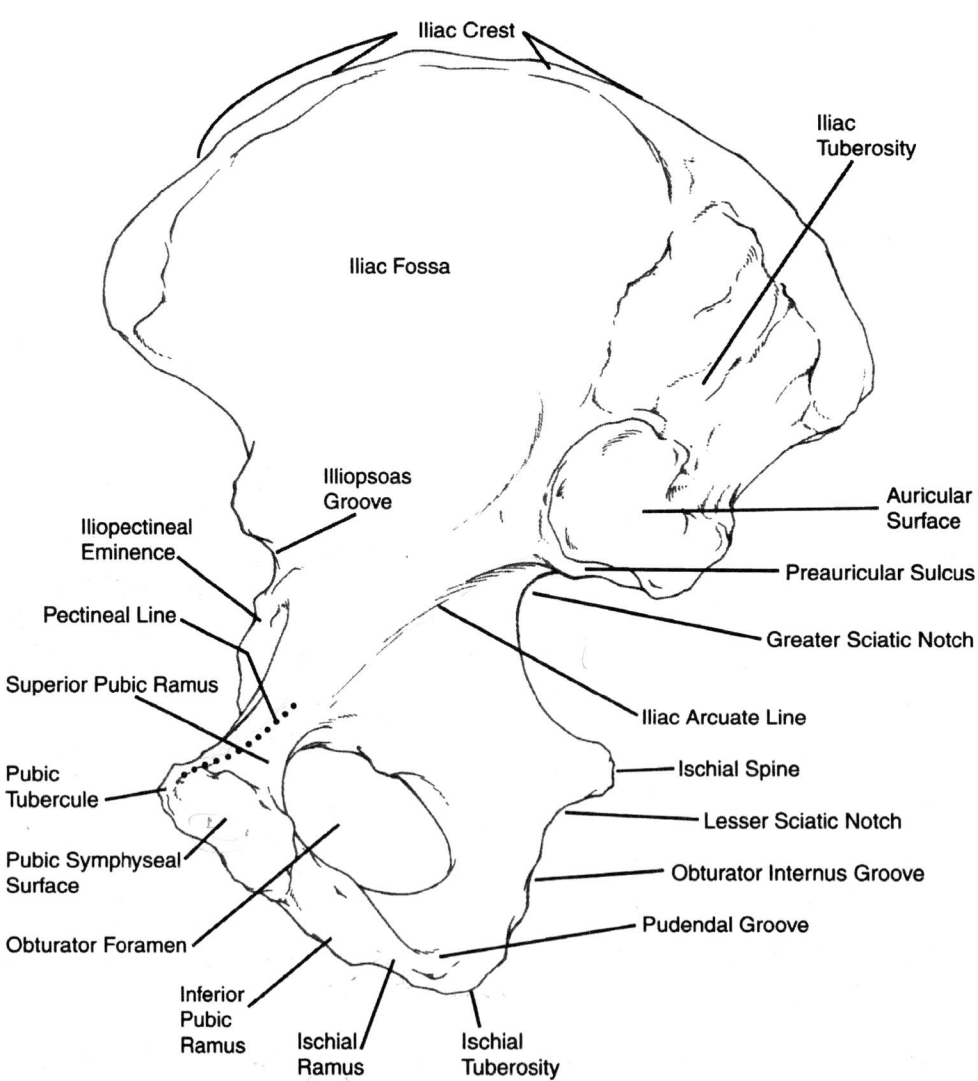

Figure 7-1 The right os coxae (innominate or hip bone), medial (pelvic) aspect.

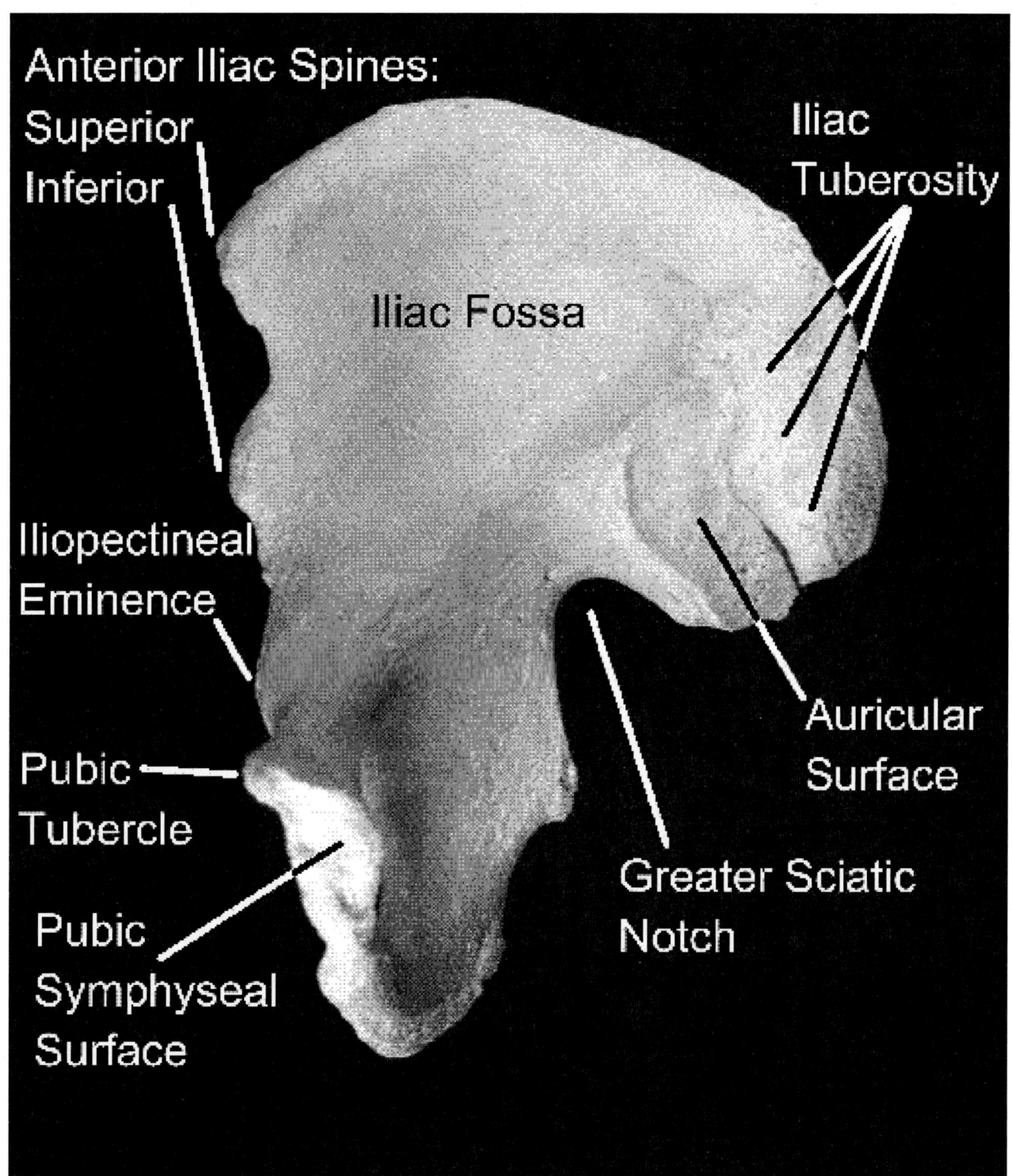

Figure 7-1 *(Continued)* Photographs of anterior and posterior views of the right os coxae (innominate or hip bone), medial (pelvic) aspect. See accompanying line drawing for identification of additional landmarks.

anterior portion, with its attachment outlined by the anteriormost extent of the ilium, the anterior iliac crest and a line which originates from the iliac crest and passes posteroinferiorly toward the greater sciatic notch (see below). This is the *anterior gluteal line*.

The medius is attached posterior to the anterior gluteal line from an almost circular area outlined by the crest and a second line, the posterior gluteal line. The *posterior gluteal line* is a short continuation of the iliac crest toward the *greater sciatic notch*. The posterior terminus of the iliac crest is the *posterior superior iliac spine*. Note that the posterior gluteal line and posterior part of the crest (anterior from the posterior superior spine for a distance about 1-2 inches) forms a rugose triangular area. To this area is attached a portion of the gluteus maximus muscle (the bulk of its origin is from the sacrum). A vague line may run from the *anterior inferior iliac spine* to the greater sciatic notch delineating the inferior extent of the gluteus minimus origin. This is the *inferior gluteal line*.

Also note that the iliac blade is bent back relative to the massive portion of the innominate which houses the acetabulum. The ilium's inferior surface thus forms the *greater sciatic notch*, which is shaped in males very much like an inverted fishhook, and in females very much like a boomerang. That is, the walls of the notch are much more divergent from one another in

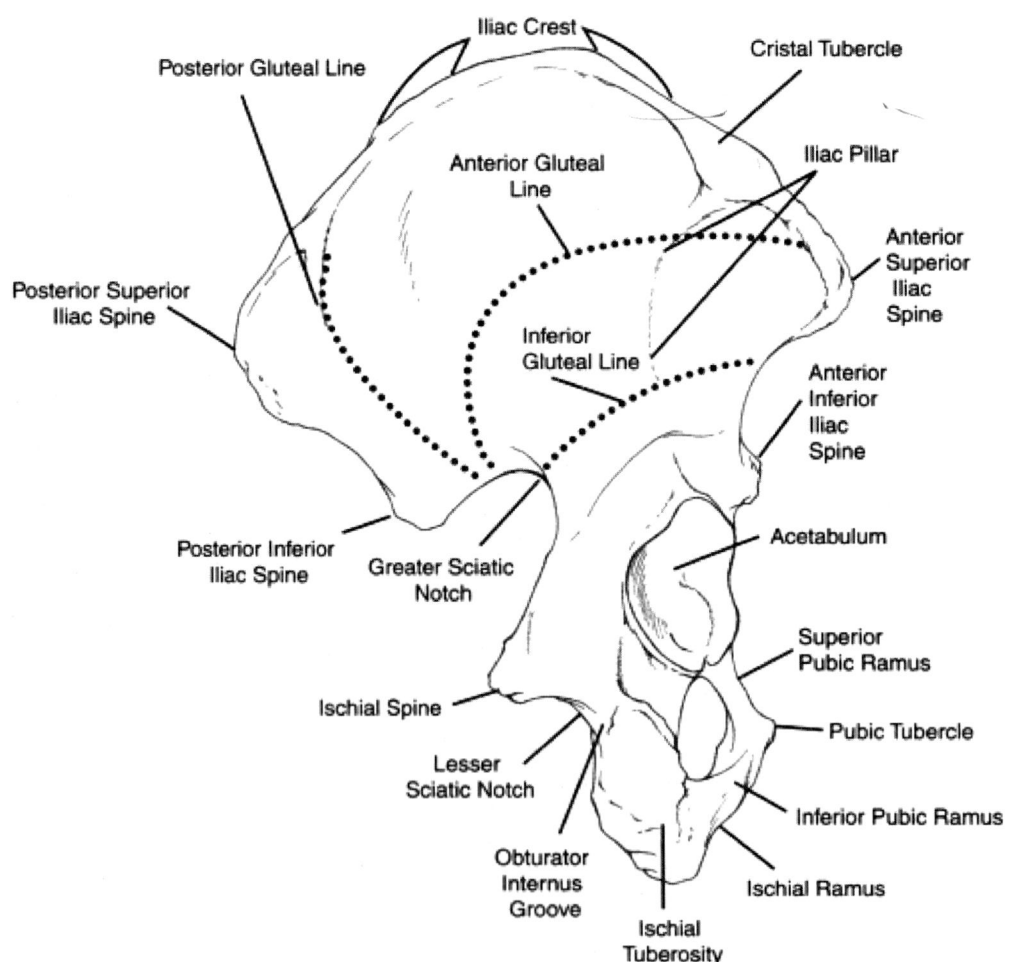

Figure 7-2 The right os coxae (innominate or hip bone), lateral aspect.

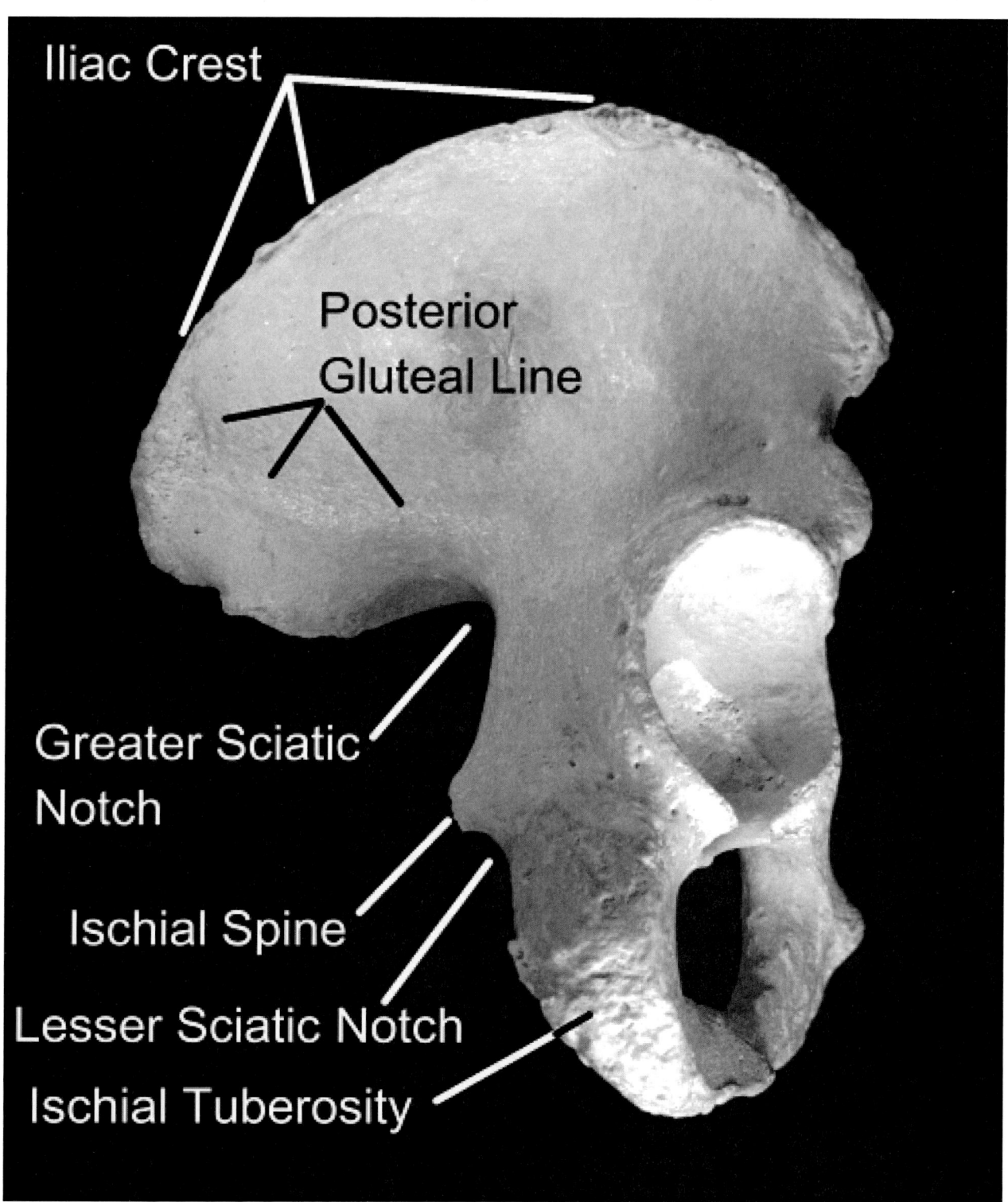

Figure 7-2 *(Continued)* Photographs of anterior and posterior views of the right os coxae (innominate or hip bone), lateral aspect. See accompanying line drawing for identification of additional landmarks.

females than in males. The posterior terminus of this notch is the *posterior inferior iliac spine*. The anterior terminus of the notch is actually located on the ischium and is called the *ischial spine*.

In external view the ilium will display a distinct thickening of its body about one third the distance from the anterior superior iliac spine to the posterior superior iliac spine. This thickened region runs from the iliac crest to the acetabular lip. It is called the *iliac pillar*. At the juncture of the pillar with the crest, there is a thickening of the latter called the *cristal tubercle*. The pillar is crossed by the anterior gluteal line.

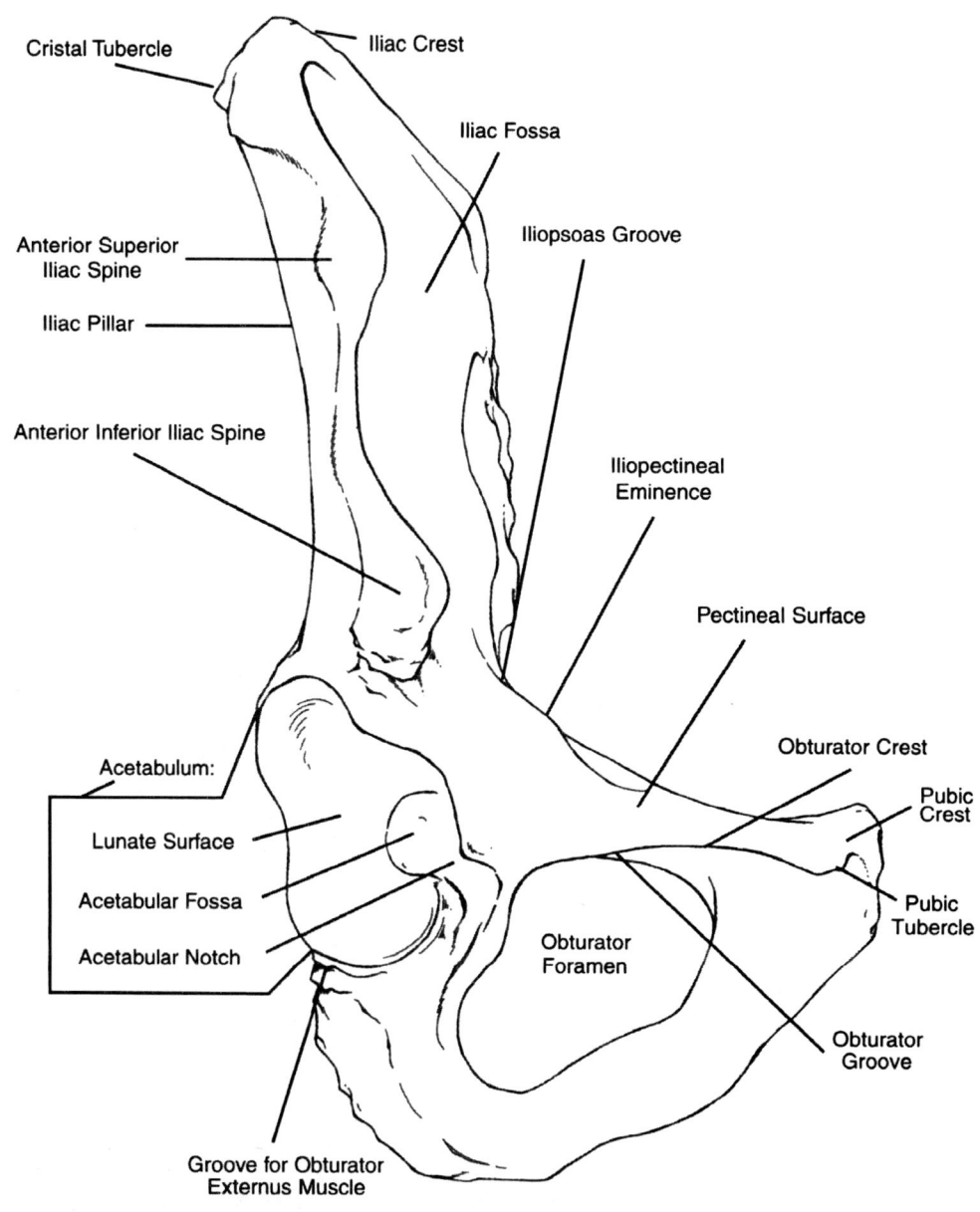

Figure 7-3 The right os coxae (innominate or hip bone), anterior view.

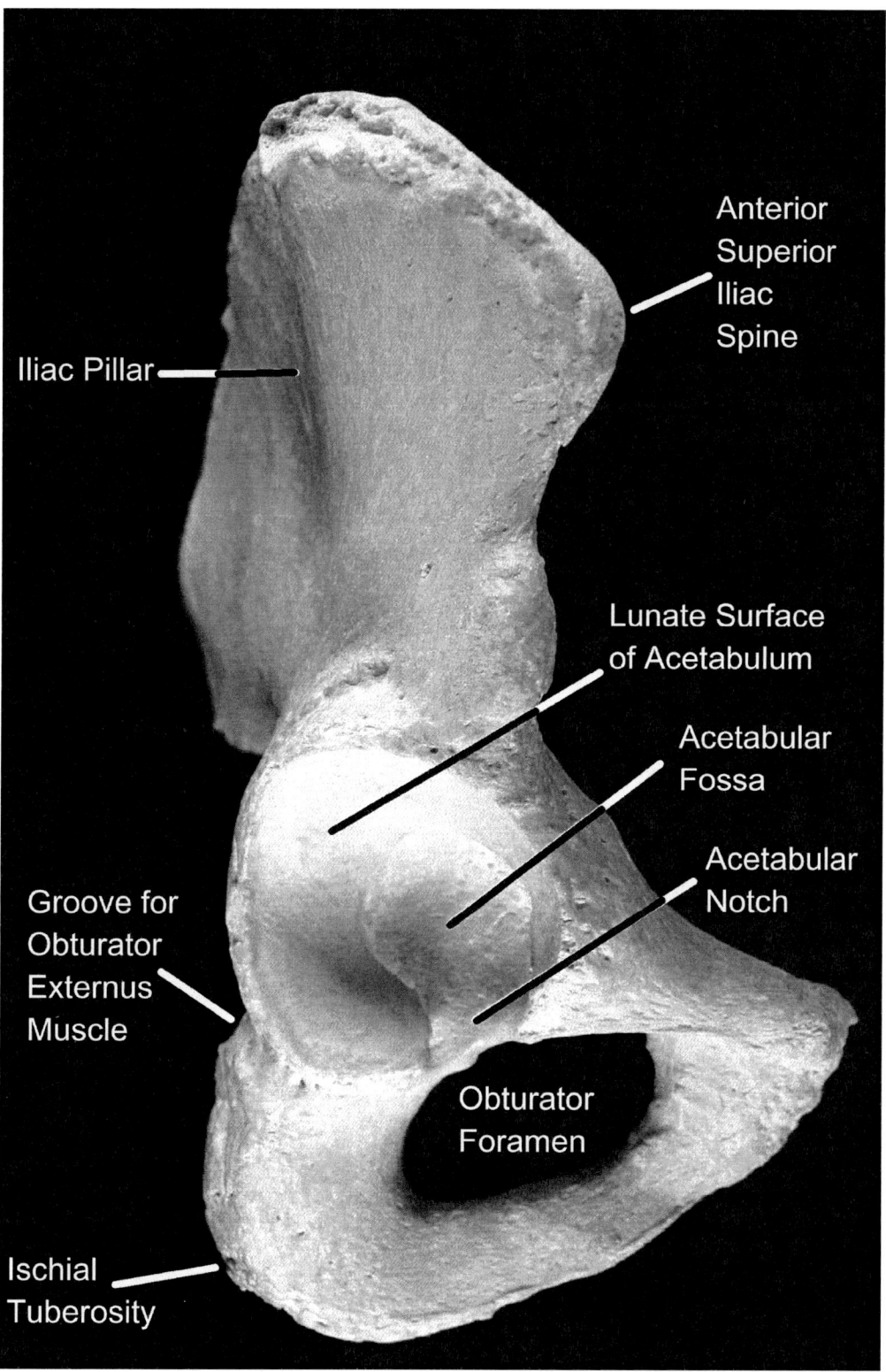

Figure 7-3 *(Continued)* Photographs of anterior and posterior views of the right os coxae (innominate or hip bone), anterior view. See accompanying line drawing for identification of additional landmarks.

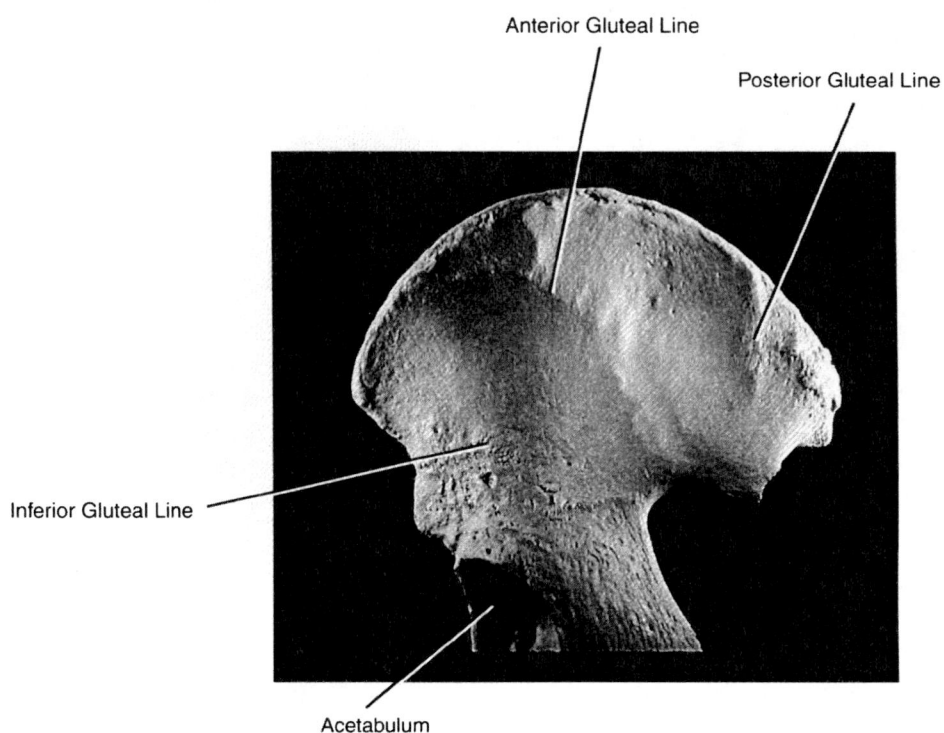

Figure 7-4 The left os coxae, lateral aspect of the ilium demonstrating the gluteal lines.

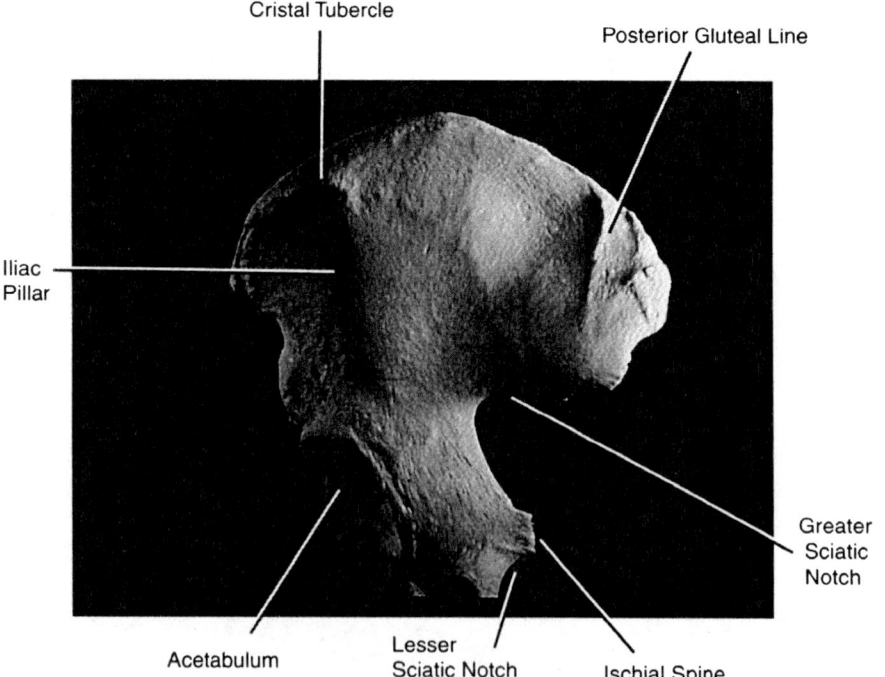

Figure 7-5 The left os coxae, lateral aspect of the ilium demonstrating the iliac pillar.

The lower portion of the acetabulum is formed by the *corpus* or *body* of the *ischium* (see Figures 7-1 throughout 7-3 throughout the discussion of the ischium). From this a large *ischial ramus* projects anteriorly and joins the inferior ramus of the pubis. The inferior portion of the corpus is surmounted by a large rugose region, called the *ischial tuberosity*. The tuberosity is formed by the posterior terminus of a separate apophysis which surmounts the length of the ischiopubic ramus. The ischial tuberosity's surface is highly rugose and faceted. These facets are for the origin of the so-called hamstring muscles. The superiormost facet gives origin to the semimembranosus muscle. The next facet is for the biceps femoris muscle, and the most inferior is for the origin of the semitendinosus muscle. These three muscles arise from the more vertical portion of the tuberosity, before it becomes sharply angled to provide a horizontal surface for the attachment of the adductor magnus muscle.

A distinct groove lies between the ischial tuberosity and the remaining surface of the ischium on its internal surface. This is the *pudendal groove* for the pudendal nerve and the internal pudendal vessels. Between the ischial tuberosity and the *ischial spine* is a second involution of the lesser pelvis, called the *lesser sciatic notch*. A groove may be occasionally palpated just inferior to the ischial spine for the obturator internus muscle: the *obturator internus groove*. A second groove is formed by the anterosuperior edge of the ischial tuberosity and the lower lip of the acetabulum. This is the *obturator externus groove*. The pubis and ischium together surround a large opening called the *obturator foramen*. It is closed by a membrane during life. Its outer surface gives attachment to the obturator externus muscle, while its internal surface gives attachment to the obturator internus muscle.

The *pubis* (see Figures 7-1 throughout 7-3 throughout the discussion of the pubis) is formed of two separate portions called *rami* (the *superior and inferior rami*) and a central portion which provides the attachment to the two rami and to the pubic symphysis called the *corpus* or *body*. The arcuate line of the ilium is continuous onto the superior ramus of the pubis, and may be followed anteriorly to a large anterior projection called the *pubic tubercle*. The continuation of the *iliac arcuate line* on the pubis is called the *pectineal line* or the *pecten* (Figures 7-1).

The iliac arcuate line and the pectineal line together form the *linea terminalis*. The pectineal line together with the iliac arcuate line and the *sacral promontory* form the brim of the lesser or "true" pelvis, which is the area inferior to this brim. Medial to the pubic tubercle is the *pubic crest*. The surface of the superior ramus just anterior to the pectineal line is called the *pectineal surface*. A second ridge or line runs from the pubic tubercle to the acetabular rim. This is the *obturator crest*. It gives attachment to the pubofemoral ligament.

The inferior ramus of the pubis joins the ramus of the ischium. It gives origin to the three adductor muscles of the thigh. At the apex of the obturator foramen on the inferior and posterior aspect of the superior ramus of the pubis is the *obturator groove* for the passage of the obturator nerve and vessels. In life this is converted into the obturator canal by the obturator membrane.

The acetabulum bears a *lunate (articular) surface*, an *acetabular notch*, a *rim*, and a central nonarticular portion, the *acetabular fossa*. The acetabulum is traditionally described as being composed of two-fifths ilium, two-fifths ischium, and one-fifth pubis. In life, a fibrocartilagenous labrum attaches to and deepens the acetabular rim. The acetabular notch is a gap in the acetabular rim and is actually closed in life by the *transverse acetabular ligament*. Additionally, a small ligament of the head of the femur runs from this transverse acetabular ligament to the fovea capitis, a small depression in the center of the head of the femur.

The innominate bone and pelvis are very important in the identification of both sex and age in unidentified skeletal remains and in radiographs. There are a number of sexual differences in the morphology of the innominates (Figure 7-6). As noted before, in females the greater sciatic notch is broad and shallow, while it is deep and narrow in males. Among females, the *superior pubic ramu*s is relatively long, and the *subpubic angle* (formed by the meeting of the two inferior pubic rami at the pubic symphysis) is obtuse in females, while in males it is acute.

Important age related changes which occur at the pubic symphysis and the auricular surface of the ilium. In young adulthood both of these surfaces are covered with ridges and furrows which progressively fill and become smoother with age. Other additional degenerative changes occur with age as well. These changes occur with some regularity, and can be used by trained observers to estimate an age at death with reasonable certainty.

Additionally, throughout the skeleton, the appearance and fusion of ossification centers occurs with a high degree of regularity. Therefore, the appearance and fusion of apophyses and epiphyses are an excellent means of determining skeletal age in subadults, either in x-rays or in unidentified skeletal remains. Likewise, the appearance of tooth buds and the eruption sequence of teeth can be used to determine age in subadult skeletons. Anyone interested in the subject of diagnosis of age and sex in human skeletal remains would do well to consult as a starting point *The Human Skeleton in Forensic Medicine* (1st edition, Krogman, 1962; 2nd edition, Krogman and Iscan, 1986). See the references and further readings at the end of the text for additional information.

Bones of the Lower Limb

The *lower limb* is divisible into a more superior *thigh (femur)*, which runs from the hip to the knee, and a more inferior *leg (crus)*, which runs

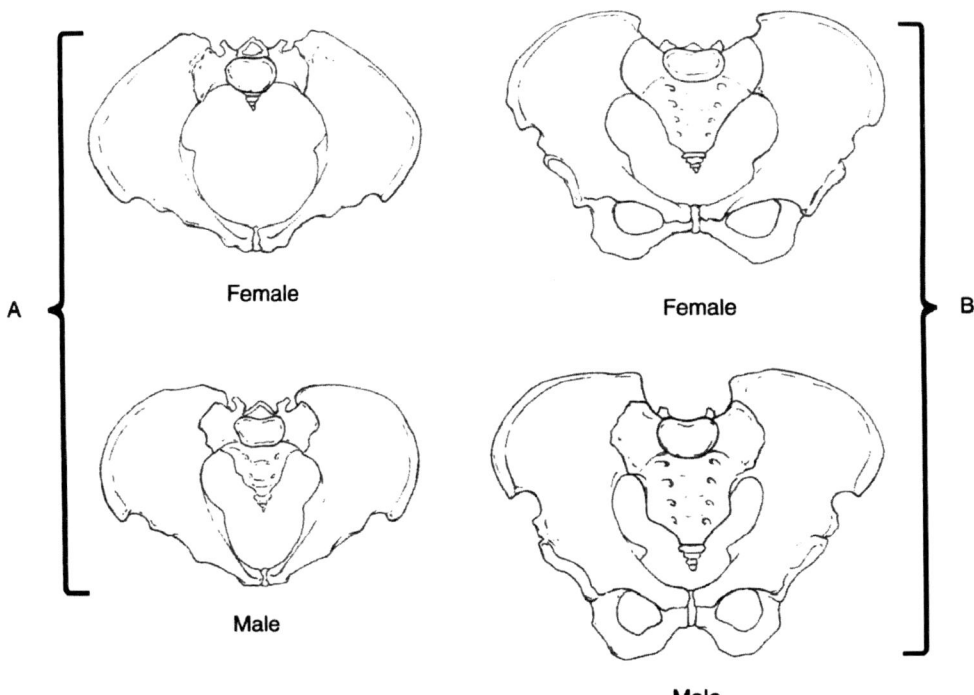

Figure 7-6 Gender-related differences in the male and female pelvis: (*A*) superior views, (*B*) anterior views. Note the much greater relative mediolateral diameter of the pelvic inlet and outlet in the female and the absolutely greater subpubic angle of the female.

from the knee to the ankle. The *femur (os femoris)* (Figure 7-7) is the bone of the thigh, while the *tibia* and *fibula* are the two bones of the leg (Figure 7-8). Strictly speaking, the term *femur* refers to the thigh, and *os femoris* is the bone of the thigh. Convention is to use *thigh* to refer to bone and soft tissue, while the term *femur* is generally used to mean the bone alone.

The Femur

The *femur* is the largest bone of the body, and is the skeleton of the thigh. It consists of a proximal head, a neck, two large proximal protuberances: the greater and lesser trochanters, a shaft or body, and a distal extremity composed of medial and lateral articular condyles.

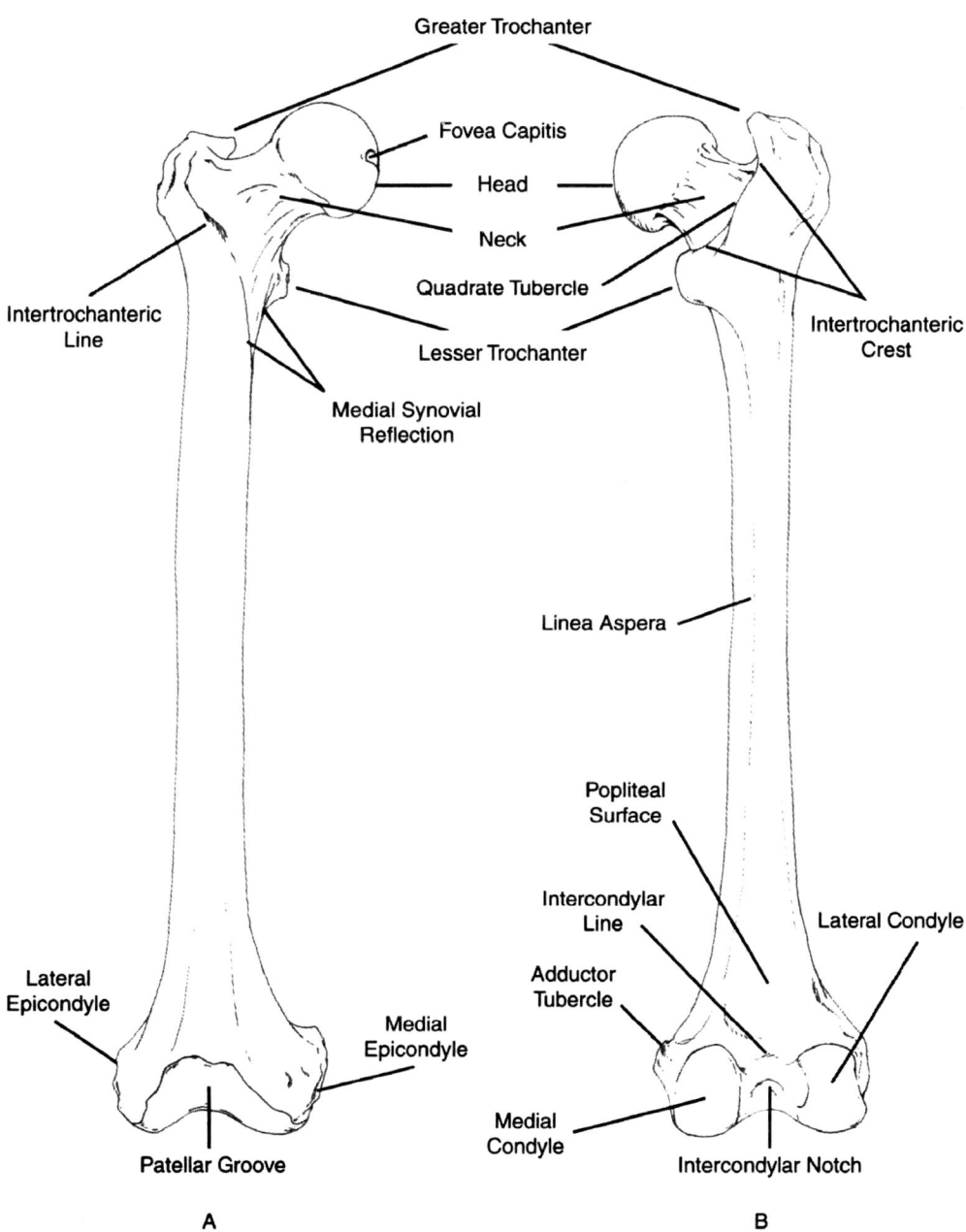

Figure 7-7 The right femur: (*A*) anterior and (*B*) posterior views.

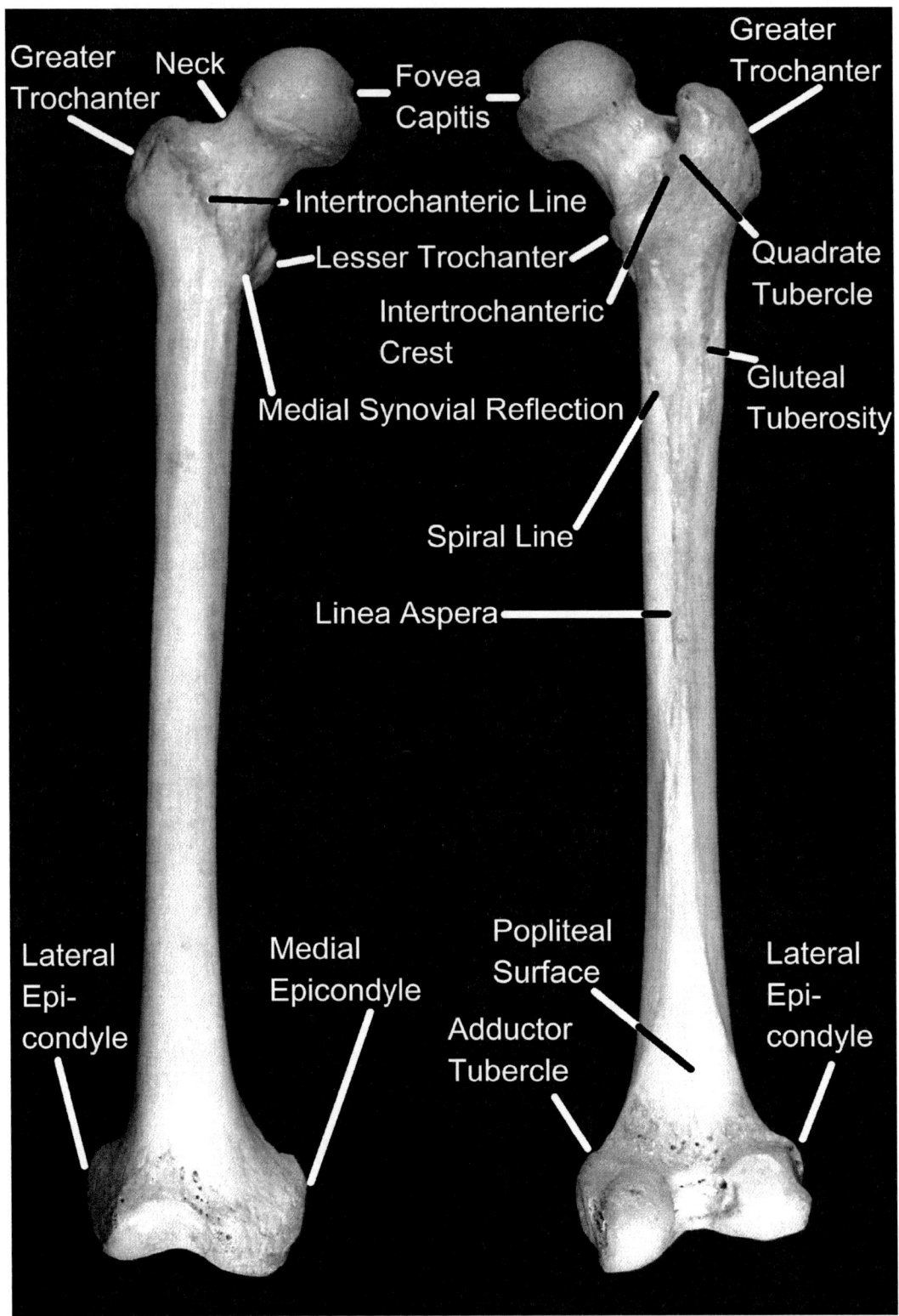

Figure 7-7 *(Continued)* The right femur, anterior (left) and posterior views. See accompanying line drawing for identification of additional landmarks.

Bones of the Postcranial Skeleton: Lower Limb

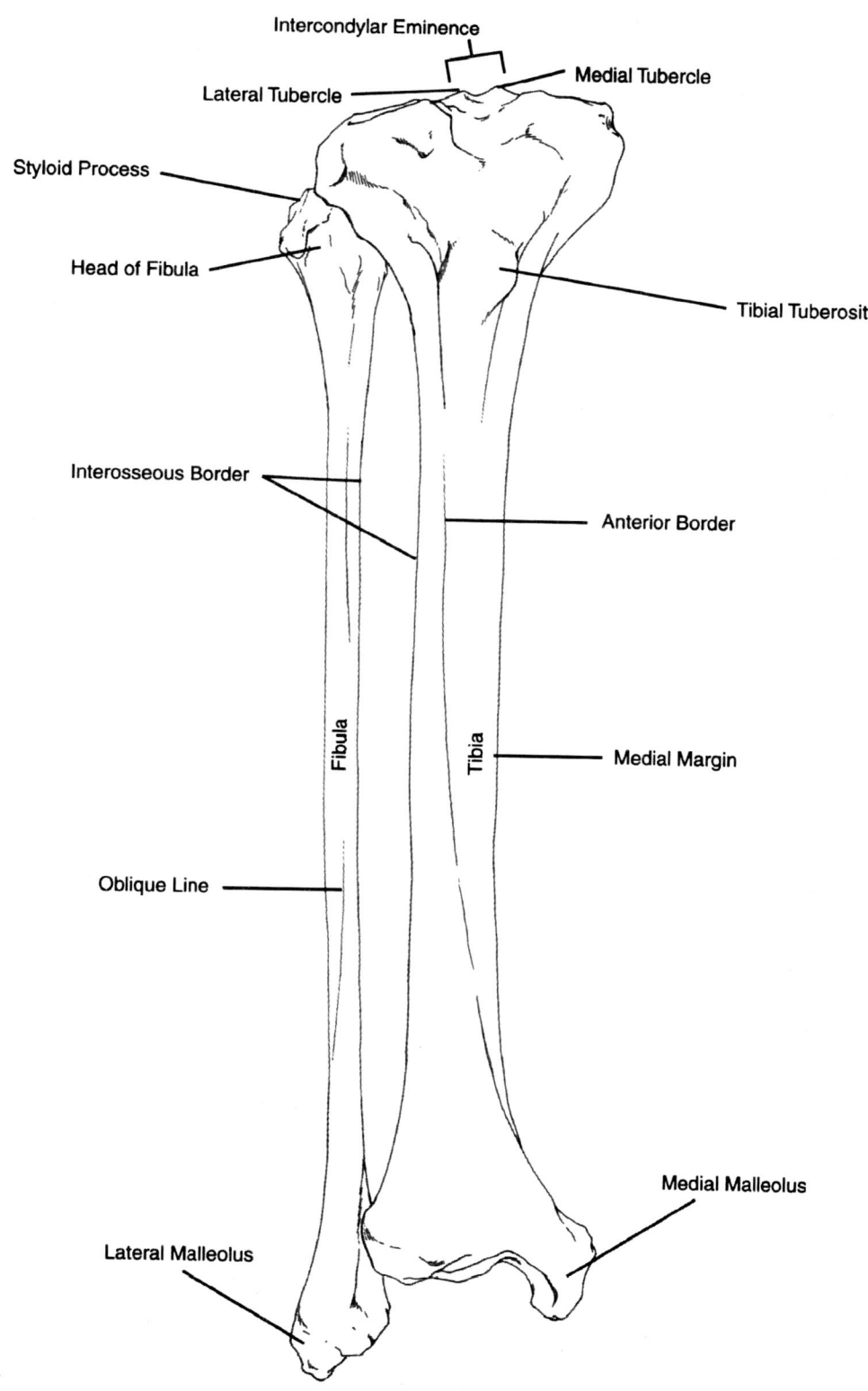

Figure 7-8 The right fibula and tibia: (*A*) anterior.

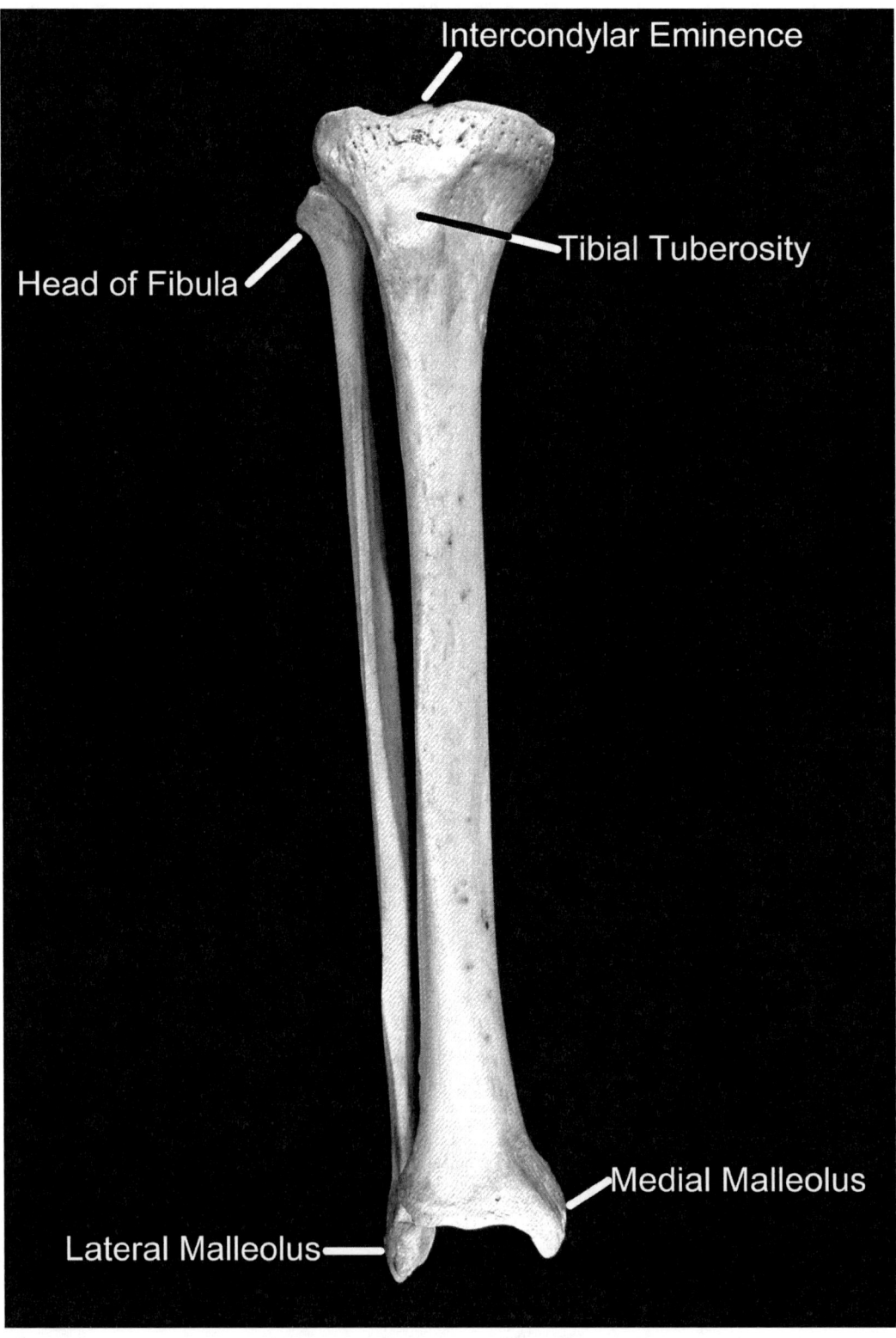

Figure 7-8 A *(Continued)* Photograph of anterior view of the right tibia and fibula, articulated. See accompanying line drawing for identification of additional landmarks.

Bones of the Postcranial Skeleton: Lower Limb

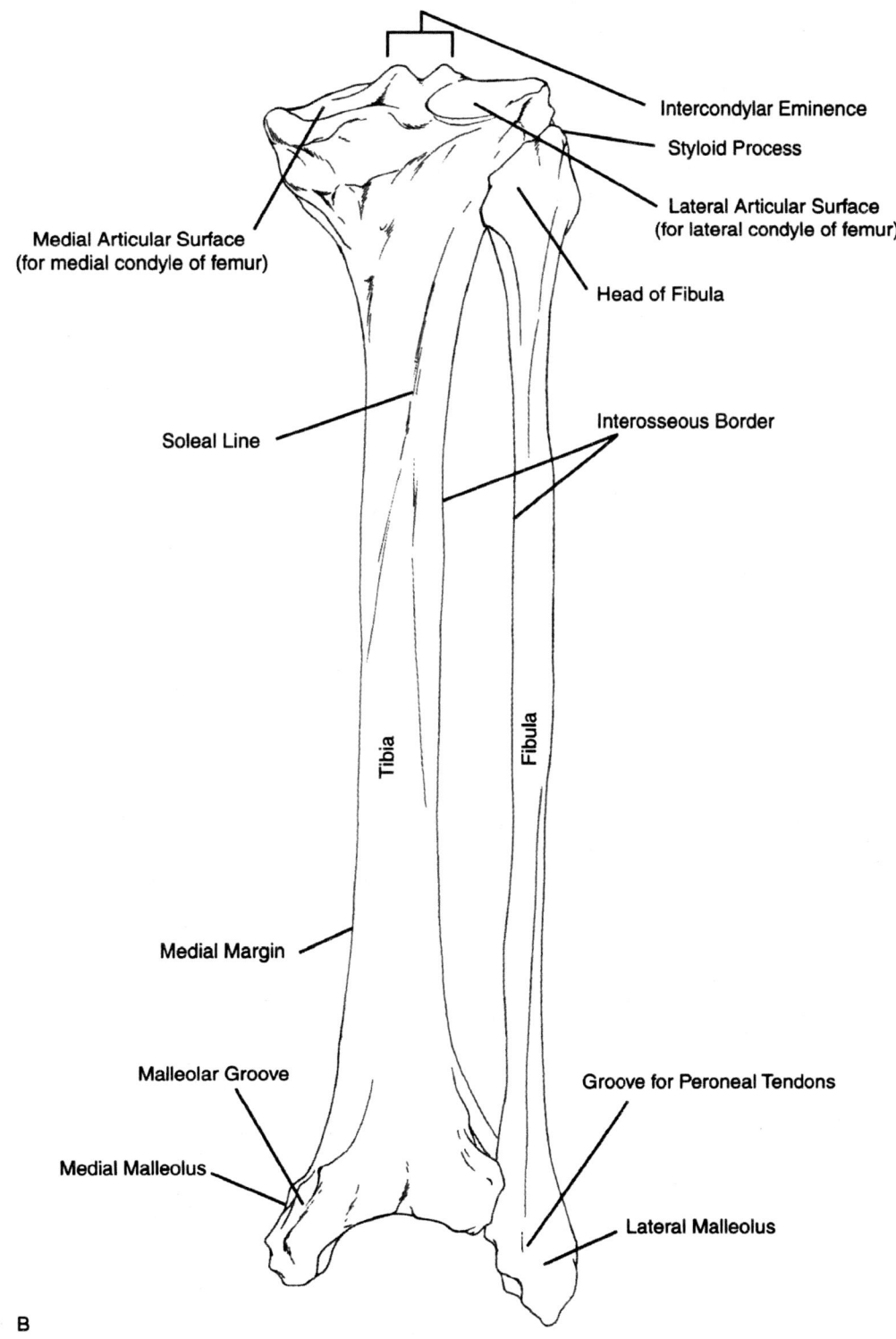

Figure 7-8 (*Continued*) (*B*) posterior.

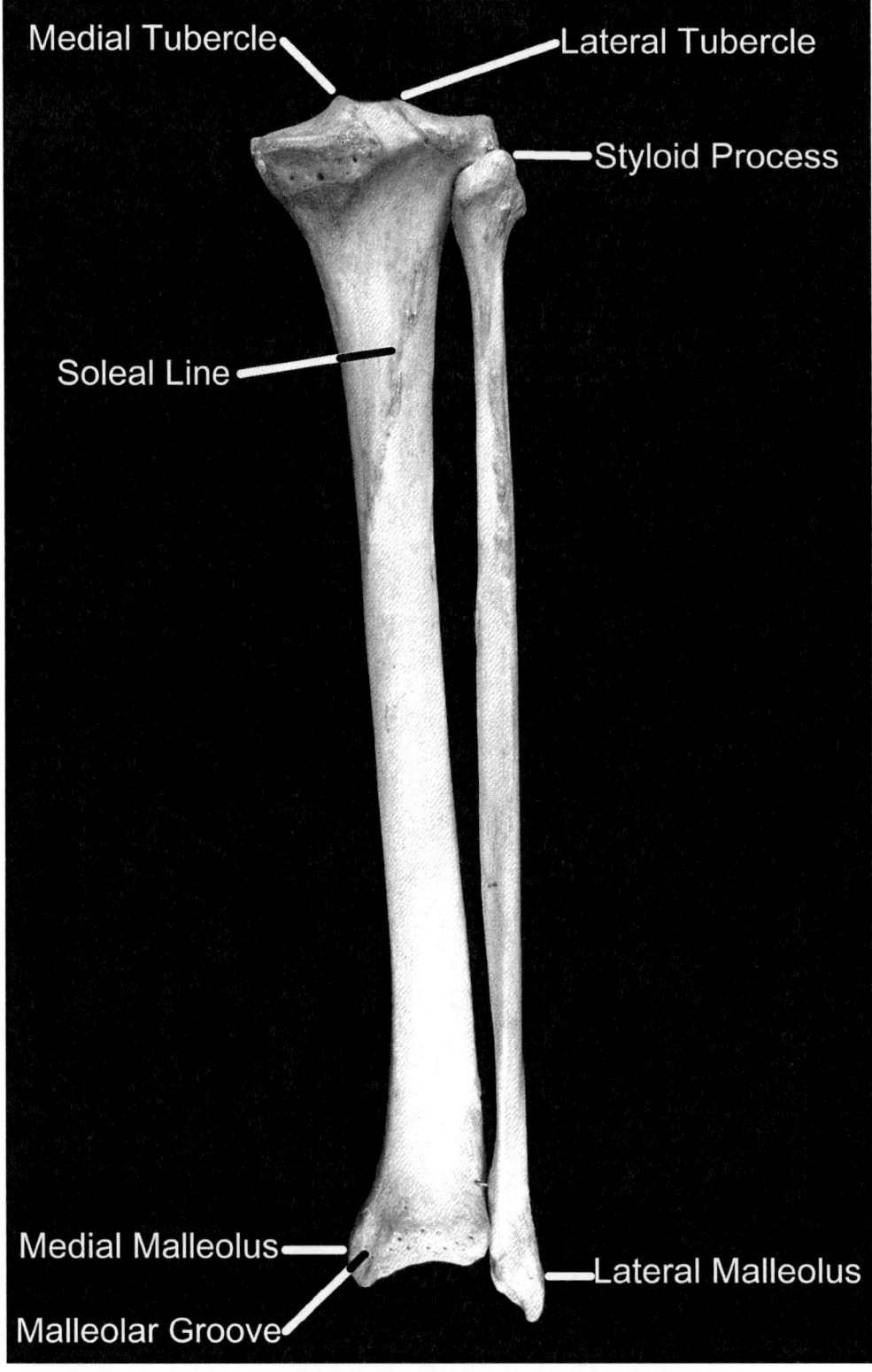

Figure 7-8 B *(Continued)* Photograph of posterior view of the right tibia and fibula, articulated. See accompanying line drawing for identification of additional landmarks.

Femur: The Proximal End

The *head* of the femur approximates a sphere. It is covered with hyaline articular cartilage and articulates with the acetabulum of the os coxae (innominate bone). The *fovea capitis* is a superomedial depression in the head, and lacks articular cartilage. The fovea capitis receives the ligament of the head of the femur. The head is joined to the shaft by the *femoral neck*. There are two *trochanters*, or projections, on the proximal femur, and they are formed by means of separate apophyses. The larger and more lateral is the *greater trochanter*, while the smaller, more medial is the *lesser trochanter*.

Viewed laterally, the greater trochanter can be seen to approximate a square. The square is divided into two triangular areas by a transverse ridge that begins at its superoposterior corner and runs to its anteroinferior corner. The ridge is the insertion of the gluteus medius. The upper of the two triangles thus formed is also covered by the gluteus medius. The area may also underlie a bursa. The lower of the two triangles also underlies a bursa that aids movement of the gluteus maximus which covers this region. The greater trochanteric surface continues onto the anterior surface of the femoral shaft. There an inferior ovoid region may be distinguished which receives the insertion of the gluteus minimus.

At the juncture of the femoral neck and shaft on the anterior surface is a linear rugosity that runs from th*e greater trochanter inferomedially toward the lesser trochanter.* Before the rugose area reaches the lesser trochanter, however, it begins to ascend superomedially, forming a V. The entire line is called the *intertrochanteric line*. The V shaped portion is referred to as the *medial synovial reflection*. The path of the line marks the insertion of a portion of the hip joint capsule called the iliofemoral ligament. The intertrochanteric line continues around the posterior aspect of the femur as the *spiral line*, and eventually reaches the linea aspera just inferior to the pectineal line.

The medial surface of the greater trochanter displays a deep fossa, the *trochanteric fossa*, which receives the insertion of the obturator externus muscle. A groove may be palpated running across the posterior aspect of the neck superolaterally in the direction of the fossa. This is the *obturator externus groove*. Just anterior to the fossa the insertions of the obturator internus and gemelli may be detected. Just above them, on the superior surface of the trochanter is an oval region to which the piriformis attaches.

On the posterior aspect of the proximal femur, the posterior surface of the greater trochanter is connected via a thick broad elevation to the lesser trochanter. This ridge is called the *intertrochanteric crest*. Midway between the two trochanters on this crest is the *quadrate tubercle*. The quadratus femoris muscle inserts onto this tubercle and the portion of the intertrochanteric crest between it and the lesser trochanter. The lesser trochanter is the site of insertion of the combined iliopsoas muscle. The apex of the lesser trochanter receives the psoas major portion of the muscle, just distal to this is the insertion of the iliacus portion of the muscle.

Femur: The Shaft

The most prominent feature of the *shaft* of the femur is the rugose line which runs two-thirds the length of the shaft on its posterior aspect. This raised ridge is called the *linea aspera* (literally, the "rough line"). The linea aspera may be thought of as the arcs of two circles of very large diameter that intersect at about the center of the shaft. As the linea aspera approaches either end of the shaft it divides into two crests or lips, a medial and a lateral. Proximally the lateral crest passes directly into a very long and thick rugosity, the *gluteal tuberosity*. The gluteal tuberosity receives the insertion of a part of the gluteus maximus.

An additional apophysis may occasionally be present at this site, which is then termed the *third trochanter*. (While the third trochanter is only occasionally present in *Homo sapiens*, it is a constant feature in some quadrupeds.) The medial divergence of the proximal linea aspera receives the tendon of the pectineus muscle, and is called the *pectineal line* (superior to the *spiral line*).

The central third of the linea aspera is densely occupied by numerous origins and insertions of thigh and hip muscles. Inserting upon it are the adductor brevis, adductor magnus, and adductor longus. Originating from it are the vastus medialis, vastus intermedius, and the short head of the biceps femoris.

Femur: The Distal End

As the distal extremity of the bone is approached, the linea aspera again divides into medial and lateral portions that descend to *medial* and *lateral condyles* of the femur. In doing so, they define a distinct triangular area on the posterior surface of the distal femur. This is called the *popliteal surface* of the femur. The less distinct medial portion of the divided linea aspera may be seen to terminate just proximal to the medial condyle at a small roughening called the *adductor tubercle*. The "hamstring" part of the adductor magnus muscle inserts on this tubercle. At the base of the triangular popliteal surface superior to the medial and lateral condyles are two local elevated areas for the origins of the medial and lateral heads of the gastrocnemius muscle.

The dominant feature of the distal femur is its articular surface, composed of the large *medial* and *lateral femoral condyles*. The two condyles are confluent anteriorly and proximally to form the *patellar groove* or *surface*, but are separated distally and posteriorly by the *intercondylar notch*. Each of the condyles bears a slight depression that may be easily palpated on the dry bone. These represent the anterior limits of the intraarticular menisci of the knee joint.

Within the intercondylar notch may be found two distinctly smooth areas, one on either condyle. The one on the lateral condyle is more posterior and is the site of attachment of the *anterior cruciate ligament*, while that on the medial condyle is more anterior and is the site of attachment of the *posterior cruciate ligament*. A slight ridge may be seen to join the two condyles at the posterior limit of the intercondylar notch. This is called the *intercondylar line*.

The medial surface of the medial condyle and the lateral surface of the lateral condyle both bear *epicondyles* above the articular surfaces. The *medial epicondyle* is the larger of the two (about the size of a quarter). The tibial (or medial) collateral ligament of the knee takes origin from it. The *lateral epicondyle* is smaller (about the size of a nickel) and gives origin to the fibular (or lateral) collateral ligament. A smooth depression may be located posteroinferior to the lateral epicondyle. This is the origin of the popliteus muscle.

Other Important Facts About the Femur

There are important age- and sex-related variations in the femur. Relative to femoral length, women have greater interacetabular widths than do men. As a result of this, women have a greater *bicondylar angle*, the angle formed between the shaft of the femur and a vertical parasagittal plane.

Because the proximal femur is an area particularly affected by age-related osteoporosis, x-rays of the proximal femur can aid in the establishment of age in undocumented human skeletal material. Such features, together with other features of the skull and pelvis, are useful in medicolegal contexts in the identification of sex in unidentified or archeological human skeletal remains. The femur is also a particularly frequent site for the application of various histological and other methods of determination of age at death from bone. (See Walker and Lovejoy, 1985, Walker, et al., 1994 and other references at the end of the text for further information.)

The Patella

The *patella* (Figure 7-9) is a flat, rounded, triangular bone located in the front of the knee joint. It is generally regarded as a large sesamoid bone in the

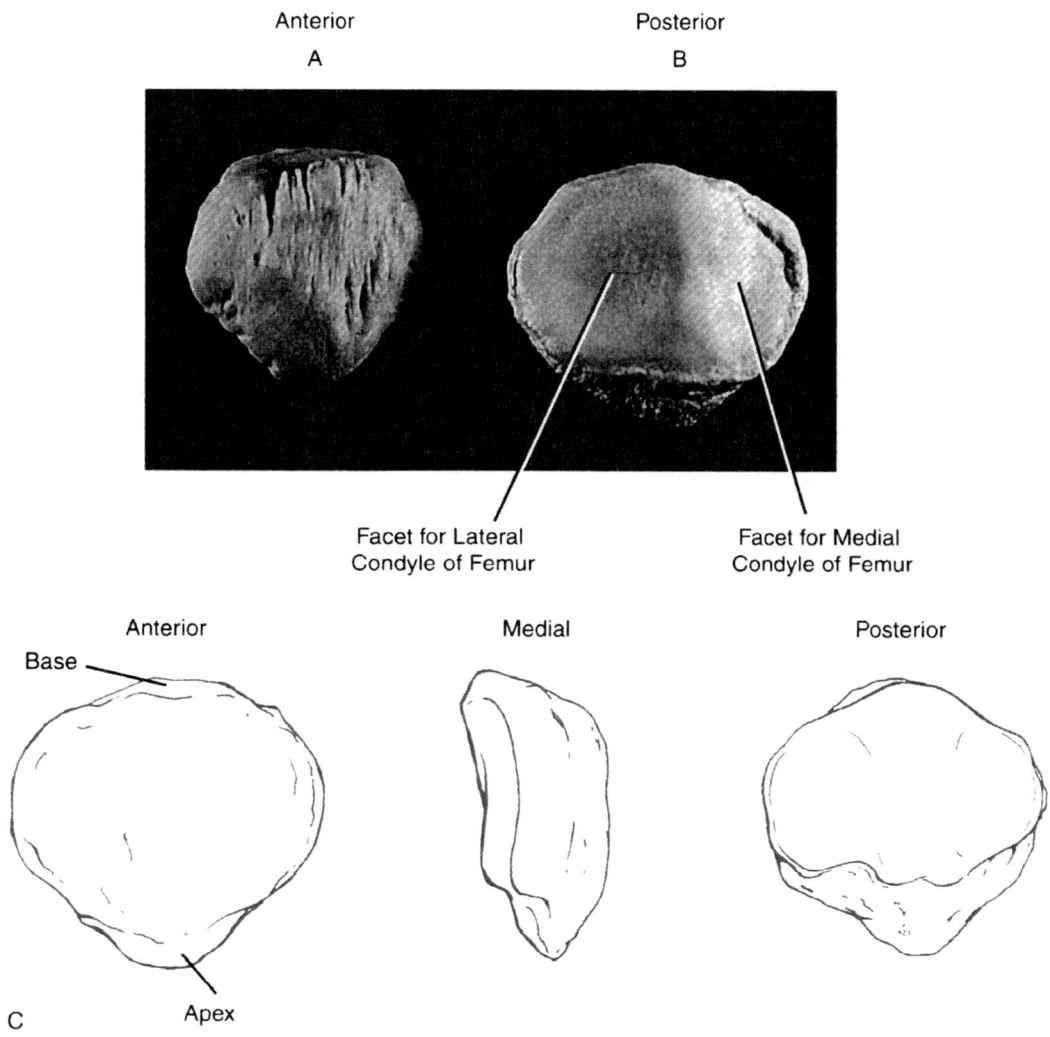

Figure 7-9 The right patella: (A) anterior surfaces and (B) articular (posterior) surfaces. (C) Diagrammatic representation from anterior, medial, and posterior aspects.

tendon of the quadriceps femoris muscle. It serves to protect the front of the knee joint and increases the effectiveness of the quadriceps femoris muscle as an extender of the leg on the knee by increasing its moment arm about the knee. It has an anterior surface, a posterior surface, a base, and an apex.

The anterior surface is subcutaneous, while the posterior surface is possessed of a smooth oval articular area divided into two facets by a vertical ridge. The ridge contacts the patellar groove of the femur in extension, while the two demifacets on either side of this ridge contact the articular surfaces of the medial and lateral condyles of the femur in flexion. The base is proximal, and receives the quadriceps tendon proper. The apex is distal.

Running between the apex of the patella and the tibial tuberosity on the proximal anterior tibia is the *patellar ligament* (patellar tendon). It is properly termed a *ligament* because it connects bone to bone, but functionally it is a part of the tendon of the quadriceps, and possesses the histological features and receptors that characterize a tendon.

The Tibia

The *tibia* (see Figure 7-8) is the main weight-bearing bone of the leg. The fibula transmits only ten percent of the force through the leg. With the exception of the femur, the tibia is the largest bone of the skeleton. It has five articulations, three (superior, middle, and inferior) for the fibula, and one each for the talus and the femur.

Tibia: The Proximal End

The femur rests upon the superior surface of the tibia. This superior surface is called the *tibial plateau*. The plateau supports two articular surfaces on its superior aspect, the *medial* and *lateral articular facets*, and a narrow nonarticular area that separates them. (The medial and lateral aspects of the tibial plateau that support the articular surfaces are often incorrectly referred to as *condyles*. A condyle is a convex, somewhat knuckle-shaped articular surface that articulates with a complementary concave surface on another bone. Thus, the femur possesses *medial* and *lateral condyles* that articulate with medial and lateral *articular surfaces* on the *tibial plateau*.) The medial articular surface of the tibial plateau is ovoid and usually concave while the lateral articular surface of the tibial plateau is usually (but not always) convex.

The area between the articular surfaces is divided into anterior and posterior regions. The anterior region is triangular in form and called the *anterior intercondylar area*. There is typically a deep groove where it borders the lateral articular surface. The anterior attachment of the lateral meniscus lies in this groove.

The anterior attachment of the medial meniscus is less well defined. A depression may usually be found near the anteromedial edge of the anterior interarticular area. This is the site of attachment of the anterior cruciate ligament. (The names of the cruciate ligaments are based on their attachments to the tibia).

In the center of the superior surface of the tibia are the *medial* and *lateral tubercles* that form the *intercondylar eminence*. This structure slopes posteriorly into the *posterior interarticular area*. The medial and lateral menisci are attached in the depressions that occur at the junctures of the medial and lateral tubercles of the intercondylar eminence within this area. The posterior cruciate ligament attaches in the center of this area.

The anterior portion of the proximal tibia is marked by a large triangular area. The inferior apex of this triangle is marked by the *tibial tuberosity*. It has a superior smooth portion and an inferior rough portion. The rough portion is the site of insertion of the patellar ligament, while the smooth portion underlies the infrapatellar bursa. Lateral and superior to the tibial tuberosity is a ridge marking the site of attachment of the iliotibial tract.

The thick shelf-like form of the two tibial "condyles" is broken posteriorly by a gap beneath the posterior interarticular area. Just medial to this gap, on the border of the medial articular surface, is a distinct groove, the site of attachment of the semimembranosus tendon. Just lateral to the gap, on the lateral aspect of the proximal tibia, is the *articular facet for the fibula*. Ligaments binding these two bones are attached around the circumference of this facet. Inferomedial to the tibial tuberosity is the site of the combined insertions of the sartorius, semitendinosus, and gracilis muscles. The marking for this triple-tendon complex is called the *pes anserinus* (Latin meaning "goose foot").

Tibia: The Shaft

Descending from the fibular facet and coursing medially is the *soleal line* (spiral line). It marks the origin of the soleus muscle from the fibular head to the proximal 2/5 of the tibia. Superior to this line is a broad undefined triangular region that gives origin to the popliteus.

Almost joining the soleal line at the fibular facet is a distinct angular elevation that runs the length of the lateral surface of the tibia. This is the *interosseous border* (or *interosseous margin*). It separates the lateral surface of the shaft on its anterior side from the *posterior surface* of the shaft on its posterior side, and gives attachment to the interosseous membrane that runs between the tibia and the fibula. The *anterior border* of the tibial shaft proceeds distally from the tip of the tibial tuberosity along the length of the shaft. It terminates on the anterior surface of the medial malleolus.

The *medial margin* is ill-defined. It develops from the expanded buttress of the medial side of the tibial plateau and descends to the posterior surface of the medial malleolus. It is more defined in its lower half. Much of its proximal portion is the soleal line. Just anterior to the medial margin, where the buttressing of the tibial plateau blends with the shaft, there will usually occur a roughened poorly defined region that is the insertion of the tibial (medial) collateral ligament of the knee.

The proximal two-thirds of the lateral surface of the tibial shaft gives rise to the tibialis anterior. The posterior surface of the tibial diaphysis gives rise to the tibialis posterior laterally and flexor digitorum longus medially (as well as the soleus and popliteus). Their origins are variable but occur along the middle third of the shaft. The tibialis posterior origin extends up to the lower portion of the soleal line and bears the nutrient foramen (in most cases). The tibialis posterior and anterior are separated from one another by the interosseous membrane.

Tibia: The Distal End

The prominent projection of the distal end is the *medial malleolus*. Its inferior surface is occasionally pierced by vascular channels. Its posterior surface bears a deep groove which upon palpation can usually be felt to be double. These grooves are for the tibialis posterior and flexor digitorum longus tendons. It is called the *malleolar groove*.

Another groove may also be detectable on the lateral portion of the posterior surface of the distal tibia. This is *groove for the flexor hallucis longus tendon*. The lateral surface of the distal end bears the *fibular notch*. The inferior end of the medial malleolus, when viewed medially, usually appears notched. The deltoid ligament originates in this notch. The inferior articular surface, for the talus, frequently bears on its anterior edge articular extensions that contact the talus in extreme dorsiflexion of the foot. These are called *squatting facets*. They are not a constant feature, and their frequency varies from population to population.

The Fibula

The *fibula* (see figure 7-8) bears about ten percent of the force transmitted through the leg. It has three articular surfaces: two (proximal and distal) for the tibia, and one distally for the talus. The joint between the distal fibula and the talus is synovial, while the joints between the fibula and tibia are more fibrous, though at least in part synovial.

Fibula: The Proximal End

The proximal enlargement of the fibula is the *head*. It bears a flat *articular facet* for the corresponding facet on the lateral aspect of the tibial plateau. Surrounding it are the insertions of the tibiofibular joint capsule. The head has a distinct superiorly directed apex, the *styloid process*. The lateral surface of the head receives the tendon of the biceps femoris muscle. The posterior surface of the head receives the origin of the soleus muscle. The lateral aspect also receives the fibular (lateral) collateral ligament of the knee joint and gives origin to the fibularis (peroneus) longus muscle. The extensor digitorum longus arises from the anterior aspect of the head. *None of these origins and insertions are clearly defined osteologically.*

Fibula: The Shaft

The shaft is highly variable and its muscular markings are usually difficult to interpret. The shaft is prismatic in cross-section. The lateral surface is usually the widest and smoothest. Note that it twists as it descends distally and begins to face more posteriorly. It terminates on the lateral surface of the distal end of the fibula which is called the *lateral malleolus*. Its proximal 2/3 gives origin to the fibularis (peroneus) longus muscle. Its distal two-thirds gives origin to the fibularis (peroneus) brevis muscle. The tendons of these muscles pass through a groove on the posterior surface of the lateral malleolus.

Directly opposite the lateral surface of the shaft is the *interosseous crest* of the fibula, affording attachment to the interosseous membrane. It divides the remainder of the fibular surface into a *medial surface* and a *posterior surface*. The *medial surface* is usually grooved and gives rise to the extensor digitorum longus, peroneus tertius, and extensor hallucis longus. The posterior surface is frequently divided by a prominent line, the *oblique line*. It divides the posterior surface into two parts, one facing more medially, which gives rise to the tibialis posterior, and the other facing more posteriorly, which gives rise to the soleus and the flexor hallucis longus.

Fibula: Distal End

The distal extremity of the fibula is the *lateral malleolus*. Its medial surface is composed of the more anterior *malleolar articular surface* and the more posterior *lateral malleolar fossa*. In the latter are the origins of the posterior tibiofibular ligament and posterior talofibular ligament. The distal extension of the posterior face of the lateral malleolus bears a *groove for the tendon of the fibularis (peroneus) longus and brevis muscles*. The anterior border of the lateral malleolus gives attachment to the anterior tibiofibular ligament, the anterior talofibular ligament, and the calcaneofibular ligament. The *articular surface* of the lateral malleolus is for articulation with the talus. The distal tibiofibular joint is a syndesmosis, and therefore has no subchondral bone.

The Foot

The skeleton of the *foot* (figure 7-10) is composed of 26 bones which consist of three groups: the *tarsal bones*; the *metatarsal bones*; and the *phalanges*.

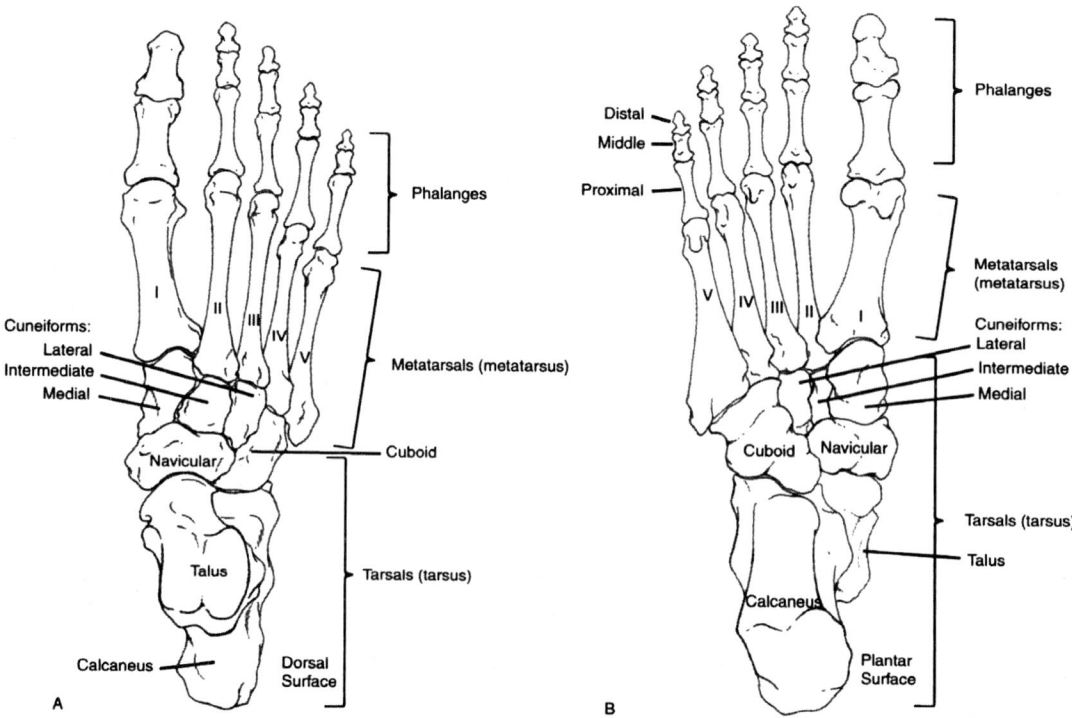

Figure 7-10 The skeleton of the right foot: (*A*) dorsal, (*B*) plantar.

The *tarsal bones* are most proximal, form the ankle and heel, and articulate proximally with the tibia and fibula. The *metatarsals* are a set of five small long bones. They are numbered I through V from the medial (big toe) side. Their bases articulate with the distal row of tarsal bones, and their heads articulate

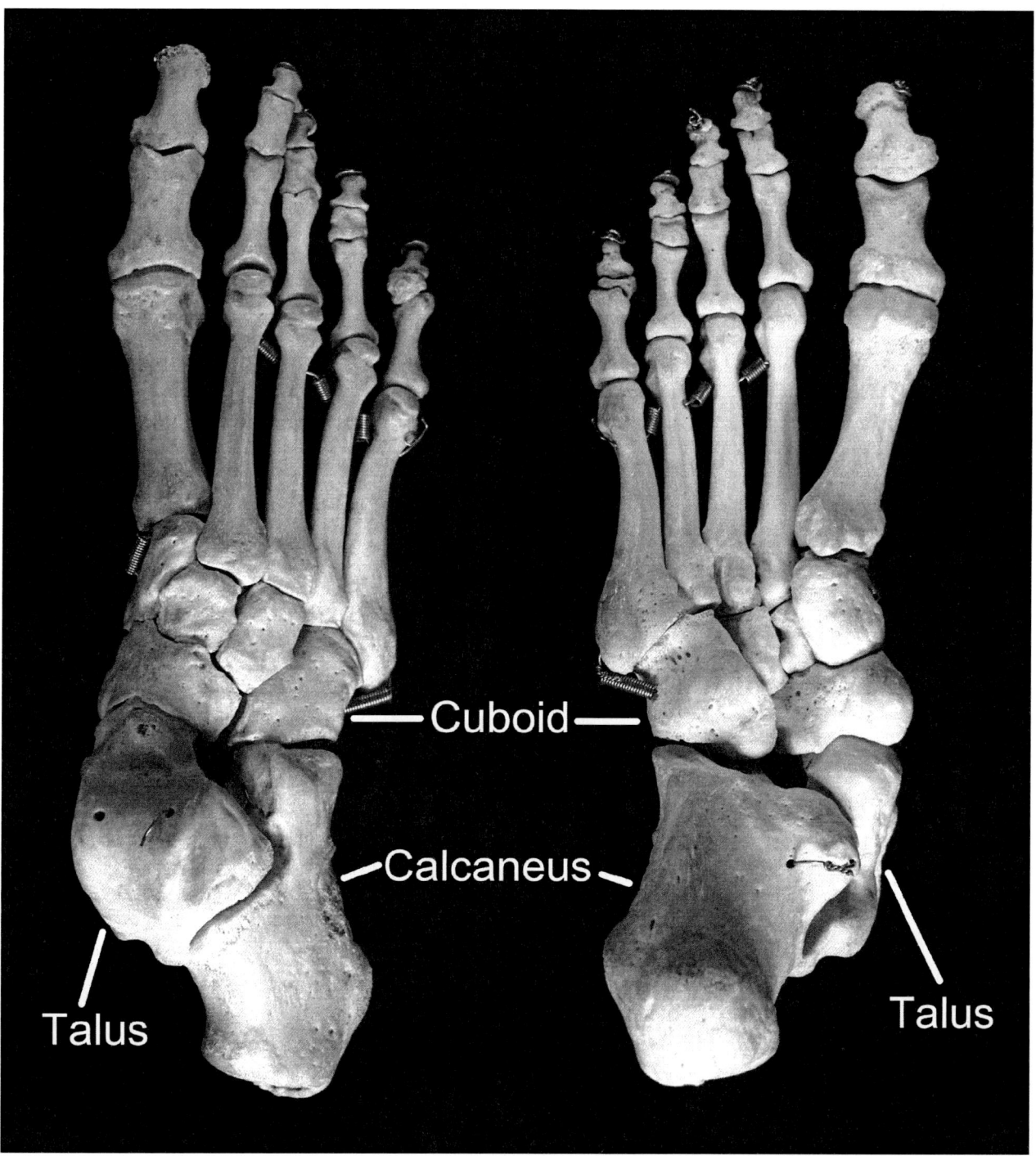

Figure 7-10 *(Continued)* Photograph of the skeleton of the right foot, dorsal surface on left, plantar surface on right. See accompanying line drawing for identification of additional landmarks.

with the proximal phalanges of each of the digits. Each digit has three *phalanges*, a proximal, a middle, and a distal, with the exception of the big toe or hallux. The hallux has only two phalanges, termed proximal and distal.

Foot: The Tarsus

The *tarsus* (see Figure 7-10) consists of seven small, irregularly shaped bones, arranged in three rows, a proximal, an intermediate, and a distal. They are closely applied to one another and are held firmly in their rows by strong ligaments. As a whole they form a strong fibrous and bony mass. The *proximal row* consists of the talus and the calcaneus. The talus rests on the superior surface of the calcaneus, between the latter and the medial and lateral malleoli. The *intermediate row* consists of a single bone, the navicular, situated between the head of the talus and the distal row of tarsals. Medially to laterally, the *distal row* consists of the medial cuneiform, the intermediate cuneiform, the lateral cuneiform, and the cuboid.

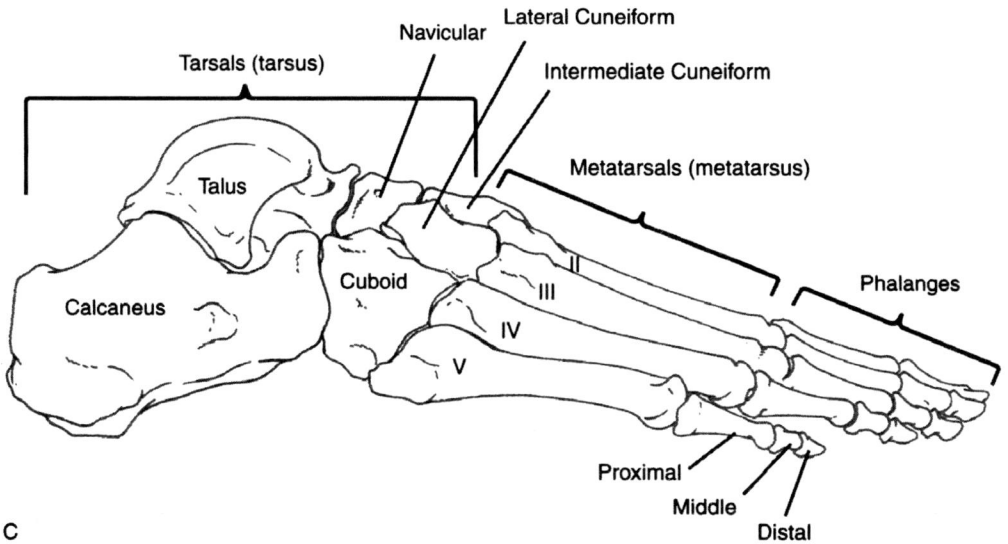

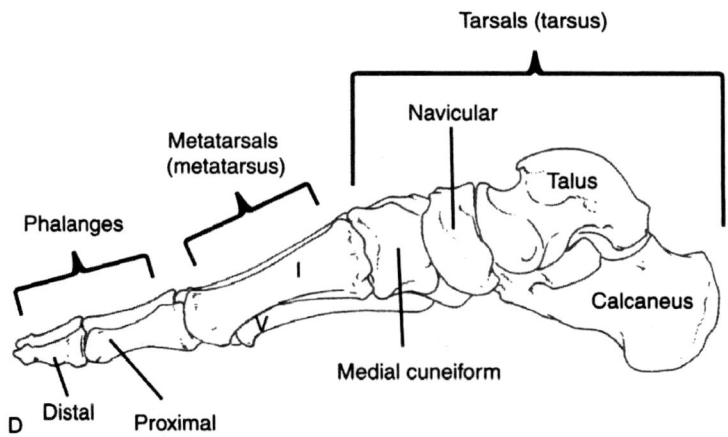

Figure 7-10 *(Continued)* (C) lateral and (D) medial views.

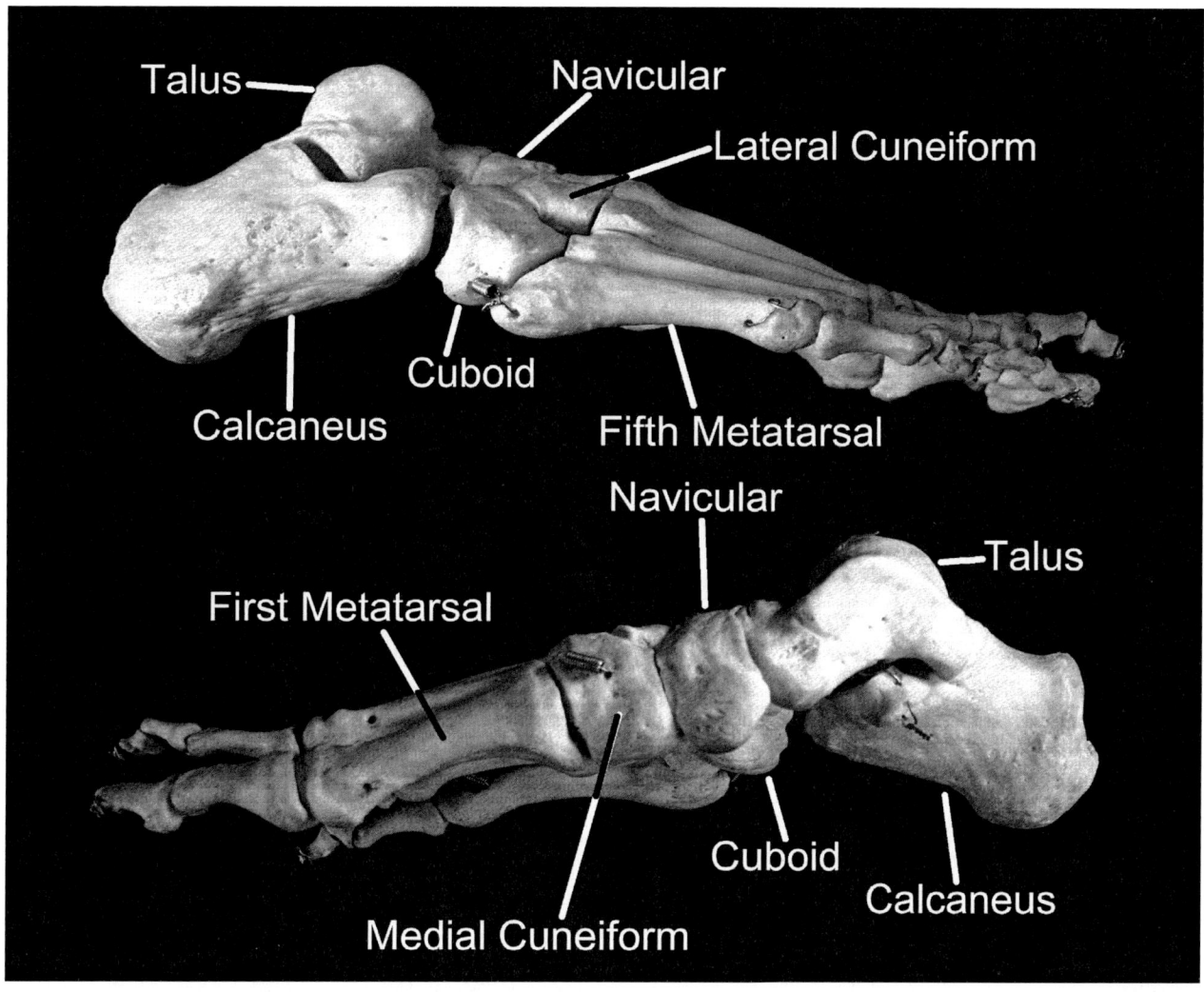

Figure 7-10 *(Continued)* Photograph of the skeleton of the right foot, lateral aspect above, medial aspect below. See accompanying line drawing for identification of additional landmarks.

The talus (Figure 7-11 A, see also Figure 7-10) is the largest bone of the body that lacks any muscle attachments. The three major portions are the *corpus, neck* and *head*. The *superior articular surface (trochlear surface)* (on which rests the tibia) and the two *malleolar surfaces (medial malleolar surface* and *lateral malleolar surface)* constitute the *trochlea* of the talus. Immediately posterior to the medial malleolar facet is a vascularized area; behind this is a roughened area for the insertion of the deltoid ligament that firmly binds the medial malleolus with the talus and calcaneus.

The posterior surface of the talus is marked by a deep *groove for the tendon of the flexor hallucis longus*. On either side are the *lateral and medial tubercles of the talus*. The *lateral tubercle* receives the posterior talofibular ligament. The *medial tubercle* receives the medial talocalcaneal ligament.

The inferior surface of the corpus is occupied by the *posterior calcaneal articular surface*. The inferior surface of the neck of the talus is occupied by the *anterior calcaneal articular surface*. (Occasionally this is divided into a smaller *anterior* and larger *middle calcaneal articular surface*, to correspond with a like division on the calcaneus). Between these two articular areas is the deep *talar groove*. This groove is occupied by the interosseous talocalcaneal ligament that is a dense band filling the entire length of the groove.

Anterior to the neck of the talus is the *talar head*. Its articular surface is usually distinguishable into three planes. The most superior of these is the *navicular articular surface*. The other two planes, if present, define the anterior and middle parts of the anterior calcaneal articular surface mentioned above.

The *calcaneus* (see Figures 7-10 and 7-11 B) is the only bone of the tarsus with an *apophysis*, which develops at the site of insertion of the gastrocnemius-soleus (calcaneal or Achilles) tendon. It appears before the age of 10 and unites during the late teens. The large posteroinferior buttress is called the *tuberosity of the calcaneus*. The superior surface usually has three facets for articulation with the talus, the *anterior, middle,* and *posterior talar articular surfaces*. The anterior and middle facets are often conjoined. The anterior surface of the bone is defined by the *cuboidal articular surface*.

When, you articulate the talus and calcaneus, note that the talar groove has a corresponding floor on the calcaneus. This channel created between the talus and calcaneus is called the *tarsal sinus*. The floor is called the *calcaneal groove*. The sinus and groove contain the *interosseous talocalcaneal ligament*, which also attaches in the talar groove.

The rough inferior surface ends posteriorly in an abrupt fashion as it develops into the *tuberosity of the calcaneus*. A well defined anteriorly projecting spike of bone from this tuberosity is the site of origin of the long plantar ligament, which passes anteriorly to attach to the cuboid and four lateral metatarsal bases. The superior part of the calcaneal tuberosity is smooth while the inferior part presents definite striations. The superior part underlies the bursa deep to the calcaneal tendon, while the rough inferior part receives the calcaneal tendon.

When the talus and calcaneus are articulated, the lateral malleolar surface of the talus comes to a point and this "points" inferiorly to a small but well defined area of bone that is the *fibular (peroneal) trochlea*. Posterior to this is a slight *groove for the fibular (peroneal) muscles*. Superior to the groove, at about the midpoint along the length of the bone is a roughened area for the insertion of the calcaneofibular ligament.

The process on which is found the middle talar articular facet is called the *sustentaculum tali*. The sustentaculum tali projects medially and is easily identified in even fragmentary specimens by the presence of the *sustentaculum tali*. Note the deep groove on the under surface of this structure. This is the *groove for the flexor hallucis longus tendon*.

The *navicular bone* (Figure 7-12) separates the talus and the three cuneiform bones. Its *posterior surface* is deeply concave for articulation with the head of the talus. Its *anterior surface* has three distinct facets for the three cuneiform bones. The largest of these is for the medial or first cuneiform. Extending out medially from the navicular is the *navicular tuberosity*, which receives the tendon of the tibialis posterior muscle. The superior surface is convex.

The *medial (first) cuneiform* (see Figure 7-12) is the largest of the three cuneiform bones. Its largest facet is kidney-shaped and defines the anterior surface. It articulates with the first metatarsal. Opposite it, on the posterior surface, is a concave facet for the navicular. The lateral surface has an L-shaped facet for articulation with the second cuneiform. On the medial surface is a smooth area that is actually a groove for the tendon of the tibialis anterior. The bone is wedge-shaped. The inferior surface is narrow while the superior surface is broad. The inferior surface receives the insertion of the tibialis anterior, tibialis posterior, and peroneus longus muscles.

The *intermediate (second) cuneiform* (see Figure 7-12) is the smallest of the cuneiforms. Two of its surfaces, the anterior and posterior, are completely covered by subchondral bone. The posterior concave surface articulates with the navicular, the anterior convex surface with the second

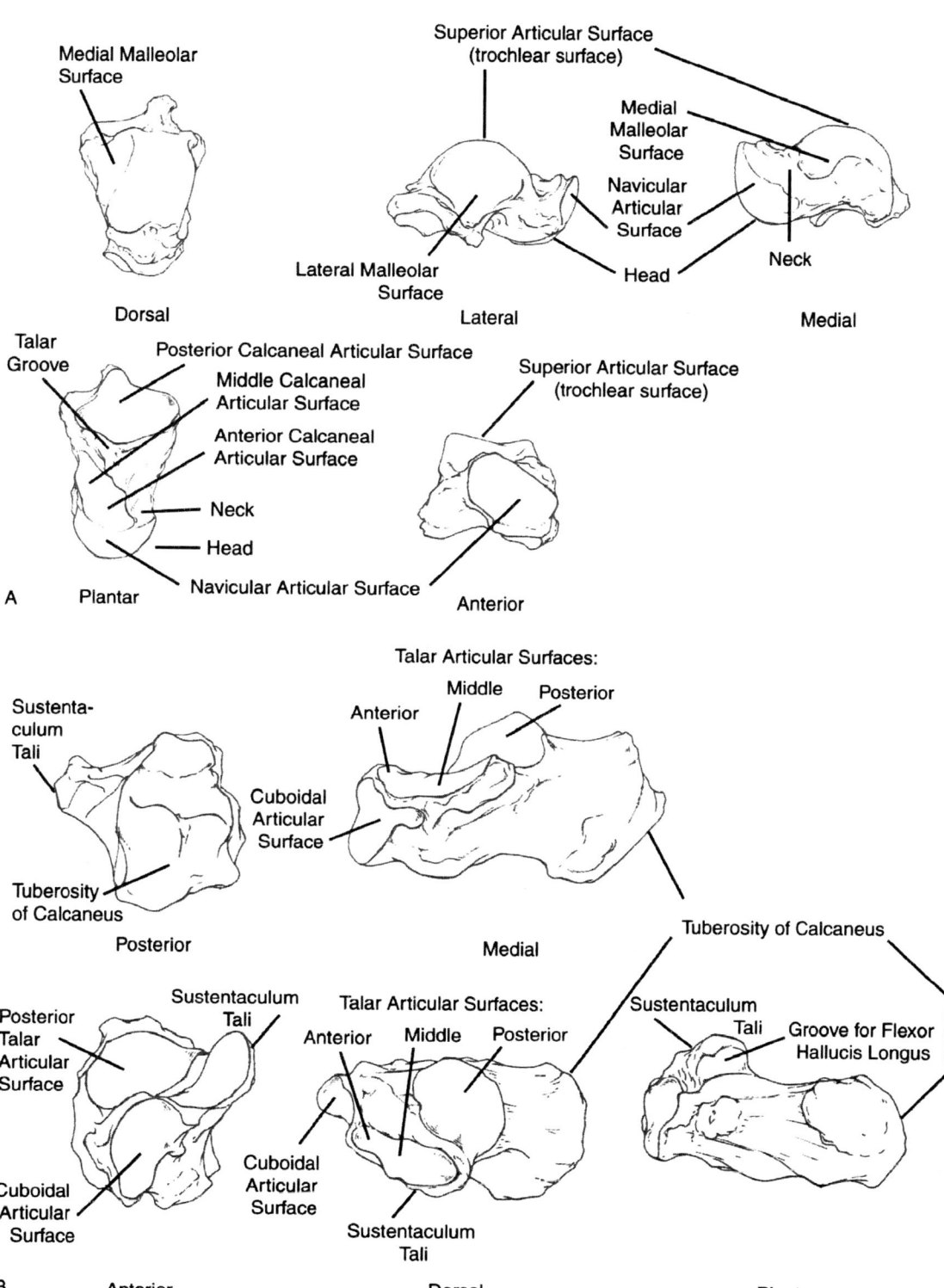

Figure 7-11 (A) Right talus and (B) right calcaneus.

metatarsal. An L-shaped articular facet is located on its medial surface and matches the facet on the first cuneiform. The inferior surface receives an insertion of the tibialis posterior.

The *lateral (third) cuneiform* (see Figure 7-12) also has anterior and posterior surfaces completely covered by subchondral bone. The large anterior surface articulates with the third metatarsal. The smaller posterior surface articulates with the navicular. The navicular facet continues on to the medial surface of the bone to form the articular surface for the second cuneiform.

The *cuboid* (see Figure 7-12) is quite distinctive in shape. Note that it has only three articular facets. One of these is small and isolated. This is for the third cuneiform, and matches the corresponding facet on that bone. The superior surface is broad. The inferior surface has a prominent ridge. The smaller of the two remaining articular facets is anterior and is subdivided for the fourth and fifth metatarsals. The posterior surface is saddle-shaped with a prominent projection. It articulates with the calcaneus. The prominent ridge on the inferior surface is for the *attachment of the long plantar ligament*. Anterior to this is a deep *groove for the tendon of the*

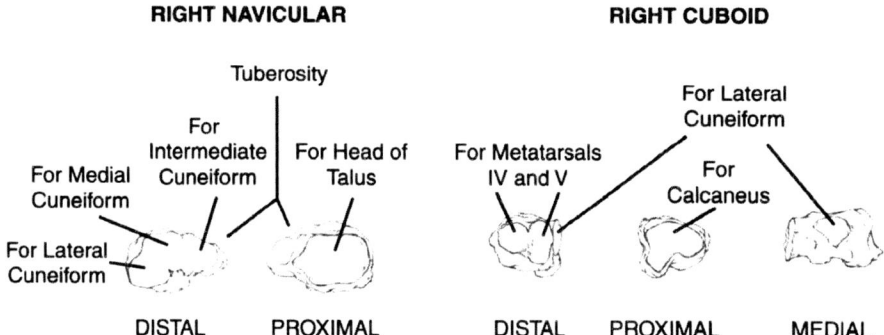

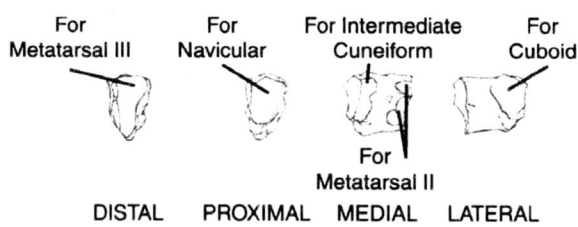

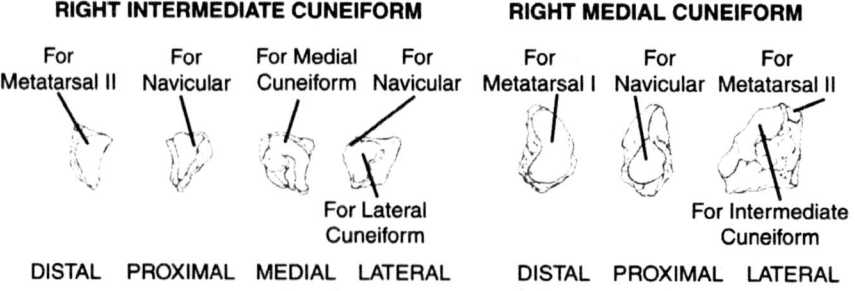

Figure 7-12 The remaining tarsal bones.

peroneus longus muscle. The notch on the very small lateral surface of the bone is an artifact of the peroneus longus groove.

The Foot: The Metatarsus

The *metatarsus* (see Figure 7-10) is formed by the five *metatarsal bones*. Each of the five metatarsal bones has a *shaft*, a proximal *base* and a distal *head*. The first metatarsal is the most robust and the shortest. The peroneus longus and tibialis anterior insert at the inferior surface of its base. There is occasionally a facet for the articulation with the second metatarsal. The *head* is subdivided into *two grooves* by a prominent inferior ridge. These grooves are for the *two sesamoid bones* in the tendons of the flexor hallucis brevis. The nutrient foramen is usually clearly identifiable.

The rest of the metatarsals may be easily identified as follows if all are present. The *second* is the longest. The *third, fourth,* and *fifth* successively decrease in length. The fifth is especially distinctive because of the presence of a prominent inferolateral *tuberosity* that receives the tendon of the fibularis (peroneus) brevis muscle. The bases of the second and third metatarsals have distinctive triangular articular facets for the second and third cuneiforms. The second is much longer and its facet is more concave, well rounded, and distinctly non-angular.

The Foot: The Phalanges

There are three phalanges (see Figure 7-10) in each toe, a *proximal,* a *middle,* and a *distal*, with the exception of the hallux, which possesses only two: a proximal and a distal, just as does the *pollex*. The phalanges of the toes are similar to the phalanges of the hand in general morphology. They are, however, generally shorter. The phalanges of the hallux are very robust, much more robust than those of the pollex.

The *distal phalanx of the hallux* is distinctive. It is very large and the bone shows considerable torsion along its long axis. The flexor digitorum brevis inserts upon the middle phalanges of digits 2 to 5. The distal phalanges receive the tendons of the flexor digitorum longus and extensor digitorum longus. Unlike the palmar surfaces of the phalanges of the hand, the plantar surfaces of the phalanges of the foot are rounded.

Articulations of The Lower Limb

The joints of the lower limb consist of the joints of the hip, knee, ankle, and foot. The pelvic girdle is intimately associated with the hip joint, and so we shall consider it here with the lower limb because it is one of the few structures that directly link the axial skeleton to the pelvic girdle.

Articulations and Ligaments of the Pelvis

The bony pelvis (which includes the sacrum) is joined to the lumbar vertebrae via the anterior and posterior longitudinal ligaments, the intervertebral disc between L5 and S1, the ligamenta flava, the interspinous and supraspinous ligaments, and articular capsules of the zygapophyseal joints (refer to figures 7-13 to 7-15 throughout the discussion of the articulations and ligaments of the pelvis). Additionally, the 5th lumbar vertebra (and occasionally the 4th) is joined with the pelvis via one additional ligament on each side, the *iliolumbar ligaments* (figures 7-13, 7-14). The *iliolumbar ligament* consists of a *superior band* and an *inferior band* (also known as the *lumbosacral ligament*).

Both bands attach medially to the transverse process of L5 and sometimes L4 as well. The *superior band* runs from the transverse process of L5 to crest of the ilium immediately lateral to the sacroiliac articulation. The *inferior band* runs from the transverse process of L5 to the base of the sacrum, where it blends with the *anterior sacroiliac ligament*. These two bands limit lateral flexion of the lumbar spine.

Other articulations of the pelvis can be divided into four groups: (1) those connecting the

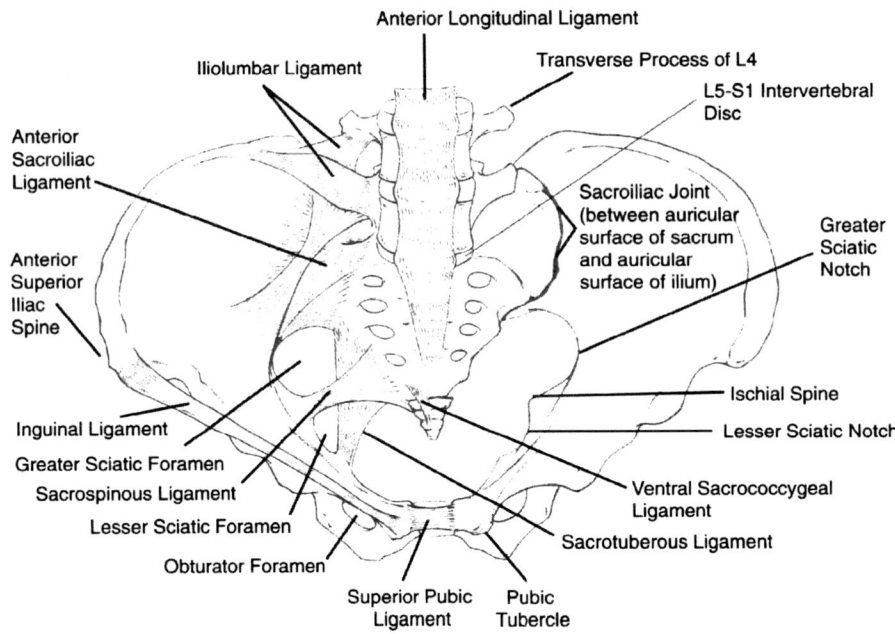

Figure 7-13 Ligaments of the pelvis, anterior view.

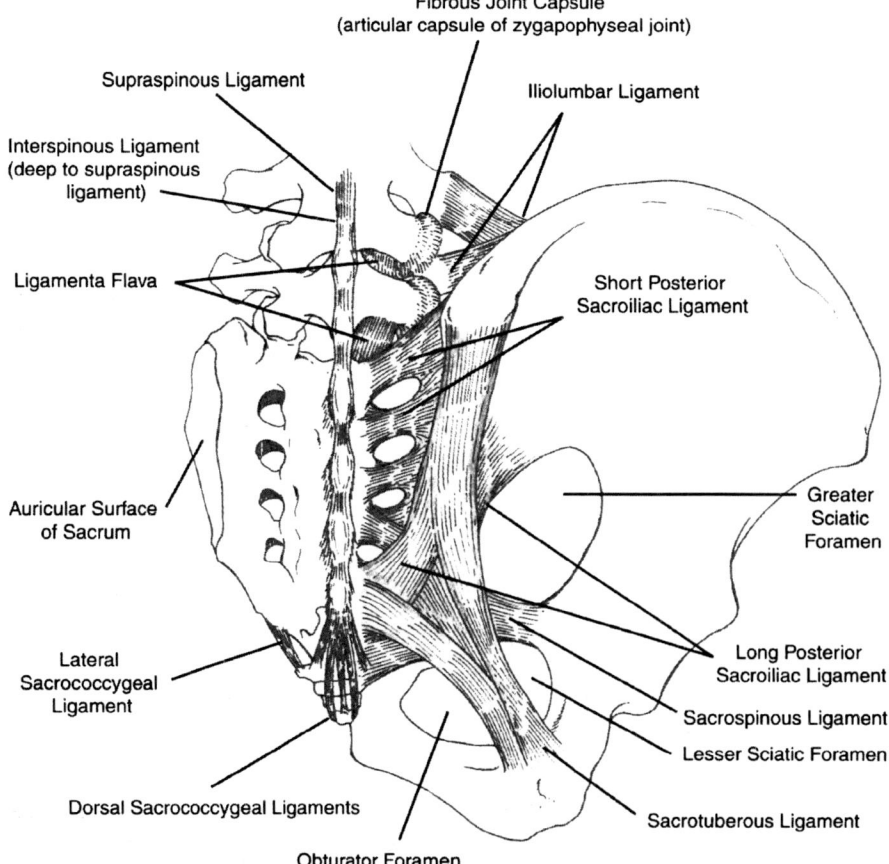

Figure 7-14 Ligaments of the pelvis, posterior view.

sacrum and ilium; (2) those passing between the sacrum and ischium; (3) those uniting the sacrum and coccyx (discussed with the vertebral articulations); and (4) those between the two pubic bones.

The Sacroiliac Articulation

The *sacroiliac articulation* occurs between the *auricular surface of the ilium* and the matching *auricular surface of the sacrum*. This joint is a synchondrosis. The joint has a very special character because the sacral portion develops from somitic mesoderm while the innominate portion develops from lateral plate mesoderm. The sacral portion presents a typical chondral surface and is also a growth center, or epiphysis. The surface of the innominate portion has a different texture and includes a significant amount of fibrocartilage. The two parts of the joint are united by patches of softer fibrocartilage. Over a considerable extent of the joint, especially in later life, there may be a synovial cavity, so that it exhibits some characteristics of a plane synovial joint. There are three ligaments of this joint:

- The Anterior Sacroiliac Ligament
- The Posterior Sacroiliac Ligament
- The Interosseous Sacroiliac Ligament

The *anterior sacroiliac ligament* extends from the base and lateral parts of the ventral surface of the sacrum to the margins of the auricular surface of the ilium.

The *posterior sacroiliac ligament* consists of both long and short fibers. The *short posterior sacroiliac ligament* runs from intermediate and lateral sacral crests to the posterior superior iliac spine and internal lip of the iliac crest at its dorsal end. The *long posterior sacroiliac ligament* runs from the lower part of the lateral sacral crest to the posterior superior iliac spine, where some of its fibers merge with the fibers of the *sacrotuberous ligament*.

The *interosseous sacroiliac ligament* is located between the sacroiliac joint and the posterior sacroiliac ligaments, uniting the tuberosities of the ilium and the sacrum. It is a large syndesmosis lying deep to the posterior sacroiliac ligament.

During flexion and extension of the trunk, some rotation may occur at the sacroiliac articulation (Soames and Atha, 1981). The axis of rotation is located 5 to 10 cm. inferior to the sacral promontory. The greatest amount of motion of the sacrum relative to the innominate bones occurs in rising from a sitting position. The sacral promontory moves forward, a movement called *nutation,* as much as 5 or 6 mm. as body weight pushes down on the top of the sacrum. Less posterior motion occurs at the sacral apex.

The range of motion is the same in males and nonpregnant females, but *nutation* and *counternutation* increase in pregnancy. The sacroiliac joint may be injured by falls, automobile accidents, childbirth, and other causes, and can result in pain in walking, climbing stairs, going from a sitting to a standing position and vice versa. The patient may walk with a limp. (See Edwardson, 1995 for more details.)

Ligaments Connecting the Sacrum and Ischium

The two ligaments connecting the sacrum and the ischium are the sacrotuberous ligament and the sacrospinous ligament.

The *sacrotuberous ligament* extends from the posterior inferior iliac spine, the 4th and 5th transverse tubercles and caudal part of the lateral margin of the sacrum and coccyx to the inner margin of the tuberosity of the ischium. Its more proximal fibers fuse with the long posterior sacroiliac ligament.

The *sacrospinous ligament* is a thin triangular sheet attached by its broad base to the lateral margins of the sacrum and coccyx. Its apex is attached to the ischial spine. The sacrotuberous and sacrospinous ligaments close the *greater and lesser sciatic notches* of the ilium to create the *greater and lesser sciatic foramina*.

The Pubic Symphysis

The pubic symphysis is the joint between the bodies of the two pubic bones. As in all symphyses, a fibrocartilagenous disc is interposed between the articular surface of the two bones. Each of these two surfaces is covered with a thin layer of hyaline cartilage. The ligaments of this articulation are the superior pubic ligament, and the arcuate pubic ligament or inferior pubic ligament.

The *superior pubic ligament* connects the two pubic bones superiorly and extends laterally as far as the pubic tubercles. The *arcuate pubic ligament* is a thick band of fibers which unite the two pubic bones and forms the boundary of the pubic arch (subpubic angle). Its fibers blend with those of the interpubic fibrocartilagenous disc. It attaches to the inferior pubic rami.

Other Ligaments of the Pelvis

In addition to the ligaments associated with the various articulations of the pelvis, there are also important ligaments associated with the attachment of the anterior abdominal wall to the innominate bone. These include:

- *The Inguinal Ligament*
- *The Lacunar Ligament*
- *The Pectineal Ligament.*
- *The Iliopectineal Band.*

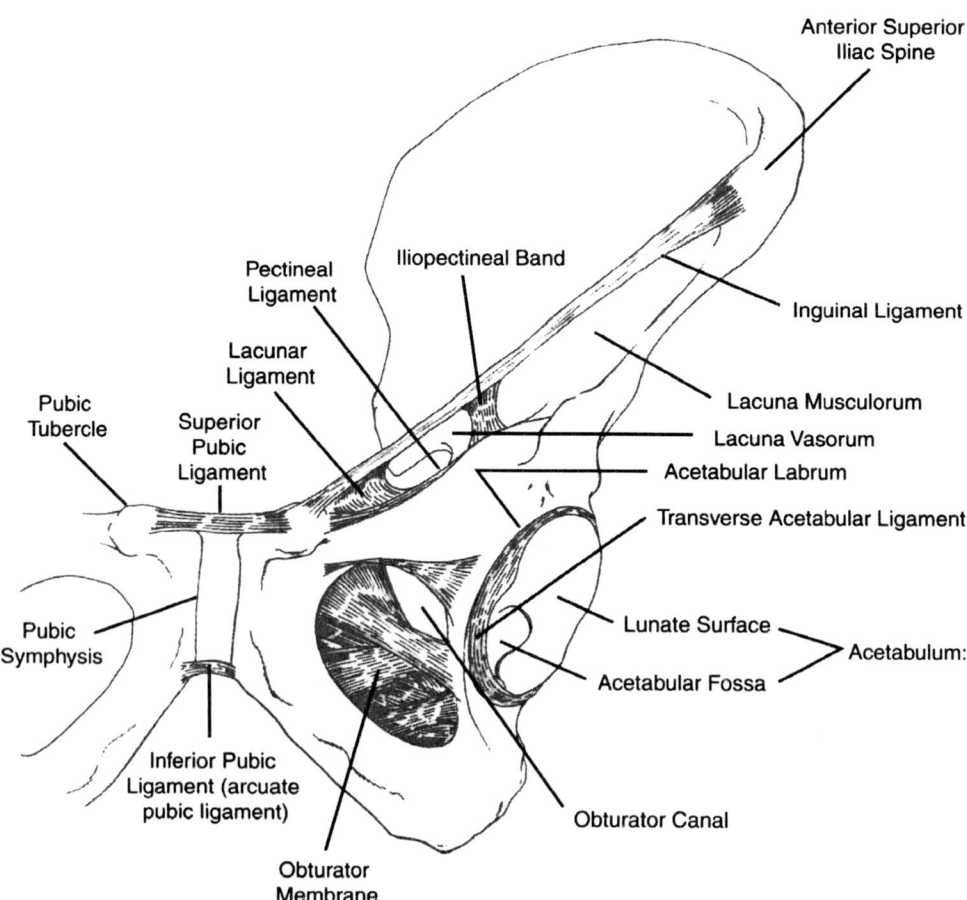

Figure 7-15 Inguinal and associated ligaments, pubic ligaments, obturator membrane, transverse acetabular ligament, and acetabular labrum, left anterior view.

The *inguinal ligament* is formed by the aponeurosis of the external oblique muscle. It attaches to the anterior superior iliac spine and the pubic tubercle. The medial part of the inguinal ligament is reflected horizontally back and is attached to the pecten pubis as the *lacunar ligament*. Some of the fibers of the lacunar ligament continue laterally along the pectineal line of the pubis. These fibers are known as the *pectineal ligament*.

Connecting the inferior surface of the middle portion of the inguinal ligament to the superior surface of the iliopectineal eminence is the *iliopectineal band*. This iliopectineal band divides the space beneath the inguinal ligament into a more lateral *lacuna musculorum* (for the passage of the iliopsoas muscle and femoral nerve – the floor of this lacuna is the iliopsoas groove of the innominate) and a more medial *lacuna vasorum* (for the passage of the femoral vessels and lymphatics). The *lacunar ligament* is so called because it demarcates the medial extent of the lacuna vasorum.

The Hip Joint

The *hip (acetabulofemoral) joint* (Figure 7-16) is a spheroidal, or ball-and-socket joint. As such, it has a great range of motion. In fact, only the glenohumeral joint has a greater range of motion. However, the hip joint has a much deeper bony socket for the reception of the femoral head than does the shoulder joint for the head of the humerus. The hip joint is also surrounded by massive amounts of muscle and a very strong fibrous joint capsule with well-developed intrinsic ligaments relative to the glenohumeral joint. As a result, while the hip joint is more stable and much less subject to displacement/dislocation than is the shoulder joint, its range of motion is somewhat more limited.

The simplest definition of the hip joint is the articulation between the spheroidal head of the femur with the cup-shaped acetabulum. While that is certainly a straightforward definition, the joint is not quite that simple. The convex femoral head and the concave acetabulum are not quite congruent, in that

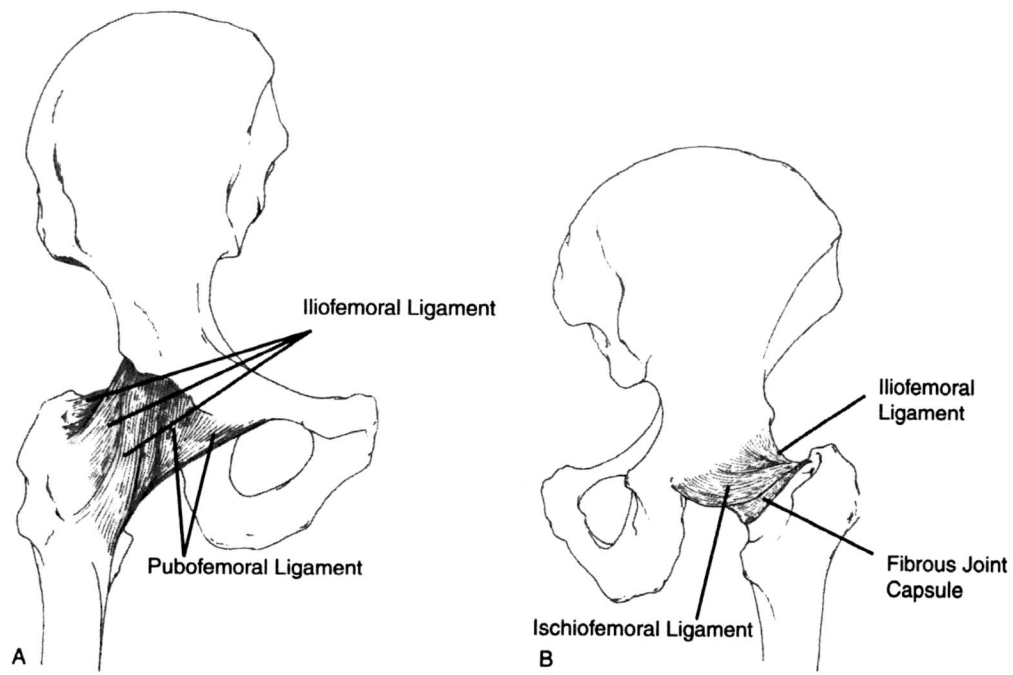

Figure 7-16 The fibrous capsule and ligaments of the right hip joint: (*A*) anterior and (*B*) posterior views.

they do not have exactly the same curvature. Further, the femoral head is not quite spheroid, but is slightly ovoid. The joint surfaces approach a nearly spherical form in older age.

Additionally, the femoral head does not present a smooth, uninterrupted articular surface. In fact, near the center of the head of the femur is the roughened *fovea capitis* for the reception of the *ligament of the head of the femur (ligamentum teres capitis femoris)*, a small ligament that runs from the transverse acetabular ligament to the fovea capitis. The articular surface of the femoral head covers more than half a sphere.

The acetabulum does not present a complete complementary articular surface. The articular cartilage of the acetabulum covers only the crescent-moon shaped *lunate surface*. Surrounding the margin of the acetabulum and deepening it is a fibrocartilagenous *acetabular labrum* which is similar in structure and function to the shoulder's glenoid labrum. The labrum reaches over the head of the femur to the point that reaches beyond the greatest circumference of the femoral head. Thus the greatest diameter of the femoral head exceeds the diameter of the opening through the acetabular labrum. This has the effect of making it especially difficult for the head of the femur to be dislocated from the acetabulum.

The entire hip joint is surrounded by a very thick and strong *fibrous joint capsule* and is reinforced by a number of ligaments. These include:

- The Transverse Acetabular Ligament
- The Ligament of the Head of the Femur
- The Fibrous Joint Capsule
- The Iliofemoral Ligament
- The Pubofemoral Ligament
- Ischiofemoral Ligament

The transverse acetabular ligament and the ligament of the head of the femur are intracapsular ligaments, in that they are contained inside the joint's fibrous capsule. The fibrous joint capsule itself contains three very prominent and strong intrinsic ligaments: the pubofemoral, ischiofemoral and iliofemoral ligaments (see Figure 7-16).

As noted, most of the acetabulum is surrounded by a fibrocartilagenous labrum. The labrum continues across the gap of the acetabular notch as the *transverse acetabular ligament*, which completes the circumference of the acetabulum. The *ligament of the head of the femur (ligamentum teres capitis femoris)* arises from the transverse acetabular ligament, and follows the acetabular notch to reach the fovea capitis of the femoral head. In living persons it is ordinarily compressed into the pad of fat that fills the acetabular fossa. The ligamentum teres is too weak to play an important role in reinforcing the hip joint, but it does carry a small artery, a branch of the obturator artery, to the head of the femur. This small artery may sometimes spare the head of the femur from avascular necrosis in the event of a fracture of the neck of the femur.

The fibrous joint capsule of the hip joint is thick, dense, and strong. It attaches around the margins of the acetabulum, extending 5 or 6 mm. beyond the margins of the acetabular labrum. It also attaches to the transverse acetabular ligament and to the margin of the obturator foramen. Inferiorly and laterally, it attaches to the neck of the femur. *Anteriorly* on the femur it attaches to the intertrochanteric line, *superiorly* to the base of the neck of the femur, *posteriorly* a cm. or so medial and superior to the intertrochanteric crest and *inferiorly* to the inferior part of the neck close to the lesser trochanter.

The capsule has two zones: an inner *zona orbicularis*, which consists of circular fibers that surround the neck of the femur; and an outer longitudinal zone. The outer longitudinal fibers run between the acetabulum and femur, and are reinforced by three strong ligamentous bands, the iliofemoral, pubofemoral, and ischiofemoral ligaments.

The *iliofemoral ligament* is a strong, triangular ligamentous band. It lies anterior to the joint and its fibers blend with those of the joint capsule. Its apex attaches to the anterior inferior iliac spine and to the body of the ilium. It then divides into two bands of fibers. One passes laterally and inferiorly to attach to the lower part of the intertrochanteric line of the femur, while the other passes more laterally to insert into the more superior part of the intertrochanteric line. Because of the two bands of fibers, this ligament is sometimes called the **Y**-shaped ligament, as it resembles an inverted **Y**. The two bands of the ligament are joined by thinner material, and often the two bands blend tightly together. The iliofemoral ligament defines the limits of hip extension, as it tightens in extension. It reinforces the anterior part of the fibrous capsule when body weight in the standing position tends to want to push the trunk backwards on the two hip joints.

The *pubofemoral ligament* is likewise a strong band of ligamentous fibers that helps reinforce the fibrous joint capsule of the hip joint. Medially it attaches to the pubis near the acetabulum and to the adjoining superior pubic ramus. Its fibers then pass laterally, anterior to the femoral head, to attach to the femoral neck. Its fibers blend with those of the joint capsule and with the nearby medial portion of the iliofemoral ligament. The pubofemoral ligament defines the limits of hyperextension and hyperabduction of the femur

The *ischiofemoral ligament* is a triangular band of fibers arising from the body of the ischium inferior and posterior to the acetabulum. It blends with the zona orbicularis of the capsule. Its more superior fibers are oriented horizontally across the hip joint, while its lower fibers spiral superiorly and laterally to attach to the neck of the femur medial to the greater trochanter. This ligament becomes tense near the limits of extension of the hip joint.

Unique features of the human hip

The human hip is structured to accommodate human habitual bipedality (walking on the hind limbs only as the principal means of locomotion). Half of the weight above the hip joint is transmitted through each of the femoral heads in the upright position. As a result, there are some unique structural components to the human hip.

Already noted in the discussion of the hip bone is the iliac pillar. The iliac pillar is the thickened column of trabecular bone that runs from the cristal tuberosity of the iliac crest to the superior margin of the acetabulum. It becomes increasingly prominent as it approaches its interface with the cristal tubercle. This is a natural consequence of the structure being the "bony history" of the growth of the cristal tubercle, which is a thickening of the iliac crest.

The lateral surface of the ilium resists compressive forces generated by the contraction of the hip abductors (the gluteus medius and minimus muscles) in the *single stance* (one-legged) *phase* of human locomotion. Body weight is transferred through the acetabulum and then on to the neck of the femur. A very well developed group of bony trabeculae lead from the femoral head to the inferior cortical bone of the neck of the femur. This is called the *calcar femorale*, and serves to transmit body weight and joint reaction forces from the head of the femur to its shaft.

Likewise, the cortical bone of the femoral neck is uniquely arranged in humans. In many other animals, the cortex of the femoral neck is of uniform thickness all around its perimeter. In human beings, the inferior cortex of the femoral neck is much thickened, while the superior cortex of the femoral neck is very thin (Lovejoy, 1988). The inferior aspect of the femoral neck carries most of the weight from the head of the femur to the shaft of the femur.

Ranges of Motion of the Hip Joint

The hip can circumduct, and thus movements at the hip joint include flexion/extension, abduction/adduction, and rotation. The range of motion is extensive, but not as extensive as those of the glenohumeral joint, primarily being restricted by the presence of other body parts. Motions of the hip joint are referred to in relation to the femoral neck. The femoral neck itself is relatively long, and meets the shaft of the femur at an angle of approximately 125 degrees. In order for the thigh to flex and extend, the head and neck of the femur must rotate about their long axes. Likewise, during medial and lateral rotation of the thigh, the junction of the neck and shaft of the femur must swing forward and backward.

Disorders of the Hip Joint

In severe cases of *osteoporosis*, the femoral neck is at great risk of fracture because of its important weight bearing role. When such fractures occur in the elderly, they are often slow to heal, and complications frequently prove fatal. The head of the femur does not have a rich blood supply. Often in fractures of the neck of the femur, the femoral head loses its blood supply and the bone of the femoral head dies, resulting in *avascular (aseptic) necrosis*.

The hip joint can be subject to severe *osteoarthritis*. Osteoarthritis can result in excessive joint pain because of degeneration of the articular cartilage and the formation of *osteophytes,* pathological bony spicules growing in and around the joint; and mobility is often severely limited. Extreme cases may require *joint replacement surgery*.

Snapping hip is a condition in which the iliotibial band or anterior aspect of the gluteus maximus muscle slips over the greater trochanter with an audible popping sound. It may or may not be associated with pain.

Multiple muscles and their tendons are associated with the hip joint, as are multiple bursae. As a result, *tendonitis* and *bursitis* are frequent problems of the hip.

Traumatic dislocation of the hip joint is much less frequent than dislocations of the shoulder due to the much stronger and much more restrictive construction of the hip joint. Dislocation of the hip joint occurs most commonly in automobile accidents when a person is sitting with his/her thighs abducted and laterally rotated.

Congenital dislocation is more common at the hip than elsewhere.

The Knee Joint

The knee is a compound joint (Figure 7-17) consisting of a bicondylar joint between the distal femur and proximal tibia (tibiofemoral joint) and a sellar joint between the patella and the patellar surface of the femur (patellofemoral joint). The knee also involves several different ligaments and complex motion and is closely associated with the proximal tibiofibular articulation.

Tibiofemoral Articulation

The principal joint of the knee is the tibiofemoral articulation. This joint is sometimes incorrectly referred to as a ginglymus or hinge joint. It is, in fact, a bicondylar joint between the condyles of the distal femur and the articular surfaces of the proximal tibia. As with all bicondylar joints, the motion that occurs between surfaces is not a simple hinge motion (as in a true ginglymus), but a combination of hinge-type and rotatory motion. As has been described, the distal femur possesses two condyles, the medial and the lateral. The medial condyle is longer than the lateral condyle. That incongruity in length of the two condyles results in lateral rotation of the tibial shaft in extension of the knee and medial rotation of the tibial shaft in flexion of the knee joint. (Motions of the knee will be discussed later.) A number of ligaments are associated with the tibiofemoral joint and reinforce the fibrous joint capsule. Additionally, since the femoral and tibial joint surfaces are incongruent with one another, *menisci* (articular disks) separate femoral condyles from the articular facets of the proximal tibia.

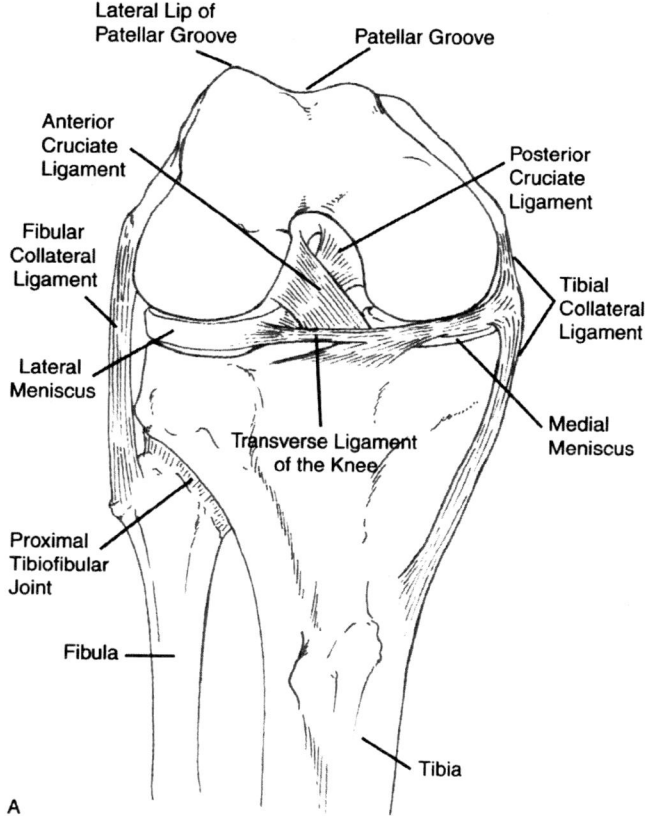

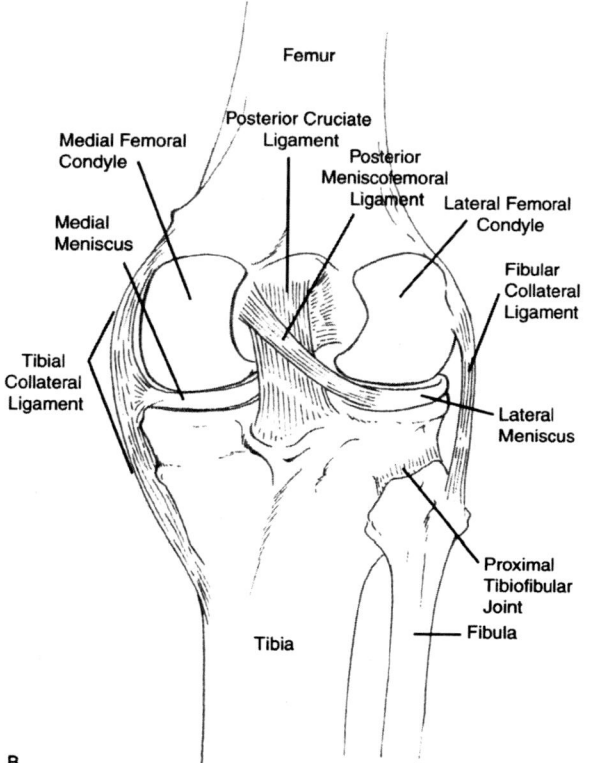

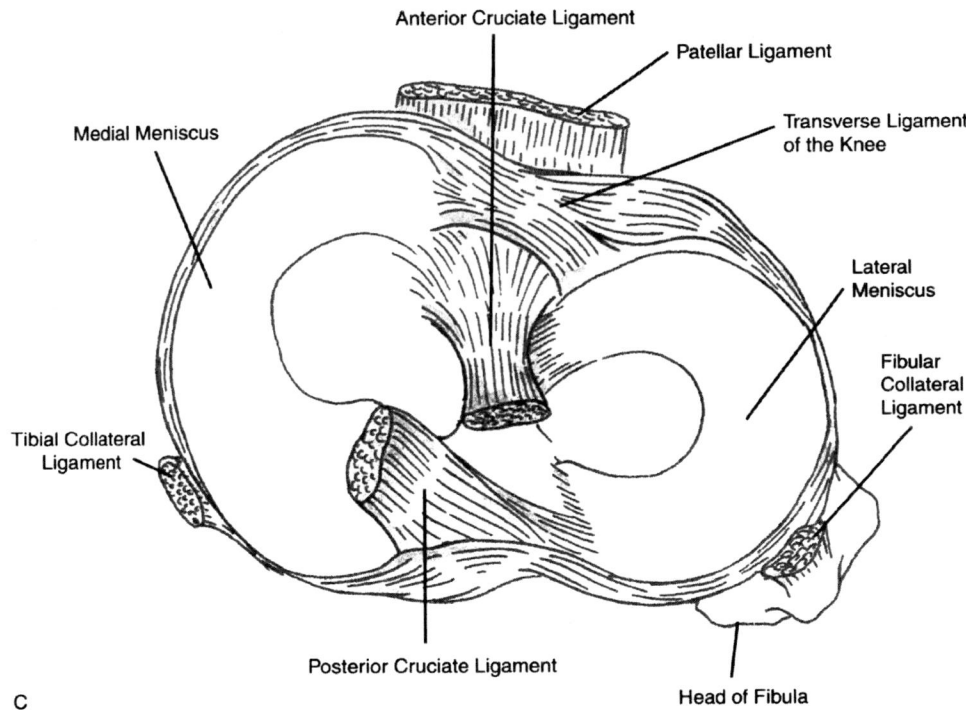

Figure 7-17 Ligaments and menisci of the knee: (A) anterior, (B) posterior, and (C) superior views.

The Fibrous Joint Capsule

The tibiofemoral joint is surrounded by a fibrous joint capsule. The capsule is partly deficient, and blends with the ligaments and tendons that surround the joint. Anteriorly, superior to the patella the fibrous capsule is deficient, and the synovial lining of the cavity is in contact with the tendon of the quadriceps femoris muscle.

At the level of the patella and more inferiorly, the capsule blends with the tendons of the vastus lateralis and vastus medialis muscles and with the *fascia lata* (the broad, thick fascia of the thigh) and its *iliotibial band* (a thickening of the fascia lata laterally that attaches the tensor fascia lata muscle above with the lateral aspect of the proximal tibia below).

The attachments of the vastus muscles' tendons with the medial and lateral aspects of the fibrous capsule of the tibiofemoral joint and to the patella form the *medial* and *lateral patellar retinacula*. The posterior part of the joint capsule consists of vertical fibers that arise from the posterior portions of the femoral condyles and the margins of the intercondylar notch. These fibers proceed inferiorly and attach along the posterior margins of the tibial plateau. Because the synovial membrane of the joint capsule proceeds around the margins of the intercondylar notch in such a way that it passes anterior to the *anterior cruciate ligament*, the cruciate ligaments remain outside the synovial cavity, but within the fibrous joint capsule.

On the *lateral* aspect of the joint, the fibers of the capsule arise from the border of the lateral femoral condyle above the origin for the popliteus muscle and attach inferiorly to the lateral aspect of the tibial plateau and to the fibular head. *Medially*, the fibrous capsule attaches to the inferior aspect of the medial femoral condyle superiorly and to the medial aspect of the tibial plateau inferiorly.

The Synovial Membrane

The *synovial membrane* of the knee joint lines the internal aspect of the fibrous joint capsule. The synovial membrane of the knee joint is the most complicated in the human body. It attaches to the superior border of the patella and then extends upward for a distance beneath the tendon of the quadriceps femoris tendon. This extension is frequently referred to as the *suprapatellar bursa,* but it is not a true bursa; it is actually a continuation of the synovial cavity of the joint. It has its own muscle, the *articularis genu,* to support it during movements of the knee.

The synovial membrane extends to either side of the patella deep to the aponeuroses of the vastus lateralis and vastus medialis muscles, especially under the vastus medialis. Inferiorly, it is separated from the patellar ligament by a fat pad, the *infrapatellar fat pad.* Two folds, the left and right *alar folds,* of the synovial membrane project from the medial and lateral aspects of the articular surface of the patella. They converge inside the joint to form the *infrapatellar synovial fold* that attaches to the anterior aspect of the intercondylar notch of the femur.

From the superior and posterior aspect of the joint, the synovial membrane forms two pouches, one of which reaches beneath each head of the gastrocnemius muscle as the heads rise from the femur. The synovial membrane covers the superior and inferior surfaces of the menisci. The internal free borders of the menisci are not covered by the synovial membrane. The synovium then continues from the inferior surfaces of the menisci and attaches to the proximal tibia.

A recess of the synovial capsule continues as a small sac between the tendon of the popliteus muscle and the lateral meniscus. This is the *subpopliteal recess.* The synovial membrane in general follows the course of the fibrous joint capsule but is reflected anterior to the cruciate ligaments, which therefore remain outside the synovial cavity. A large number of bursae surround the knee joint to reduce friction between the skin and underlying tissue as the knee moves.

Ligaments

A number of very strong ligaments reinforce the knee joint. Since the two knee joints bear all the body weight above the knees, and since the highly convex femoral condyles rest somewhat precariously on the relatively shallow articular facets of the proximal tibia, the joint would be highly unstable without a number of strong, cordlike ligaments to help maintain the femur upon the tibial plateau. Even with the presence of these ligaments, the knee is still highly susceptible to injury. Four well developed ligaments, the tibial (medial) and fibular (lateral) collateral and the anterior and posterior cruciate ligaments are particularly important to the maintenance of joint integrity. Other ligaments include the oblique popliteal and arcuate popliteal ligaments, the coronary ligament, and the transverse ligament of the knee.

THE FIBULAR COLLATERAL LIGAMENT: The lateral or fibular collateral ligament is a round, dense band, about the diameter of a pencil, that extends from the lateral epicondyle of the femur directly inferiorly to the head of the fibula. (The *superior tibiofibular joint* lies just lateral and inferior to the knee joint.) The tendon of insertion of the popliteus muscle separates the lateral collateral ligament from the lateral meniscus. The lateral collateral ligament reinforces the lateral side of the knee joint capsule, assisted by the iliotibial band as it passes the lateral aspect of the knee to attach to the proximal tibia.

THE TIBIAL COLLATERAL LIGAMENT: The *tibial (medial) collateral ligament* is composed of two broad bands of fibers that arise off the medial epicondyle of the femur just below the adductor tubercle. The deeper band passes inferiorly and posteriorly to insert onto the more posterior and medial part of the border of the tibial plateau. The more superficial band passes directly inferiorly to attach to the medial and superior aspect of the tibial shaft. The tibial collateral ligament is firmly attached to the margin of the *medial meniscus* of the tibiofemoral joint. This limits the freedom of the medial meniscus to move about in the joint cavity. Because the medial meniscus is firmly attached to the medial collateral ligament (unlike the lateral meniscus and collateral ligament, which *are not* joined together), damage to the medial collateral ligament regularly leads to damage to the medial meniscus.

THE CRUCIATE LIGAMENTS: The anterior and posterior cruciate ligaments are two very strong intraarticular ligaments, each about the diameter of a pencil or larger. They are located in the middle of the joint, but closer to the posterior margin of the tibial plateau than the anterior. They are called cruciate ("cross-shaped") because they cross one another in a manner resembling a letter **X** or tilted cross. The cruciate ligaments help affix the proximal tibia to the distal femur, and help reinforce the knee joint. They are named according to their points of connection on the tibia.

The *anterior cruciate ligament* attaches to a depression anterior to the intercondylar eminence on the tibial plateau, where it blends with the anterior horn of the lateral meniscus. It passes superiorly, posteriorly and laterally to insert into the posterior part of the medial surface of the lateral condyle of the femur. The anterior cruciate ligament helps limit hyperextension of the knee and helps limit the anterior displacement of the tibia relative to the femur. This limits extension of the lateral condyle of the femur, and causes medial rotation of the femur on the tibia as the medial condyle continues to extend until it reaches full extension. The collateral ligaments and the oblique popliteal ligament become taut as this occurs, and eventually limit any other rotation from occurring. This rotation with full extension is often referred to as the "screw home" mechanism of the knee.

The *posterior cruciate ligament* is stronger, but shorter and more oblique than is the anterior ligament. It attaches to the tibia in a depression posterior to the intercondylar eminence of the tibial plateau, and has some fibers that proceed onto the posterior aspect of the tibia. It attaches to the posterior horn of the lateral meniscus. It then proceeds superiorly, anteriorly, and medially to insert onto the lateral surface of the medial condyle of the femur, passing medial to the anterior cruciate ligament as it does so. The posterior cruciate ligament defines the limits of hyperflexion of the knee and defines the limits of anterior motion of the femur relative to the surface of the tibia. This ligament is especially important in walking down steep inclines or steps, where body weight is transferred to the tibia at the same time that the knee is flexed.

THE OBLIQUE POPLITEAL LIGAMENT: The *oblique popliteal ligament* blends with and reinforces the posterior portion of the fibrous joint capsule of the tibiofemoral articulation. It attaches laterally and proximally to the margin of the intercondylar fossa and to the posterior aspect of the lateral condyle of the femur. It then passes distally and obliquely medially to insert onto the posterior and medial aspect of the proximal tibia just above the soleal line. Its more superficial fibers blend with the semimembranosus muscle, which passes nearby.

THE ARCUATE POPLITEAL LIGAMENT: The *arcuate popliteal ligament* attaches superiorly to the posterior aspect of the lateral condyle of the femur and then passes medially and inferiorly to pass over the posterior aspect of the knee joint articular capsule, passing over the tendon of the popliteus muscle. It attaches to the tibial plateau just posterior to the attachment of the posterior cruciate ligament. It also sends fibers to the posterior aspect of the apex of the fibular head.

The Menisci

The medial and lateral menisci (see Fig. 7-17) are two C-shaped fibrocartilagenous articular discs between the femoral condyles and the articular facets on the tibial plateau. The internal borders of the menisci are free, but the external surfaces are bound to the proximal tibia by the *coronary ligament*. The coronary ligament, as the name implies *(coronary* means "crown"), surrounds the top of the proximal tibia. Its fibers are blended with the fibrous joint capsule.

Additionally, the *medial meniscus* is tightly bound to the medial collateral ligament. The two menisci, as noted, are shaped like the letter C. The opening of each C is directed inward toward the center of the joint, and the free ends of each meniscus are attached to the anterior and posterior ends of the tibial intercondylar eminence. The medial meniscus is broader posteriorly than anteriorly. Its anterior end is pointed and attaches anteriorly to the intercondylar eminence of the tibia, in front of the anterior cruciate ligament. Posteriorly it attaches behind the intercondylar eminence, between the attachment of the lateral meniscus and the posterior cruciate ligament. Its outer surface attaches to the fibrous joint capsule of the tibiofemoral joint. It is also firmly attached to the internal aspect of the medial collateral ligament.

The *lateral meniscus* is almost circular, but it is open medially where its anterior and posterior ends attach anteriorly and posteriorly to the intercondylar eminence of the tibia. The anterior end of the lateral meniscus attaches to the tibia lateral and posterior to the anterior cruciate ligament. Its posterior end attaches anterior to the posterior attachment of the medial meniscus. Its outer surface attaches to the fibrous joint capsule of the tibiofemoral joint. The *transverse ligament of the knee* attaches the convex anterior margin of the lateral meniscus to the anterior end of the medial meniscus. It is variable in size and sometimes absent.

The superior surfaces of the two menisci are concave to receive the convex surfaces of the femoral condyles. The inferior surfaces are flat and rest upon the tibial plateau. As a result, the menisci are almost triangular in cross-section. The fibrocartilagenous menisci are somewhat deformable and increase the congruity between the articular surfaces of the femoral condyles and the articular facets of the proximal tibia. They deepen the receptacles for the femoral condyles on the tibial plateau.

Additionally, the curvatures of the femoral condyles are not uniform. Because the menisci are somewhat compliant, they can change shape throughout the range of motion of the joint to maintain their congruency, which is important to joint lubrication. The presence of the menisci in the joint also helps to more evenly distribute the transfer of weight from the femur to the tibia.

Patellofemoral Articulation

The second component of the knee joint is the patellofemoral joint. The patella itself is a large sesamoid bone embedded in the tendon of the quadriceps femoris muscle. It is customary to refer to the band of collagenous tissue that joins the inferior aspect of the patella to the tibial tuberosity as the *patellar ligament*, following the old adage that *ligaments connect bone to bone* (patella to tibia), while *tendons connect muscle to bone* (quadriceps femoris to patella).

The patellar ligament is, in fact, the distal continuation of the tendon of the quadriceps femoris muscle, the patella simply being a large sesamoid bone that interrupts that tendon. It is therefore also referred to as the patellar tendon. However, we will follow convention and refer to it as the *patellar ligament*.

The superficial fibers of the patellar ligament are continuous over the anterior surface of the patella and blend with the quadriceps tendon. The tendons (or aponeuroses) of insertion of the vastus medialis and vastus lateralis give rise to the *medial* and *lateral patellar retinacula* that attach along the medial and lateral aspects of the patella.

The posterior aspect of the patella is its articular surface, which is divided into *medial* and *lateral articular facets*. These articulate with the patellar groove on the anterior and inferior aspect of the femur. The joint itself takes the form of a sellar articulation. The synovial capsule of the patellofemoral joint is simply a continuation of the synovial capsule of the tibiofemoral joint, described previously. During flexion and extension of the knee joint, the patella tracks along the patellar groove of the femur.

A principal function of the patella is to move the quadriceps tendon further anterior to the center of rotation of the knee joint. This increases the moment arm of the quadriceps femoris and gives it greater mechanical advantage in extension of the knee. If the patella is absent, about 30 percent more force is required to extend the knee (Jenkins, 1991).

Unique Features of the Human Knee.

Human habitual bipedal locomotion has led to some unique features of the human knee. During normal walking, the body must be supported on one leg while the other is swung forward to take the next step. The body's weight can be better balanced on a single knee joint if that knee joint is under the center of gravity, in the midline. In order to bring the knees under the midline of the body, the shafts of the femurs angle inward from the hip to the knee, so that the medial aspects of the two femurs nearly meet at the knee. Thus, the shafts of the femurs are not vertical, but are angled slightly.

The tibial plateaus are approximately horizontal, as are the inferior aspects of the medial and lateral condyles of the femurs when in anatomical position. The angle thus formed between the shaft of the femur and the vertical is the *bicondylar angle*. In humans, this angle is on average about 9 or 10 degrees, being higher in females due to the greater relative width of the hips (Heiple and Lovejoy, 1971).

The bicondylar angle develops in growing children as they begin to walk. Prior to that, the condition in human children is much like that in quadrupedal mammals, where the knees are vertically aligned with the hips. Because of the presence of a large bicondylar angle, the line of action, or direction of pull, of the quadriceps femoris muscle on the patella is angled, generating both a Y-component force (vertical) and an X-component of force (horizontal). As such, it has a tendency to pull the patella laterally.

The patella is prevented from slipping laterally out of the patellar groove by the presence of the medial and lateral patellar retinacula, the lowermost fibers of the vastus medialis muscle, and by the anteriorly projecting lateral lip of the patellar groove on the distal femur. Because women have a larger bicondylar angle than men, the force vector generated by the quadriceps femoris muscle will have a relatively greater **X** (horizontal) component than in men. As a consequence, women have a higher incidence of lateral dislocation of the patella than do men.

Ranges of Motion of the Knee Joint

Movements of the knee (tibiofemoral and patellofemoral) are described as *flexion, extension,* and *medial* and *lateral rotation*. As has been noted, the knee is not a simple hinge joint, but is a bicondylar joint. As with all *bicondylar joints* (such as that between the occipital condyles and the atlas, or the temporomandibular joint), hinge motions are coupled with rotation.

As the knee goes through flexion and extension, the tibial articular surfaces do not simply slide fore and aft on the femur. During extension, the tibial shaft undergoes lateral rotation, and during flexion, it undergoes medial rotation. With the foot fixed, the last 30 degrees of extension involve medial femoral rotation, while early flexion involves lateral femoral rotation. From a vertical femorotibial axis, the extreme range of extension is about 5 to 10 degrees beyond that axis, while extreme flexion is about 120 degrees with the hip extended, or 140 degrees with the hip flexed. Conjunct rotation at the knee with flexion and extension is about 20 degrees, but passive rotation can be 60 or 70 degrees.

Disorders of the Knee Joint

Like the hip joint, the knee is a frequent site of *osteoarthritis*. Severe cases can result in complete joint replacement surgery, which involves prosthetic replacement of the femoral, tibial, and patellar articular surfaces. Because of the lack of congruity between the articular surfaces of the tibia and femur, joint stability relies solely on the surrounding soft tissues. In ordinary circumstances, surrounding musculature and ligaments maintain the joint, but these can be damaged in trauma.

The tibial (medial) collateral ligament is frequently damaged when a force is applied to the lateral aspect of the knee when the foot is firmly fixed on the substrate and the knee is slightly flexed ("clipping injury"). If the force is strong enough, the medial collateral ligament will rupture. If the force is even greater, it may rupture both the medial collateral and the anterior cruciate ligaments. Because the *medial meniscus* is tightly bound to the medial collateral ligament, damage to that ligament frequently also results in damage to the meniscus.

If a force is applied to the anterior aspect of the knee in full extension, driving it into hyperextension, the anterior cruciate ligament may rupture. If the force is great enough, the posterior cruciate ligament may rupture as well. A force applied to the anterior aspect of the tibia in a flexed knee may result in rupture of the posterior cruciate ligament.

Runner's syndrome (iliotibial band syndrome) is an overuse syndrome in which the iliotibial band becomes inflamed after repetitive slipping back and forth over the lateral epicondyle of the femur (see Edwardson, 1995 for more details).

Tibiofibular Articulations

The fibula does not participate in the knee joint (except as the site of attachment of the lateral collateral ligament), but the proximal tibiofibular articulation lies just below the lateral aspect of the knee joint on the lateral and inferior aspect of the tibial plateau. There are three tibiofibular articulations, the proximal, middle, and inferior tibiofibular joints.

Proximal Tibiofibular Articulation

The *proximal (superior) tibiofibular joint* (see Figure 7-17 A and 7-17 B) is a plane synovial joint. The synovial capsule may communicate with the tibiofemoral joint posterolaterally through the popliteal bursa which lies deep to the popliteus tendon. Little movement occurs at this joint in human beings, as the fibula's primary role is as a site of muscle attachment and to contribute to the lateral wall of the ankle joint. The articular surface of the tibia is located on the lateral and inferior aspect of the tibial plateau, and that of the fibula is located on the fibular head. Both the fibular and tibial articular surfaces are flat and oval and covered with hyaline cartilage. The joint is surrounded by a fibrous joint capsule which is reinforced by the *anterior* and *posterior ligaments of the head of the fibula*. These run from the lateral borders of the tibial plateau to the anterior and posterior aspects of the head of the fibula.

Middle Tibiofibular Articulation

The *middle tibiofibular joint (tibiofibular syndesmosis)* joint consists of the *crural interosseous membrane (crural* = leg below the knee). Like the interosseous membrane between the radius and ulna, this membrane runs between the interosseous borders (crests) of the two adjacent bones, but in the leg rather than the forearm. Its superior margin does not quite reach the proximal tibiofibular joint, but presents a free concave superior edge over which the anterior tibial artery and vein pass. Distally, it becomes continuous with the interosseous ligament of the distal tibiofibular joint. The membrane helps keep the fibula firmly in articulation with the tibia, separates the anterior and posterior muscular compartments of the *leg (crus)* and gives attachment to the deep dorsiflexor and plantarflexor muscles.

The Distal Tibiofibular Articulation

The *distal (inferior) tibiofibular articulation* joint (Figure 7-18) exhibits very little motion, and is a combination syndesmosis and diarthrosis. The medial aspect of the distal end of the fibula bears a rough, triangular, convex surface that articulates with the rough, concave, fibular notch on the lateral aspect of the distal tibia. Most of the joint surfaces are joined to one another by a fibrous interosseous ligament.

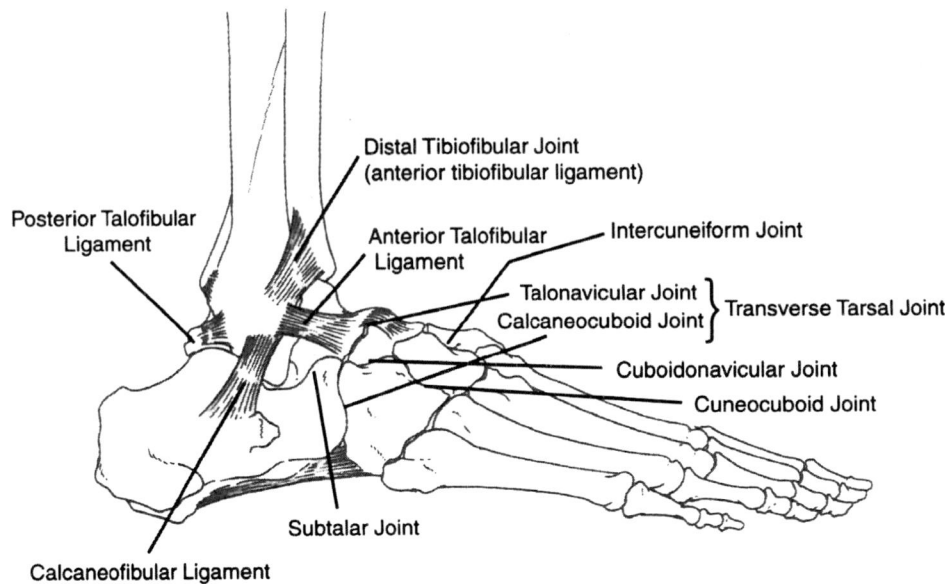

Figure 7-18 Ligaments of the ankle (talocrural joint), lateral view.

However, the distal half centimeter or so of the joint consists of a plane synovial joint. Its synovial capsule is continuous with that of the ankle joint. In human beings, this joint binds the lateral malleolus of the fibula tightly to the distal tibia to form the bony mortise of the ankle joint.

The distal tibiofibular joint in human beings is less mobile than in other hominoid (human-like) primates, but does have a special function in humans. During the heel strike phase of locomotion, the fibula moves downward relative to the tibia, which serves to deepen the mortise of the ankle. This puts tension on the interosseous membrane of the middle tibiofibular joint and the interosseous ligament of the distal tibiofibular joint. This draws the fibula and tibia more tightly together reinforcing the ankle mortise. This in turn dissipates some of the compressive force of heel strike by means of ligament distention (Weinert et al., 1973).

The Talocrural Joint

The *talocrural (ankle) joint* (Figure 7-19; refer also to Figure 7-18 throughout the discussion of the talocrural joint and its ligaments) is the joint between the distal ends of the *tibia* and *fibula* superiorly, laterally and medially, and the superior surface of the *talus* inferiorly. This is principally a uniaxial joint classified as a ginglymus.

The articular surfaces superiorly are located on the medial aspect of the lateral malleolus of the fibula, the inferior end of the tibia, and on the lateral aspect of the medial malleolus of the tibia. The malleoli, the inferior tibia, and the inferior transverse (tibiofibular) ligament together complete a deep concavity for reception of the superior surface of the talus. The superior or proximal surface of the talus is highly convex and spool-like in form. This proximal surface articulates with the large facet at the inferior surface of the tibia. Likewise, the talus possesses medial and lateral articular facets which articulate with the facets on the inner aspects of the medial and lateral malleoli.

Capsules and Ligaments.

The bones involved in the talocrural joint are joined by a number of ligaments, as well as the articular capsule.

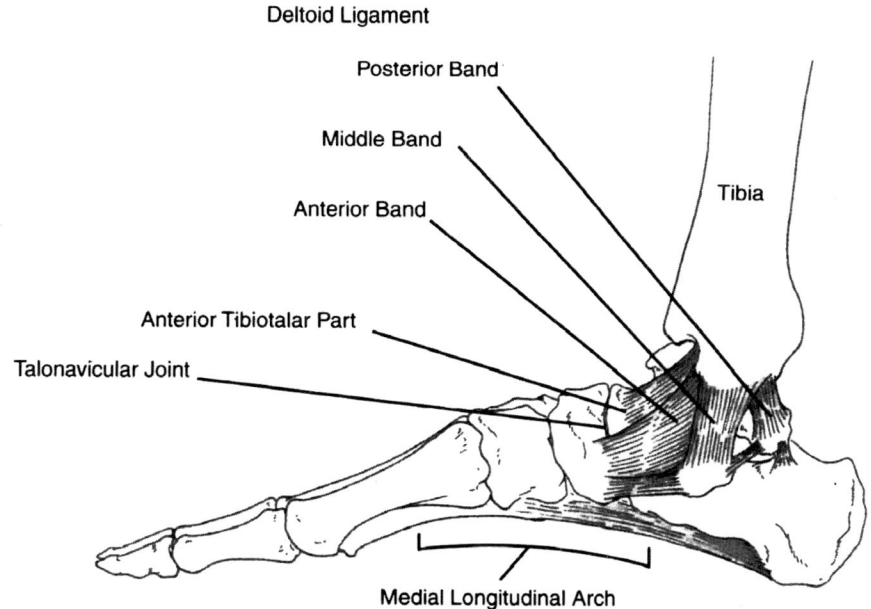

Figure 7-19 Ligaments of the ankle (talocrural joint), medial view.

Articular Capsule

The fibrous *articular capsule* of the talocrural joint attaches superiorly along the margins of the articular surfaces of the tibia and malleoli. Inferiorly, it attaches along the margins of the involved articular facets of the talus. The synovial membrane lines the internal aspects of the capsule, and sends a small extension superiorly between the distal ends of the tibia and fibula at the inferior tibiofibular articulation. The articular capsule is thin anteriorly and posteriorly, but is reinforced medially and laterally by a number of ligaments.

The Anterior Talofibular Ligament

Some describe this and the following two ligaments as three parts of a single lateral ligament of the talocrural joint. The *anterior talofibular ligament* is the shortest of these three and runs from the anterior margin of the lateral malleolus to the talus in front of its lateral articular surface.

The Posterior Talofibular Ligament

The *posterior talofibular ligament* is both the strongest and deepest of the three lateral ligaments of the ankle. It runs almost horizontally from the lateral malleolar fossa to the posterior surface of the talus lateral to the groove for the tendon of the flexor hallucis longus.

The Calcaneofibular Ligament

The *calcaneofibular ligament* is the longest of the three lateral ligaments of the ankle. It is a narrow rounded cord that extends from the apex of the fibular malleolus. It runs posteriorly and inferiorly to insert into a tubercle on the lateral surface of the calcaneus. The tendons of the fibularis (peroneus) longus and brevis muscles cross over and cover this ligament.

The Deltoid Ligament

The *deltoid (or medial) ligament* is a broad flat band of ligamentous tissue, triangular in shape. It attaches superiorly to the apex, anterior and posterior borders of the medial malleolus, and inferiorly attaches by three bands of fibers to the tarsal bones. It has both *superficial* and *deep fibers*.

The *superficial fibers* can be divided into three bands, an *anterior (tibionavicular) band*, *middle (tibiocalcaneal) band*, and *posterior (posterior tibiotalar) band*, which join with the bones indicated by their names. The *deep fibers* form the *anterior tibiotalar part*. The tendons of the tibialis posterior and flexor digitorum longus muscles cross the deltoid ligament passing from the posterior crural compartment into the sole of the foot.

The Intertarsal Joints

There are a considerable number of intertarsal joints. As with the intercarpal joints of the wrist, most of these joints are of the plane synovial type. We will not discuss all the details of each one of these joints, but will touch upon some of the more important aspects of their anatomy.

The Talocalcaneonavicular joint

The *talocalcaneonavicular joint* is a compound, multiaxial, spheroidal synovial joint. The head of the talus is rounded, and is received by a socket formed by the posterior surface of the navicular, the anterior articular surface of the calcaneus, and by the *plantar calcaneonavicular (spring) ligament* and *calcaneonavicular part of the bifurcated ligament*. Gliding and rotatory movements are permitted at this and at the subtalar joint.

Subtalar (talocalcaneal) joint

The talus and calcaneus have two articulations, an anterior and a posterior. The anterior joint takes part in the *talocalcaneonavicular joint*, discussed in the previous section. The posterior, or *subtalar (talocalcaneal) joint*, forms between the posterior concave calcaneal articular facet on the inferior surface of the talus and the convex articular facet on the posterior superior calcaneus. This is a plane synovial joint, surrounded by an articular capsule and reinforced by *anterior, posterior, medial, lateral,* and *interosseous talocalcaneal ligaments*. The interosseous talocalcaneal ligament is the chief union between the talus and calcaneus and forms from the union of the fibrous capsules of the talocalcaneonavicular and subtalar joints. It fills the tarsal sinus, as described with the calcaneus. The synovial capsule of this joint is unique among the intertarsal joints in that it does not communicate with the synovial capsules of the other intertarsal joints.

The Calcaneocuboid joint

The *calcaneocuboid joint* between the calcaneus and cuboid has the characteristics of a sellar joint, but does not produce the wide range of motions usually associated with sellar joints. Ligaments reinforcing the fibrous joint capsule between the more posterior calcaneus and the more anterior cuboid are the *dorsal calcaneocuboid ligament; bifurcated ligament which consists of the calcaneocuboid part and the calcaneonavicular part; the long plantar ligament (Figure 7-20); and the plantar calcaneocuboid ligament*. Movements at this joint accompany those at the subtalar and talocalcaneonavicular joints. They are basically translatory and rotational in nature.

The Transverse Tarsal Joint

The *transverse tarsal joint* (Chopart's joint; see Figure 7-18) is a term used to describe collectively the *calcaneocuboid* and *talonavicular joints*. Together the joint capsules of these two joints extend transversely, though somewhat irregularly, across the foot. The talus and calcaneus lie posterior to the transverse tarsal joint, while the navicular and cuboid lie anterior to it. Some plantarflexion and dorsiflexion of the anterior part of the foot is permitted at the transverse tarsal joint, as well as some slight inversion and eversion. *Pronation and supination of the foot* take place principally at this joint.

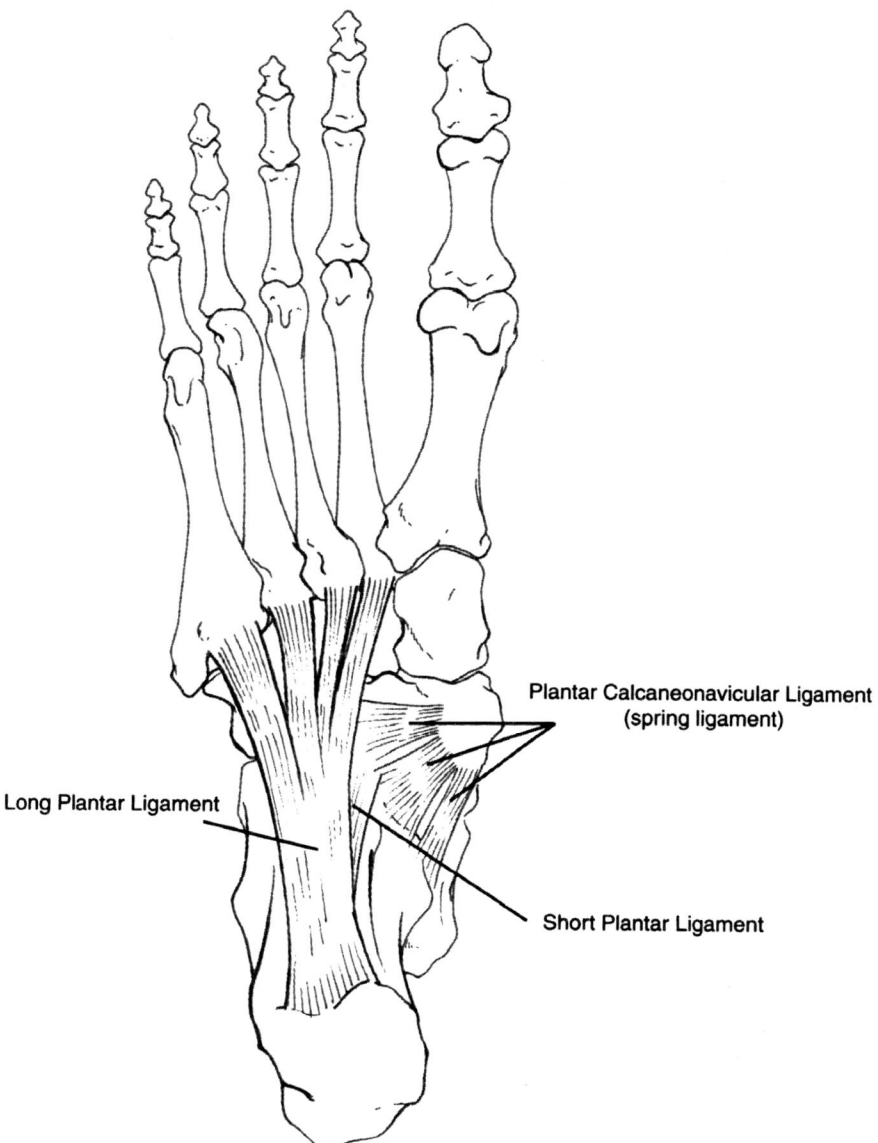

Figure 7-20 Plantar ligaments of the foot.

The Calcaneonavicular Joint

The calcaneus and navicular do not directly articulate, but are joined by ligaments. Important ligaments associated with this joint are the *calcaneonavicular part of the bifurcated ligament*, from the anterior calcaneus to the lateral navicular, and the *plantar calcaneonavicular (spring) ligament*. These ligaments will be further discussed below. By supporting the head of the talus, the spring ligament helps maintain the *medial longitudinal arch* of the foot.

Cuboidonavicular Joint

The *cuboidonavicular joint* is a syndesmosis and consists of *dorsal, plantar, and interosseous cuboidonavicular ligaments*. This joint is occasionally synovial, in which case, its synovial cavity is continuous with the cuneonavicular joint.

The Intercuneiform Joints

The intercuneiform joints are plane synovial joints. *Dorsal, plantar*, and *interosseous ligaments* join adjacent bones with one another, reinforcing the joint capsules.

The Cuneocuboid Joint

The cuneocuboid joint is a plane synovial joint. The bones are united by *dorsal, plantar,* and *interosseous cuneocuboid ligaments*.

The Cuneonavicular Joint

The distal surface of the navicular bone articulates with the three cuneiform bones in a joint that is typically described as a plane synovial joint. However, the navicular's articular surface is transversely convex, and divided by faint ridges into three facets, one for each cuneiform bone. Its articular capsule and synovial cavity are continuous with those of the intercuneiform and cuneocuboid joints. *Dorsal* and *plantar cuneonavicular ligaments* join the dorsal and plantar surfaces of the navicular with each of the three cuneiform bones.

Ligaments in the Foot Associated with the Tarsal Bones

Plantar Calcaneonavicular Ligament

The *plantar calcaneonavicular (spring) ligament* (see Figure 7-20) is broad and thick and connects the anterior margin of the sustentaculum tali of the calcaneus to the plantar surface of the navicular. The dorsal surface of the ligament has a triangular fibrocartilagenous facet upon which rests the head of the talus. By supporting the head of the talus the ligament helps maintain the medial longitudinal arch of the foot and thus keeps the "spring" in the step. The plantar surface of the ligament is supported by the tendons of the tibialis posterior, flexor hallucis longus and flexor digitorum longus muscles. The medial border of the ligament blends with the deltoid ligament.

The Bifurcate Ligament

The bifurcate ligament is so called because it divides into two major bands or parts. These are the (lateral) *calcaneocuboid ligament and the (medial) calcaneonavicular ligament*. The bifurcate ligament is **Y** shaped, with the stem of the **Y** attaching to the anterior superior (or dorsal) calcaneal surface. Distally it divides into its *calcaneocuboid* and *calcaneonavicular* parts. The *calcaneocuboid ligament* is the principal connection between the proximal and distal rows of tarsal bones. The *calcaneonavicular ligament* attaches to the dorsolateral part of the navicular.

The Long Plantar Ligament

The long plantar ligament is the longest ligament of the tarsus. It connects the plantar aspect of the calcaneus to the ridge and tuberosity of the plantar surface of the cuboid (deep fibers) and continues on to the bases of the second, third, fourth, and sometimes fifth metatarsals (superficial fibers). With the groove on the plantar aspect of the cuboid, this ligament forms a tunnel for the passage of the tendon of the fibularis (peroneus) longus muscle. The ligament and the tendon of fibularis (peroneus) longus help maintain the *lateral longitudinal arch* of the foot.

The Plantar Calcaneocuboid Ligament

The *plantar calcaneocuboid (short plantar)* ligament is deeper than the long plantar ligament, and is separated from it by areolar connective tissue. It runs from the anterior calcaneal tubercle and depression anterior to it to the plantar surface of the cuboid. It also helps maintain the *lateral longitudinal arch of the foot*.

The Arches of the foot

The foot has three arches, the medial longitudinal arch, the lateral longitudinal arch, and the transverse arch. The medial longitudinal arch, as alluded to earlier, is maintained principally by the spring ligament, while the lateral arch is maintained by the long and short plantar ligaments.

The *medial longitudinal arch* is comprised of the calcaneus, talus, navicular, the three cuneiform bones and the first, second and third metatarsal bones. The concavity of the arch is supported by *posterior (calcaneal) and anterior (metatarsal) pillars*. The anterior pillar is formed by the heads of the three medial metatarsal bones.

The flatter *lateral longitudinal arch* is composed of the calcaneus, cuboid, and fourth and fifth metatarsals. Its *posterior pillar* is the tuberosity of the calcaneus and its *anterior pillar* is the heads of the two lateral metatarsals.

The *transverse (metatarsal) arch* is formed by the navicular, the three cuneiforms, the cuboid, and the bases of the five metatarsal bones. It is maintained by the ligaments that bind the cuneiforms and the bases of the metatarsal bones together. All three arches are concave inferiorly. The resilience and elasticity of the medial arch serves to absorb reaction forces applied to the foot such as those received in jumping. The lateral arch, which is lower and less resilient, is constructed to transmit forces to the ground, along the lateral aspect of the foot, such as in walking and running.

The Tarsometatarsal Joints

The *tarsometatarsal joints* are plane synovial joints. They include articulations between the medial, intermediate and lateral cuneiforms and the cuboid proximally and the bases of the five metatarsal bones distally. The *first metatarsal* articulates with the medial cuneiform. The *second metatarsal* articulates with adjacent surfaces of both the intermediate cuneiform and on either side with small areas of the medial and lateral cuneiforms. The *third metatarsal* articulates with the lateral cuneiform. The *fourth metatarsal* articulates with the lateral cuneiform and the cuboid, while the *fifth metatarsal* articulates only with the cuboid. The five tarsometatarsal joints share only three synovial cavities.

The *medial tarsometatarsal joint cavity* is restricted to the joint between the first metatarsal and the medial cuneiform. The *intermediate tarsometatarsal joint cavity* is between the intermediate and lateral cuneiforms and the second and third metatarsals. The *lateral tarsometatarsal joint cavity* is between the cuboid bone and the bases of the fourth and fifth metatarsals. *Dorsal and plantar tarsometatarsal ligaments* join the dorsal and plantar surfaces of the bases of the metatarsals to the adjacent surfaces of the tarsal bones. *Interosseous cuneometatarsal ligaments* pass from the cuneiform bones to the bases of adjacent metatarsals. They are variably present, but the first, between the medial cuneiform to the second metatarsal is constant.

Intermetatarsal joints

The *bases* of the second to the fifth metatarsals articulate at plane synovial joints and are also joined by *dorsal, plantar and interosseous intermetatarsal ligaments*. The *heads* of all the metatarsals are joined syndesmotically to one another on their plantar surfaces by the *deep transverse metatarsal ligaments*.

Metatarsophalangeal Joints

The metatarsophalangeal joints are ellipsoid synovial joints. They have loose fibrous joint capsules. The capsules are reinforced by a *plantar ligament* and *medial* and *lateral collateral ligaments*. The dorsal surface of the capsules are strengthened by the extensor hoods of the extensor muscles of the toes. The joints are capable of flexion, extension, abduction and adduction.

Interphalangeal Joints

These joints are *ginglymi*, or synovial hinge joints. The distal joint surface of each phalanx is pulley-shaped, and the proximal joint surface of the articulating phalanx has a double concavity to fit these pulley-shaped surfaces. Each joint has an articular capsule and a *plantar ligament* as well as *medial* and *lateral collateral ligaments*.

Ranges of Motion of the Joints of the Ankle and Foot

Motions of the ankle and foot include *plantarflexion* and *dorsiflexion, inversion* and *eversion,* and *pronation* and *supination,* as well as flexion and extension of the phalanges and metatarsophalangeal joints. Movements at the talocrural joint include plantarflexion and dorsiflexion. Dorsiflexion is the action of bringing the top of the foot closer to the shin; plantarflexion is the opposite movement. The human talocrural joint allows about 55 degrees of plantarflexion and 35 degrees of dorsiflexion.

Movements at the subtalar joint and at the talocalcaneonavicular joints permit a wide range of gliding and rotatory movements. Movements at these joints result in inversion and eversion of the plantar surface of the foot. In inversion, the medial border of the foot is raised, whereas in eversion, the lateral border of the foot is raised. Movements at the calcaneocuboid joint accompany those of inversion and eversion at the previously mentioned joints.

Pronation and supination of the foot take place principally at the transverse tarsal joint (composed of the calcaneocuboid and talonavicular joints). Supination of the foot occurs when the anterior aspect of the plantar surface of the fully inverted foot is rotated so that the heads of the metatarsals remain in contact with the substrate. Relative to the inverted foot, this involves lifting the lateral border of the foot. Inversion and supination increase as the feet are spread apart, as in standing with the feet widely separated.

Pronation of the foot is the opposite motion; it also occurs at the transverse tarsal joint. During walking, the suspended foot is held in supination. This maximizes the contact of the plantar surface of the foot following heel strike. As the heel rises, and the foot heads toward toe off (the phase of walking in which the foot pushes off with the big toe), the foot pronates to transmit body weight to the head of the first metatarsal.

Disorders of the Joints of the Ankle and Foot

A *sprained ankle* results from forced inversion (turning inward) or from forced eversion (turning outward) of the foot. Most sprains are of the inversion type. This causes the tearing of ligaments at the lateral aspect of the joint. Eversion sprains are less frequent and cause damage to the deltoid ligament. More frequently, the medial malleolus fractures before the ligament ruptures.

Flat foot results from the loss of the longitudinal arches. In this situation, forces are transmitted through the entire plantar surface of the foot. The flattening is brought on by the excessive elongation of the ligaments of the talonavicular and cuneonavicular joints.

Ossification of the Lower Limb Bones

The Os Coxae

The os coxae (Figure 7-21) ossifies from *three primary centers*, one each in the *ilium*, the *ischium*, and the *pubis*. There are *five principal secondary centers* of ossification: the *iliac crest*, the *ischial tuberosity*, the *anterior inferior iliac spine*, the dorsal rampart of the *pubic symphyseal surface*, and the *triradiate cartilage*, the zone of hyaline growth cartilage in the acetabulum that separates the three primary centers from one another. (There may be additional secondary centers of ossification for the pubic tubercle, pubic crest, and ischial spine.)

The primary centers appear in this order: (1) ilium, (2) ischium, (3) pubis. The primary center for the ilium appears at about 8 weeks *in utero*, for the ischium at about three months *in utero*, and for the pubis at about between four and five months *in utero*. The centers first appear near the acetabulum, that is. the lower ilium, the body of the ischium, and

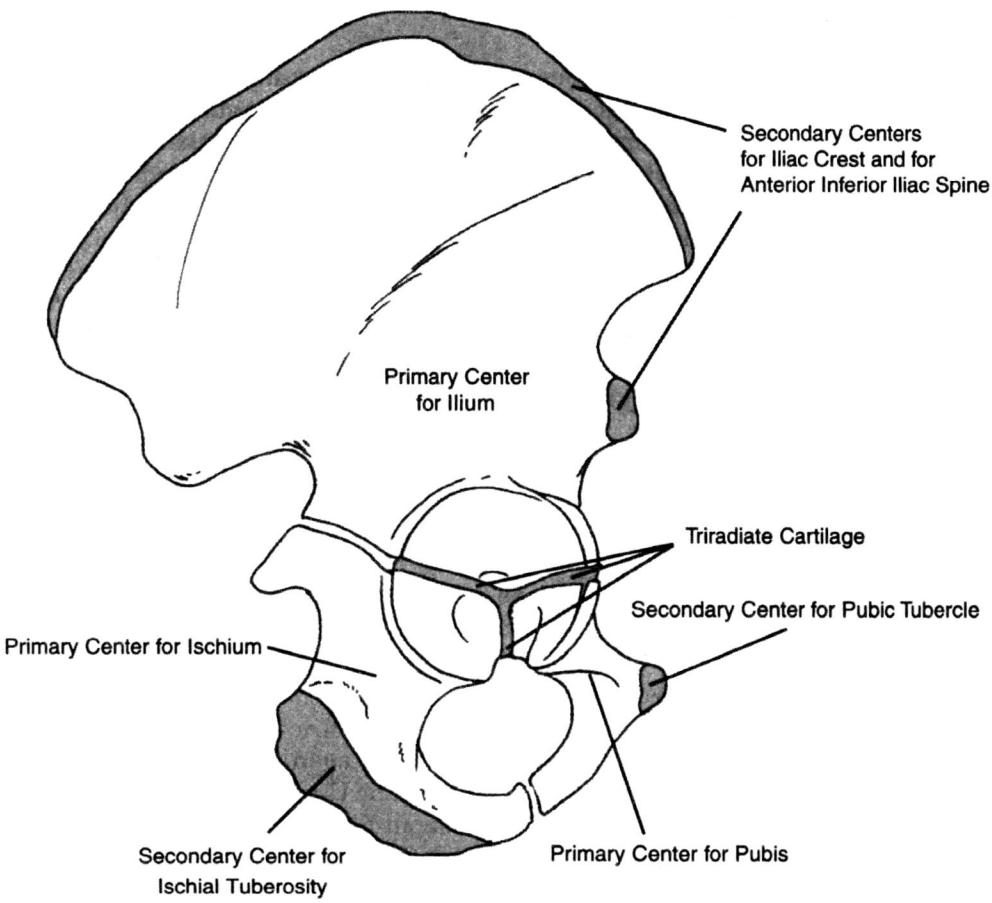

Figure 7-21 Centers of ossification of the os coxae.

the superior pubic ramus. By birth, the portions of the three elements of the hip bone nearest the acetabulum have undergone ossification, and only the triradiate cartilage remains cartilaginous. Between 8 and 10 years the ischial ramus and the inferior pubic ramus have met (earlier in girls than boys).

Immediately afterward, secondary centers of ossification develop within the triradiate cartilage. The number of these centers may be multiple, and some may fuse with one another as the *os acetabuli* before fusing with the three primary elements of the innominate. The os acetabuli usually appears at about the age of 12 between the ilium and pubis. It fuses with them by age 18, and forms the pubic part of the acetabulum. The other secondary centers of ossification begin to fuse with the remainder of the innominate at or just after puberty, and all are completely synostosed with the remainder of the bone by age 25 (Breathnach, 1965; Krogman, 1962).

The Femur

The femur (Figure 7-22) ossifies from a single primary center for the diaphysis, which begins to ossify at two months or less *in utero*, and four secondary centers: the femoral head, the greater trochanter, the lesser trochanter, and the distal end. There may occasionally be an apophysis for the gluteal tuberosity. The secondary center in the distal epiphysis appears at about age 1 year, and fuses with the distal end of the shaft at about 17 years in girls and at about 18 to 19 years in boys.

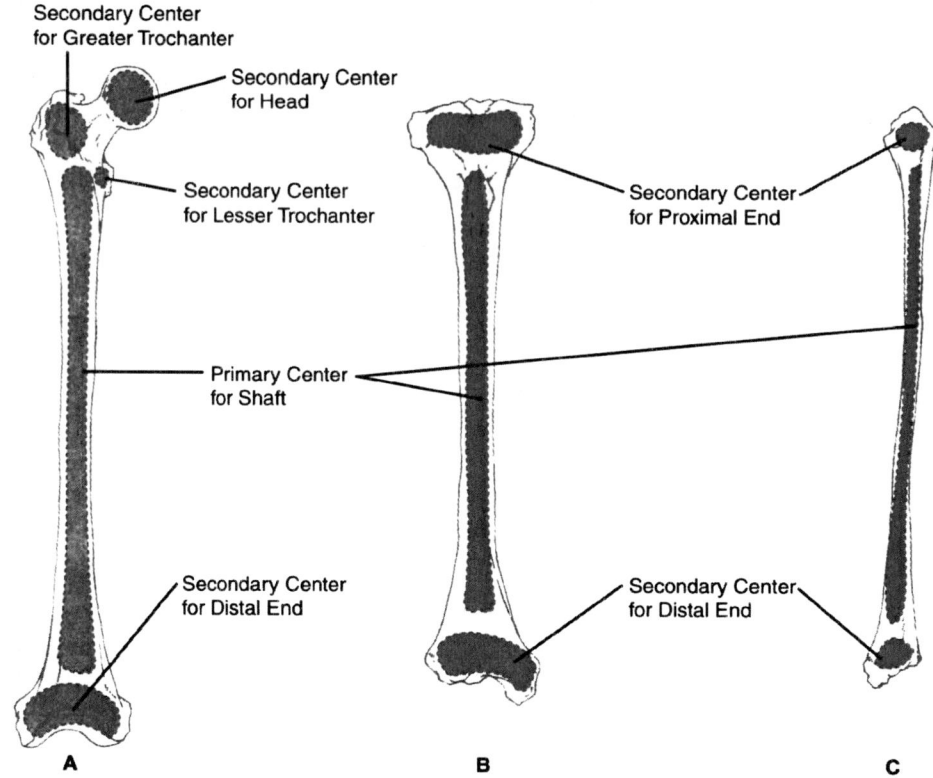

Figure 7-22 Centers of ossification of the (A) femur, (B) tibia, and (C) fibula.

Ossification of the femoral head begins at about one year, and fuses with the femoral neck between 17 and 18 years, earlier in girls than in boys. The center for the greater trochanter appears at about three years in girls and at about four years in boys. That for the lesser trochanter appears between 10 and 11 years. Both the greater and lesser trochanters fuse with the shaft of the femur between 17 and 18 years, usually a little before the head fuses with the neck. Again, this is slightly advanced in girls relative to boys (Breathnach, 1965; Krogman, 1962).

The Patella

The patella is cartilaginous at birth. By age two or three (but as late as age six) several bony centers have appeared. The centers rapidly fuse, and by puberty ossification of the bone is complete (Breathnach, 1965; Gray, 1985; Krogman, 1962).

The Tibia

The tibia (Figure 7-22) is the classic example of ossification in a long bone. It ossifies from three centers: a primary center for the diaphysis, and a secondary center in each of the proximal and distal epiphyses. The center for the diaphysis appears at about seven weeks *in utero*, that for the proximal epiphysis shortly before or after birth, and that for the distal epiphysis at about age two. The distal epiphysis fuses with the shaft of the bone at about 18 years, and that for the proximal epiphysis at about age 20 (Breathnach, 1965; Gray, 1985; Krogman, 1962).

The Fibula

The fibula (figure 7-22), like the tibia, ossifies from three centers: a primary center for the diaphysis, and a secondary center in each of the proximal and distal epiphyses. The center for the diaphysis appears at about eight weeks *in utero*, that for the proximal epiphysis at about age three or four, and that for the distal epiphysis at about age two. The distal epiphysis fuses with the shaft of the bone at about 20 years, and that for the proximal epiphysis at about age 25 (Breathnach, 1965; Gray, 1985; Krogman, 1962).

The Tarsals

All the tarsals ossify from a single center with the exception of the calcaneus, which has a secondary center for its posterior end. The centers of ossification appear in the following order: *calcaneus* (4-7 months *in utero*); *talus* (6 months); *cuboid* (9 months); *lateral cuneiform* (1 year); *medial cuneiform* (2 years); *intermediate cuneiform* and *navicular* (3 years). The posterior epiphysis of the calcaneus appears at 8 years and unites with the body of the calcaneus at about puberty (Breathnach, 1965; Gray, 1985; Krogman, 1962).

The Metatarsals

Metatarsals II through V all ossify from two centers, one for the shaft and one for the head (distal end). Metatarsal I (for the hallux) ossifies from two centers as well, but its epiphysis is located at its proximal rather than its distal end. In this regard it follows the pattern of the phalanges (as with the metacarpal for the pollex). The shafts of the metacarpals begin ossification at the ninth or tenth week *in utero*. At about age three the center of ossification for the epiphysis of the base of the first metatarsal appears, while the centers of ossification for the heads of the other four metatarsals appear between ages five and eight. These epiphyses unite with the shafts of the metatarsals between ages 18 and 20 (Breathnach, 1965; Gray, 1985; Krogman, 1962).

The Phalanges

The phalanges, like the first metatarsal, all ossify from two centers: one for a proximal epiphysis and one for the shaft. The ossification centers for the shafts of the proximal and distal phalanges appear between the eighth and twelfth weeks *in utero*, those for the middle phalanges at about 16 weeks *in utero*. The proximal epiphyses, or phalangeal bases, ossify between ages two and ten, and fuse with the phalangeal shaft by age 18 (Breathnach, 1965; Gray, 1985; Krogman, 1962).

References Cited

Breathnach, AS, ed. (1965) *Frazer's Anatomy of the Human Skeleton,* Sixth Edition. J and A Churchill, London.

Edwardson, BM (1995) *Musculoskeletal Disorders: Common Problems.* Singular Publishing Group, Inc., San Diego.

Gray, Henry (1985) *Gray's Anatomy,* 30th American Edition. C.D. Clemente, ed. Lea and Febiger, Philadelphia.

Gray, Henry (1995) *Gray's Anatomy,* 38th British Edition. P.L. Williams, L.H. Bannister, M.M. Berry, P. Collins, M. Dyson, J.E. Dussek and M.W.J. Ferguson, eds. Churchill Livingstone, New York.

Heiple, KG and Lovejoy, CO (1971) The distal femoral anatomy of *Australopithecus*. *American Journal of Physical Anthropology* 35: 75-84.

Krogman, W.M. (1962). *The Human Skeleton in Forensic Medicine.* Charles C. Thomas, Springfield, Illinois.

Krogman, W.M. and M. Y. Iscan (1986) *Human Skeleton in Forensic Medicine*, 2nd Edition. Charles C. Thomas, Springfield, Illinois.

Jenkins, DB (1991) *Hollinshead's Functional Anatomy of the Limbs and Back,* Sixth Edition. WB Saunders Company, Philadelphia.

Lovejoy, CO (1988) Evolution of human walking. *Scientific American* 259: 82-89.

Soames, RW and Atha, J (1981) The role of the antigravity musculature during quiet standing in man. *Eur. J. Appl. Physiol.* 47: 159-167.

Walker, R.A., and C.O. Lovejoy (1985) Radiographic Changes in the clavicle and proximal femur and their use in the determination of skeletal age at death. *American Journal of Physical Anthropology* 68:67-78.

Walker, R.A., C.O. Lovejoy and R.S. Meindl (1994) The histomorphological and geometric properties of human femoral cortex in individuals over 50: Implications for histomorphological determination of age-at-death. *American Journal of Human Biology* 6: 659-667.

Weinert, CR, McMaster, JH, and Furgeson, RJ (1973) Dynamic function of the human fibula. *American Journal of Anatomy* 138: 145-150.

CHAPTER
|Eight|

Osteology of The Human Cranium

The skull (figure 8-1) is divided into two portions: the *neurocranium* (cranial vault); and the *visceral cranium* or *splanchnocranium* (facial skeleton). The bones of the skull are joined to one another by immovable joints called *sutures*. For all practical mechanical purposes, sutures effectively block motion at the sutures and unify the cranium into a rigid structural unit that houses and protects the brain, the sense organs, and the structures associated with speech and mastication. (There is a school of thought in chiropractic and osteopathic medicine that physiologically important amounts of motion continue at the sutures and that adjustment of the sutures can affect cerebrospinal fluid (Adams et al., 1992; Heisey and Adams, 1993). See Rogers and Witt (1997) for a contrary view.)

The *neurocranium* is composed of eight bones that form the cranial vault. Some are paired (one is present on each side) and some are unpaired (present only in the midline). These eight bones are:

Ethmoid - unpaired

Frontal - unpaired

Occipital - unpaired

Sphenoid - unpaired

Temporal - paired

Parietal - paired.

The *visceral cranium* is composed of fourteen bones. Like those of the neurocranium, some are paired bilaterally, while some are present singly in the midline. These bones are:

Vomer - unpaired

Mandible - unpaired

Maxilla - paired

Zygomatic (Malar) - paired

Nasal - paired

Lacrimal - paired

Palatine - paired

Inferior Concha - paired.

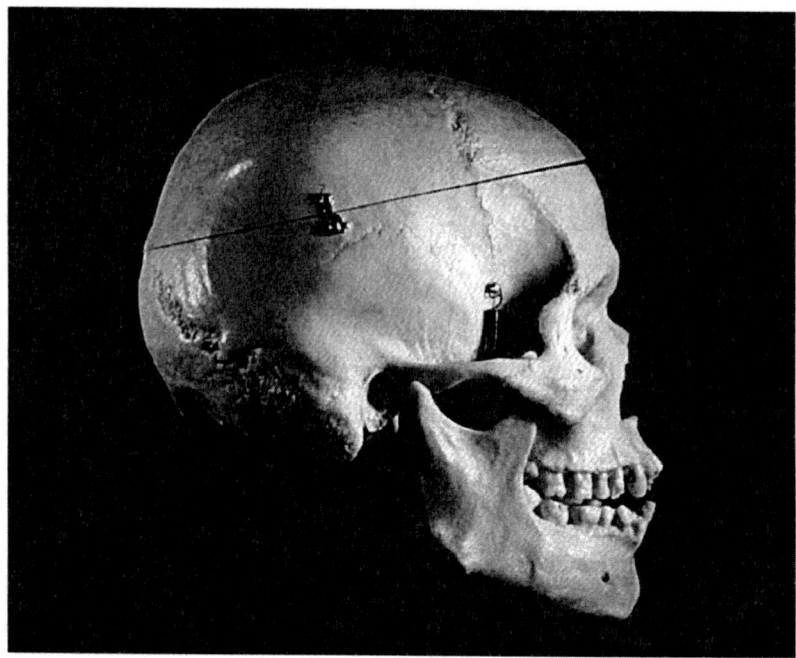

Figure 8-1 Photograph of the human skull: norma lateralis, or lateral view.

The mandible will be considered in detail later. In addition to these 22 bones that compose the skull proper, there are several associated bones that also must be considered. These include the bones associated with the organs of hearing (the *auditory ossicles: malleus, incus and stapes)* and speech (the *hyoid* bone).

The Auditory Ossicles

The auditory ossicles (Figures 8-2 and 8-3) are housed in the petrous portion of the temporal bone and transmit sound impulses from the tympanic membrane (eardrum) to the cochlea. They are six in number, three on each side. They are named as follows:

Malleus - paired.

Incus - paired.

Stapes - paired.

The handle of the *malleus* (colloquially: "hammer") is connected to the tympanic membrane (colloquially: "ear drum"). The malleus in turn articulates with the *incus* (colloquially: "anvil"). The incus articulates with the *stapes* (colloquially: "stirrup"). The base of the stapes rests upon the fenestra vestibuli (oval window) of the cochlea. The joints between the auditory ossicles are typical, although very small, synovial joints. Through this chain of small bones sound vibrations received by the tympanic membrane are transferred to the fenestra vestibuli of the cochlea.

The Hyoid Bone

The *hyoid* (Figure 8-4) is situated in the anterior part of the neck inferior to the mandible. Many of the muscles and ligaments of the tongue and larynx attach to it. It does not articulate with any other bone, but is best considered with the skull.

It is a small bone that consists of a central *body;* small superior *lesser cornu* projecting posterosuperiorly from the superolateral aspects of each side of the body; and larger, inferior, *greater cornu*, also projecting posteriorly from the lateral aspect of each side of the body.

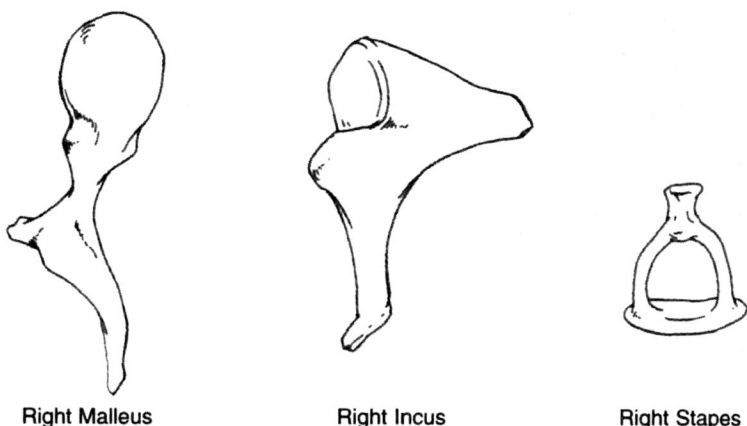

Figure 8-2 The auditory ossicles. The largest dimension is approximately 0.5 cm.

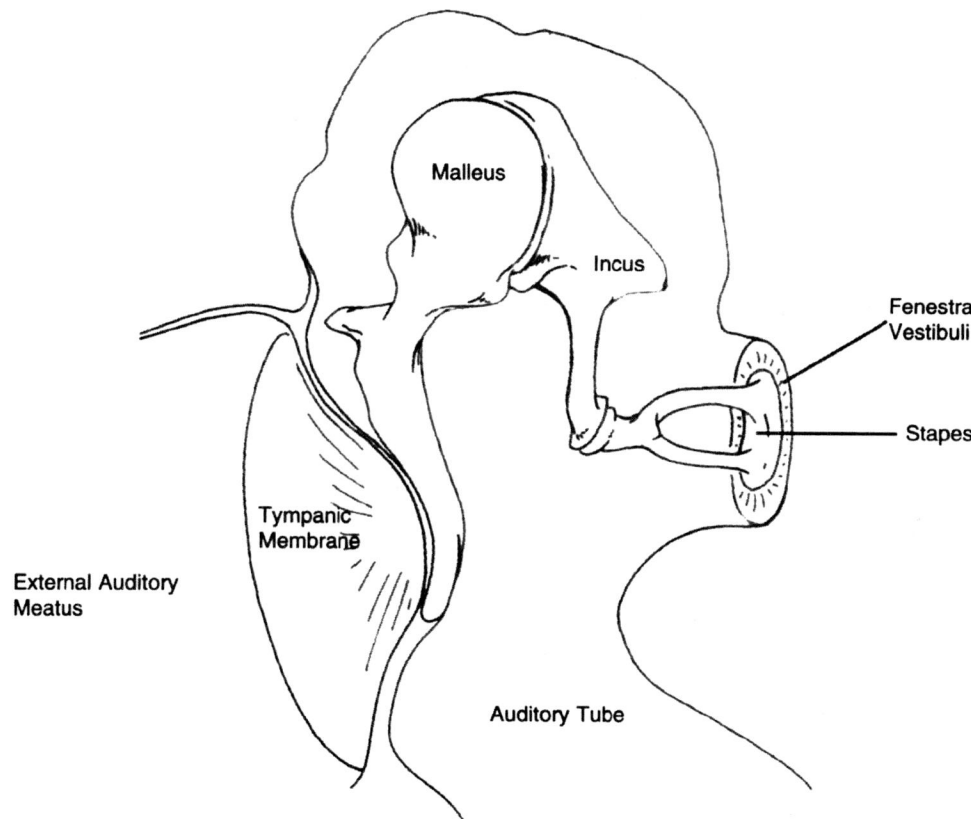

Figure 8-3 Diagrammatic representation of the middle ear, right side, anterior view.

The stylohyoid ligament attaches to the lesser cornu. The thyrohyoid ligament attaches to the greater cornu, as do several muscles of the neck and pharynx (the hyoglossus, the middle pharyngeal constrictor, digastric, thyrohyoideus and stylohyoideus muscles). Additional neck, pharyngeal, and oral muscles attach to the body (the genioglossus, geniohyoideus, mylohyoideus, sternohyoideus, and omohyoideus muscles).

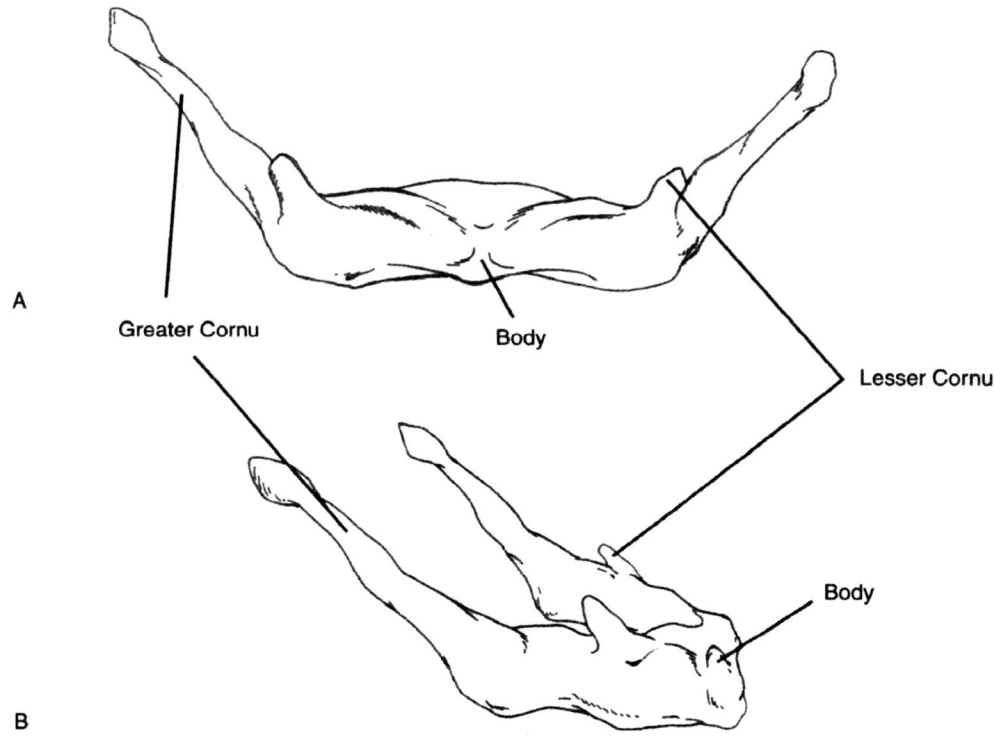

Figure 8-4 The hyoid bone: (A) anterior and (B) right lateral views.

The Skull

For ease of studying, the features of the skull will be broken down by view: superior, lateral, anterior, posterior, basal (inferior), and interior.

Superior View of the Skull: *Norma Verticalis*

The major bones seen in the *superior view of the skull, or norma verticalis* (figure 8-5) view of the skull are the *parietals*, the *occipitals*, and the *frontal*. Major landmarks are discussed next.

Landmarks and Features

The *sagittal suture* lies in the midline between the two *parietal bones* that form the roof of the *cranial vault*. The *coronal suture* is located between the two *parietal bones* posteriorly and the *frontal bone* anteriorly. The *lambdoidal suture* is found between the two *parietals* anteriorly and the *occipital* bone posteriorly.

Bregma is the point at the junction of the coronal and sagittal sutures. In infancy it is the location of the *anterior fontanelle*. The *fontanelles* are the so-called soft spots of the infant cranium, representing areas where the cranial vault bones have not yet grown together at the sutures. *Lambda* is the junction of the sagittal and lambdoidal sutures. In infancy it is the location of the *posterior fontanelle*.

Vertex is the most superior point on the skull when it is oriented in the Frankfort Horizontal Plane, in which the upper margin of the external acoustic meatus and the lower margin of the left orbit are aligned in a horizontal plane. Vertex usually lies on the sagittal suture about halfway between bregma and lambda. Near the center of each parietal bone, lateral to the sagittal suture, is the *parietal eminence*. It is the point of maximum convexity of the parietal bone, and marks the spot where ossification of the parietal bone began.

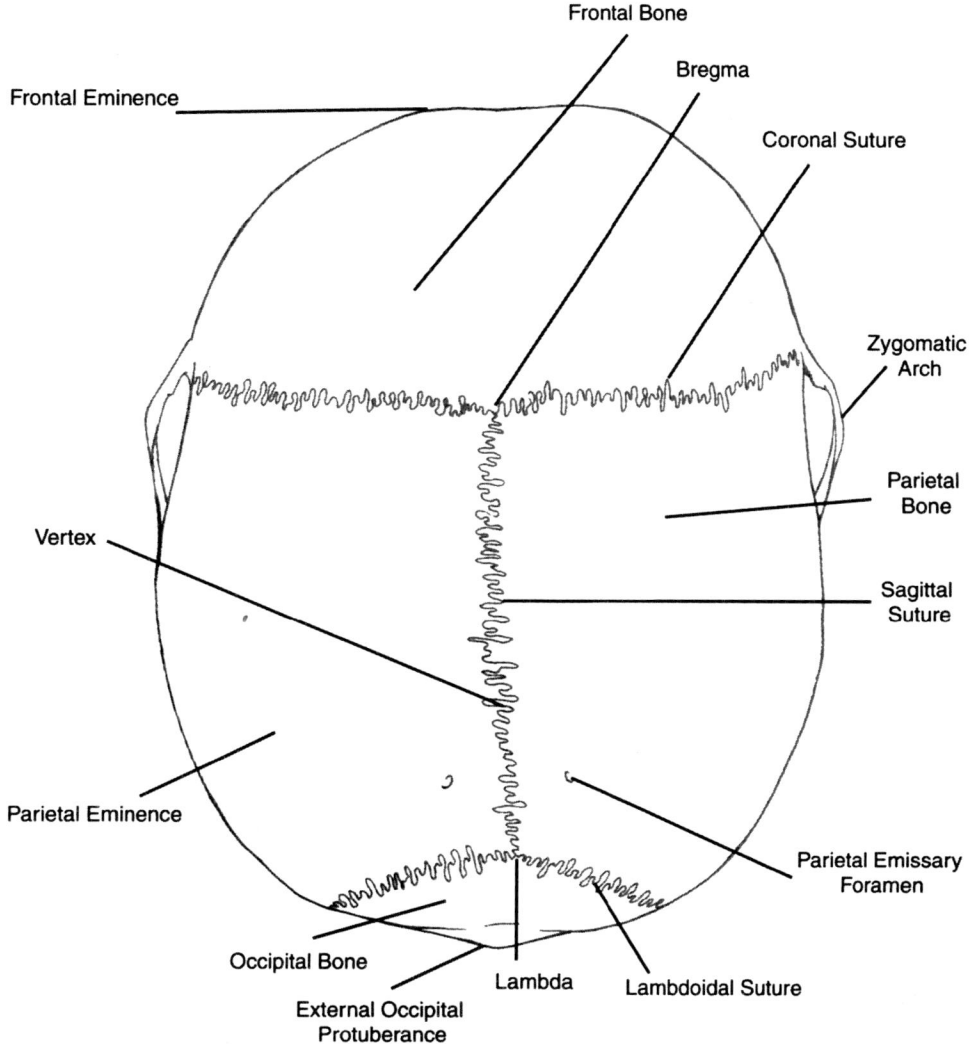

Figure 8-5 The adult human cranium: norma verticalis, or superior view.

Lateral View of the Skull: *Norma Lateralis*

Bones visible in the lateral view of the skull, or *norma lateralis* (Figure 8-6), are the frontal, zygomatic, temporal, parietal, occipital, sphenoid, nasal, maxilla, and mandible. The latter three will be discussed later. Visible landmarks are discussed next.

The *coronal* and *lambdoidal* sutures were discussed above The *squamosal suture* is present between the *squamous part of the temporal bone* and the *parietal bone*. The *occipitomastoid suture* is located between the occipital bone and the mastoid part of the temporal bone. The *parietomastoid suture* is between the posteroinferior parietal bone and the mastoid part of the temporal bone.

Pterion is the junction of the greater wing of the *sphenoid*, the *frontal*, the *parietal*, and the *temporal* bones, located in the region of the temple, superior to the zygomatic arch. It usually represents a region, and not a single point. It is the weakest region of cranial vault. It is the site of the *anterolateral* or *sphenoidal fontanelle* in the skull of the infant.

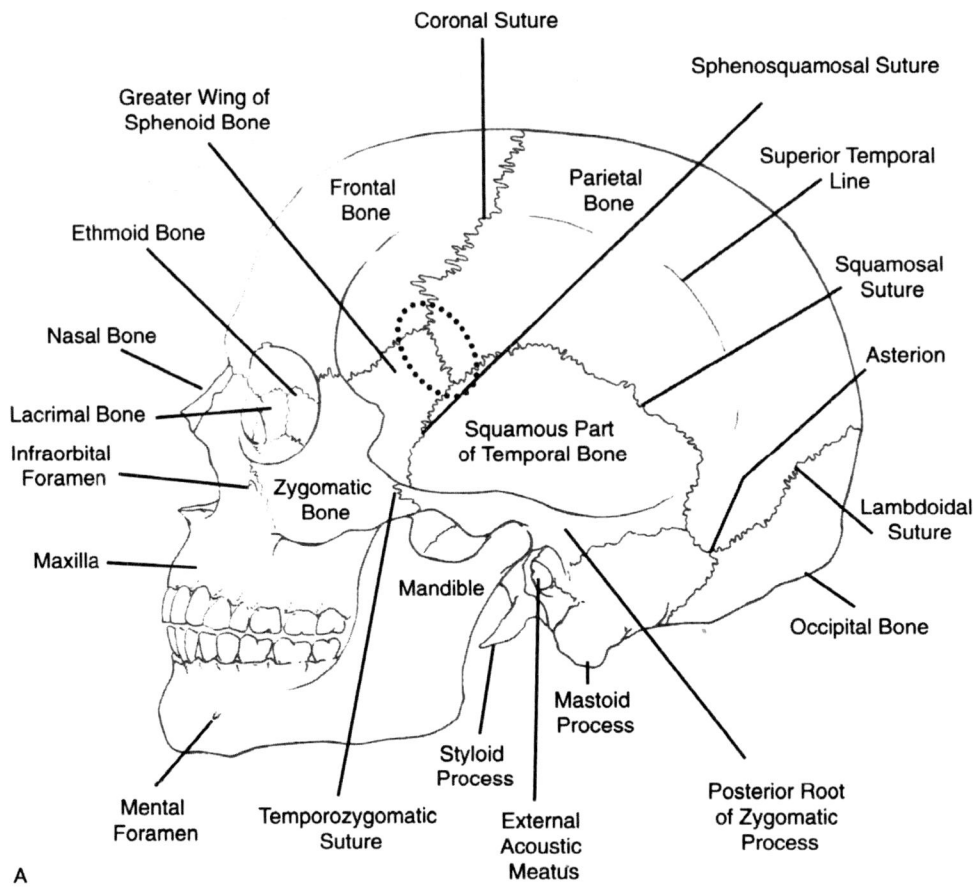

Figure 8-6 The adult human cranium: norma lateralis, or lateral view. The figure is divided into *A* and *B* portions to make the labeling of structures clearer.

Landmarks and Features

Asterion is the point marking the junction of the *occipitomastoid suture, parietomastoid suture,* and *lambdoidal suture.* It is the site of the *posterolateral fontanelle* in the skull of the infant. *Bregma, lambda,* and *vertex* are all landmarks visible laterally as well as superiorly, and were described earlier.

The *frontal bone* is the bone of the forehead. It has *zygomatic processes* on either side, projecting to articulate with the zygomatic bones. The *frontal eminences* are the areas of greatest convexity.

The parietal bones each exhibit *superior* and *inferior temporal lines* that arch anteroposteriorly above the squamosal suture. The superior and inferior temporal lines (Figure 8-7) mark the origin of the temporalis muscle, and extend from the *zygomatic process of the frontal bone,* across the frontal and parietal bones, and then curve anteriorly to become continuous with the *supramastoid crest* and the *posterior root of the zygomatic arch* (figures 8-6, 8-8). The *parietal eminences* are the areas of greatest convexity of the parietal bones.

The *temporal bone* demonstrates several prominent features in the lateral view. The *external acoustic (auditory) meatus* (Figure 8-9) is the bony opening into the ear. In life it is closed by the tympanic membrane which articulates with the malleus (see Figure 8-3). The *squamous part* of the temporal bone is the flat part just inferior to the parietal bone.

Osteology of The Human Cranium — Page 247

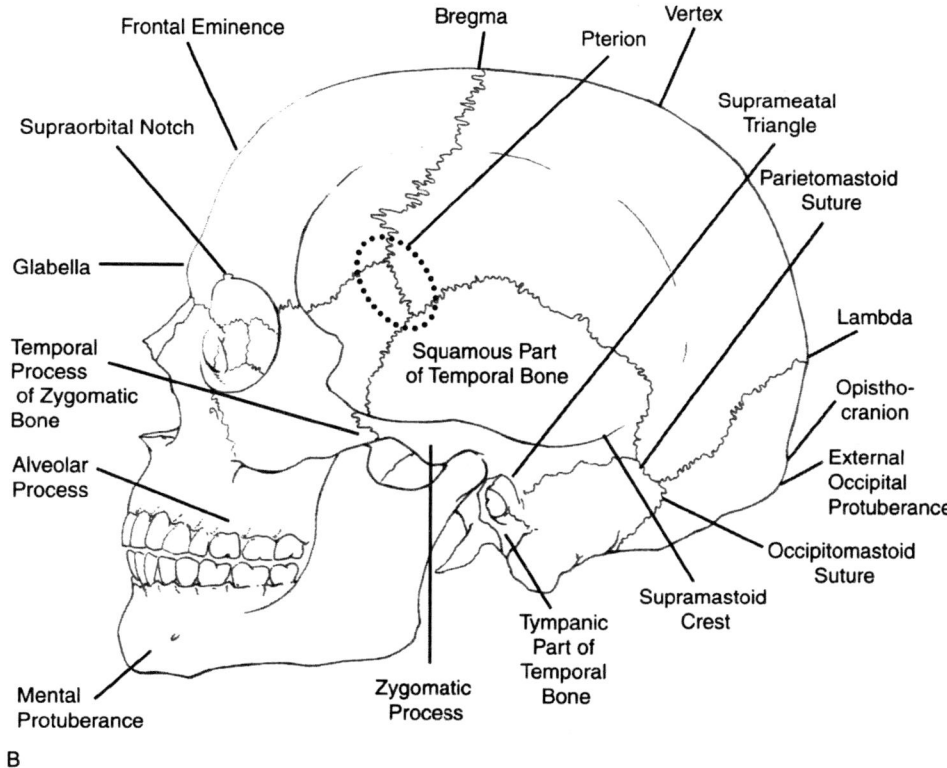

B

Figure 8-6 (*Continued*)

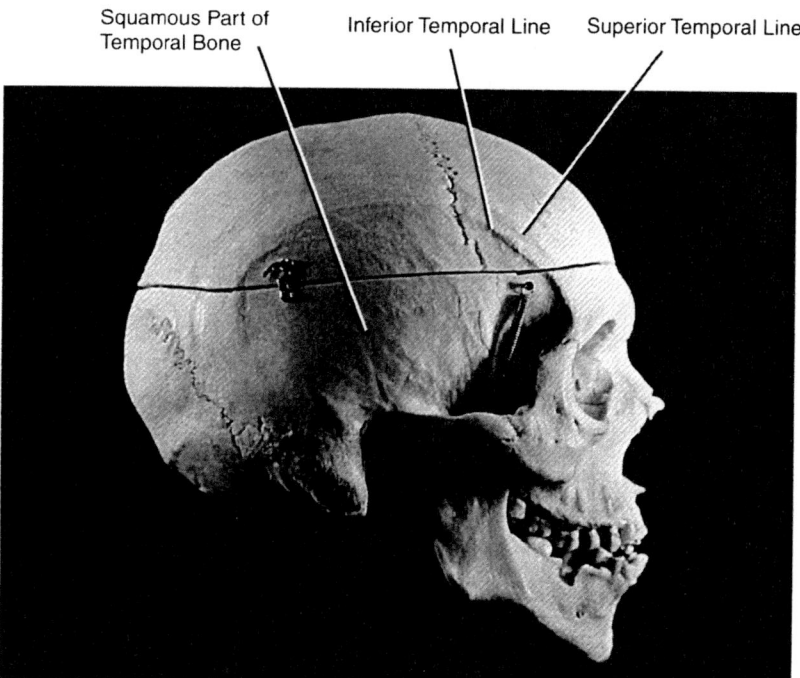

Figure 8-7 Photograph of the adult human cranium from the right side, demonstrating the superior and inferior temporal lines.

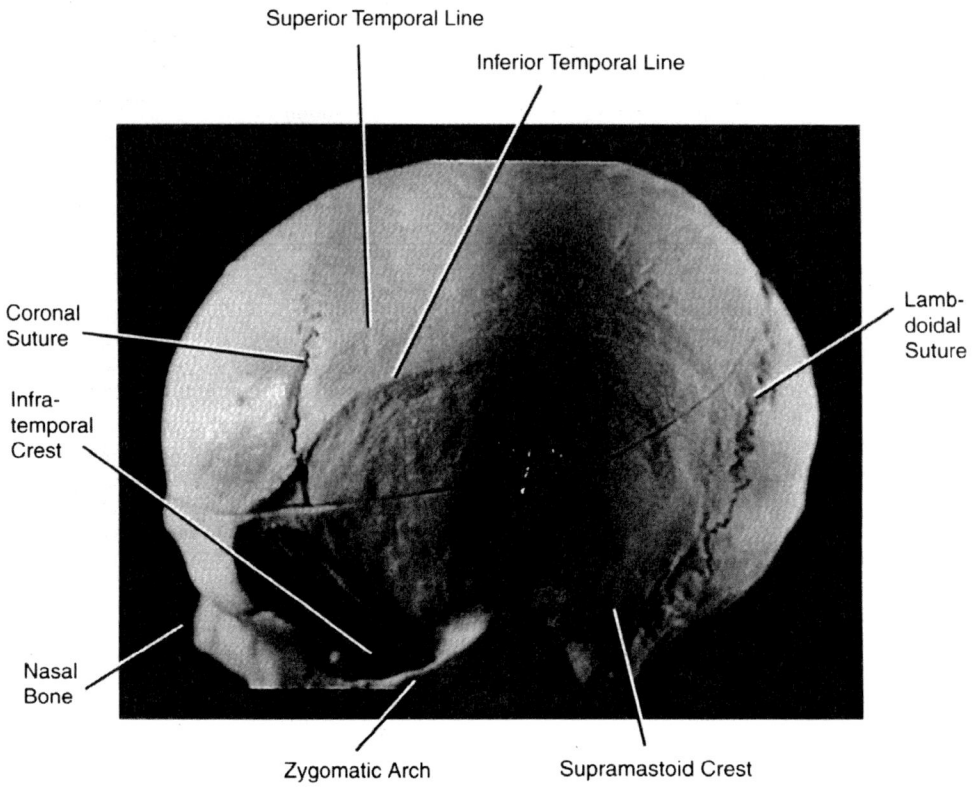

Figure 8-8 Photograph of the adult human cranium from the left superior side.

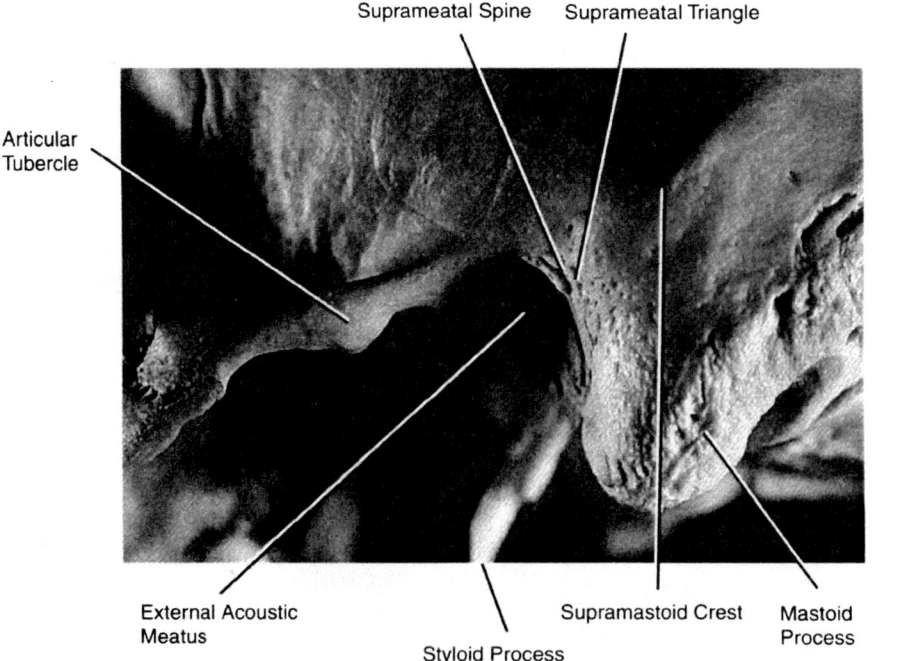

Figure 8-9 The region of the external acoustic meatus, left side.

It exhibits a *supramastoid crest*, which extends posteriorly from the posterior root of the zygomatic process and superior to the mastoid process. A s*uprameatal triangle and spine* can also be found between the posterior root of the zygomatic arch and the posterosuperior part of the external acoustic meatus.

The *tympanic part* (Figures 8-10 and 8-11) of the temporal is a curved plate of bone lying inferior to the squamous part, anterior to the mastoid process, and posterior to the mandibular fossa. The *Mandibular fossa (glenoid fossa)* is located between the *articular tubercle* and the tympanic part of the temporal bone. It is divided in two by the *squamotympanic* fissure. The anterior and larger part of the fossa articulates with the condyle of the mandible, which forms the *temporomandibular joint*.

The *zygomatic process* projects anteriorly from the temporal squama above the mandibular fossa and articular tubercle. Forming the anterior boundary of the mandibular fossa is the bump-like articular tubercle. Posterior to the mandibular fossa is the *postglenoid process*, which is a small, pointed projection of the temporal bone just anterior to the tympanic part of the temporal bone. The *styloid process* is a long, pointed projection that extends inferiorly and anteriorly from the tympanic part and gives attachment to several muscles and ligaments.

The *mastoid portion* of the temporal bone is posterior to the external acoustic meatus. Two structures of interest can be found on this part of the temporal bone. The *Mastoid Process* (see Figures 8-6 and 8-9) is a blunt projection for muscle attachments found immediately posterior to the tympanic part and the external auditory meatus. A *mastoid foramen* may be found posterior and/or superior to the mastoid process. It transmits an emissary vein from sigmoid venous sinus inside the skull.

The *zygomatic bone* (see Figures 8-6 and 8-7) is the bone of the cheek. It presents a *temporal process* that meets the zygomatic process

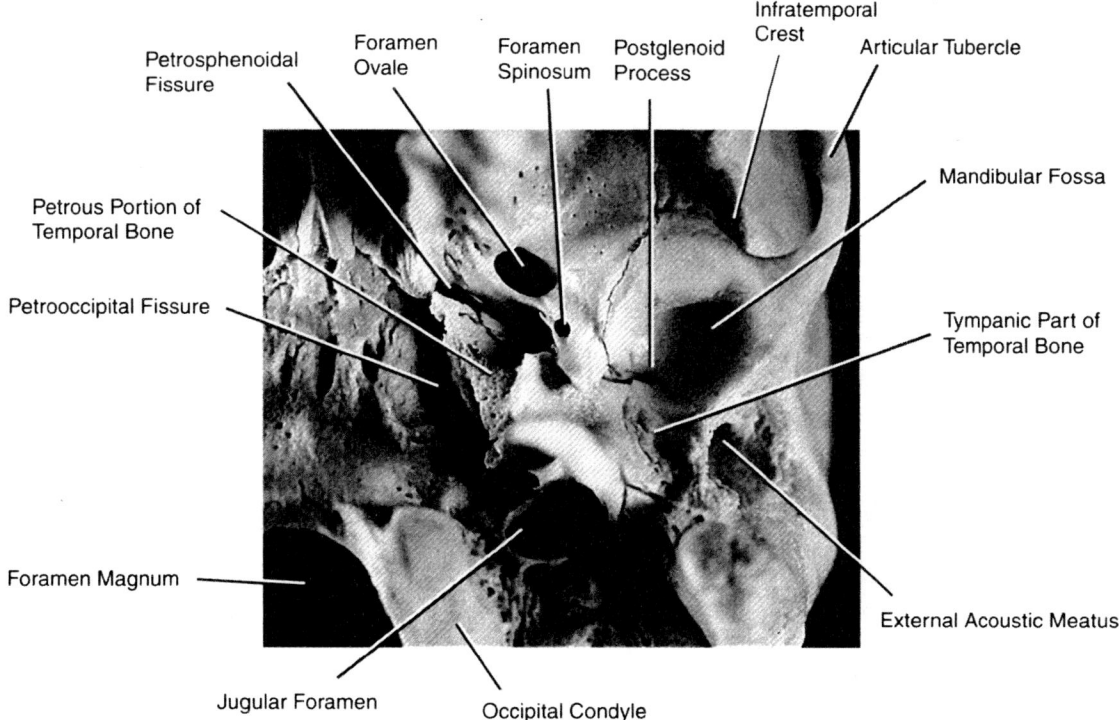

Figure 8-10 The region of the external acoustic meatus, right inferior view.

of the temporal bone, a *zygomaticofacial foramen* on its anterolateral aspect, and a *zygomaticotemporal foramen* posteriorly. These foramina transmit nerves to the skin of the lateral face and temple, respectively.

The *zygomatic arch* is formed by the *zygomatic process of the temporal* and the *temporal process of the zygomatic*. The zygomatic process of the temporal arises from *two roots*. The *anterior root* is directed inward anterior to the mandibular fossa where it expands and forms the *articular tubercle* of the temporal bone. There is also a *posterior root* which runs posteriorly above the external acoustic meatus and is continuous with the *supramastoid crest*.

The *sphenoid bone* is found between the frontal and zygomatic bones anteriorly and the parietal and temporal bones posteriorly. The *greater wing of the sphenoid bone* (see Figure 8-6) is a superolateral extension of bone that helps form the temple region. At its inferior aspect is a linear roughness or ridge called the *infratemporal crest*.

The *infratemporal crest* is located at about the level of the zygomatic arch. This crest marks the point at which the greater wing of the sphenoid bone takes a sharp turn and becomes a part of the base of the skull. The infratemporal crest marks the inferior extent of the temporal fossa and marks the superior border of the *infratemporal fossa*. The *lateral pterygoid plate* of the sphenoid is a thin plate of bone posterior to the maxilla and inferior to the greater wing of the sphenoid. The lateral pterygoid plate will be better observed from the *norma basalis*, or inferior view of the skull.

The *temporal fossa* is formed by parts of the frontal, temporal, sphenoid, and zygomatic bones. This fossa occupies a large part of the lateral aspect of the parietal bone. The temporal fossa is the region bounded by the temporal lines superiorly and posteriorly, the frontal and zygomatic bones anteriorly, and the zygomatic arch and infratemporal crest (on the greater wing of the sphenoid) inferiorly. The bones making up the medial wall of the temporal fossa are the squamous part of the temporal bone, the greater

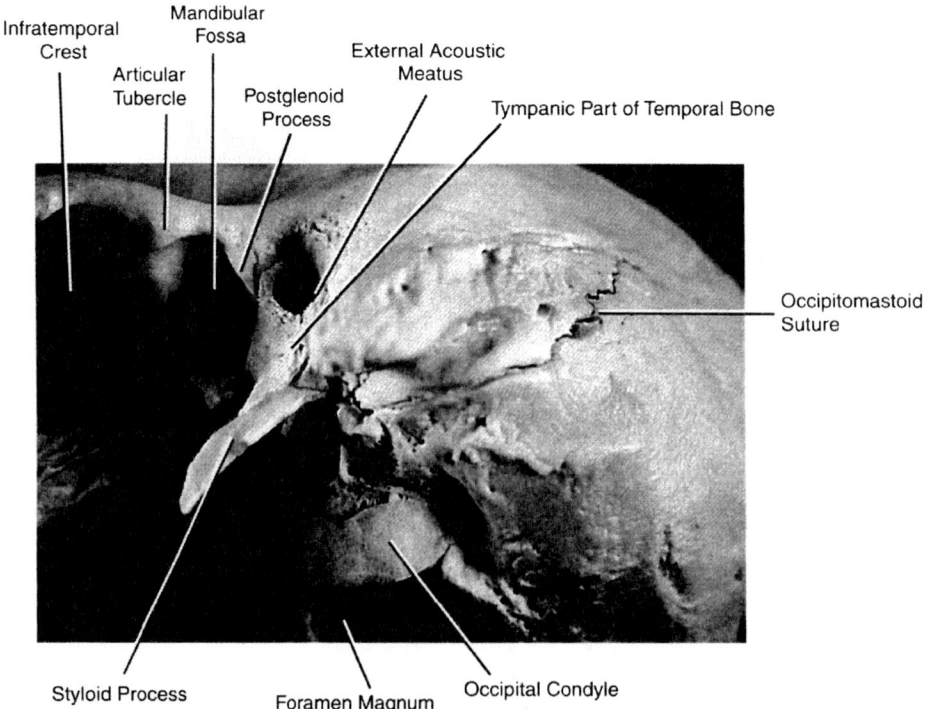

Figure 8-11 The region of the external acoustic meatus, left side, oblique posteroinferior view.

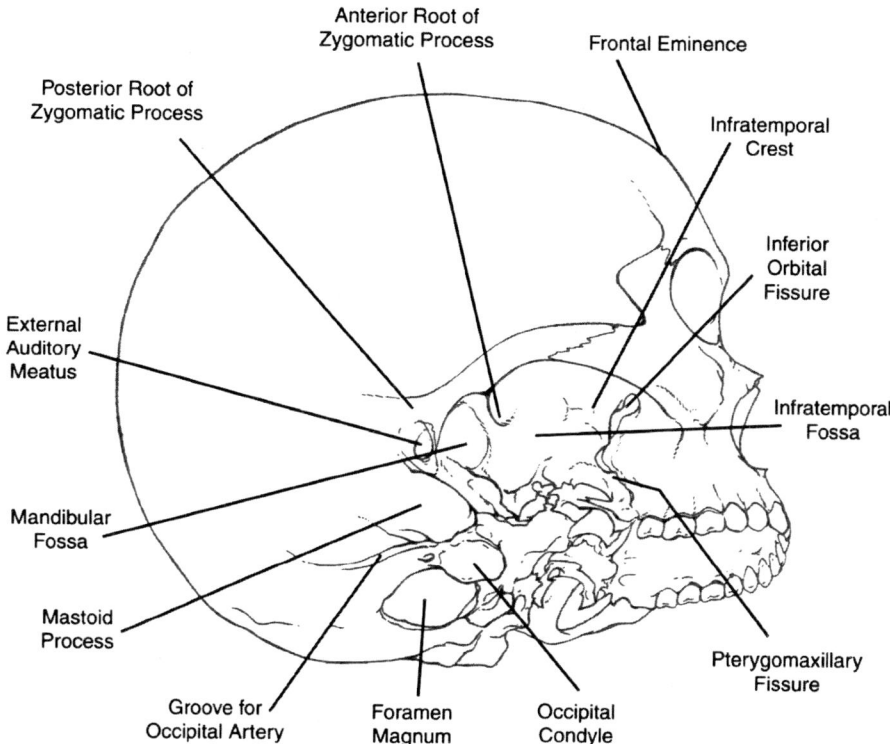

Figure 8-12 The adult human cranium, right side, oblique posteroinferior view.

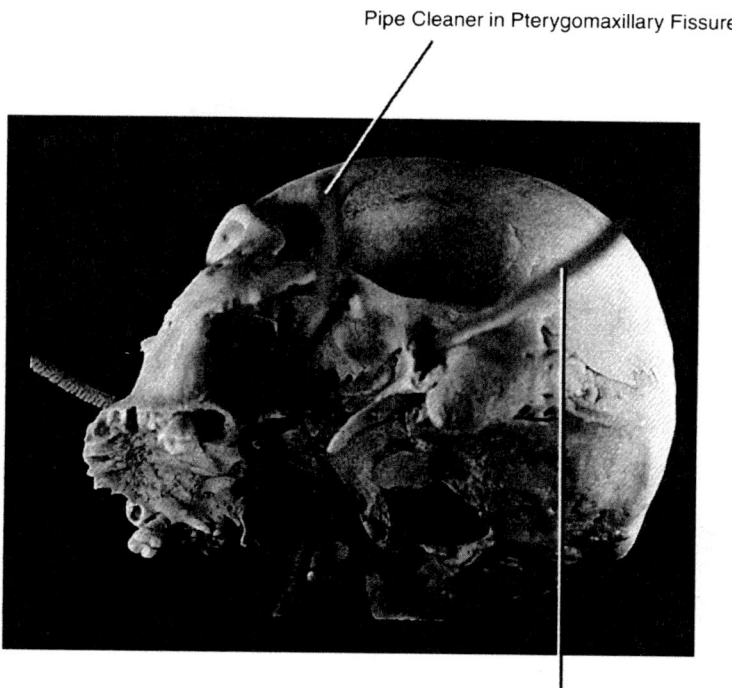

Figure 8-13 Photograph of the adult human cranium, left side, oblique posteroinferior view.

wing of the sphenoid bone, and parts of the frontal and parietal bones.

The *infratemporal fossa* (Figures 8-12 and 8-13) is an irregular space behind the maxilla formed by the infratemporal surface of the greater wing of the sphenoid and part of the temporal squama. This irregularly shaped cavity lies medial or deep to the zygomatic arch. It is bounded anteriorly by the alveolar process and infratemporal surface of the maxilla and the ridge from the maxilla's zygomatic process; posteriorly by the articular tubercle of the temporal and the spine of the sphenoid; superiorly by the greater wing of the sphenoid and the inferior surface of the temporal squama; medially by the *lateral pterygoid plate;* and laterally by the zygomatic arch.

The *foramen ovale, foramen spinosum*, and the *alveolar canals* open into it (discussed later). Two fissures open into the superior and medial part of the infratemporal fossa and meet at right angles: the *inferior orbital fissure* and the *pterygomaxillary fissure.*

The *inferior orbital fissure* opens into the posterolateral part of the orbit. It is bounded by the greater wing of the sphenoid, lateral border of the orbital surface of the maxilla, the orbital process of the palatine bone, and a small portion of the zygomatic bone.

The *pterygomaxillary fissure* is formed by the divergence of the lateral pterygoid plate of the sphenoid from the maxilla. This fissure connects the pterygopalatine fossa with the infratemporal fossa and transmits the terminal part of the maxillary artery and veins.

The Anterior View of the Skull: *Norma Frontalis*

Major bones visible in the anterior view of the skull, or *norma frontalis* (Figure 8-14) include the frontal, nasals, maxillae, zygomatics, ethmoid, vomer, mandible (discussed in detail below), and inferior nasal concha. Major visible landmarks are discussed next.

Landmarks and Features

The *frontal bone* (Figures 8-15 and 8-16; see also Figure 8-14) forms the forehead. The *frontal eminences* are the areas of greatest convexity on the frontal bone. One is located superior to each superciliary arch. The *superciliary arches* are the bony arches underlying the eyebrows. *Glabella* is the raised area between the two superciliary arches. The *supraorbital margin* is the superior margin of the orbit, formed by the frontal bone. The medial third of the supraorbital margin is rounded, the lateral two thirds are thin and prominent. At the junction of the medial one third and the lateral two thirds is the *supraorbital notch or foramen* for the transmission of the supraorbital nerve and vessels.

Facial Skeleton

The *frontonasal suture* (see Figure 8-14) is located below glabella. This suture is formed at the junction of the frontal bone with the two small *nasal bones*. The midpoint of this suture is a point termed *nasion*. The *anterior nasal aperture* (figure 8-14) is the external opening of the nasal cavity in the skull. The external opening in the living person is the external nares.

At the inferiormost point of the anterior nasal aperture in the midline is the *anterior nasal spine*. Within the nasal cavity, the *bony nasal septum* is visible. The superior and anterior part of the septum is formed by the *perpendicular plate of the ethmoid bone*, while the inferior and posterior part is formed by the *vomer*. Projecting medially from the lateral nasal wall on each side is a thin bone called the *inferior nasal concha*.

Below the orbit on the maxilla is the *infraorbital foramen* (figure 8-14). The part of the maxilla containing the roots of the upper teeth is the *alveolar process of the maxilla*.

Facial Skeleton: The Orbit

The orbits (see Figures 8-15 and 8-16) are the conical cavities which contain the eyes and associated structures. They are so placed that the medial walls of the two orbits are approximately parallel with each other, while their lateral walls are widely divergent. The orbit has a roof, a floor, a medial wall, a lateral wall, a base, and an apex.

The *roof* of the orbit is formed by the orbital plate of the frontal bone and the lesser wing of the sphenoid. The *lacrimal fossa* is located laterally and in life houses the lacrimal gland. The *floor* (figure 8-14) is formed by the orbital surface of the maxilla, the orbital process of the zygomatic bone, and the orbital process of the palatine bone.

The *medial wall* of the orbit is nearly vertical and formed by the frontal process of the maxilla, the lacrimal bone, the lamina orbitalis of the ethmoid, and a small part of the body of the sphenoid. Anteriorly is the *lacrimal sulcus* that houses the lacrimal sac. The sulcus is bounded by an *anterior*

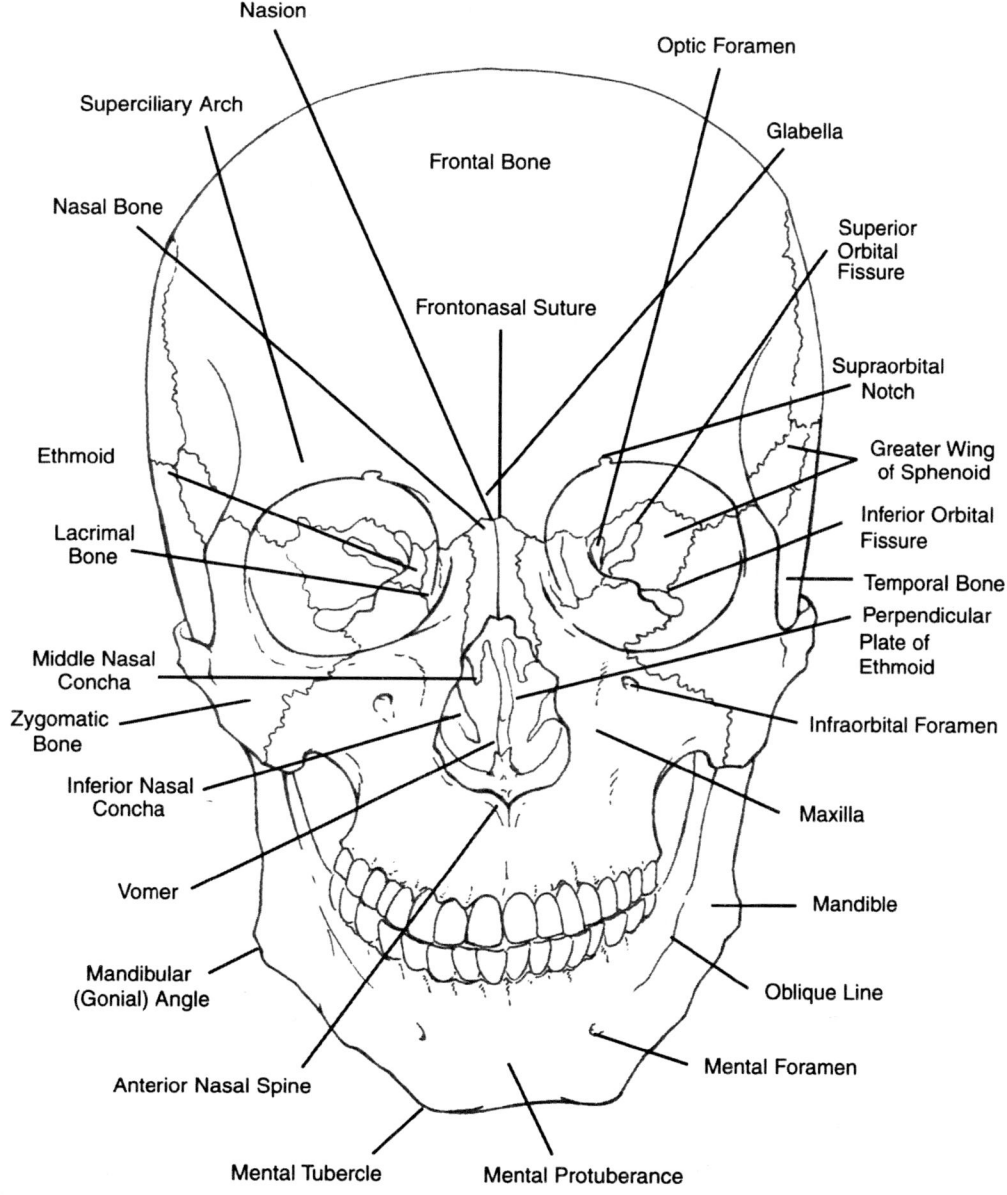

Figure 8-14 The adult human cranium: norma frontalis, or anterior view.

lacrimal crest on the frontal process of the maxilla and a sharper *posterior lacrimal crest* on the lacrimal bone.

In or near the *frontoethmoidal suture* (between the frontal and ethmoid bones) are the anterior and posterior ethmoidal foramina. The *anterior ethmoidal foramen* transmits the anterior ethmoidal nerve, one of the terminal branches of the nasociliary nerve, and anterior ethmoidal vessels. The *posterior ethmoidal foramen* transmits the posterior ethmoidal nerve and vessels.

The *lateral wall* of the orbit is formed by the orbital process of the *zygomatic* and the orbital surface of the *greater wing of the sphenoid*. Between the roof and the lateral wall near the apex of the orbit — or more specifically, between the greater and lesser wings of the sphenoid bone — is the *superior orbital fissure*. The oculomotor, trochlear, opthalmic division of the trigeminal, and abducent nerves enter the orbital cavity through this fissure The *inferior orbital fissure* lies between the lateral wall and the floor of the orbit, inferior to the greater wing of the sphenoid.

The *base* of the orbit is formed by the orbital margin, while the *apex* of the orbit is formed by the *optic foramen*. The *optic foramen* is the anterior opening of the *optic canal*, which transmits the optic nerve and the opthalmic artery. This is an opening that runs through the sphenoid bone.

The zygomatic arch has already been discussed, and the mandible will be discussed later in a separate section.

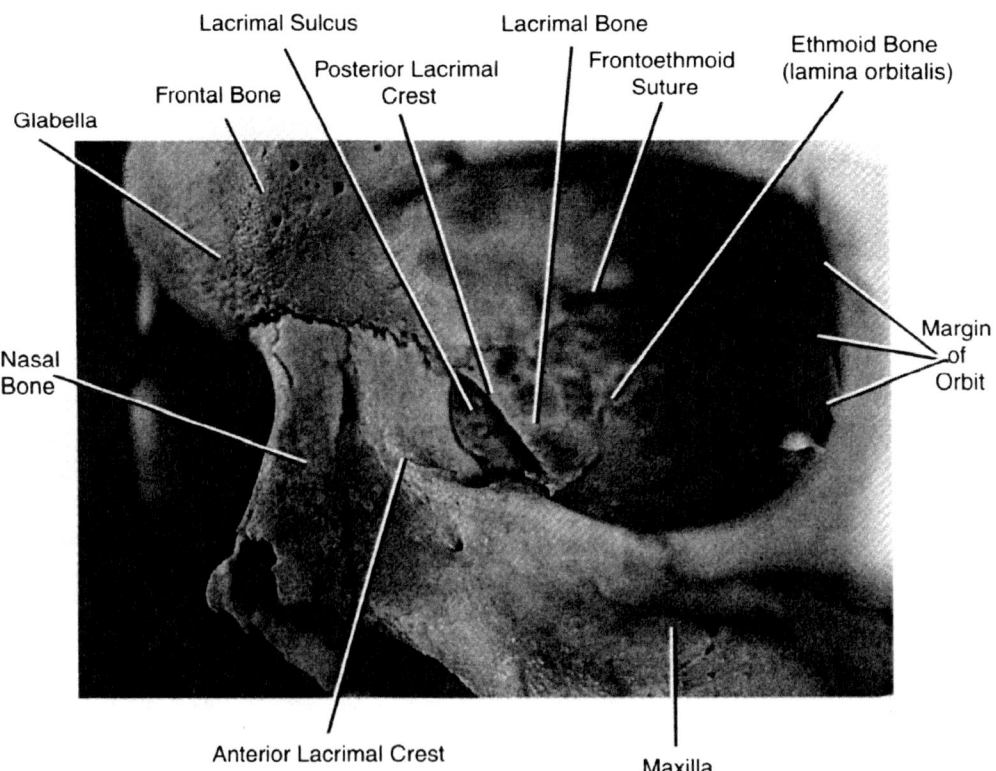

Figure 8-15   Orbit I: photograph of the left orbit, demonstrating the medial orbital wall and the orbital margin.

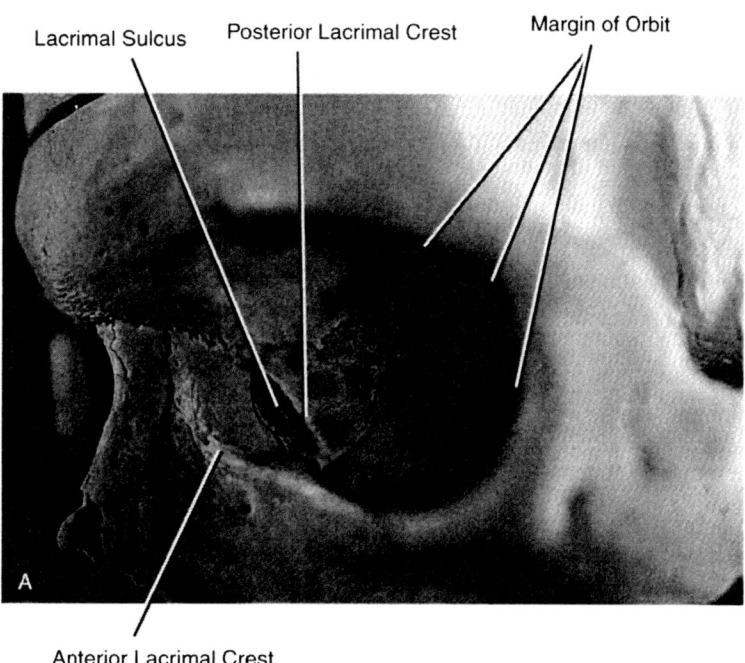

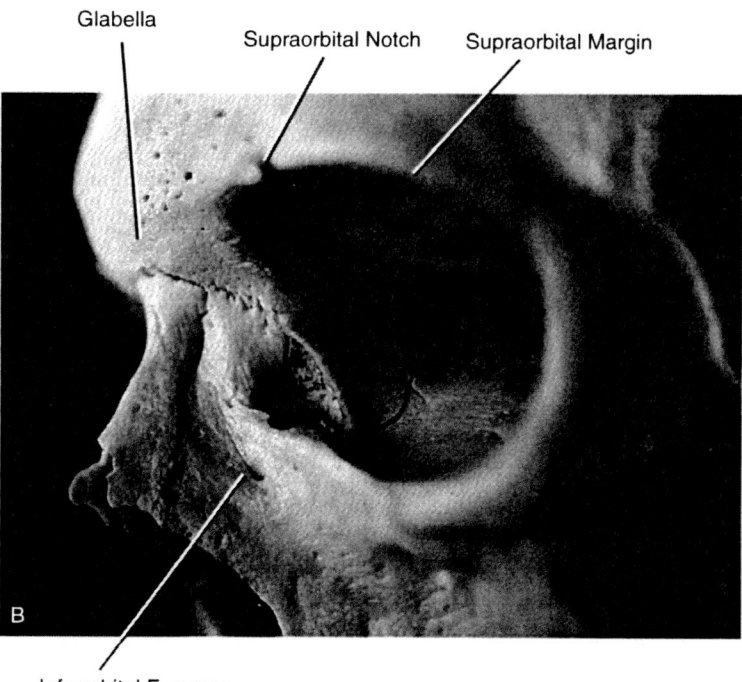

Figure 8-16 Orbit II: photograph showing a broader view of the left orbit, including (*A*) the lacrimal sulcus and associated structures and (*B*) the orbital margin and associated structures.

The Posterior View of the Skull: *Norma Occipitalis*

The major bones visible in the posterior view of the skull, or *norma occipitalis* (Figures 8-17 through 8-19), are the occipital, the parietals, and the temporals. Landmarks are examined next.

Landmarks and Features

The sagittal and lambdoidal sutures are visible in this view as well as in norma verticalis. Occasional small bones, the *sutural* (or *Wormian* or *Inca*) bones, may be present in the lambdoidal suture between the parietals and the occipital. They are not a constant feature, and their frequency may vary from population to population. Sutural bones may occur in other sutures of the cranium as well, but are far more common in the lambdoidal suture than elsewhere. The *parietomastoid suture,* between the parietal bone and the mastoid portion of the temporal bone, is visible in this view as well as in the norma lateralis. The same is true for the *occipitomastoid suture*, which is between the *occipital bone* and the *mastoid portion of the temporal bone.*

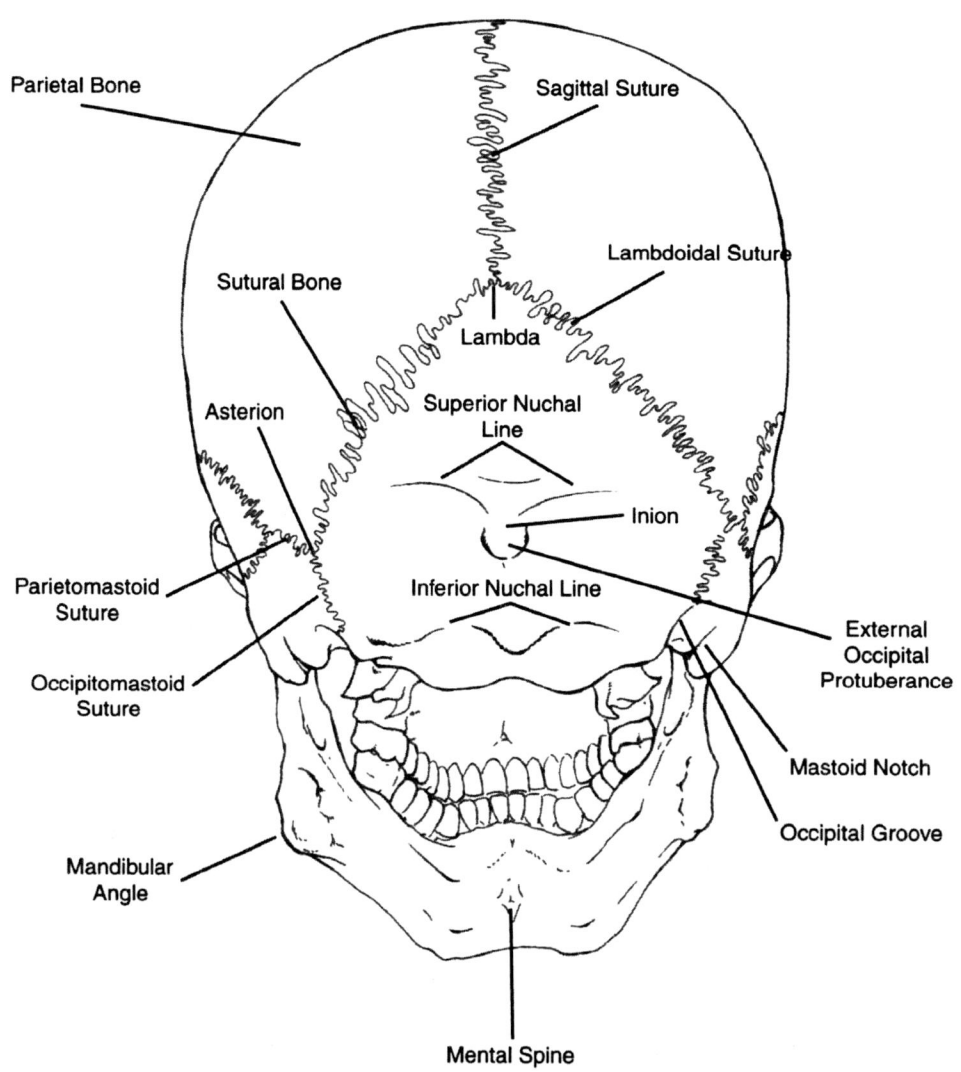

Figure 8-17 The adult human cranium: norma occipitalis, or posterior view.

Osteology of The Human Cranium   Page 257

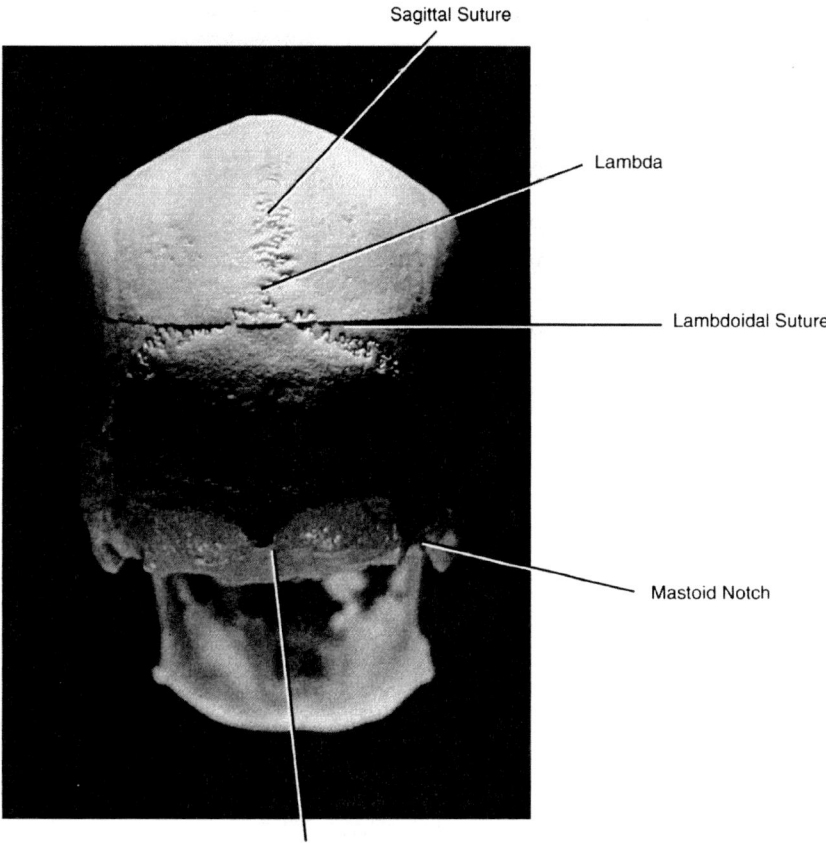

Figure 8-18 Photograph of the adult human cranium in norma occipitalis, or posterior view.

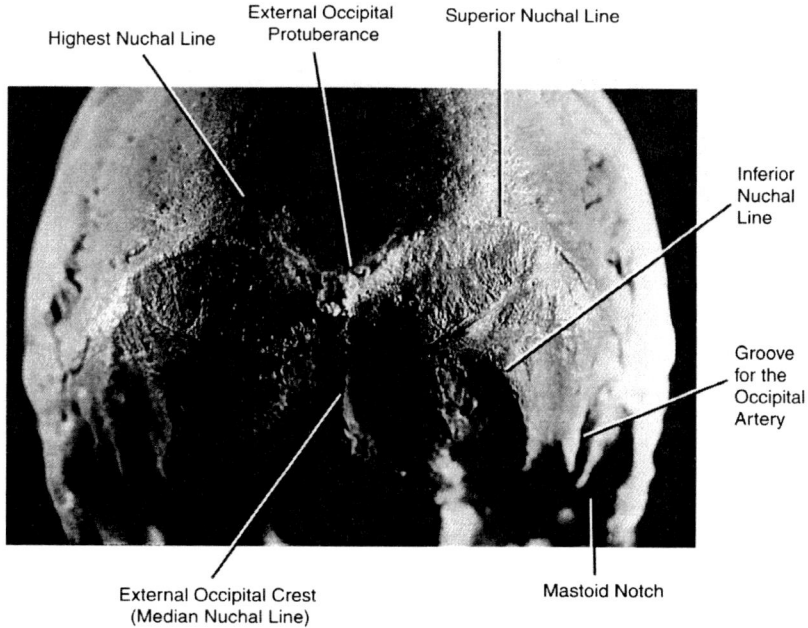

Figure 8-19 Photograph of the nuchal region of the adult human cranium.

Asterion is visible from the posterior view, as well as from the lateral view. Lambda is also visible posteriorly. *Opisthocranion* is the most posterior point on the skull when it is oriented in the Frankfort Horizontal (defined as when the superior border of the external auditory meatus and the inferior margin of the orbit are oriented in the same horizontal plane). It may occasionally lie on the external occipital protuberance, but is more often located superior to the protuberance. *Inion* is the most posterior point on the external occipital protuberance. Very occasionally opisthocranion may occupy this same location.

The *external occipital protuberance* is a prominent raised area at the back of the skull near the midline of the occipital squama. Extending laterally from it on either side are slightly arching ridges that are called the *superior nuchal lines*. Approximately one cm. above the superior nuchal line there may be present the faint *highest nuchal line*, for the

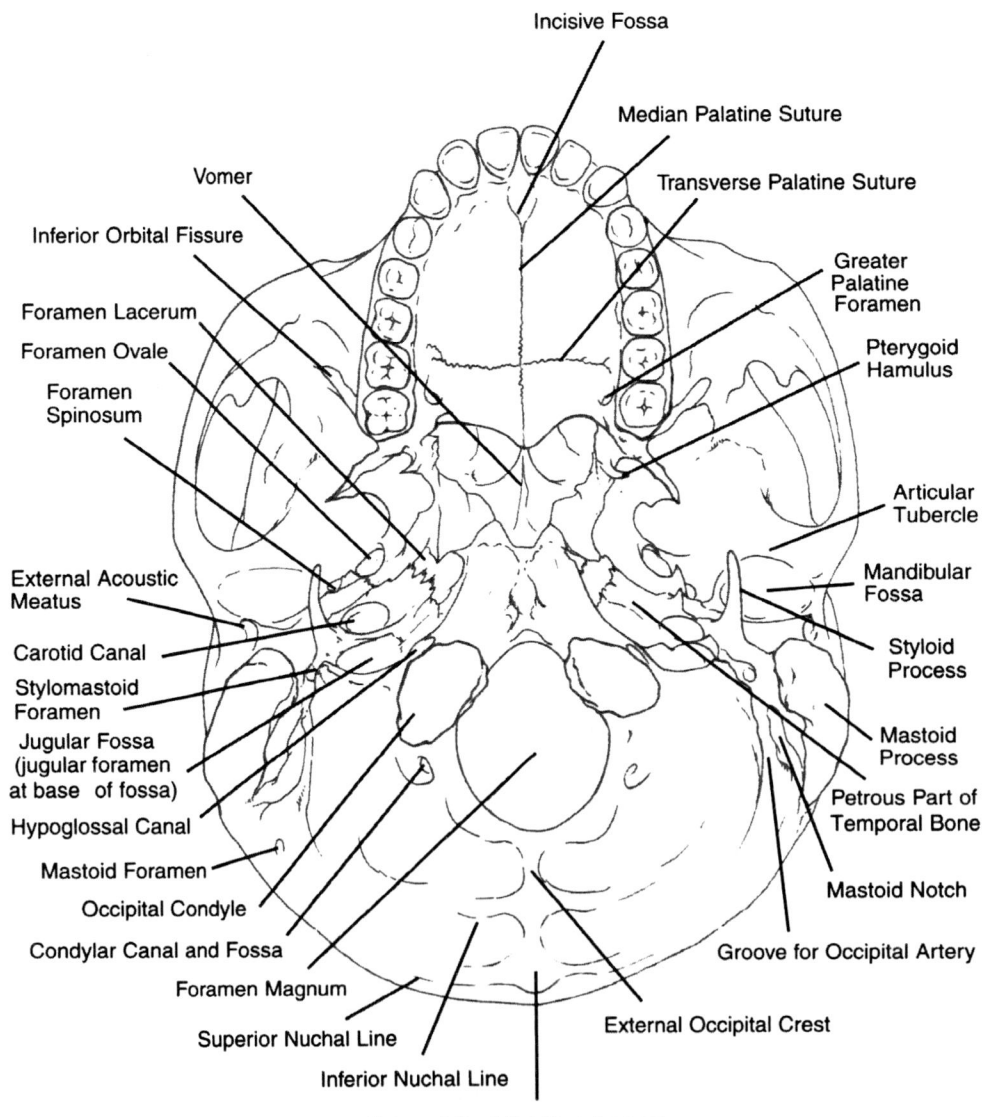

Figure 8-20 The adult human cranium: norma basalis, or inferior view. The figure is divided into *A* and *B* portions to make the labeling of structures clearer.

attachment of the galea aponeurotica. The *external occipital crest (median nuchal line)* is a median ridge that extends from *inion* to the *foramen magnum*, and gives attachment to the *ligamentum nuchae*. The *inferior nuchal line* begins at the midpoint of the external occipital crest and runs laterally across the occipital bone inferior to the superior nuchal lines.

The *mastoid notch* can be seen on the medial side of the *mastoid process*. This notch is for the posterior belly of the digastric muscle. Medial to the notch is the *occipital groove* for the occipital artery.

The Basal View of the Skull: *Norma Basalis*

The major bones visible in the basal (inferior) view of the skull, or *norma basalis* (Figures 8-20 and 8-21), are the occipital, the temporals, the sphenoid, the maxillae, the palatines, the vomer, and the mandible. Visible landmarks are discussed next.

Landmarks and Features

The *hard palate* is formed by the *palatine processes of the maxillae* and the *horizontal plates of the*

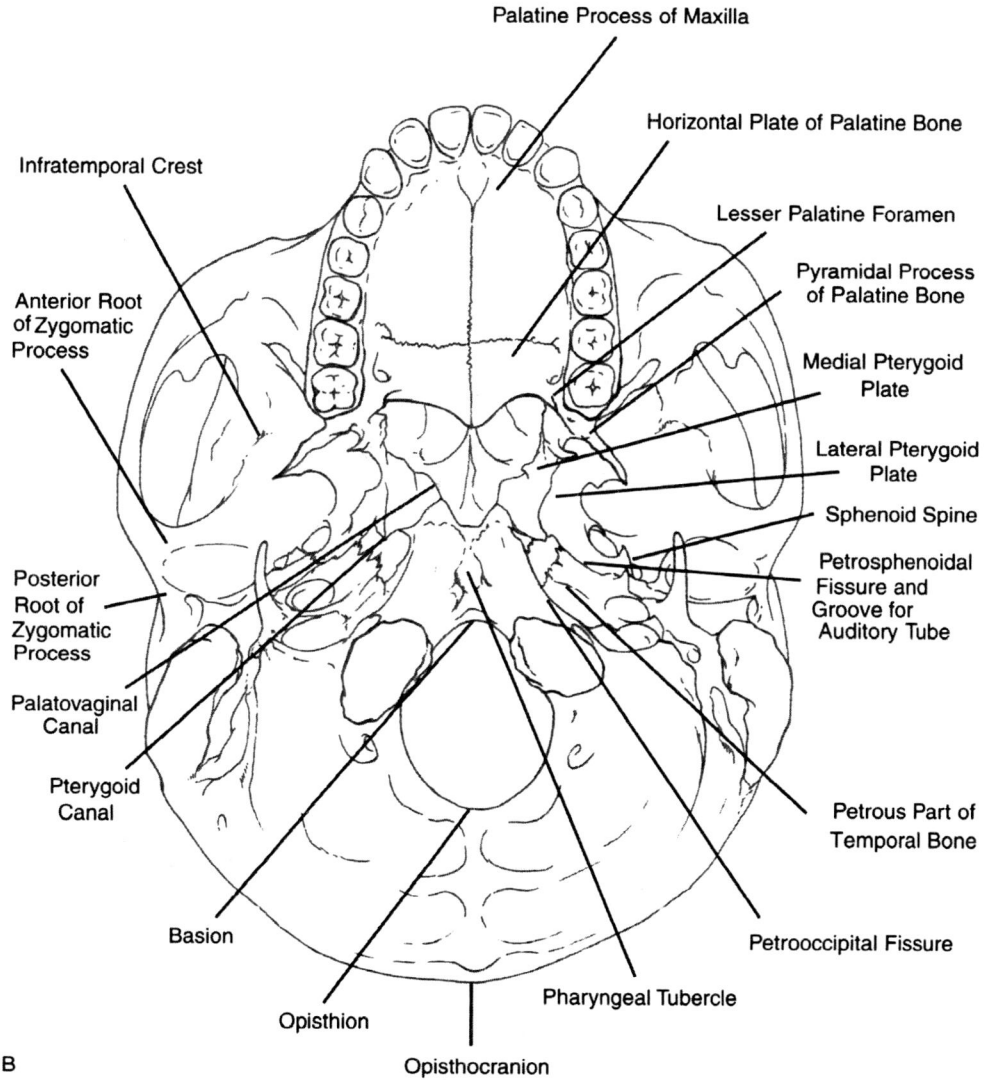

Figure 8-20 *(Continued)*

palatine bones. At the posterior angles of the hard palate, in the palatine bone, are the *greater palatine foramina*. Posterior to these foramina are the *pyramidal processes* of the palatine bones, which are perforated by one or more *lesser palatine foramina*. The greater palatine foramen transmits the greater palatine vessels and nerve.

Posterior and superior to the hard palate are the *choanae* or internal nasal apertures. At the anterior aspect of the hard palate in the midline just posterior to the central incisor teeth is the *incisive fossa*. Deep in the incisive fossa are the paired *incisive foramina* that transmit branches of the nasopalatine nerve. Lateral and anterior to the hard palate are the *alveolar*

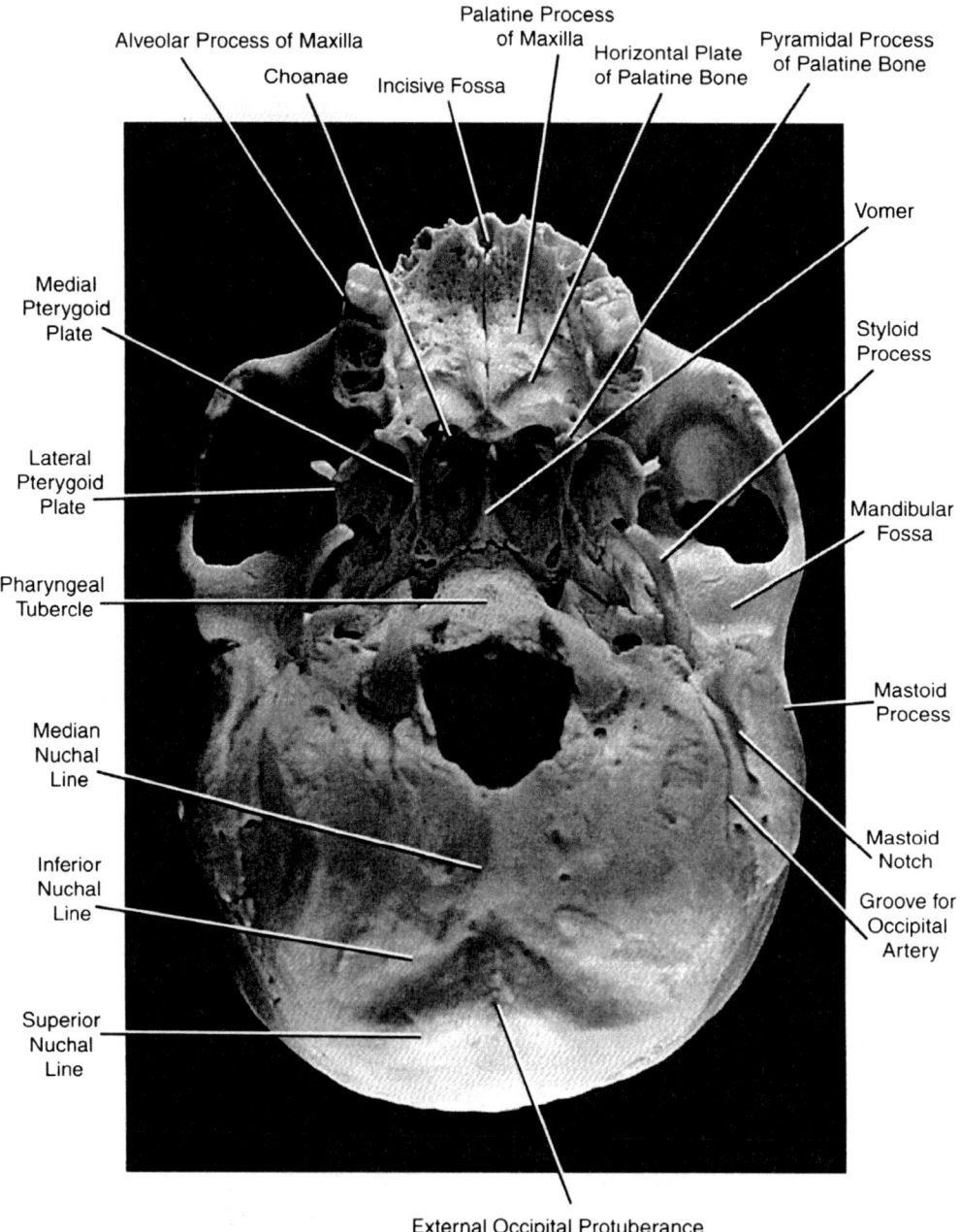

Figure 8-21 Photograph of the adult human cranium: norma basalis, or inferior view.

processes of the maxillae, holding the roots of the upper teeth. On the posterior aspect of the lateral wall of the nasal cavity is the *perpendicular plate of the palatine bone*, at the top of which is an opening called the *sphenopalatine foramen* (figure 8-22).

The *pterygoid canal* is a bony tunnel through the sphenoid bone that eventually opens into the *pterygopalatine fossa* (described in detail later). The posterior opening of the canal is located superiorly above the pterygoid fossa at the point where the medial

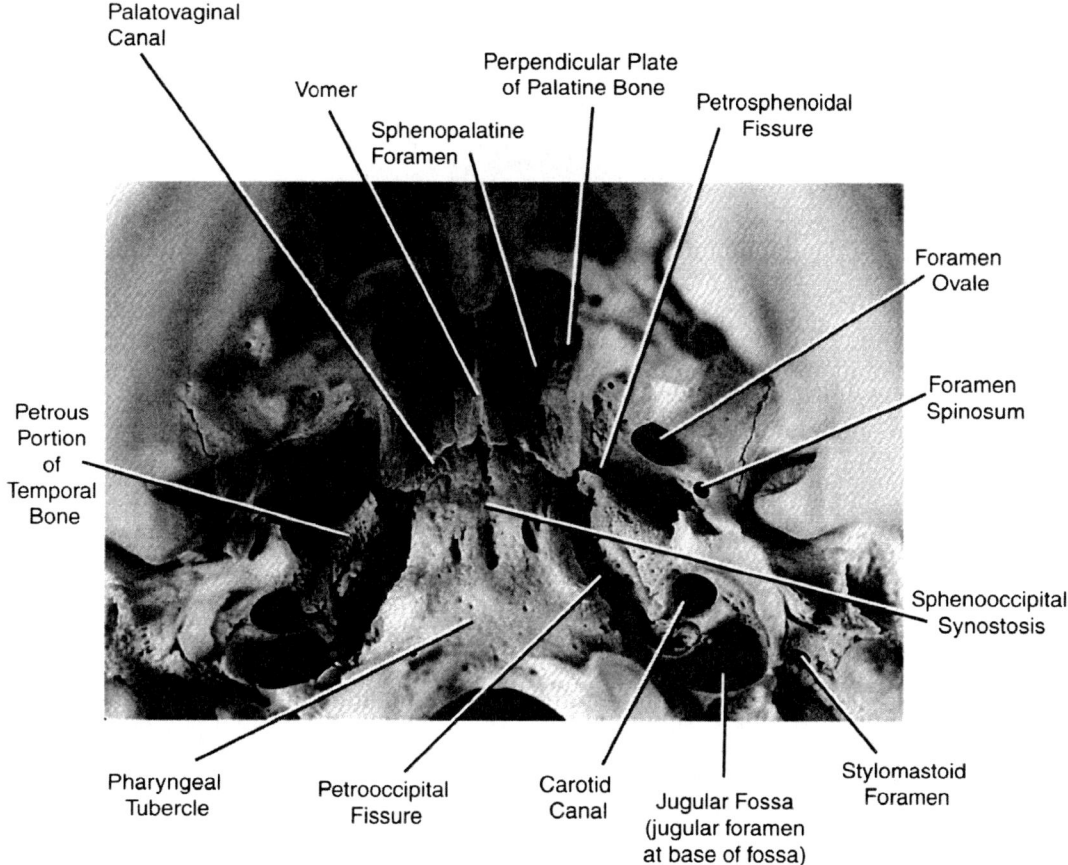

Figure 8-22 Photograph of the adult human cranium: norma basalis, or inferior view. Close-up of the region of the sphenooccipital synostosis.

Different parts of the sphenoid bone can be seen inferiorly, and include the medial and lateral pterygoid plates, the hamulus and the pterygoid fossa. The *medial pterygoid plate* is posterior to the palatine bone, and the hook-shaped *pterygoid hamulus* is attached to the posterior part of the free inferior border of the medial pterygoid plate. More laterally is the *lateral pterygoid plate*. The space between the medial and lateral pterygoid plates is the *pterygoid fossa*.

and lateral pterygoid plates meet (Figure 8-23; see also Figure 8-20 B). Another tiny opening to the pterygopalatine fossa is the *palatovaginal (pharyngeal) canal*, which passes anterolaterally through the sphenoid just lateral to where the vomer articulates with the sphenoid. The vomer forms the posterior and inferior part of the bony nasal septum and can be seen through the choanae, separating the nasal cavity into two parts.

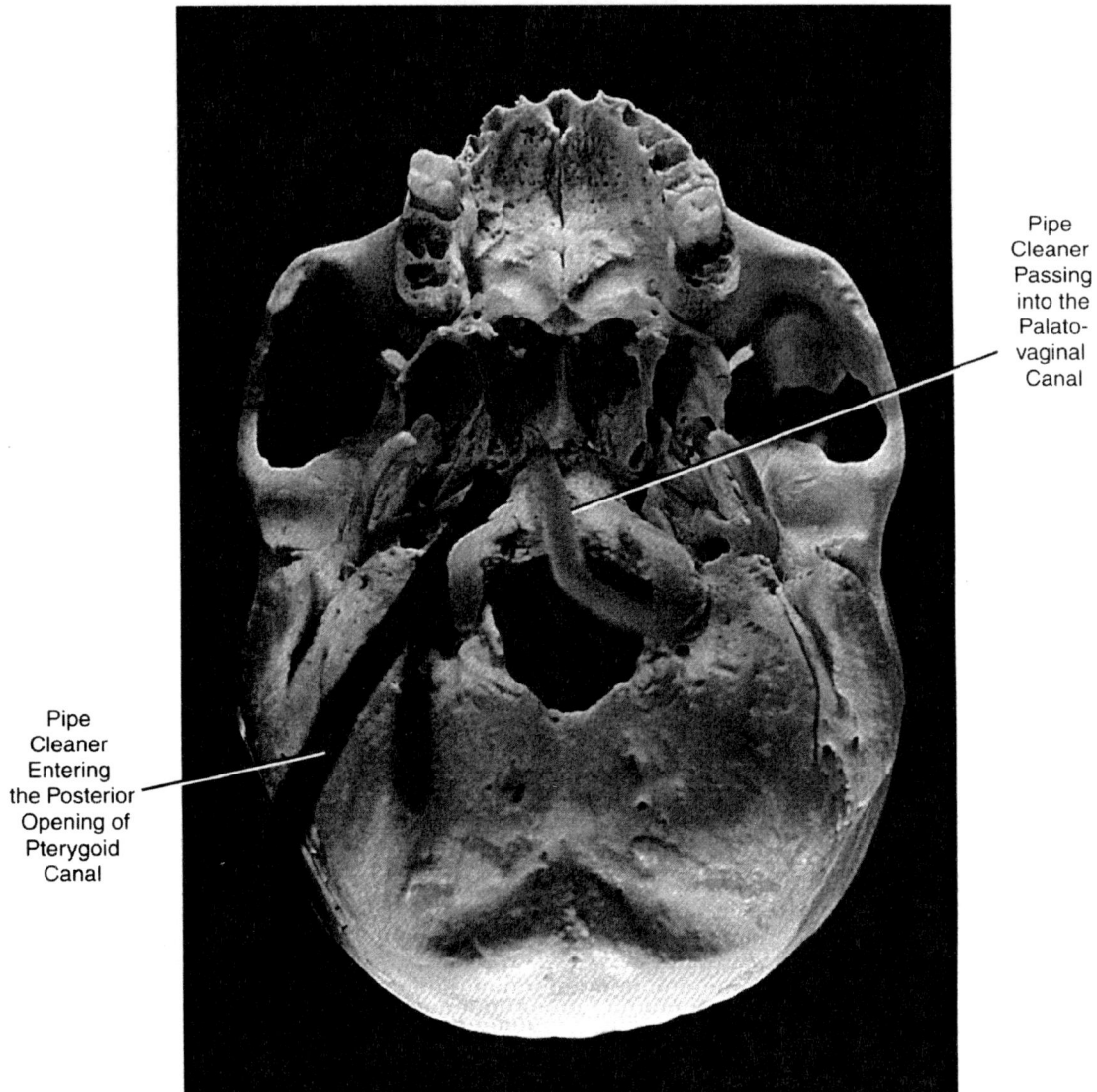

Figure 8-23 Photograph of the adult human cranium: norma basalis, or inferior view. A pipe cleaner is in the posterior opening of the pterygoid canal.

More laterally on the sphenoid is the *foramen ovale*. It is an oval opening located at the base of the lateral pterygoid plate for the transmission of the mandibular division of the trigemenal nerve.

The *sphenoidal spine* is a pointed projection located lateral to the foramen ovale. It gives attachment to the sphenomandibular ligament and tensor veli palatini muscle. The sphenoidal spine is pierced by the *foramen spinosum*.

The *pharyngeal tubercle* is located posterior to vomer on the basilar portion of the occipital bone. This tubercle is for the attachment of the fibrous raphe of the pharynx.

Various parts of the temporal bone are seen in this view. These include the squamous and tympanic parts, which were also seen laterally. The *mastoid process and notch* and the *occipital groove* (see Figures 8-20 A and 8-21) for the occipital artery are visible in this view as well as in the norma occipitalis.

The *styloid process* (Figure 8-24) is visible in this view as well. Lying at its base and separating it from the mastoid process is the *stylomastoid foramen.* Visible in the norma basalis but not in previous views is the *petrous part of the temporal bone* (see Figures 8-20 and 8-22). This is a wedge-shaped part of the bone that is very dense and "rock hard" (to which "petrous", meaning "rocklike" refers). It extends between the sphenoid and occipital bones.

Several foramina and fissures in addition to those mentioned above are visible on the base of the skull. The *foramen lacerum* (Figures 8-20) is at the base of the medial pterygoid plate, between

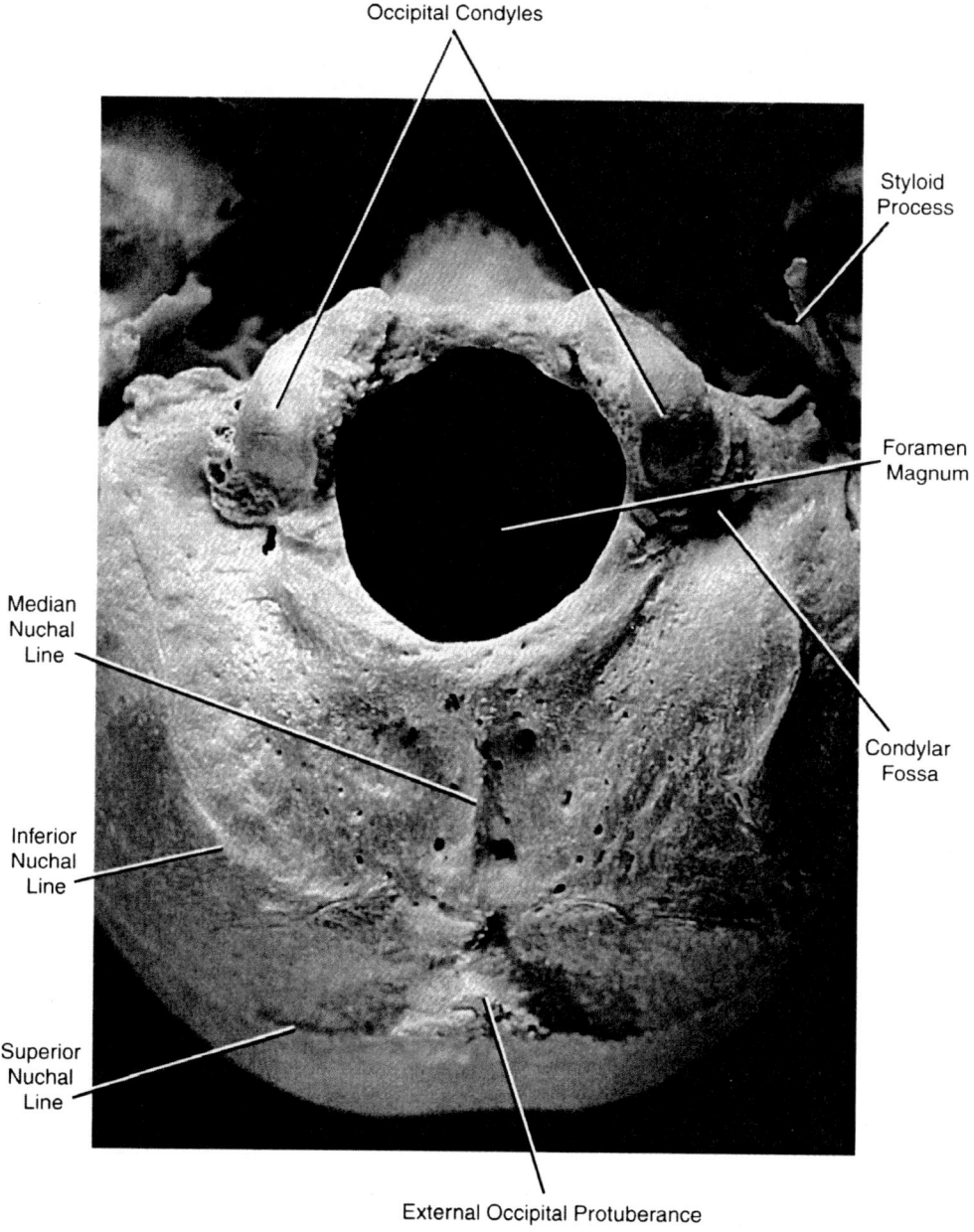

Figure 8-24 Photograph of the adult human cranium: norma basalis, or inferior view. Close-up of the region of the foramen magnum and the nuchal region of the occipital bone.

it and the tip of the petrous part of the temporal bone. It usually has very irregular edges. The *carotid canal* (Figures 8-20 A and 8-22) is a relatively large bony tunnel for the internal carotid artery. Its inferior opening is usually circular and it passes through the petrous part of the temporal bone.

The *groove for the auditory tube (sulcus tubae auditivae;* see Figures 8-20 B) is located between the petrous portion of the temporal and the greater wing of the sphenoid. This groove houses the cartilaginous part of the auditory tube. The *petrosphenoidal fissure* (see Figures 8-20 B and 8-22) is at the "bottom" of this groove. In fact, in

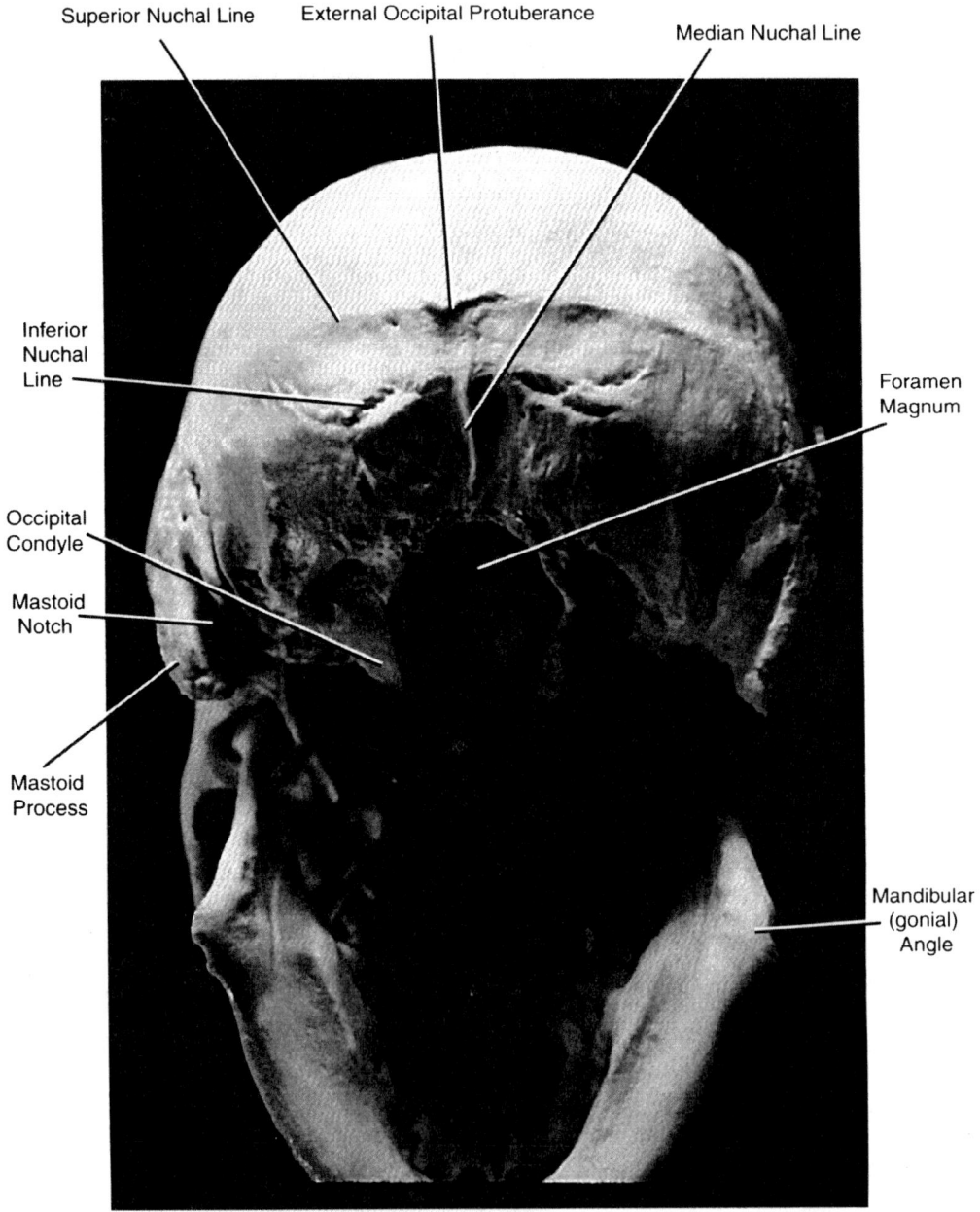

Figure 8-25 Photograph of the adult human cranium, oblique inferoposterior view, demonstrating the region of the foramen magnum and the nuchal region of the occipital bone.

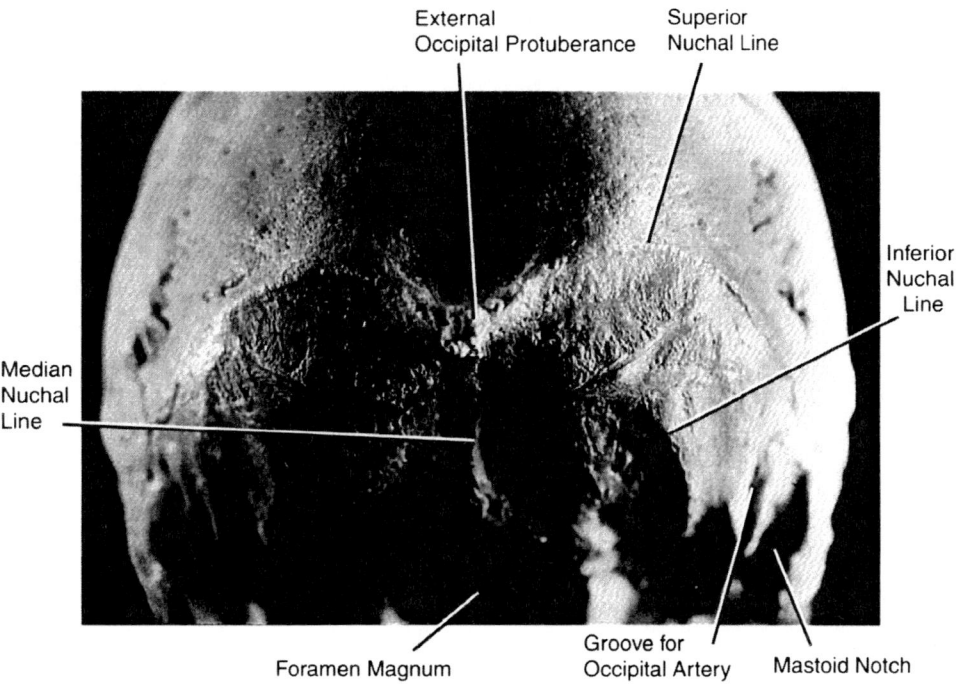

Figure 8-26 Photograph of the adult human cranium, oblique inferoposterior view, demonstrating the region of the external occipital protuberance, the median nuchal line, and the superior and inferior nuchal lines.

anatomical position, the fissure is in the superior aspect or "roof" of the groove. This fissure is the actual space between the petrous part of the temporal bone and the sphenoid, whereas the groove extends a little to either side, making it wider than the fissure.

The jugular fossa and foramen (see Figures 8-20 A and 8-22) are located posterior to the *carotid canal*. The *jugular fossa* is a depression seen posterior to the carotid canal on the temporal bone, while the *jugular foramen* is the opening into the interior of the skull located medially in the fossa. It is located between the temporal and occipital bones. The jugular process of the occipital bone is an extension of the occipital lateral to the occipital condyles, and its anterior border forms the posterior boundary of the jugular foramen. The *petrooccipital fissure* extends anteriorly from the jugular foramen to the foramen lacerum, separating the petrous part of the temporal bone from the occipital.

The *foramen magnum*, (Figure 8-25; see also Figures 8-20 A and 8-24) which transmits the medulla oblongata (part of the brainstem) and its associated structures, is a very large opening located posterior to the basilar portion of the occipital bone. It is bounded laterally by the oval and convex *occipital condyles* (see Figure 8-25; see also Figure 8-20 A). These articulate with the superior articular facets of the atlas. Posterior to each condyle is an opening, the *condylar canal*, or a depression, the *condylar fossa* (See Figures 8-20 A and 8-24). The canal transmits an emissary vein which connects venous sinuses on the inside of the skull to veins outside the skull. *Basion* is the anteriormost point on the margin of the foramen magnum, while *opisthion* is the posteriormost point on its margin. The *hypoglossal canals* (see Figure 8-20 A) are openings located superior to each occipital condyle. Each transmits a hypoglossal nerve and a meningeal artery.

The external occipital crest (median nuchal line) and the superior and inferior nuchal lines (Figure 8-26, see also figures 8-20 A, 8-24 and 8-25) are visible in the basilar view as well as from the posterior view of the cranium.

The Interior View of the Skull

Bones visible from the interior of the skull (Figures 8-27 and 8-28) include the *frontals, parietals, occipital, temporal, sphenoid,* and *ethmoid*. Landmarks are discussed next.

Landmarks and Features

The interior of the skull is divided into three basins or fossae — the *anterior, middle, and posterior cranial fossae* (see Figures 8-27 and 8-28).

The Anterior Cranial Fossa

The floor of the *anterior cranial fossa* (see Figures 8-27 and 8-28) is formed by the *orbital plates* of the frontal, the small and much perforated *cribriform plates* of the ethmoid, and the lesser wings and body of the sphenoid. The floor supports the frontal lobe of the cerebrum. The olfactory nerves run to the nasal cavity through the perforations of the cribriform plates, and the olfactory bulbs of the brain sit on these plates. The *crista galli* is a small perpendicular plate of bone that extends superiorly

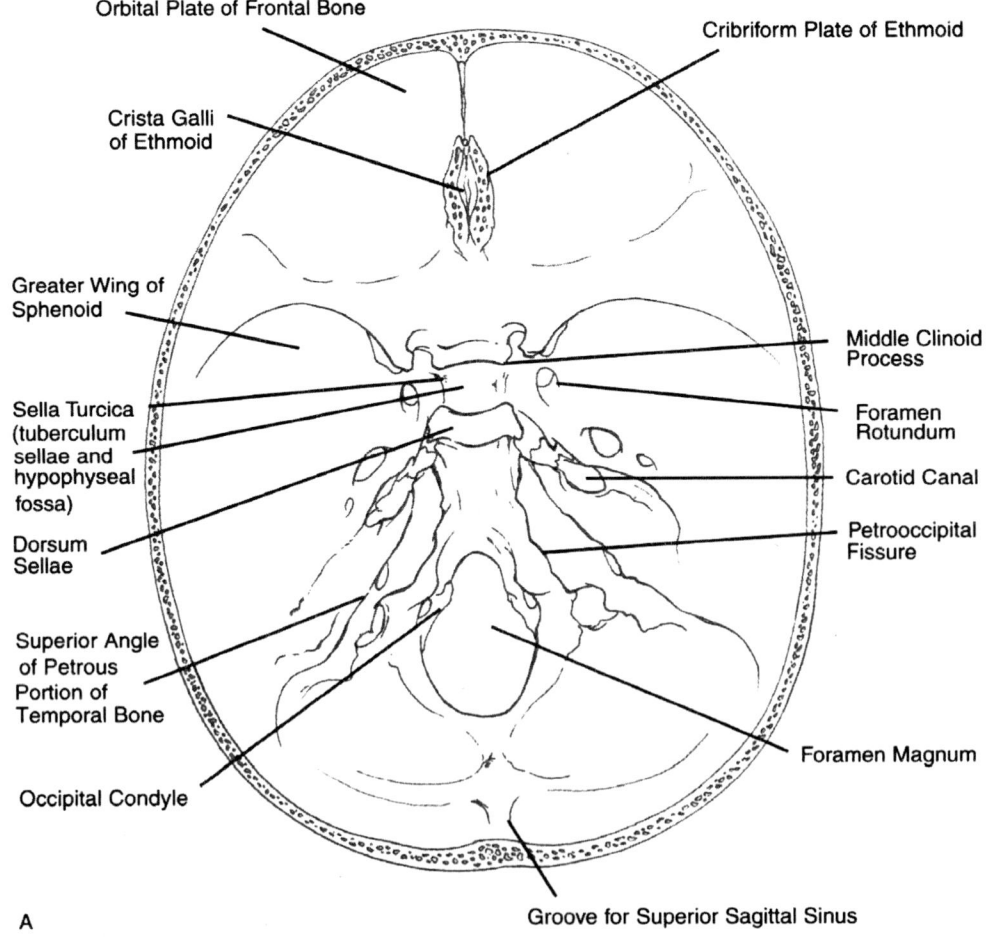

Figure 8-27 Interior of the adult human cranium, superior views, demonstrating the cranial fossae. The figure is divided into *A* and *B* portions to make the labeling of structures clearer.

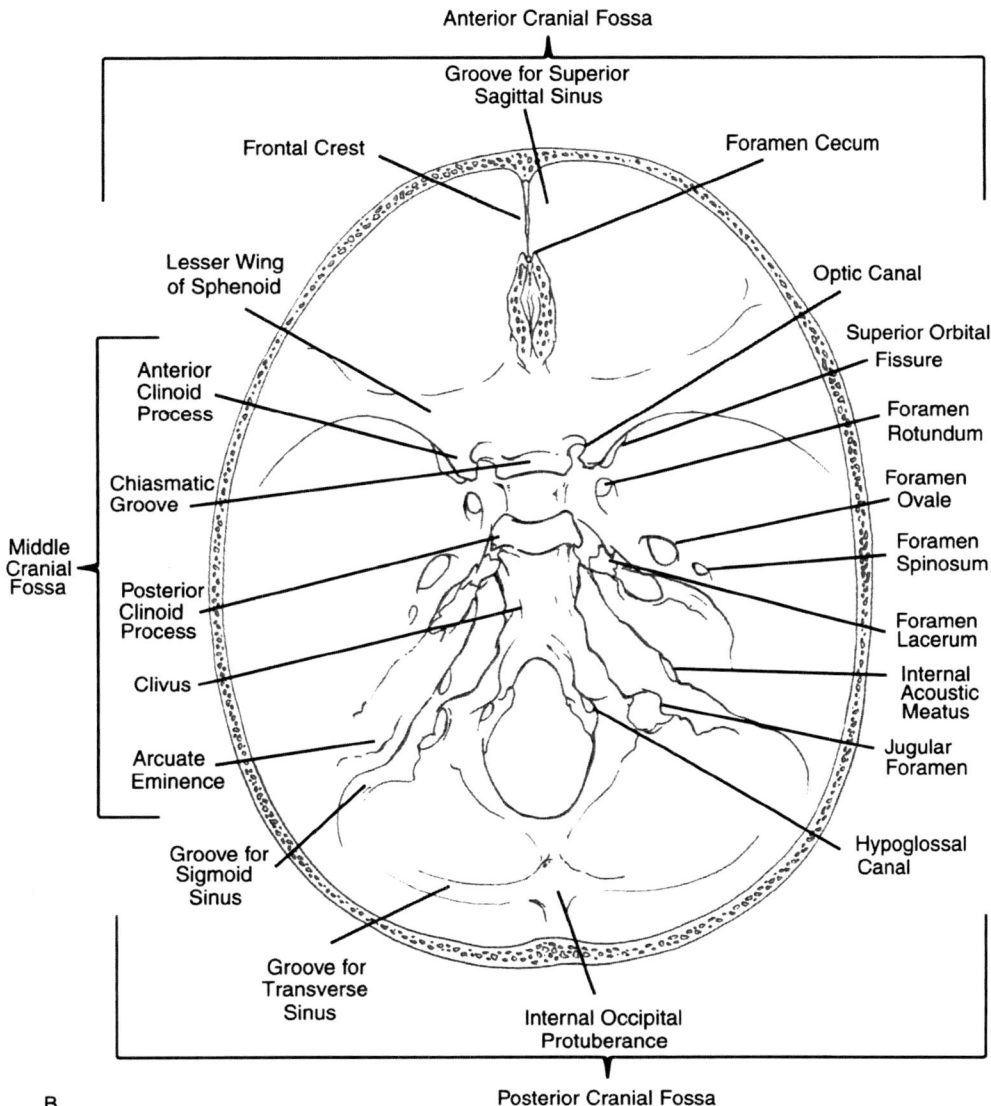

Figure 8-27 (Continued)

from the median portion of the superior aspect of the ethmoid bone between the two cribriform plates. It serves as an anchoring point for the cranial meninges.

The Middle Cranial Fossa

The middle cranial fossa (Figure 8-29; see also Figures 8-27 and 8-28) is bounded anteriorly by the *posterior margins of the lesser wings of the sphenoid*, the *anterior clinoid processes*, and the ridge forming the *anterior margin of the chiasmatic groove*. This boundary also defines the posterior margin of the anterior cranial fossa. The posterior margin of the middle cranial fossa is formed by the *superior angles of the petrous portions of the temporal bones* and the *dorsum sellae*. This boundary also defines the anterior margin of the posterior fossa.

In the median portion of the middle cranial fossa are the *chiasmatic groove* and the *tuberculum sellae* (Figure 8-30, see also Figures 8-27 through 8-29). The *chiasmatic groove* extends across the midline and ends on either side at the *optic canal*. The tuberculum sellae is a mild prominence inferior to the groove.

The *anterior clinoid processes* are projections of the lesser wings of the sphenoid located posterior and lateral to the optic canals. Posterior to the tuberculum sellae is the highly concave *fossa hypophyseos* (see Figure 8-30), which houses the pituitary gland. This fossa is at the center of a structure termed the *sella turcica* (see Figures 8-29 and 8-30),

which is bounded anteriorly by the tuberculum sellae, and posteriorly by the dorsum sellae.

To either side on its anterior wall may be the *middle clinoid process*. The middle clinoid processes are highly variable in their presence or absence. They may be entirely absent, present as very small bumps, or as very well developed projections. The posterior margin of the sella turcica is formed by the wall-like dorsum sellae. At the superolateral corners of the dorsum sellae are small projections called the *posterior clinoid processes* (see Figures 8-27 B and 8-30).

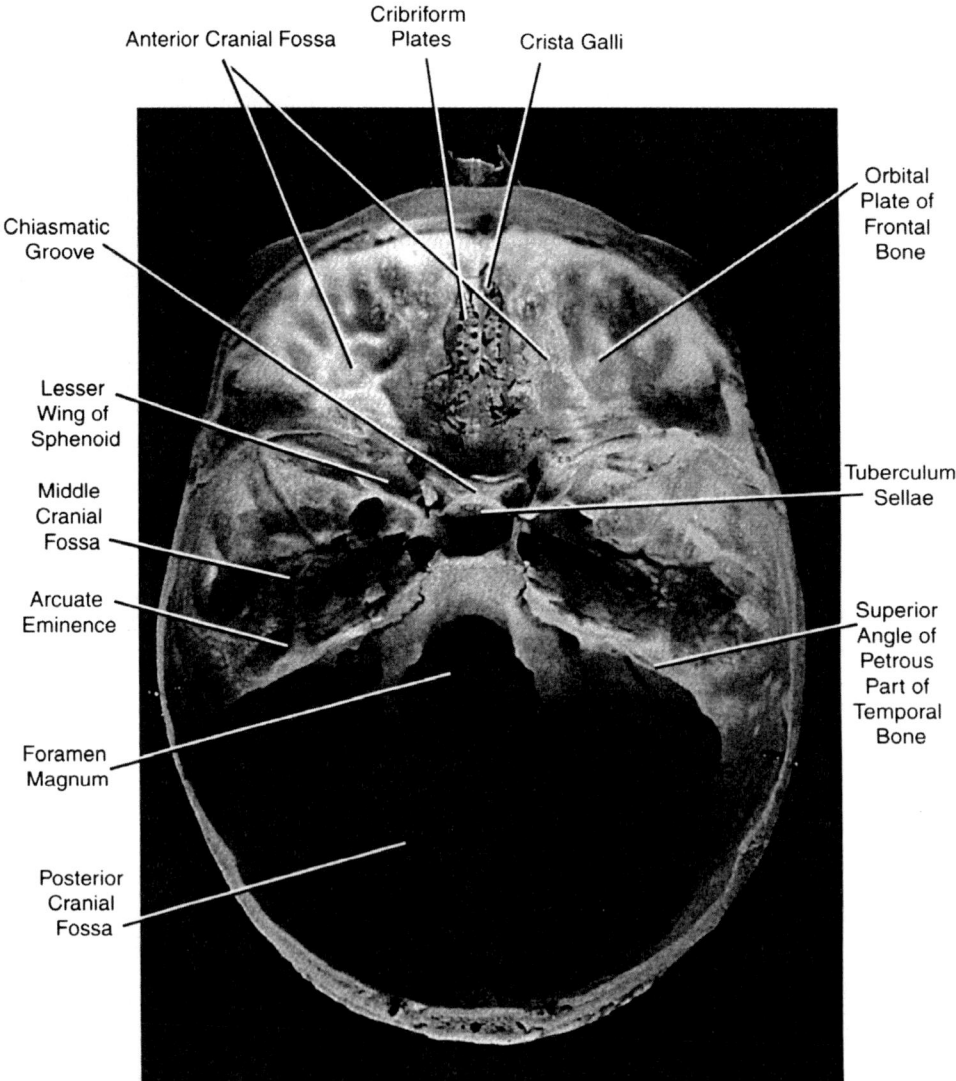

Figure 8-28 Photograph of the interior of the adult human cranium, superior view, demonstrating the cranial fossae.

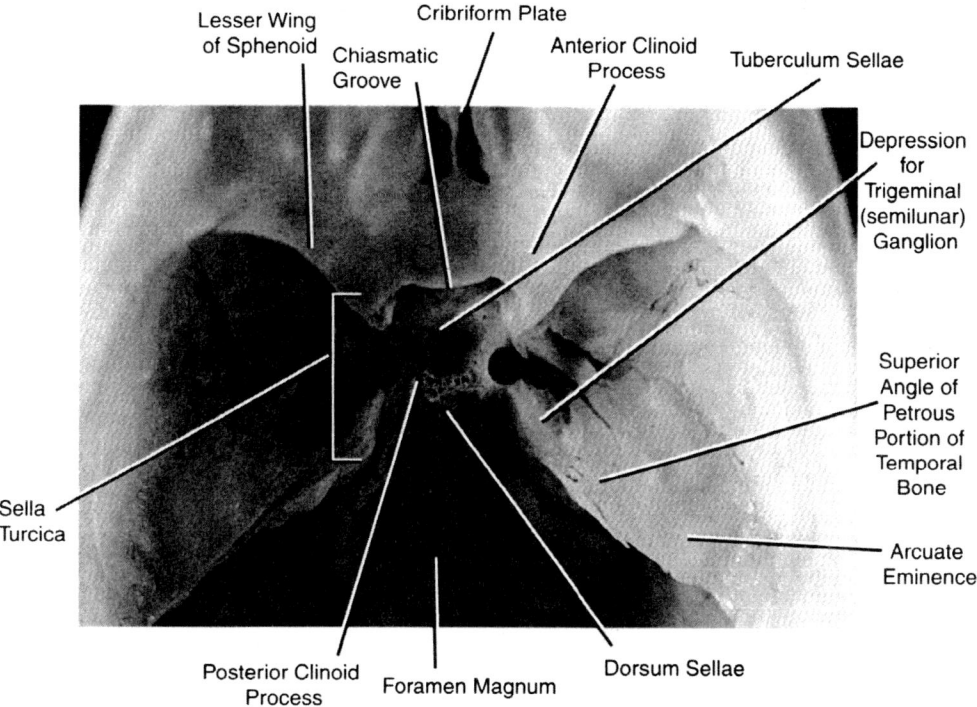

Figure 8-29 Photograph of the interior of the adult human cranium, superior view, demonstrating the middle cranial fossa.

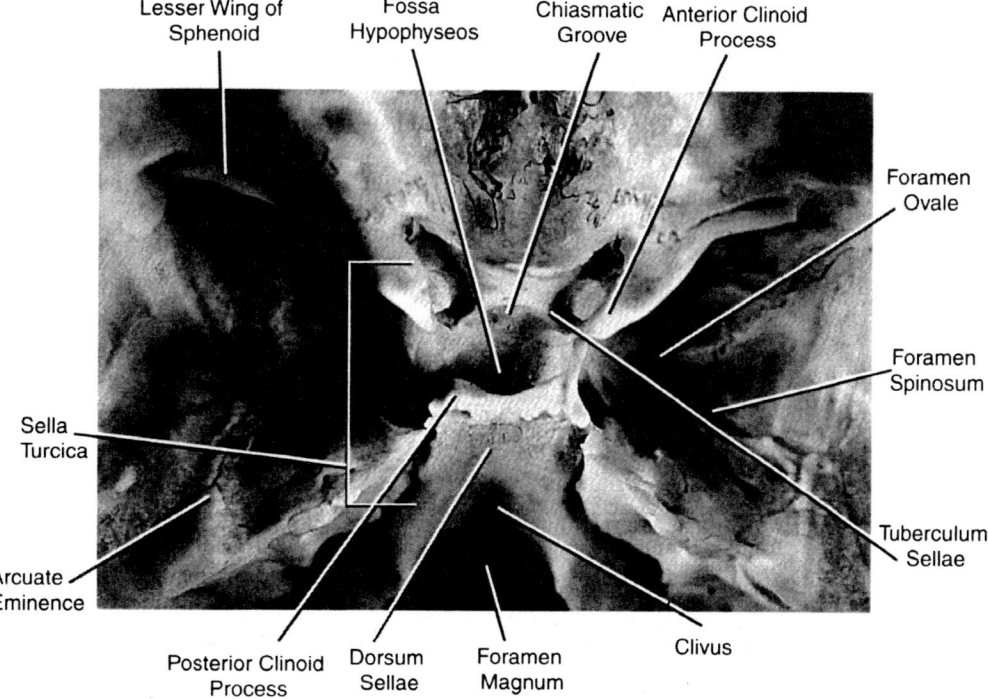

Figure 8-30 Photograph of the interior of the adult human cranium, superior view, demonstrating the sella turcica and associated structures in the middle cranial fossa.

The anterior and posterior clinoid processes provide attachment to a meningeal membrane called the tentorium cerebelli.

The lateral deeper portions of the middle fossa house the temporal lobes of the brain. Opening into the middle cranial fossa are the *superior orbital fissure, foramen rotundum, foramen ovale, foramen spinosum, foramen lacerum* and the *carotid canal* (see Figures 8-27 and 8-28), most of which are seen in the inferior view.

The internal opening of the carotid canal usually lies more medially, next to the occipital, than the external opening. The foramen rotundum was not visible inferiorly. It is a small, usually round, opening between the foramen ovale and the superior orbital fissure.

Figure 8-31 demonstrates the isolated sphenoid bone. It is a highly complex bone which is surrounded by other bones of the cranium. It is, however, a key feature of the middle cranial fossa. Its superior and inferior aspects have been seen. Its principal features

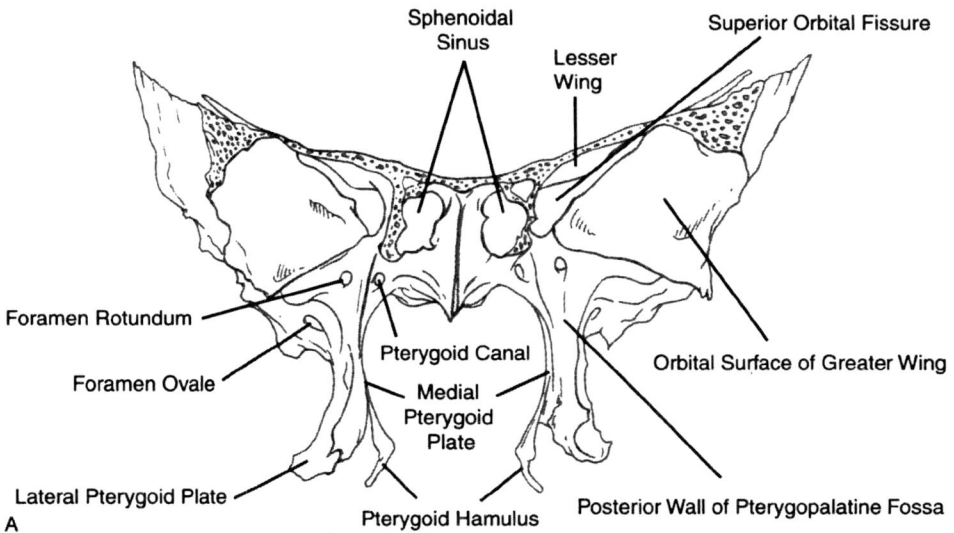

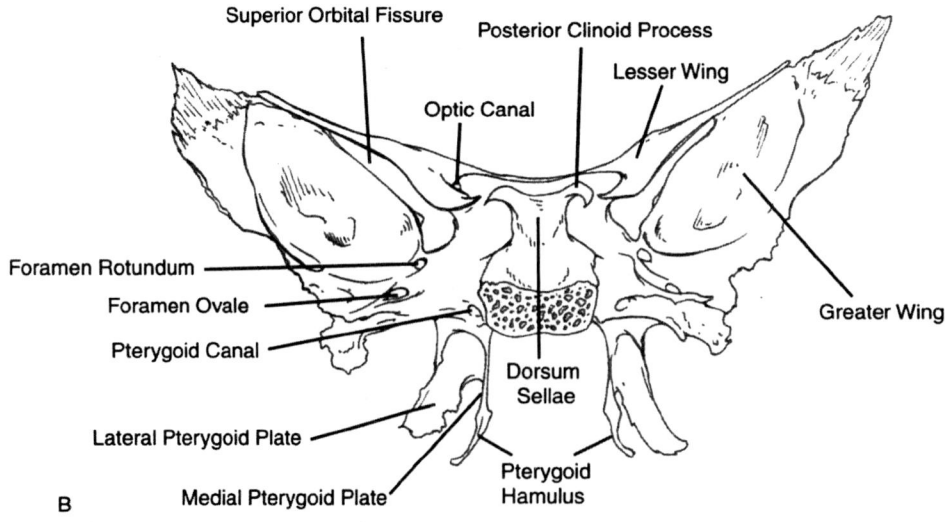

Figure 8-31 The isolated sphenoid bone: (*A*) anterior and (*B*) posterior views.

visible from its anterior and posterior aspects, including the dorsum sellae, greater and lesser wings, and its major foramina are shown in this figure.

Visible on the anterior surface of the *petrous portion of the temporal bone* (See Figure 8-27 B), within the middle cranial fossa, is a ridge or swelling called the *arcuate eminence*. It is caused by the projection of the superior semicircular canal. Also in the middle cranial fossa is the *depression for the trigeminal (semilunar) ganglion* of the trigeminal nerve (figure 8-29). This depression is located on the anterior surface of the petrous portion of the temporal bone just lateral to the orifice of the carotid canal.

The Posterior Cranial Fossa

The posterior cranial fossa (Figure 8-32, see also Figures 8-27 and 8-28) is the largest and deepest of the three fossae. It is formed by the *dorsum sellae*, the anterosuperior slope of the *clivus of the sphenoid* and *occipital bones* (see Figures 8-27 and 8-30), the *petrous* and *mastoid portions of the temporal bone,* the *internal aspect of the squama* (the flat, interparietal portion) *of the occipital bone,* and the *mastoid angles of the parietals*. The latter are the corners of the parietal bones that sit next to the partietomastoid sutures. This fossa houses the cerebellum, pons, and medulla oblongata. The *foramen magnum* pierces the center of this fossa.

Also present in the posterior cranial fossa are the *petrooccipital fissure,* the *jugular foramen*, the *hypoglossal canal* (seen previously), and the *internal acoustic meatus* (see Figure 8-27 A and B). The internal acoustic meatus is an opening on the posterior aspect of the petrous part of the temporal bone that transmits the vestibulochoclear and facial

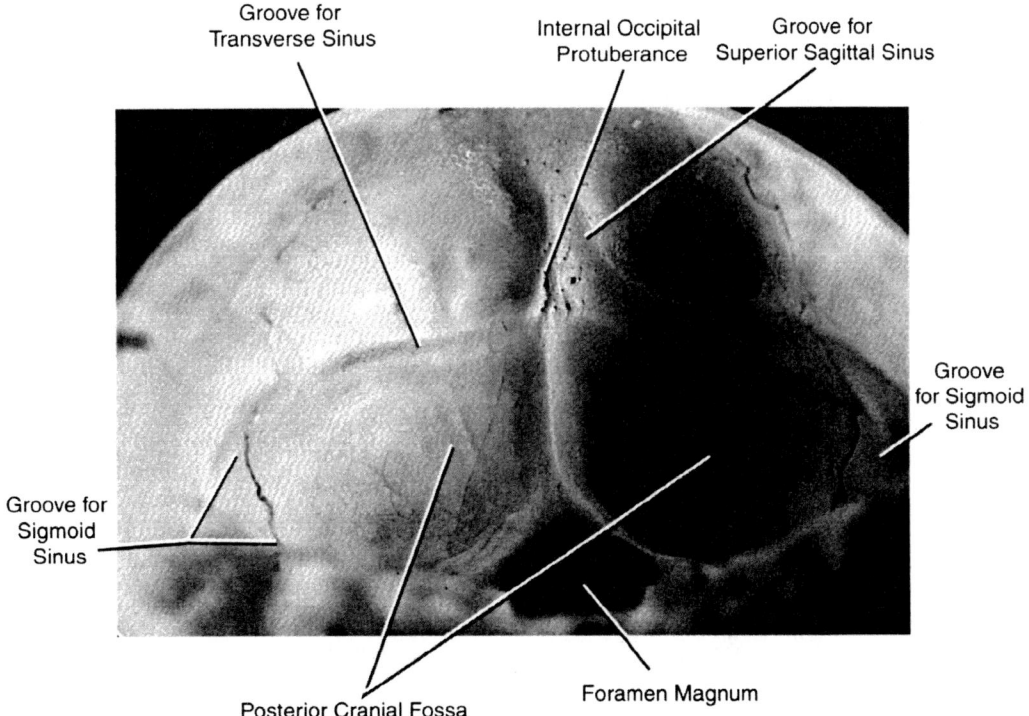

Figure 8-32 Photograph of the internal aspect of the occipital squama, demonstrating the internal occipital protuberance, the confluence of the sinuses (intersection of groove for superior sagittal sinus and grooves for transverse sinuses), and the grooves for the superior sagittal, transverse, and sigmoid sinuses.

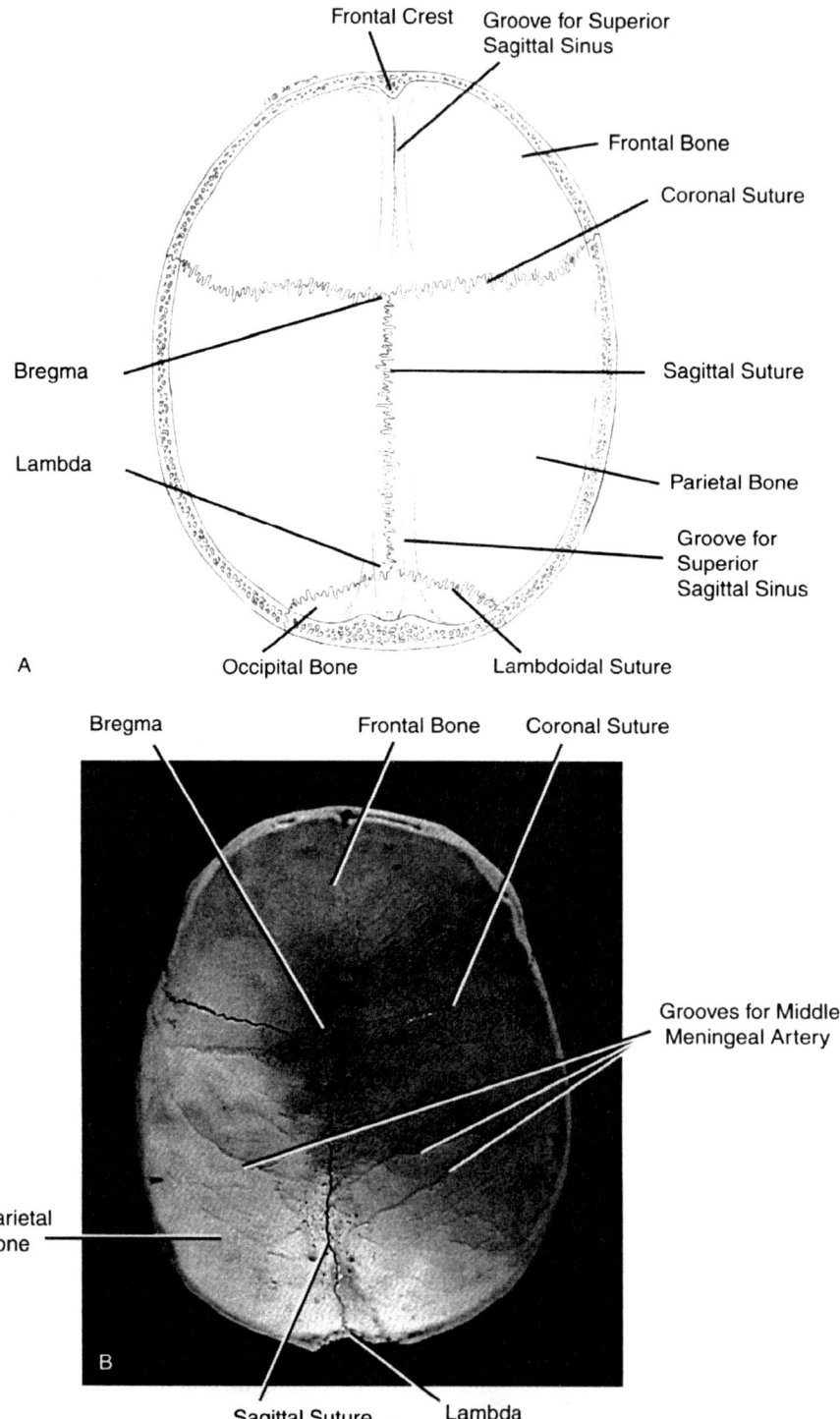

Figure 8-33 (A) Line drawing and (B) photograph of the internal aspect of the calvarium, or cranial roof.

nerves. The posterior fossa is bounded posteriorly by the deep *grooves for the transverse sinuses* (see Figures 8-27 B and 8-32). These proceed laterally and then anteriorly to meet the deep, curved *grooves for the sigmoid sinuses*, which run along the internal aspect of the mastoid part of the temporal bone and end at the jugular foramen. These are grooves for vascular sinuses, which are part of the venous system of the head, and should not be confused with the bony air sinuses of the frontal, ethmoid, sphenoid, and maxilla. The grooves for the two transverse sinuses meet in the midline posteriorly and groove the surface of the internally projecting *internal occipital protuberance*. At this structure they meet the *groove for the superior sagittal sinus,* (Figure 8-33 A; see also Figure 8-32) which lies on the internal aspect of the two parietal bones, where they meet at the sagittal suture. This groove is wider than the sagittal suture and runs posteriorly from the frontal bone to the internal occipital protuberance. The intersection of the grooves for the superior sagittal sinus and the transverse sinuses on the internal occipital protuberance is known as the *impression for the confluence of the sinuses* or *torcular Herophili*.

The Nasal Cavity and Paranasal Sinuses

Although not readily visible in a skull that has not been sagittally bisected, there are a number of important bony, yet very delicate, structures located in the nasal cavity (refer to Figures 8-34 through 8-36 throughout

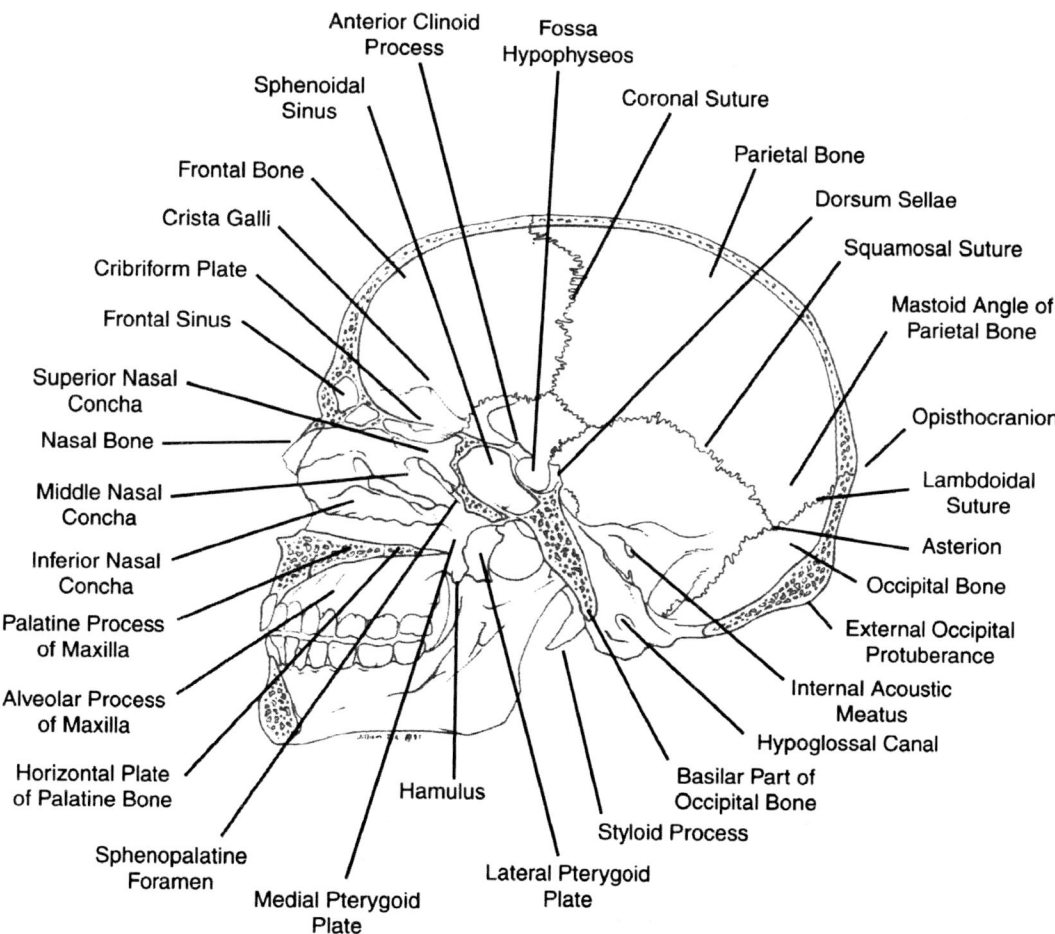

Figure 8-34 Median section of the adult human cranium, right side, demonstrating the nasal conchae as well as the frontal and sphenoid sinuses.

the discussion of the structures in the nasal cavity). On the lateral nasal wall are three scroll-like projections. These are the *nasal conchae.* They are three in number. They include the *superior* and *middle nasal conchae*, which are projections from the labyrinths of the *ethmoid bone.* They form the superiormost portion of the lateral nasal wall.

Inferior to these conchae is the *inferior nasal concha*, a separate, small bone (also known as the *turbinate bone*) that articulates with the maxilla and perpendicular plate of the palatine bone. The posterior aspect of the lateral nasal wall is also formed by the perpendicular plate of the palatine. The more anterior portion of the lateral nasal wall is formed by the *maxilla* along with the inferior nasal concha.

The lacrimal bone and the medial pterygoid plate also contribute to the lateral wall of the nasal cavity. Lateral to, and covered by, the superior, middle, and inferior nasal conchae are spaces called the *superior meatus, middle meatus,* and *inferior meatuses.* Off the superior meatus is the narrow *sphenoethmoidal recess* that separates the superior concha and the anterior aspect of the body of the sphenoid.

In some of the bones surrounding the nasal cavity are mucus membrane lined spaces called the *paranasal sinuses.* These should not be confused with the vascular sinuses discussed earlier). There are four paranasal sinuses: the *frontal sinus* in the frontal bone; the *sphenoid sinus* in the body of the sphenoid bone; the *ethmoidal air cells* in the ethmoid bone just medial to the orbit, and the *maxillary sinuses* in the maxillae just lateral to the nasal cavity and superior to the alveolar processes. All of these sinuses drain into the nasal cavity.

The *sphenoidal sinus* opens into the sphenoethmoidal recess. In the middle meatus, hidden lateral to the middle concha, is the bulging *ethmoidal bulla.* This ethmoidal bulla houses the *middle ethmoidal air cells.* Anteroinferior to the bulla is the *uncinate process of the ethmoid*, which descends posteroinferiorly. Between the uncinate process and the ethmoidal bulla is the *hiatus semilunaris,* a

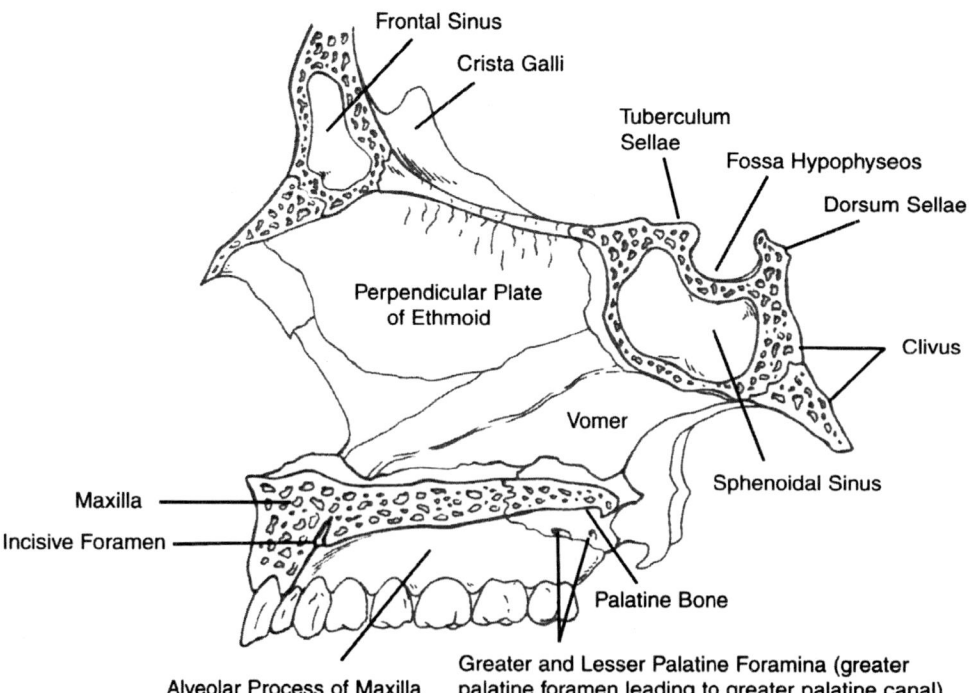

Figure 8-35 Parasagittal section of the adult human cranium, demonstrating the bony nasal septum as well as the frontal and sphenoid sinuses.

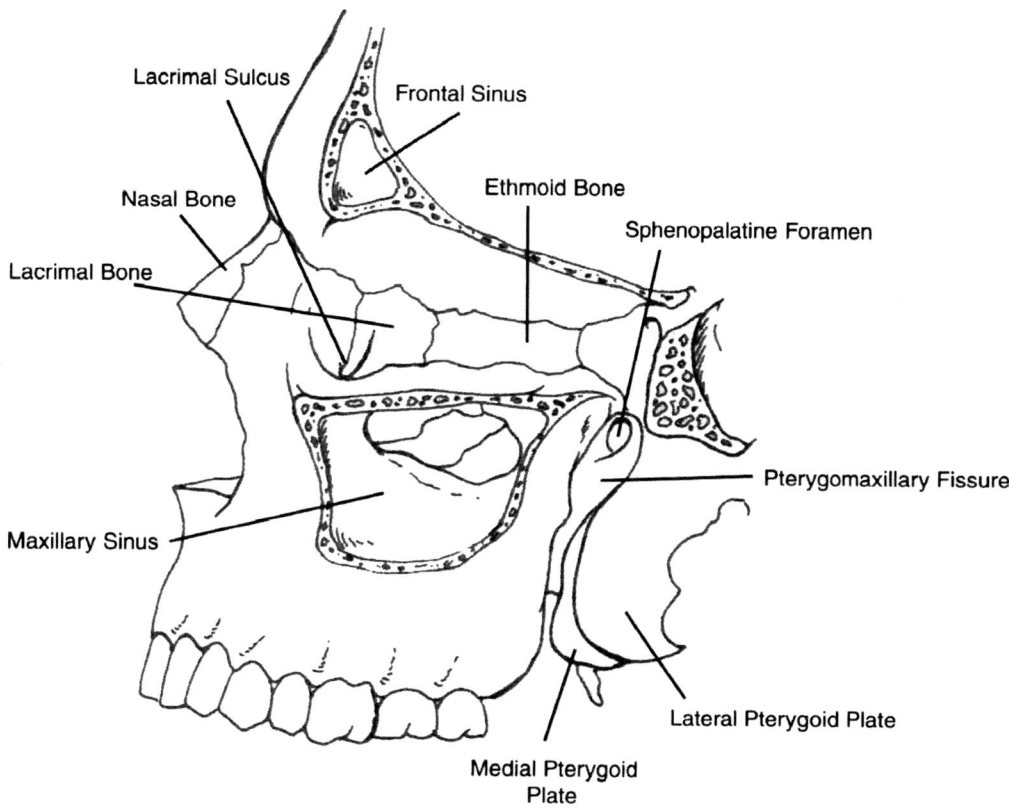

Figure 8-36 The skeleton of the face, left lateral view. The lateral wall of the orbit has been removed to expose the frontal sinus. The lateral wall of the maxillary sinus has also been removed to expose the interior of the maxillary sinus. The pterygomaxillary fissure and the sphenopalatine foramen, both opening into the pterygopalatine fossa, can also be seen.

crescent-shaped opening that allows communication between the *maxillary sinus* and the nasal cavity. At its upper end, the hiatus semilunaris is continuous with the *ethmoidal infundibulum*, a funnel-shaped space into which open the *anterior ethmoidal air cells* and sometimes the *frontal sinus*.

The inferior opening of the *nasolacrimal duct*, which is continuous with the *lacrimal sulcus* in the orbit, opens into the inferior meatus lateral to the inferior concha. This passageway drains tears from the eye into the nasal cavity.

In the midline, between the conchae of the left and right side, is the *bony nasal septum*. The bony nasal septum, as previously discussed, is formed by the vomer and the perpendicular plate of the ethmoid.

The Pterygopalatine Fossa

The *pterygopalatine fossa* (Figures 8-37 and 8-38; refer also to Figure 8-36 throughout the discussion of the pterygopalatine fossa) is a space in the skull formed at the junction of the *pterygomaxillary fissure*, the *inferior orbital fissure*, and the inferior part of the apex of the orbit. It is bounded by the body, pterygoid plates and greater wing of the sphenoid, infratemporal surface of the maxilla, and by the perpendicular plate of the palatine bone. It communicates anteriorly with the orbit via the inferior orbital fissure, medially with the nasal cavity by the sphenopalatine foramen, and laterally with the infratemporal fossa by the pterygomaxillary fissure.

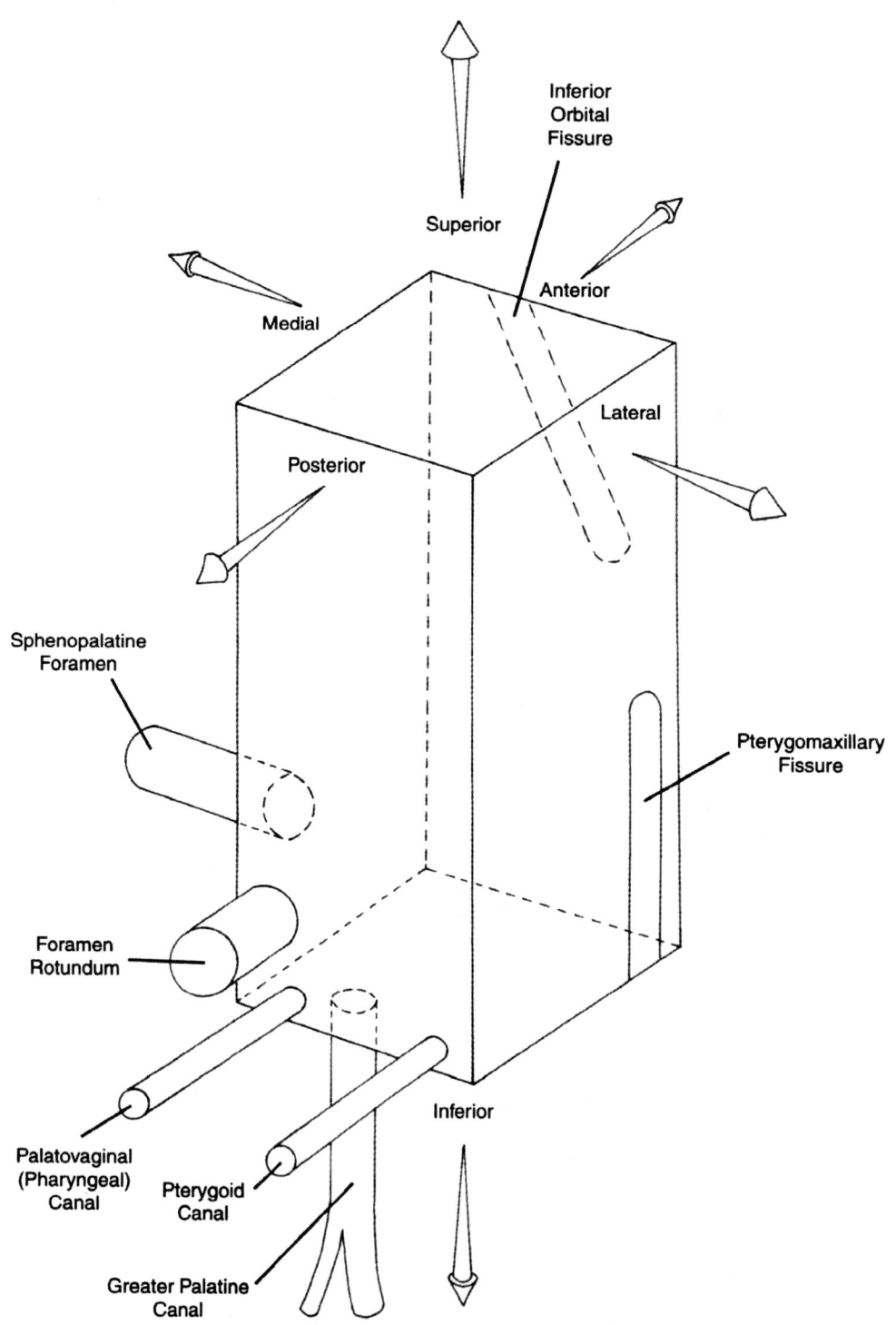

Figure 8-37 Diagrammatic representation of the pterygopalatine fossa.

Opening into it on the posterior wall are the *foramen rotundum,* the *pterygoid canal,* and the *palatovaginal (pharyngeal) canal*. Opening inferiorly into the fossa is the *greater palatine (pterygopalatine) canal* (see Figures 8-35 and 8-37). The fossa contains the maxillary nerve, the pterygopalatine ganglion, the terminal part of the maxillary artery, and various structures which connect with these through the openings into this fossa.

The pterygopalatine fossa may be visualized as a small room (figure 8-37) into which open a door (the pterygomaxillary fissure), a window (the inferior orbital fissure), and a number of ducts and pipes.

The Mandible

The *mandible* (Figures 8-39 and 8-40) consists of a u-shaped horizontal *body (corpus)* that bears the teeth, and two vertical rami that project superiorly from the posterior portion of the body.

The *body* of the mandible possesses a base or splenium that bears the *alveolar process*. The alveolar process contains the sockets for the teeth. The *external surface (buccolabial surface)* of the mandibular body exhibits several structures. A *mental protuberance* (median) and two *mental tubercles* (lateral) form the *chin*. The *oblique lines* of the mandible run from the mental tubercles to the anterior border of the ramus. An opening on either side, the *mental foramen* transmits the mental nerve and vessels. The *buccinator crest* is a faint ridge running between the oblique line and the teeth. It is for the attachment of the buccinator muscle.

The *internal surface (lingual surface;* refer to figures 8-41 to 8-43 throughout the remainder of the discussion of the mandible) presents a *mylohyoid line* on each side, which is a narrow ridge running from the ramus on the lingual surface of the body towards the midline. Anteriorly and superiorly above the mylohyoid line is the *sublingual fossa* for the sublingual salivary gland. On the inferior aspect of

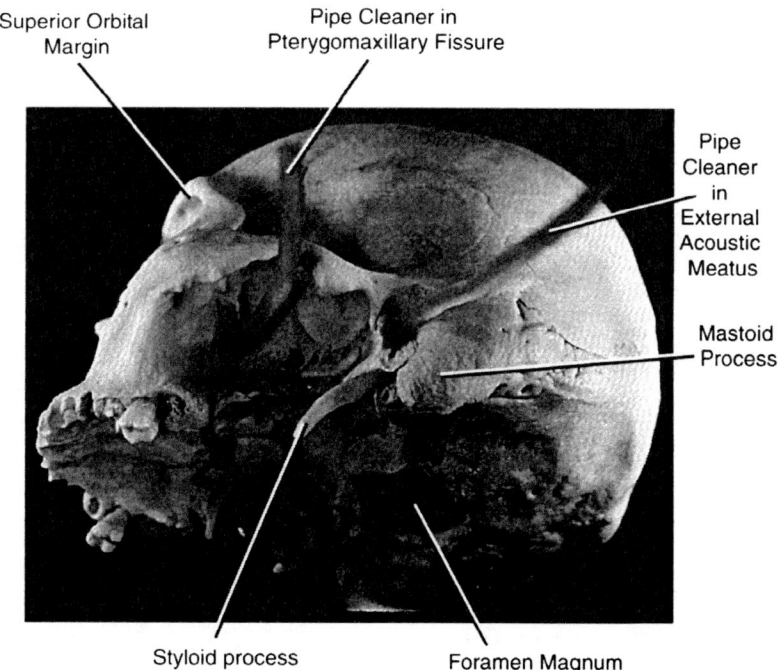

Figure 8-38 Photograph of the adult human cranium, left oblique posteroinferior view. The anterior pipe cleaner is passing into the pterygopalatine fossa through the pterygomaxillary fissure, whereas the posterior pipe cleaner is in the external acoustic meatus. Note the prominent styloid process anteroinferior to the external acoustic meatus.

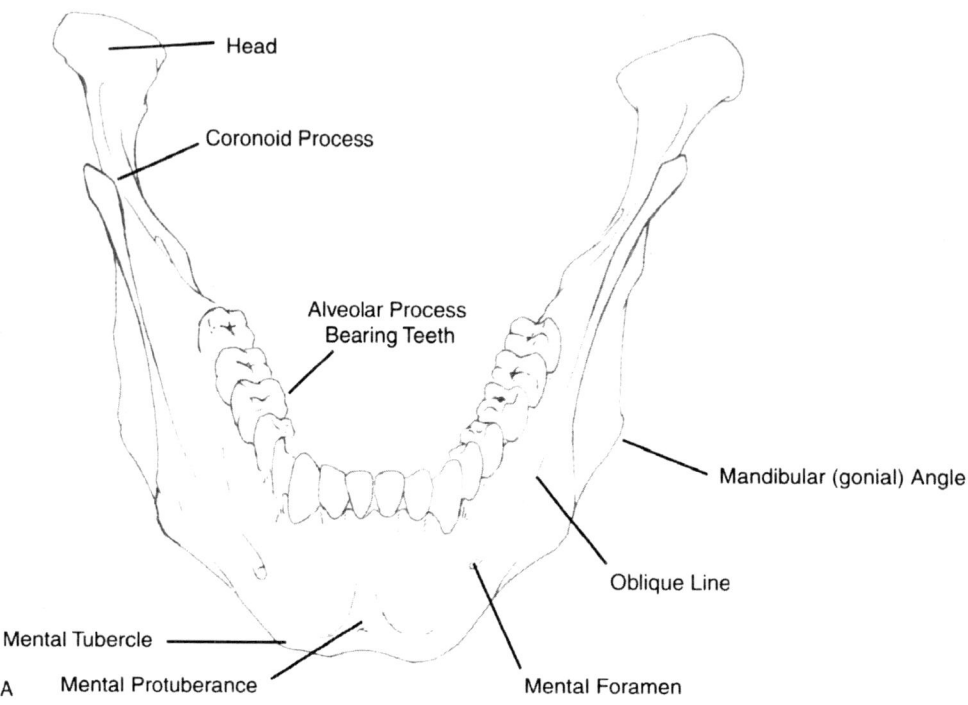
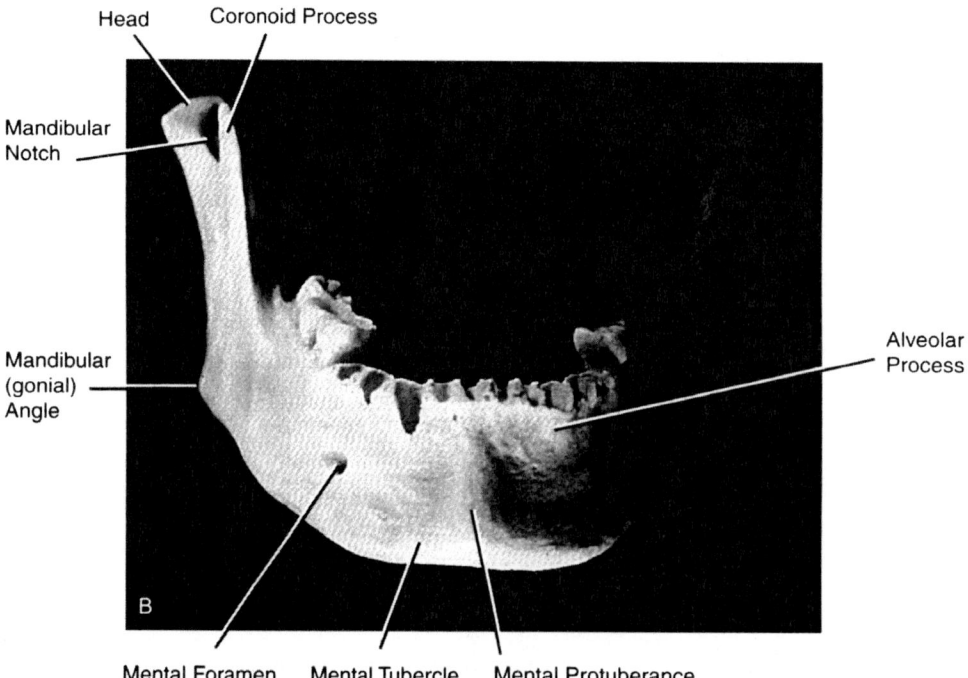

Figure 8-39 (A) Line drawing and (B) photograph of the mandible, anterior view, demonstrating the mental trigone and mental foramina.

the chin, anterior and inferior to the mylohyoid line, and just lateral to the midline, is the *digastric fossa* for the insertion of the anterior belly of the digastric muscle. Inferoposterior to the mylohyoid line is the much larger *submandibular fossa* for the submandibular salivary gland.

The midline of the mandible is the mandibular symphysis, which was a fibrocartilagenous joint early in development. At an early age the symphysis synostoses, and the two halves of the mandible become fused solidly together. Just lateral to the site of the *mandibular symphysis* are small protuberances called the *mental spines (genial tubercles)* which are attachment of the genioglossus and geniohyoid muscle (muscles of the oral cavity).

The *mandibular ramus* exhibits the following features. The *coronoid process* is a beak-like projection directed superiorly and anteriorly. It is for the insertion of the temporalis muscle. The mandibular ramus bears the mandibular *head (articular condyle)* and *neck*. The *head* lies at the superoposterior extremity of the ramus and articulates with the mandibular fossa of the temporal bone to form the temporomandibular joint. The neck is the constricted region inferior to the head that joins the head to the ramus.

The *mandibular notch* lies between the coronoid process and the mandibular head. The *medial surface* of the ramus bears the *mandibular foramen*. The medial lip of the foramen is prolonged into the *lingula*, for the attachment of the sphenomandibular ligament. The *mandibular foramen and canal* (the foramen is the opening of the canal and the canal runs through the bone) transmit the inferior alveolar nerve and vessels to the mandibular teeth.

The *mylohyoid groove* transmits the mylohyoid nerve and vessels. It originates at the mandibular

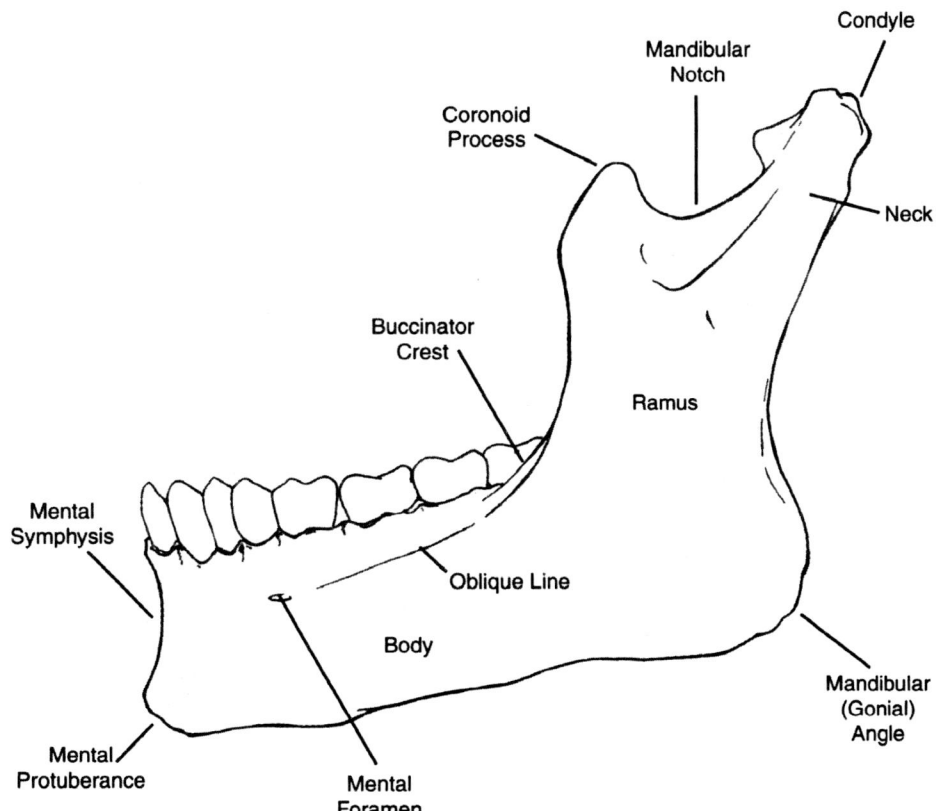

Figure 8-40 The mandible, left external view.

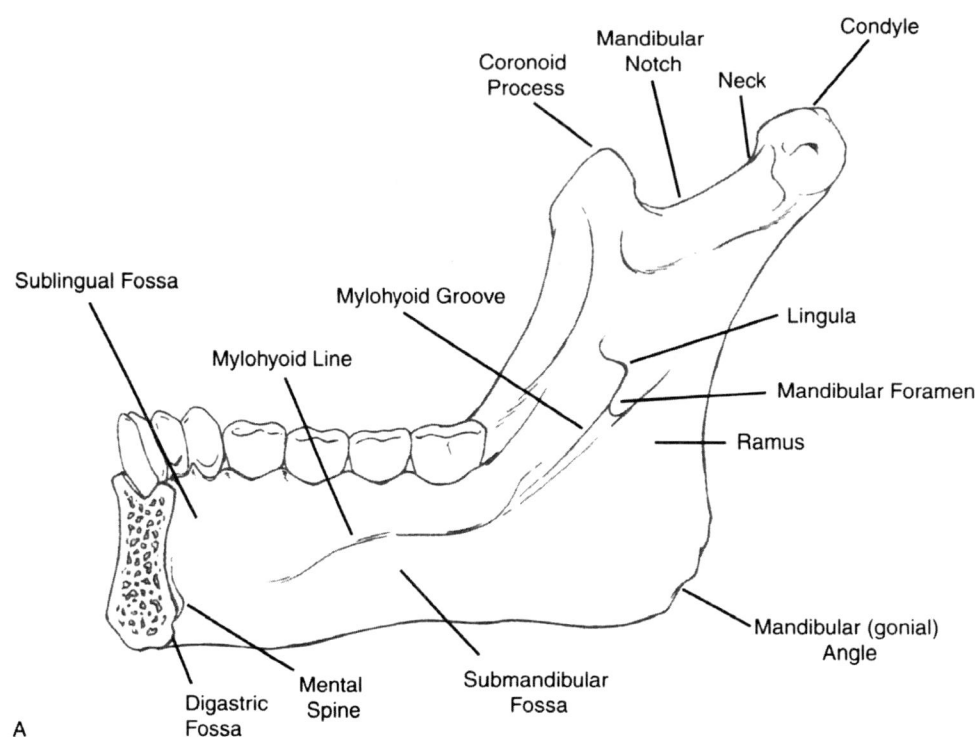
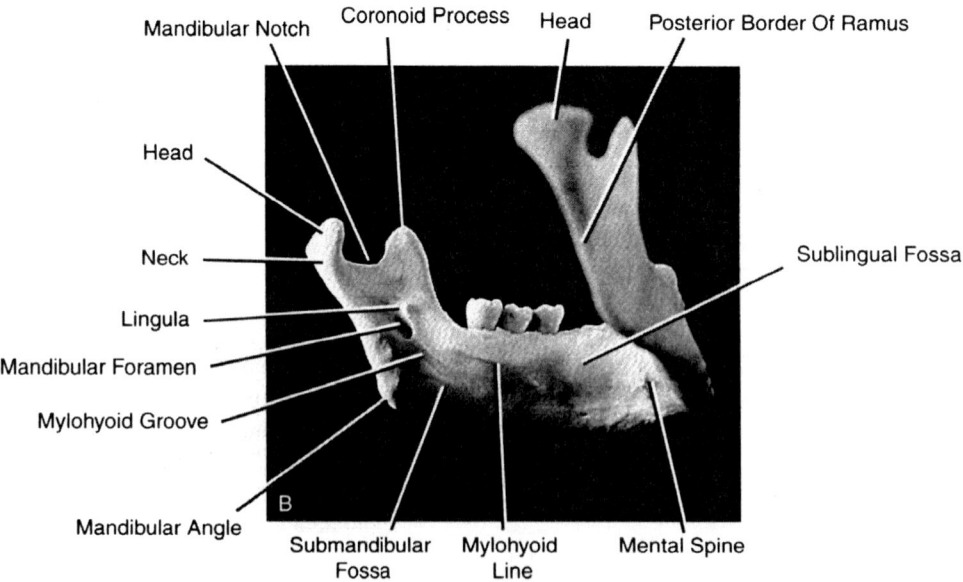

Figure 8-41 (A) Line drawing of the mandible, right internal view, and (B) photograph of left internal view.

foramen and extends inferiorly and anteriorly along the medial surface of the mandible. The *mandibular (gonial) angle* is the posteroinferior angle of the jaw, where the ramus and body of the mandible meet.

Ossification and Development of the Cranium, Mandible, and Hyoid

Vast amounts of literature and research have been devoted to craniofacial growth and development. We touch upon the high points here, concentrating especially upon the patterns of ossification within the cranium (Figure 8-45). See a work such as Sperber (1989) or Gray (1985, 1995) for a more detailed discussion of craniofacial embryology and development.

As discussed previously, the cranium is composed of two entities: the *neurocranium* and the *splanchnocranium* (the face). Both of these entities have unique developmental characteristics.

The Neurocranium

The neurocranium is composed of two elements: the *cranial vault* and the *cranial base*. The *cranial vault (calvaria)* is of recent phylogenetic origin and serves to cover the expanded brain. It is also known as the *desmocranium,* as it forms via intramembranous (dermal) ossification. It includes the frontal bone, the parietal bones, the squamous part of the occipital bone and the squamous part of the temporal bones.

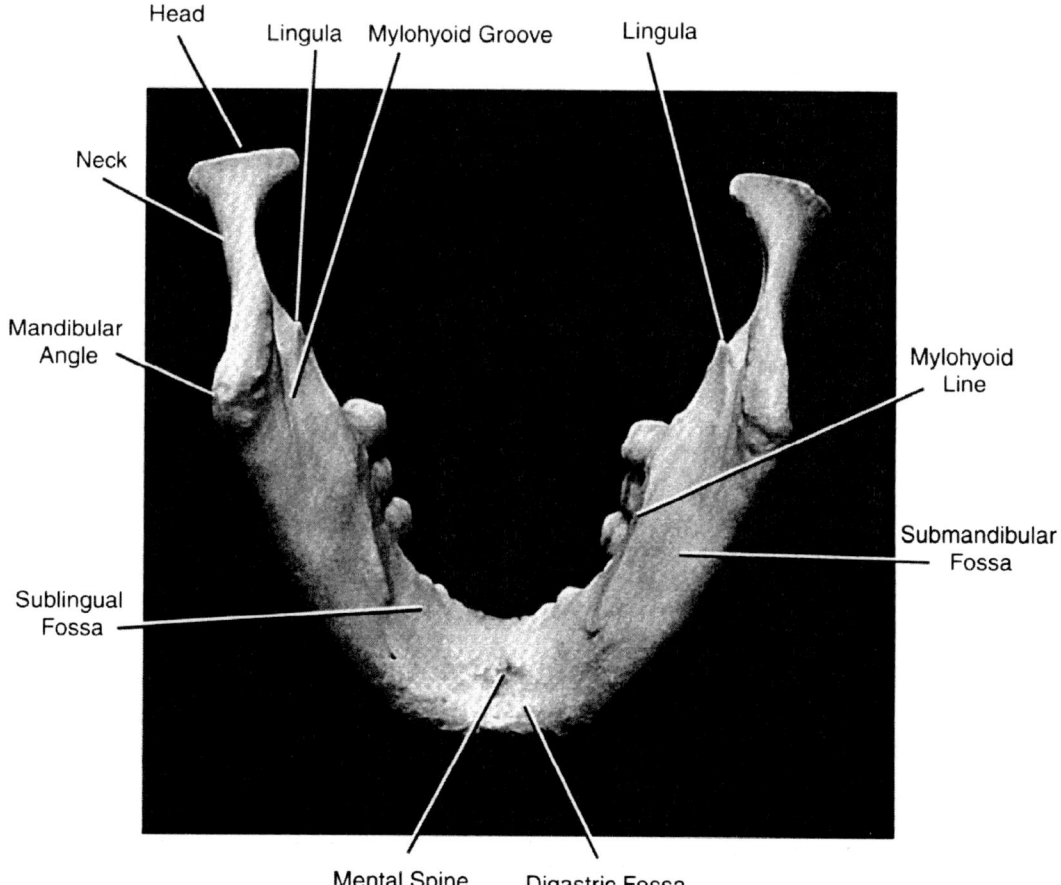

Figure 8-42 The mandible, oblique posteroinferior view, demonstrating the lingulae, mylohyoid lines, mental spines, and fossae for the sublingual and submandibular salivary glands.

The *cranial base (basicranium)* is phylogenetically much older. It includes the cranial floor and the bony capsules that surround the nasal and auditory sense organs. It forms via endochondral ossification and is hence known as the *chondrocranium*. The cranial base includes all the remainder of the cranial bones not included as part of the face or cranial vault, including most of the temporal bones, the basilar part of the occipital bone, the sphenoid and ethmoid.

The Splanchnocranium

The *splanchnocranium,* also referred to as the *viscerocranium,* is derived from the phylogenetically ancient *branchial arch* structures. A total of six branchial arches form during development (numbered in cranial-caudal sequence from one to six). However, the fifth branchial arch disappears almost as soon as it develops, leaving only five branchial arches (1- 4 and 6) to contribute to head and neck structures. Each arch possesses a cartilaginous rod. Although small portions of these cartilages will persist and form adult structures, the remainder will resorb and eventually be replaced by bone of intramembranous origin. These bones include the *maxilla, mandible, zygomatic,* and *vomer.*

Also included within the splanchnocranium are the *teeth,* which will not be considered in detail here. The teeth are also formed from specialized calcified tissue, but are not bone. They are phylogenetically derived from ectodermal placoid scales. This is reflected in their embryological origin from oral ectoderm. (For further details regarding

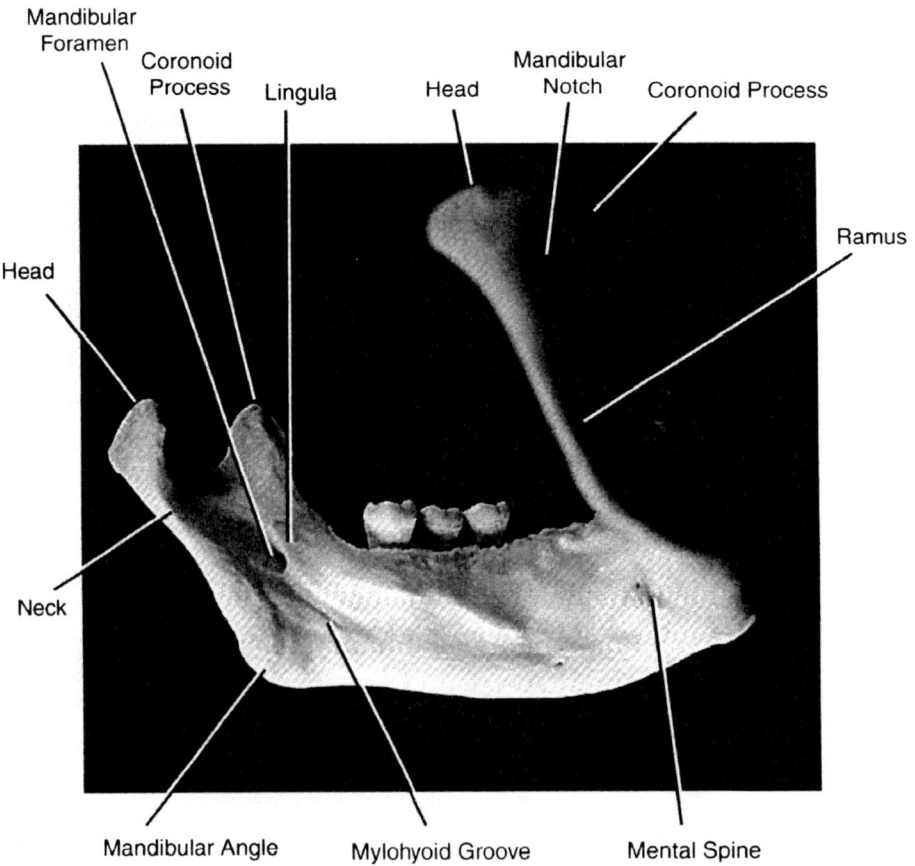

Figure 8-43 Photograph of the mandible, left internal view, demonstrating the lingula, mylohyoid line, and mylohyoid groove.

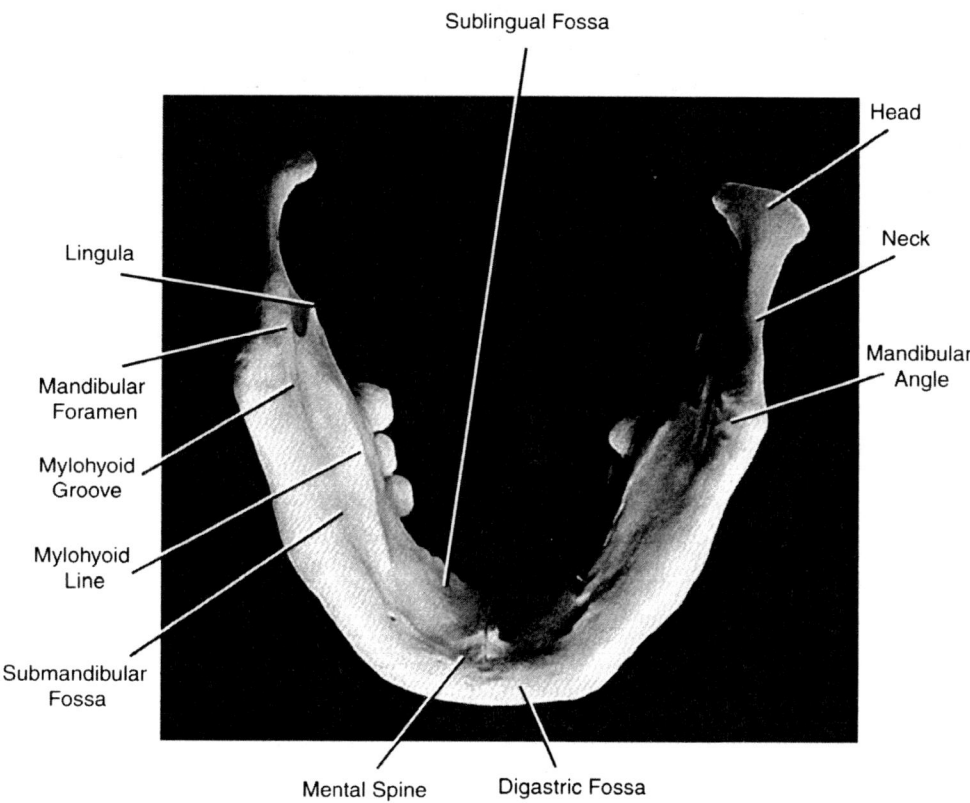

Figure 8-44 Photograph of the mandible, inferior surface, demonstrating the mylohyoid lines and grooves, digastric fossae, mental spines, and submandibular fossae.

the development and anatomy of the teeth, see Bhaskar, 1990.)

It is important to note that the alveolar processes of the mandible and maxilla bear a complex developmental relationship with the teeth, and that the presence or loss of teeth have important implications for the proper development of the maxilla, mandible, and facial skeleton in general (see Enlow and Hans, 1996). Additionally, dental problems can have serious consequences for general health.

Cranial Anlagen

As we have noted, the cranium develops both from endochondral and intramembranous ossification. We will here examine the appearance of the anlagen of these various centers.

The Chondrocranium

The portion of the cranium that develops through endochondral ossification includes the base of the skull (the *ethmoid* and *inferior nasal conchae*, and large portions of the *sphenoid, occipital* and *temporals*), as well as remnants of the *first, second* and *third branchial arch cartilages*.

The *chondrocranium* develops from cartilaginous precursors that underlie the developing brain from the interorbitonasal region to the cranial end of the vertebral column (figure 8-45). These include the following.

A pair of *parachordal cartilages* (figure 8-45) extend from the foramen magnum to the hypophyseal fossa. They are derived from the four occipital sclerotomes, the first cervical sclerotome and paraxial mesoderm (somitomeric) related to the cranial end of the notochord.

The *hypophyseal cartilages* develop on either side of the hypophyseal stem. They are derived from both paraxial mesoderm (somitomeric) and neural crest.

The *prechordal,* or *trabecular, cartilages* extend anteriorly from the hypophyseal fossa to the developing nasal capsule. At the hypophyseal fossa these cartilages become continuous with the parachordal cartilages.

As the chondrocranium begins to form, a number of capsules appear which house the developing olfactory, ocular and auditory/vestibular sensory systems.

The *cartilaginous nasal capsule* consists of a septum formed by the anterior end of the prechordal

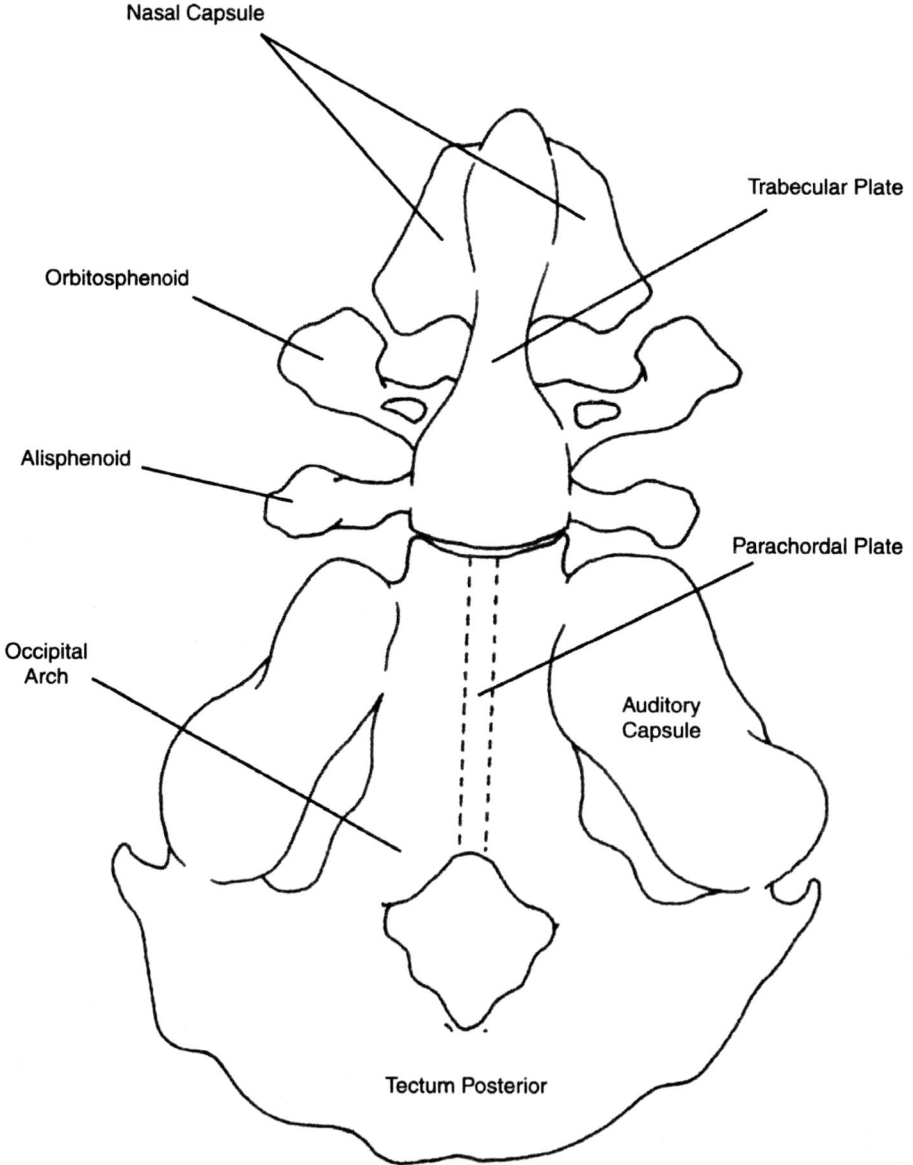

Figure 8-45 Semidiagrammatic representation of the chondrocranium and the capsules forming the anlagen of the cranial base.

plate, and a lateral cartilage on either side. Ossification of the upper part of the septum forms the perpendicular plate of the ethmoid and the crista galli. The more inferior and anterior part remains unossified to form the cartilaginous nasal septum. The cribriform plate develops in the upper part of the capsule, while the ethmoid labyrinth ossifies in the cartilage of the lateral wall of the cartilaginous nasal capsule. The inferior margin of the lateral cartilage is folded inward to form the inferior nasal concha.

The *auditory (otic) capsule* initially lies lateral to the parachordal plate. It will later ossify to contribute to the formation of the petrous and mastoid portions of the temporal bone.

The Branchial Arch Cartilages

The superior ends of the first and second arch cartilages are closely related to the underside of the auditory capsule. The *cartilage of the first arch* will form the *incus* and the *malleus*. The *cartilage of the second arch* will give rise to the *stapes*, the *styloid process* and the *lesser cornua (horns) of the hyoid bone*. The *greater cornua* and most of the *body of the hyoid* derive from the cartilage of the third branchial arch.

The *thyroid cartilage*, which sometimes ossifies, most likely derives from the *fourth branchial arch*. Some workers attribute the *cricoid cartilage* to the *sixth branchial arch*. The *fifth branchial arch* has no skeletal derivatives.

The Membranous cranium

The portion of the cranium that develops through intramembranous ossification includes the bones forming the roof and walls of the calvarium (*frontal, parietals* and *squamous portions of the occipital and the temporals*) and the bones of the face and jaws (*maxillae, palatines, nasals, lacrimals, zygomatics, vomer*, portions of the *pterygoid plates and greater wings of the sphenoid* and *mandible*), all of which are derived from the first branchial arch and frontonasal process.

Ossification of Individual Bones of the Cranium

The entirely endochondral *ethmoid bone* (figure 8-46) has three centers of ossification: one for the perpendicular plate and crista galli and one for each bony labyrinth. Ossification begins in the ethmoidal (superior and middle) nasal conchae and in the orbital plate of the ethmoid at about four months *in utero*. The bony labyrinth of the ethmoid is ossified at birth, but the septum remains cartilaginous throughout the first year. The perpendicular plate and crista galli ossify over the first year. They fuse with the labyrinths during the second year. Ethmoidal air cells develop *in utero*, and are rather narrow at birth.

An ossification center for each *inferior nasal concha* appears at four or five fetal months, and the bone rapidly ossifies. Although derived from the inferior margin of the cartilaginous ethmoid, the inferior concha will detach from the ethmoid to become an independent bone that will later synostose with the maxilla.

The *sphenoid* bone (figure 8-47) forms partly from cartilage and partly from membrane. The portions of the bone that undergo endochondral ossification are the body, the lesser wings, the medial portions of the greater wings. The rest of the bone forms either intramembranously, or endochondrally from secondary cartilages that develop in membrane (i.e., the medial pterygoid hamulus). Up to 19 individual centers of ossification coalesce to form the sphenoid. Initially the sphenoid consists of a presphenoid anterior to the tuberculum sellae, and a postsphenoid which includes the sella turcica, the dorsum sellae and basisphenoid. The pre- and postsphenoids fuse with one another shortly before birth.

Like the sphenoid, the *occipital* bone (Figures 8-48, see also Figures 8-46 and 8-47) develops via a combination of intramembranous and endochondral ossification. Most of the occipital bone above the superior nuchal line derives from two intramembranous ossification centers which appear at 8 weeks *in utero*. The remainder of the occipital

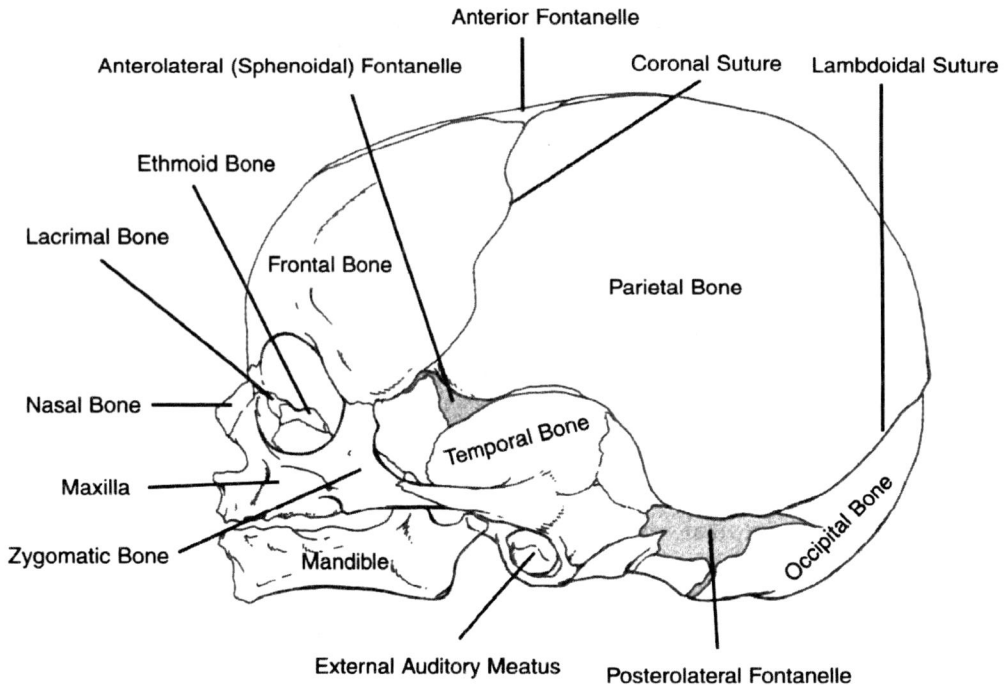

Figure 8-46 The neonatal cranium: norma lateralis, or lateral view.

bone is derived from five ossification centers which form in the parachordal cartilages. A single, unpaired ossification center will form the basioccipital bone anterior to the foramen magnum as well as the anterior one-third of the occipital condyles. This center appears at about 11 weeks in utero. All remaining ossification centers (two infranuchal centers posteriorly and two exoccipital centers laterally) appear one week later. The squamous, exoccipital and basilar portions fuse to form a single occipital bone by 7 years of age.

Anteriorly the basioccipital is separated from the basisphenoid by the sphenooccipital synchondrosis, a remnant of the chondrocranium which, like the growth cartilages of long bones, functions as a growth center throughout postnatal ontogeny. Synostosis of the sphenooccipital synchondrosis begins at age 12 to 13 in girls and age 14 to 15 in boys. Complete fusion of the synchondrosis does not occur until age 20.

The *temporal* bone (Figure 8-46) follows a complex sequence of ossification. Like the sphenoid and occipital, it is formed from both cartilage and membrane. The squamous part of the bone develops intramembranously from a single ossification center that appears above the root of the zygomatic process at 7 - 8 weeks *in utero*. The tympanic part of the temporal bone is also intramembranous in origin. It derives from four ossification centers that appear at 12 weeks *in utero* and form a ring around the external acoustic meatus. The tympanic ring will not fuse with the squamous temporal until after birth.

The remaining *petrous* and *styloid* portions of the *temporal* bone ossify in cartilage from as many as 22 ossification centers that first appear at five months *in utero*. The petrous and squamous portions of the temporal will fuse with one another during the first year. The styloid process is derived from two ossification centers that form in the second arch cartilage. The styloid process is initially attached to the temporal by cartilage and does not fuse to the bone until middle age. Occasionally the styloid process fails to fuse with the temporal bone.

Osteology of The Human Cranium

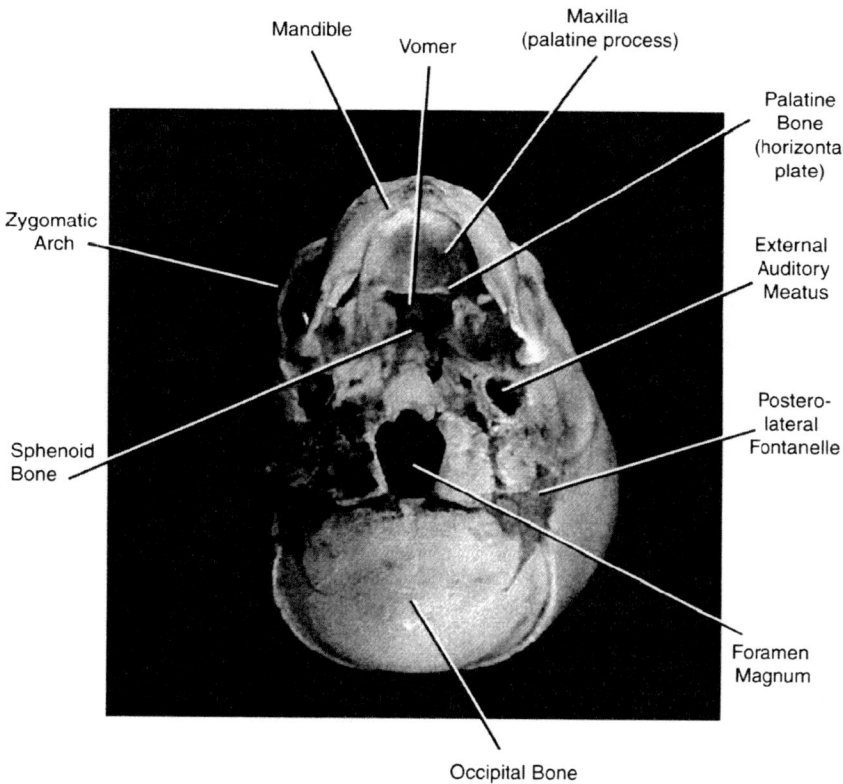

Figure 8-47 The neonatal cranium: norma basalis, or inferior view.

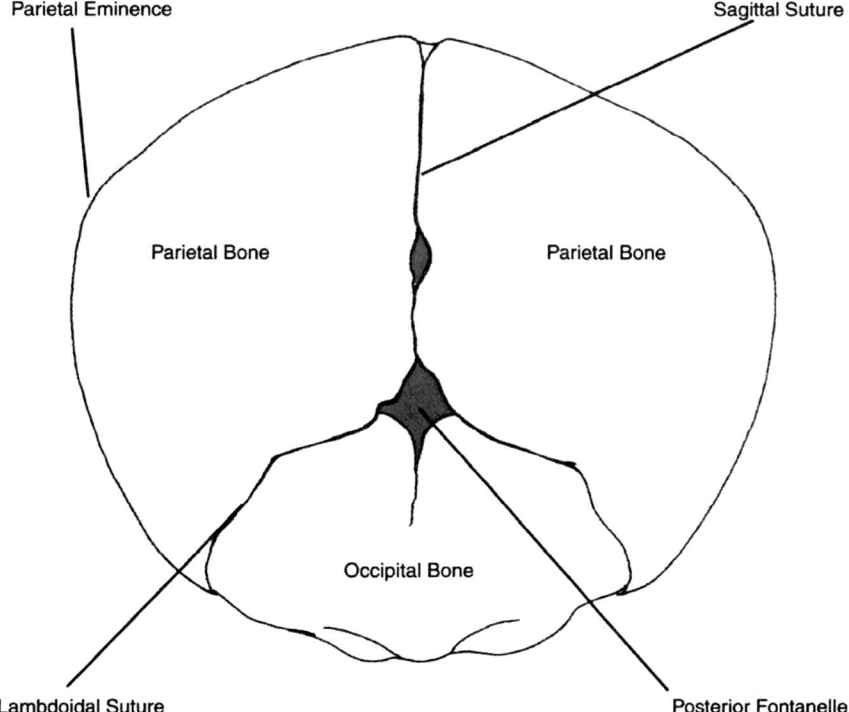

Figure 8-48 The neonatal cranium: norma occipitalis, or posterior view.

The *hyoid* bone has six centers of ossification: two for the body and one for each horn.

The *frontal* bone (Figure 8-49) derives from two primary ossification centers which appear at eight weeks *in utero* in the region of the superciliary arches. At birth the frontal bones are separated by the *metopic suture*. The middle of this suture undergoes synostosis at about two years, and by seven or eight years the two halves of the frontal are completely fused. In very rare occasions, the two halves of the frontal never fuse, and the metopic suture persists into adulthood.

The *parietal* bones (see Figures 8-46, 8-48 and 8-49) each form from two centers of ossification which appear at the parietal eminence at eight weeks *in utero*. The four corners of the parietal bones remain unossified at birth, and help form the fontanelles discussed previously. The fontanelles are completely ossified by the middle of the second year.

The *maxilla* (Figure 8-50; see also Figure 8-49) ossifies in membrane from one primary center which appears at seven weeks *in utero* at the termination of the inferior orbital nerve. Five secondary ossification centers, including two centers in the incisive region that will from the bone that encloses the four maxillary incisors, appear roughly one week later. The maxillary sinus appears as a depression in the lateral wall of the nasal cavity at about four months *in utero*, and is well formed by birth. The maxilla grows in size until the eruption of the third molars.

The palatine bones (see Figure 8-47) develop in membrane at the inferior margin of the nasal capsule. Each bone is derived from a single ossification center that appears at eight weeks *in utero* in the vicinity of the pyramidal process.

The *zygomatic, nasal* and *lacrimal* bones (see Figures 8-46, 8-49 and 8-50) each form from single centers of ossification which appear at eight weeks in utero. The lacrimal bones are visible in Figure 8-46.

The *vomer* (see Figure 8-47) derives from two centers of ossification that form on either side of the cartilaginous nasal septum. These centers, which first appear at eight weeks *in utero*, unite with each other behind the cartilaginous nasal septum where they articulate with the sphenoid.

Each half of the *mandible* (see Figures 8-46 and 8-47) begins to ossify in the sixth week *in utero* from a membranous center which appears at the future location of the mental foramina. This center, which is located lateral to Meckel's Cartilage, will spread and give rise to the body, alveolar process, and most of the ramus. During this process most of Meckel's cartilage is invaded by bone and disappears. Those portions of the cartilage that persist will give rise to the *incus* and *malleus*, the *sphenomandibular ligament* and accessory endochondral ossicles in the mental symphysis.

The head of the condyle, part of the coronoid process and the mental protuberance are formed from secondary (membrane) cartilages that appear between the 10th and 12th weeks in utero. At birth, the two halves of the mandible are separated at the mental symphysis. The two halves fuse early in the second year.

The Infant Cranium

As noted, the cranial vault is formed by intramembranous ossification. At birth, the bones of the cranial vault have not completely grown to meet one another, leaving regions of the cranium membranous. This allows the cranium to deform somewhat and makes easier its passage through the birth canal. These regions of membrane remaining at birth are termed *fontanelles,* or "soft spots." Figures 8-46 through 8-50 demonstrate the infant or neonatal cranium. Note the locations of the fontanelles at bregma, lambda, asterion, and pterion. The anterior fontanelle (see Figs. 8-49 and 8-50) is at bregma. The posterior fontanelle (see Fig. 8-48) is at lambda. The anterolateral (or sphenoidal) fontanelle (see Fig. 8-46) is at pterion. The posterolateral fontanelle (see Figs. 8-46 and 8-47) is at asterion. Also note the relatively large neurocranium, the small face, the

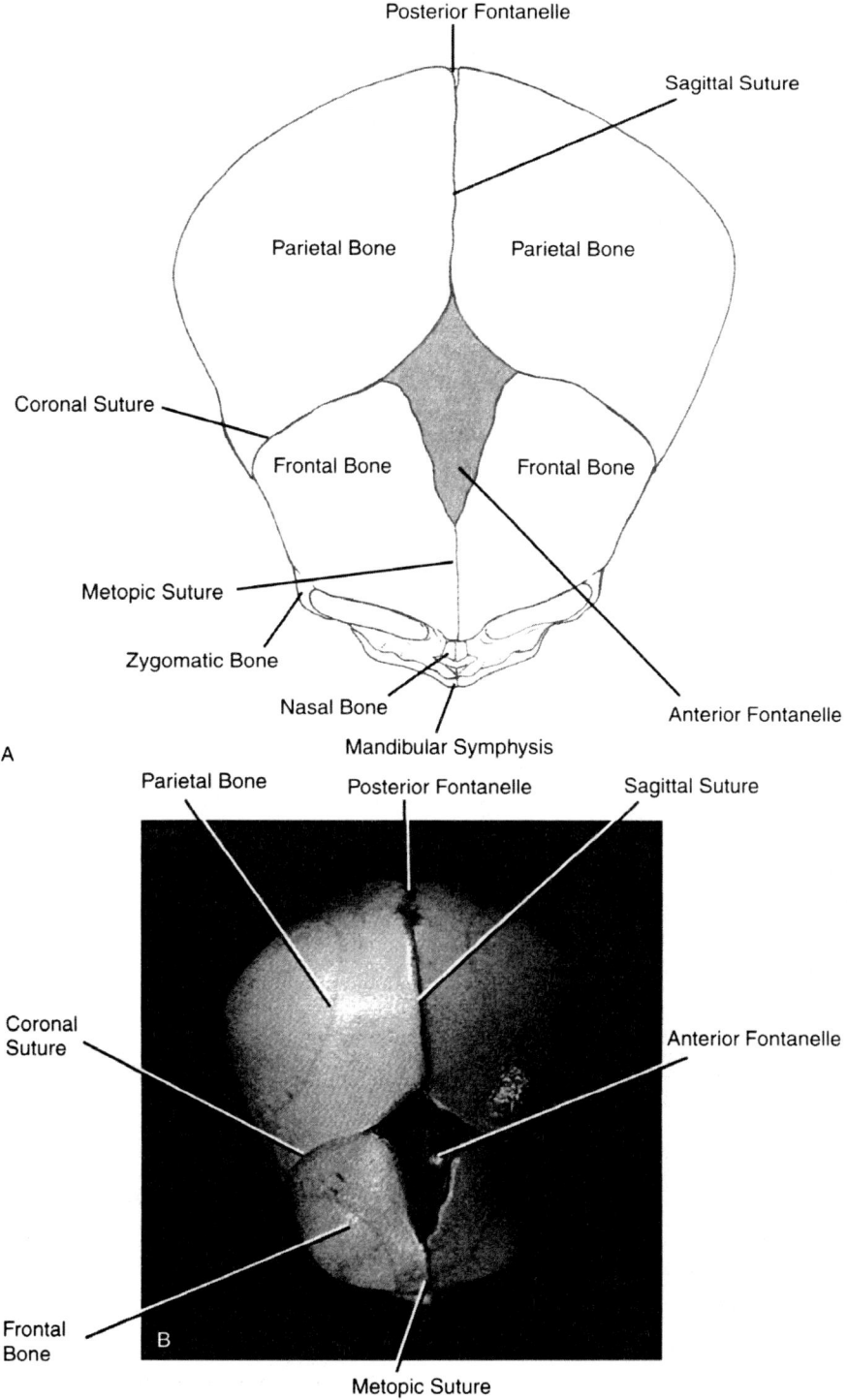

Figure 8-49 (A) Line drawing and (B) photograph of the neonatal cranium: norma verticalis, or superior view.

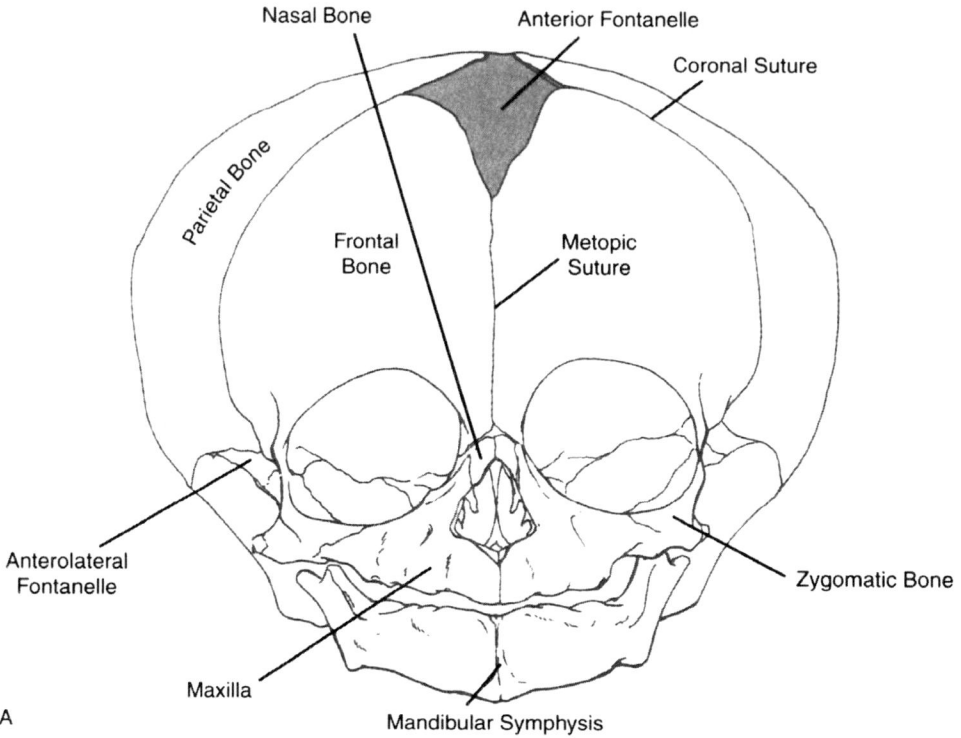

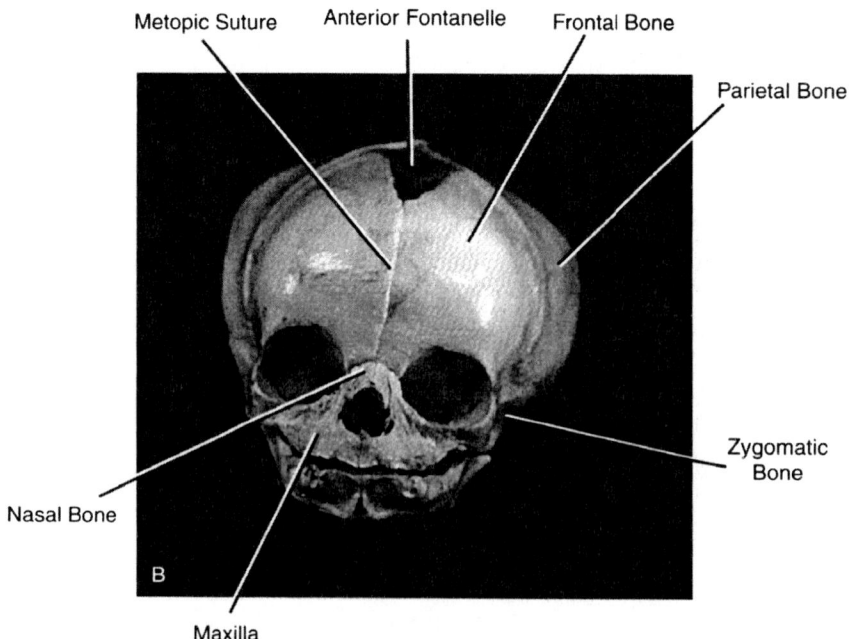

Figure 8-50 (A) Line drawing and (B) photograph of the neonatal cranium: norma frontalis, or anterior view.

edentulous mandible and maxilla, and the cartilaginous mandibular symphysis.

The process of deformation of the fetal skull during delivery is called molding. During delivery, one parietal bone slides under the other. This is referred to as the obstetrical hinge joint.

The skull of the neonate does not possess a mastoid process, because it is an apophysis that develops under the origin of the sternocleidomastoid muscle. As a result, the stylomastoid foramen and facial nerve of the infant are very superficial. During delivery, the neonate's facial nerve is very vulnerable to injury. The middle ear cavity at birth is nearly its adult size.

References Cited

Adams, T, Heisy, RS, Smith, MC, and Briner, BJ (1992). Parietal bone mobility in the anesthetized cat. *Journal of the American Osteopathic Association,* 92, 599-600, 603-610, 615-622.

Bhaskar, SN (1990). *Orban's Oral Histology and Embryology* (11th ed). Mosby-Yearbook Publishers, Chicago.

Enlow, DH, and Hans, MG (1996). *Essentials of Facial Growth.* WB Saunders, Philadelphia.

Gray, H (1985). *Gray's Anatomy* (30th Am ed). CD Clemente (Ed). Lea and Febiger, Philadelphia.

Gray, H (1995). *Gray's Anatomy* (38th Brit ed). PL Williams, LH Bannister, MM Berry, P Collins, M Dyson, JE Dussek, and MWJ Ferguson (Eds). Churchill Livingstone, New York.

Greiner, TM, and Walker, RA (1999, April). *Morphometric Variation of the Human Auditory Ossicles.* Paper presented at the 68th annual meeting of the American Association of Physical Anthropologists, Columbus, Ohio (Abstract published in *American Journal of Physical Anthropology,* Suppl 28, p 140, 1999).

Heisey, SR, and Adams, T (1993). Role of cranial bone mobility in cranial compliance. *Neurosurgery,* 33, 869-877.

Rogers, JS, and Witt, PL (1997). The controversy of cranial bone motion. *Journal of Orthopaedic and Sports Physical Therapy,* 26, 95-103.

Sperber, GH (1989). *Craniofacial Embryology.* Williams and Wilkins, Baltimore.

Further Readings and Resources

Textbooks and Anatomic Atlases:

The following texts shed further light on the human skeleton from assorted perspectives. A number of the listed references give additional information relevant to forensic use of the skeleton in identification of the unknown deceased. Others offer additional, often photographic or color, illustrations of the skeleton in relation to the soft tissue anatomy of the human body. Still others give additional clinical information about musculoskeletal disorders, or highlight specific aspects of bony anatomy and histology.

Agur, A.M.R., and Dalley, II, A.F. (2004) *Grant's Atlas of Anatomy,* 11th Edition. Lippincott, Williams and Wilkins, Baltimore.

Aiello, L. and C. Dean (2002) *An Introduction to Human Evolutionary Anatomy.* Academic Press. (Stresses the musculoskeletal anatomy of *Homo sapiens* in relationship to other primates.)

Bennett, K.A. (1993) *A Field Guide for Human Skeletal Identification.* Charles C Thomas, Springfield, Illinois.

Breathnach, A.S. (1965) *Frazer's Anatomy of the Human Skeleton,* 6th Edition. J. and A. Churchill, London. (Originally written by J. Ernest Frazer. Long out of print, this work is close in spirit to *Frazer's Anatomy*.)

Clement, J.G. and D.L. Ranson, eds. (1998) *Craniofacial Identification in Forensic Medicine.* Edward Arnold.

Clemente, C.D. (2006) *A Regional Atlas of the Human Body,* 5th Edition. Williams and Wilkins, Baltimore. (A superb anatomical atlas, edited by the editor of the most recent U.S. edition of *Gray's Anatomy*.)

Cormack, D.H., ed. (1987) *Ham's Histology*, 9th Edition. Lippincott, Philadelphia. (Originally by A.W. Ham, *Ham's Histology* has long been regarded as a classic histology text. Now out of print, the chapters on tendons and ligaments, bone, and joints still serve as an excellent introduction to the microanatomy of these structures.)

Cormack, D.H. (2001) *Essential Histology,* 2nd Edition. Lippincott-Raven, Philadelphia. (A more recent introductory histology text by the editor of the most recent edition of Ham's histology.)

Currey, J. (1984) *The Mechanical Adaptations of Bone.* Princeton University Press, Princeton, New Jersey.

Currey, J. (2006) Bones: Structure and Mechanics, 2nd Edition. Princeton University Press, Princeton, New Jersey.

Edwardson, B.M. (1995) *Musculoskeletal Disorders: Common Problems.* Singular Publishing Group, Inc., San Diego.

Gosling, J.A., P.F. Harris, I. Whitmore, P.L.T. Willan (2002) *Human Anatomy: Color Atlas with Integrated Text,* 4th Edition. Mosby. (This atlas features photographs of actual dissections accompanied by line drawings to help elucidate those figures.)

Gray, Henry (1995) *Gray's Anatomy,* 38th British Edition. P.L. Williams, L.H. Bannister, M.M. Berry, P. Collins, M. Dyson, J.E. Dussek and M.W.J. Ferguson, eds. Churchill Livingstone, New York. (The British edition of *Gray's Anatomy* is widely regarded as *the* anatomical reference. Its first edition was published in 1858. While paperback reprints of century-old editions are widely and cheaply available, they are not recommended for serious study. While the anatomy of the human body has not changed in the last 100 years, the knowledge of it has grown tremendously and much of the terminology used to describe it has changed drastically.)

Gray, Henry (1985) *Gray's Anatomy,* 30th American Edition. C.D. Clemente, ed. Lea and Febiger, Philadelphia. (The American and British versions of *Gray's Anatomy* soon diverged from one another.

The American version has remained more conservative (and slightly less expensive) than its British counterpart. The most recent American edition still retains many of the original illustrations by H. Van Dyke Carter prepared for the 1858 edition.)

Haas, J., J.E. Buikstra, and D.H. Ubelaker (1994) *Standards for Data Collection from Human* Skeletal Remains: Proceedings of a Seminar at the Field Museum of Natural History *(Arkansas Archeological Survey).* Arkansas Archeological Survey.

Hillson, Simon (2003) *Dental Anthropology.* Cambridge University Press, London.

Hutchings, R.T., S. C. Marks, Peter H. Abrahams (2003) *McMinn's Color Atlas of Human Anatomy*, 5th Edition. Mosby.

Iscan, M.Y., ed. (1989) *Age Markers in the Human Skeleton.* Charles C Thomas, Springfield, Illinois.

Jenkins, D.B. (1991) *Hollinshead's Functional Anatomy of the Limbs and Back*, 6th Edition. W.B. Saunders Co., Philadelphia.

Krogman, W.M. (1962). *The Human Skeleton in Forensic Medicine.* Charles C. Thomas, Springfield, Illinois.

(A classic text in forensic applications of human osteology. A new edition, co-authored with Yaskar Iscan was published in 1986. Not only is this text a valuable resource in itself, but it includes references to all the important literature on age, sex and "race" determination from skeletal remains published to that time. Krogman also includes detailed descriptions of the time of appearance and fusion of epiphyses in subadults.)

Krogman, W.M. and M. Y. Iscan (1986) *Human Skeleton in Forensic Medicine*, 2nd Edition. Charles C. Thomas, Springfield, Illinois.

Nordin, M, and Frankel, (2001) *Basic Biomechanics of the Skeletal System,* 3rd Edition. Lippincott, Williams and Wilkins, Baltimore.

Reichs, K.J. and W.M. Bass, eds. (1998) *Forensic Osteology : Advances in the Identification of Human Remains,* 2nd Edition. Charles C Thomas, Springfield, Illinois.

Rohen, J.W., C. Yokochi and E. Lutjen-Drecoll (2006) *Color Atlas of Anatomy*, 6th Edition. Lippincott, Williams and Wilkins, Baltimore.

Ross, H.M., L.J. Romrell, G.I. Kaye (2006) *Histology : A Text and Atlas,* 5th ed. Lippincott, Williams and Wilkins, Baltimore.

(A profusely illustrated, thorough histology text. Includes an excellent discussion of bone growth and development.)

Saunders, S.R. and M.A. Katzenberg, eds. (1992) *Skeletal Biology of Past Peoples : Research Methods.* John Wiley and Sons.

Schwartz, J.H. (2006) *Skeleton Keys : An Introduction to Human Skeletal Morphology, Development, and Analysis,* 2nd Edition. Oxford University Press.

Shipman, P., A. Walker and D. Bichell (1985) *The Human Skeleton.* Harvard University Press, Cambridge, Massachusetts and London, England.

Standring, S. (ed) (2004) *Gray's Anatomy: The Anatomical Basis of Clinical Practice,* 39th British Edition. Churchill Livingstone, New York.

(The most recent incarnation of the venerable anatomy textbook.)

Taylor, J and Resnick (2000) *Skeletal Imaging: Atlas of the Spine and Extremities.* Saunders, Philadelphia.

White, T.D. and P.A. Folkens (2005) *The Human Bone Manual.* Academic Press, Inc. San Diego.

Yochum, T.R. and L.J. Rowe (2004) *Essentials of Skeletal Radiology,* 3rd Edition. Lippincott, Williams and Wilkins, Baltimore.

Journal Articles:

American Journal of Physical Anthropology (AJPA)

The AJPA frequently publishes articles relevant to human osteology, human anatomy, and the determination of skeletal age at death, sex and "race" of unidentified skeletal remains. Most of the September, 1985 edition of the *AJPA* (Volume *68*, no. 1) was devoted to a series of articles by Lovejoy and coworkers on the determination of age and sex in unidentified human skeletal material. In subsequent years, the AJPA has published a number either confirming or inspired by the research presented by Lovejoy and coworkers. These articles include:

Lovejoy, C.O. (1985) Dental wear in the Libben population: Its functional pattern and role in the determination of adult skeletal age at death. *AJPA 68(1):* 47-56.

Lovejoy, C.O., R.S. Meindl, R.P. Mensforth and T.J. Barton (1985) Multifactorial determination of skeletal age at death: A method and blind tests of its accuracy. *AJPA 68 (1):* 1-14.

Lovejoy, C.O., R.S. Meindl, T.R. Pryzbeck and R.P. Mensforth (1985) Chronological metamorphosis of the auricular surface of the ilium: A new method for the determination of adult skeletal age at death. *AJPA 68(1):* 15-28.

Meindl, R.S., and C.O. Lovejoy (1985) Ectocranial suture closure: A revised method for the determination of skeletal age at death based on the lateral-anterior sutures. *AJPA 68(1):* 57-66.

Meindl, R.S., C.O. Lovejoy, R.P. Mensforth and L. Don Carlos (1985) Accuracy and direction of error in the sexing of the skeleton: Implications for paleodemography. *AJPA 68(1):* 79-85.

Meindl, R.S., C.O. Lovejoy, R.P. Mensforth and R.A. Walker (1985) A revised method of age determination using the os pubis, with a review and tests of accuracy of other current methods of pubic symphyseal aging. *AJPA 68(1):* 29-45.

Mensforth, R.P. and C.O. Lovejoy (1985) Anatomical, physiological, and epidemiological correlates of the aging process: A confirmation of multifactorial age determination in the Libben skeletal population. *AJPA 68(1):* 87-106.

Walker, R.A. and C.O. Lovejoy (1985) Radiographic changes in the clavicle and proximal femur and their use in the determination of skeletal age at death. *AJPA 68(1):* 67-78.

A number of other papers related to the above mentioned work (by many of the same authors):

Bedford, ME, KF Russell, CO Lovejoy, RS Meindl, SW Simpson, and PL Stuart-Macadam (1993) Test of the multifactorial aging method using skeletons with known ages-at-death from the Grant Collection. *American Journal of Physical Anthropology* 91:287-297.

Lovejoy, C.O., R.S. Meindl, TR Pryzbeck, TS Barton, KG Heiple and D. Kotting (1977) Paleodemography of the Libben site, Ottawa County, Ohio. *Science 198:* 291-293.

Meindl, R.S., M.E. Bedford, K.F. Russell, C.O Lovejoy, and S.W. Simpson (1995) Testing the test of the multifactorial aging method: a reply to Fairgrieve and Oost. *American Journal of Physical Anthropology* 97:85-87.

Walker, R.A., C.O. Lovejoy and R.S. Meindl (1994) The histomorphological and geometric properties of human femoral cortex in individuals over 50: Implications for histomorphological determination of age-at-death. *American Journal of Human Biology* 6: 659-667.

APPENDIX

Appendix: Muscle Origins, Insertions, Innervations, Blood Supplies and Principal Actions.

This appendix lists the muscles of the upper and lower limbs, vertebral column, abdomen and thorax, as well as major muscles of mastication and those attaching to the hyoid bone. These muscle origins and insertions are alluded to in the text, and are included here for reference. These muscles all leave, to a greater or lesser degree, marks upon the skeleton where their tendons of origin or insertion attach to bone. The muscles are listed regionally: first, the extrinsic muscles of the back, then the intrinsic muscles of the back, then major muscles of the abdomen and thorax, then upper limb, then lower limb, and finally the major muscles of mastication and of the hyoid region. Primary segmental innervation to most muscles are indicated in parentheses at the end of the description of innervation. The following convention is followed in listing segmental innervation. The designations C1 to C8 refer to cervical spinal nerves numbers 1 to 8, T1 to T12 refer to thoracic spinal nerves numbers 1 to 12, L1 to L5 to lumbar spinal nerves numbers 1 to 5, and S1 to S5 to sacral spinal nerves numbers 1 to 5. Innervations from cranial nerves are referred to by the names of the nerves, followed by their numeric designation, for example, Trigeminal Nerve (Cranial Nerve V).

Major Muscles of the Trunk:

The Back

The Superficial Back Muscles

The superficial back muscles are extrinsic and hypaxial, innervated by ventral rami. They act primarily upon the upper limb.

Trapezius

Origin: External occipital protuberance, medial one-third of the superior nuchal line, ligamentum nuchae, and the spinous processes of C7 to T12. *Insertion:* Superior fibers: Lateral one-third of the clavicle.

Middle fibers: Spine and acromion of scapula.

Inferior fibers: Base of spine of scapula.

Innervation: Spinal accessory nerve (motor) and ventral rami of third and fourth cervical nerves (sensory and proprioception) (C3, C4).

Blood Supply: The superficial branch of the transverse cervical artery.

Action: Raises, retracts and rotates the scapula. Bilateral contraction of superior fibers extends the head. Unilateral contraction of upper fibers laterally flexes the head and cervical vertebral column.

Latissimus Dorsi

Origin: Spinous processes of the lower six thoracic, all lumbar and the upper sacral vertebrae, from the posterior part of the iliac crest, and from the lower three or four ribs by muscular slips which interdigitate with the external oblique.

Insertion: Crest of the lesser tubercle and floor of the bicipital groove of the humerus.

Innervation: Thoracodorsal nerve (C6, C7, C8).

Blood Supply: The thoracodorsal artery.

Action: Extension, medial rotation, and adduction of the arm.

Levator Scapulae

Origin: Posterior tubercles of the transverse processes of C1 to C3 or C4.

Insertion: Medial border of scapula from superior angle to base of spine.

Innervation: Ventral rami of C3 and C4 from deep branches of the cervical plexus, twig from the dorsal scapular nerve (C5). (C3, C4, C5).

Blood Supply: The deep branch of the transverse cervical artery.

Action: Elevates and retracts scapula with trapezius, laterally flexes cervical vertebral column if scapula is fixed.

Rhomboideus Minor

Origin: Lower part of ligamentum nuchae and spinous processes of C7 and T1.

Insertion: Medial border of scapula at base (root) of the spine.

Innervation:	Dorsal scapular nerve (C4, C5).
Blood Supply:	The deep branch of the transverse cervical artery.
Action:	Elevates the medial border and retracts the scapula (same as rhomboideus major), rotation of scapula.

Rhomboideus Major

Origin:	Spinous processes of T2 to T5.
Insertion:	Medial border of scapula between base of spine and inferior angle.
Innervation:	Dorsal scapular nerve (C4, C5).
Blood Supply:	The deep branch of the transverse cervical artery.
Action:	Elevates the medial border and retracts the scapula (same as rhomboideus minor), rotation of scapula.

The Intermediate Back Muscles

The intermediate back muscles are extrinsic and hypaxial, innervated by ventral rami. They are muscles of respiration.

Serratus Posterior Superior

Origin:	Lower end of ligamentum nuchae, and spinous processes of C7 to T2 or T3.
Insertion:	Second through fifth ribs lateral to the angles.
Innervation:	Ventral rami of first four thoracic nerves (T1, T2, T3, T4).
Blood Supply:	highest intercostal artery, posterior intercostal arteries.
Action:	Elevates upper ribs, muscle of inspiration.

Serratus Posterior Inferior

Origin:	Spinous processes of T10 or T11 to L2.
Insertion:	Lower three of four ribs lateral to their angles.
Innervation:	Ventral rami of lowest four thoracic nerves (T9, T10, T11, T12).
Blood Supply:	Posterior intercostal arteries.
Action:	Depresses lower ribs, draws them down and back. Muscle of inspiration.

The Intrinsic (Deep) Back Muscles

The three layers of the intrinsic back muscles are epaxial, innervated by the dorsal rami.

The Superficial Layer of Intrinsic Back Muscles

Splenius Capitis

Origin: Caudal half of ligamentum nuchae and spinous processes of C7 to T3 or T4.

Insertion: Mastoid process and occipital bone just inferior to lateral one-third of superior nuchal line.

Innervation: Dorsal rami of middle cervical spinal nerves.

Blood Supply: Muscular and descending branches of the occipital artery, superficial branch of the transverse cervical artery.

Actions:

Unilaterally: Rotation of head and neck.

Bilaterally: Extension of head and neck.

Splenius Cervicis

Origin: Narrow tendonous band from spinous processes of T3 to T6.

Insertion: Posterior tubercles of transverse processes of upper 2 or 3 cervical vertebrae.

Innervation: Dorsal rami lower cervical spinal nerves.

Blood Supply: Muscular and descending branches of the occipital artery, superficial branch of the Transverse cervical artery.

Actions:

Unilaterally: Draws head to one side and slightly rotates.

Bilaterally: Extends head.

The Intermediate Layer of Intrinsic Back Muscles

Erector Spinae: Iliocostalis (Lateral Column of Erector Spinae)

Iliocostalis Lumborum

Origin:	Thoracolumbar Fascia.
Insertion:	Lower borders of the lower six or seven ribs near the angles.
Innervation:	Dorsal rami of lumbar spinal nerves.
Blood Supply:	Posterior intercostal arteries, lumbar arteries.
Action:	Extension and lateral flexion of vertebral column.

Iliocostalis Thoracis

Origin:	Upper borders of lower six ribs medial to insertions of iliocostalis lumborum
Insertion:	Upper six ribs and sometimes the transverse process of C7.
Innervation:	Dorsal rami of thoracic spinal nerves.
Blood Supply:	Posterior intercostal arteries.
Action:	Extension of vertebral column.

Iliocostalis Cervicis

Origin:	Upper six ribs medial to the insertion of iliocostalis thoracis.
Insertion:	Transverse processes of the fourth, fifth, and sixth cervical vertebrae.
Innervation:	Dorsal rami of spinal nerves C4 to C6 (C4, C5, C6).
Blood Supply:	Posterior intercostal arteries.
Action:	Extension and lateral flexion of the vertebral column.

Longissimus (Middle Column of Erector Spinae)

Longissimus Thoracis

Origin:	Thoracolumbar fascia and accessory processes of the lumbar vertebrae.
Insertion:	Transverse processes of the lower nine thoracic vertebrae and the lower nine ribs.
Innervation:	Dorsal rami of thoracic spinal nerves (T1 to T12).
Blood Supply:	Posterior intercostal and lumbar arteries.
Action:	Extension and lateral flexion of the vertebral column.

Longissimus Cervicis

Origin:	Transverse processes of the upper four to six thoracic vertebrae medial to the insertion of the longissimus thoracis.
Insertion:	Transverse processes of second through sixth cervical vertebrae.
Innervation:	Dorsal rami of cervical spinal nerves (C4 to C8).
Blood Supply:	Posterior intercostal arteries.
Action:	Extension and lateral flexion of the vertebral column.

Longissimus Capitis

Origin:	Medial to upper border of longissimus cervicis partly from the tendons of origin of this muscle and from the articular processes of the lower four cervical vertebrae.
Insertion:	Posterior edge of the mastoid process.
Innervation:	Dorsal rami of cervical spinal nerves (C3 to C8).
Blood Supply:	Muscular branches of occipital artery, deep cervical branch of the costocervical trunk.
Action:	Bilaterally: Extension of vertebral column. Unilaterally: Rotation of face towards same side.

Longissimus Lumborum

Although not traditionally described, a number of workers, including Bogduk (1980) and Bustami (1986), now recognize a lumbar portion of the longissimus, inserting into the transverse processes of lumbar vertebrae.

Spinalis (Medial Column of Erector Spinae)

Spinalis Thoracis

Origin:	Blends with longissimus thoracis and arises from spinous processes of T11 to L2.
Insertion:	Spinous processes of upper 4 to 8 thoracic vertebrae.
Innervation:	Dorsal rami of thoracic spinal nerves (T1 to T12).
Blood Supply:	Posterior intercostal arteries.
Action:	Extension and lateral flexion of the vertebral column.

Spinalis Cervicis

Origin:	From the lower part of ligamentum nuchae and spinous processes of C7 to T2.
Insertion:	Spinous processes of C2 to C3 and sometimes C4.

Innervation:	Dorsal rami of cervical spinal nerves.
Blood Supply:	Posterior intercostal arteries.
Action:	Extension of vertebral column.

Spinalis Capitis

May originate from upper thoracic spinous processes. Usually blends laterally with semispinalis capitis. Considered with the semispinalis capitis. Highly variable in its morphology and presence (Greiner, et al., 2004).

The Deep Layer of Intrinsic Back Muscles

Transversospinalis Group: Semispinalis

Semispinalis Capitis

Origin:	Transverse processes of upper 6 or 7 thoracic vertebrae.
Insertion:	Occipital bone between superior and inferior nuchal lines.
Innervation:	Medial branches of dorsal rami of cervical spinal nerves (C1 to C6).
Blood Supply:	Descending branch of occipital artery, deep branch of costocervical trunk.
Actions:	
Bilaterally:	Extension of the neck
Unilaterally:	Turns face toward opposite side.

Note: Spinalis capitis is frequently fused with this muscle and is represented by the more medial fibers of the combined muscle mass. It may have a separate origin from upper thoracic spinous processes (Greiner et al., 2004).

Semispinalis Cervicis

Origin:	Transverse processes of upper 5 or 6 thoracic vertebrae.
Insertion:	Second through fifth cervical spinous processes.
Innervation:	Medial branches of cervical dorsal rami (C6 to C8).
Blood Supply:	Deep branch of costocervical trunk, posterior intercostal arteries.
Actions:	
Bilaterally:	Extension of vertebral column.
Unilaterally:	Rotation of vertebral column toward opposite side

Semispinalis Thoracis

Origin: From the transverse processes of the lower six thoracic vertebrae.

Insertion: Spinous processes of upper thoracic and lower cervical vertebrae.

Innervation: Dorsal rami of thoracic spinal nerves (T1 to T6).

Blood Supply: Posterior intercostal arteries.

Actions:

Bilaterally: Extension of the vertebral column.

Unilaterally: Rotation of the vertebral column toward the opposite side.

Transversospinalis Group: Multifidus (Lumborum, thoracis, and cervicis portions)

Origin: Posterior surface of sacrum, posterior sacroiliac ligament, mammillary processes of lumbar vertebrae, transverse processes of thoracic vertebrae, and articular processes of cervical vertebrae 4 through 7.

Insertion: Spines of the vertebrae 2 to 4 segments superior to the vertebra of origin.

Innervation: Dorsal rami of all spinal nerves.

Blood Supply: Posterior intercostal and lumbar arteries, deep cervical branch of the costocervical trunk.

Actions: Extension, lateral flexion, and rotation of vertebral column, extension and lateral motions of the pelvis.

Transversospinalis Group: Rotatores Longus and Rotatores Brevis

Origin: From the transverse processes of vertebrae.

Insertion: Lamina of the vertebra immediately superior (rotatores brevis), or into the lamina of the second vertebra above the vertebra of origin (rotatores longus).

Innervation: Dorsal rami of spinal nerves.

Blood Supply: Posterior intercostal arteries, lumbar arteries.

Action: Rotation of the vertebral column.

Segmental Muscles: Levatores Costarum

Origin: 12 pairs, from transverse processes of seventh cervical and first through eleventh thoracic vertebrae.

Insertion: On external surface of rib below vertebra of origin between the tubercle and angle.

Innervation:	Dorsal rami of lowest cervical and thoracic spinal nerves (C8, T1-T11).
Blood Supply:	Posterior intercostal arteries.
Action:	Elevate ribs, assist in inspiration, rotates and laterally flexes vertebral column.

Segmental Muscles: Interspinales

Origin:	Superiorly on spine of vertebra.
Insertion:	Inferiorly on spine of vertebra immediately superior to vertebra of origin; not always present in thoracic region, frequently double in cervical region.
Innervation:	Dorsal rami of spinal nerves.
Blood Supply:	Muscular branches of the posterior intercostal arteries, lumbar arteries deep cervical branches of the costocervical trunk.
Action:	Extension of vertebral column.

Segmental Muscles: Intertransversarii

Origin:	Superiorly on transverse processes of vertebrae.
Insertion:	Inferiorly on transverse processes of vertebrae immediately superior to vertebra of origin; most highly developed in cervical and lumbar regions.
Innervation:	Dorsal rami of spinal nerves, except lateral intertransversarii in lumbar and lower thoracic, and anterior and posterior intertransversarii in cervical; these exceptions are innervated by ventral rami.
Blood Supply:	Muscular branches of posterior intercostal and lumbar arteries.
Action:	Lateral flexion of vertebral column.

Muscles of the Suboccipital Region

The muscles of the suboccipital region are intrinsic and epaxial. They are innervated by the dorsal rami.

Rectus Capitis Posterior Minor

Origin:	Posterior tubercle of the atlas.
Insertion:	Occipital bone below the inferior nuchal line.
Innervation:	Suboccipital nerve (dorsal ramus of C1).
Blood Supply:	Muscular branches of the vertebral artery, descending branch of the occipital artery.
Action:	Extension and lateral flexion of the head.

Rectus Capitis Posterior Major

Origin: Spinous process of axis.
Insertion: Lateral part of inferior nuchal line of the occipital bone.
Innervation: Suboccipital nerve (dorsal ramus of C1).
Blood Supply: Muscular branches of the vertebral artery, descending branch of the occipital artery.
Action: Extension, lateral flexion, and rotation of the head.

Obliquus Capitis Superior

Origin: Transverse process of atlas.
Insertion: Occipital bone above inferior nuchal line.
Innervation: Suboccipital nerve (dorsal ramus of C1).
Blood Supply: Muscular branches of the vertebral artery, descending branch of the occipital artery.
Action: Extension and lateral rotation of the head.

Obliquus Capitis Inferior

Origin: Spinous process of axis.
Insertion: Transverse process of atlas.
Innervation: Suboccipital nerve (dorsal ramus of C1), second cervical dorsal ramus (C1,C2).
Blood Supply: Muscular branches of the vertebral artery, descending branch of the occipital artery.
Action: Rotation of atlas and skull around dens of axis.

Major Muscles of the Trunk:
The Neck and Prevertebral Region

Sternocleidomastoid

Origin: Sternal or medial head from the superior and anterior part of the manubrium. avicular or lateral head from the superior and anterior part of the medial 1/3 of the clavicle.
Insertion: Lateral surface of the mastoid process; lateral half of the superior nuchal line.
Innervation: Spinal Accessory Nerve (Cranial Nerve XI); branches of the ventral rami of cervical spinal nerves (C2, C3).
Blood Supply: Suprascapular artery; superior thyroid artery, auricular artery.
Action: Unilaterally: rotates the head to the contralateral side; laterally flexes head and neck. Bilaterally, flexes head and neck.

Scalenus Anterior

Origin: Anterior tubercles of transverse processes of cervical vertebrae 3, 4, 5, and 6.
Insertion: Scalene tubercle of the superior surface of the first rib.
Innervation: Ventral rami of cervical spinal nerves (C5 to C8).
Blood Supply: Ascending cervical branch of the inferior thyroid artery.
Action: If neck is fixed, it elevates the first rib, if the first rib is fixed, flexes and rotates the cervical vertebral column.

Scalenus Medius

Origin: Posterior tubercles of cervical vertebrae 2, 3, 4, 5, 6 and 7.
Insertion: Superior surface of first rib posterior to the groove for the subclavian artery.
Innervation: Ventral rami of cervical spinal nerves (C3 to C8).
Blood Supply: Ascending cervical branch of the inferior thyroid artery.
Action: If neck is fixed, it elevates the first rib, if the first rib is fixed, flexes and rotates the cervical vertebral column.

Scalenus Posterior

Origin: Posterior tubercles of cervical vertebrae 4, 5, and 6.
Insertion: External surface of the second rib, deep and posterior to serratus anterior.
Innervation: Ventral rami of cervical spinal nerves (C6, 7, 8).
Blood Supply: Ascending cervical branch of the inferior thyroid artery.
Action: If neck is fixed, it elevates the first rib, if the first rib is fixed, flexes and rotates the cervical vertebral column.

Longus Colli

Origin: Vertical (vertebral) portion from the bodies of last three cervical and first three thoracic vertebrae. Inferior oblique portion from the bodies of the first three thoracic vertebrae. Superior oblique portion from the anterior tubercles of the transverse processes of cervical vertebrae 5 and 6.
Insertion: Vertical portion into the bodies of cervical vertebrae 2, 3, and 4. Inferior oblique portion into anterior tubercles of the transverse processes of cervical vertebrae 5 and 6. Superior oblique portion onto the anterior tubercle of the atlas.
Innervation: Ventral rami of cervical spinal nerves (C2 to C8).
Blood Supply: Ascending cervical, ascending pharyngeal, and vertebral arteries.
Action: Weak flexor of the neck. Slightly rotates and laterally flexes the cervical vertebral column.

Longus Capitis

Origin:	Anterior tubercles of the transverse processes of cervical vertebrae 3, 4, 5, and 6.
Insertion:	Inferior surface of the basilar part of the occipital bone lateral to the pharyngeal tubercle.
Innervation:	Ventral rami of cervical spinal nerves (C1, 2, 3, 4).
Blood Supply:	Ascending cervical, ascending pharyngeal, and vertebral arteries.
Action:	Flexes head and upper cervical vertebral column. Assists in rotating cervical vertebral column and head.

Rectus Capitis Anterior

Origin:	Anterior surface of the lateral part of the lateral mass of the atlas and root of the transverse process.
Insertion:	Inferior surface of the basilar part of occipital immediately anterior to the foramen magnum.
Innervation:	Ventral rami of first and second cervical spinal nerves (C1, C2).
Blood Supply:	Vertebral and ascending pharyngeal arteries.
Action:	Flexes the head and helps stabilize the atlantooccipital joint. May assist in rotation of the head.

Rectus Capitis Lateralis

Origin:	Superior surface of transverse process of atlas.
Insertion:	Inferior surface of jugular process of occipital.
Innervation:	Ventral rami of first and second cervical spinal nerves (C1, C2).
Blood Supply:	Vertebral, occipital, and ascending pharyngeal arteries.
Action:	Laterally flexes the head and helps stabilize the atlantooccipital joint.

Major Muscles of the Trunk:

The Thoracic Wall

Intercostal Muscles

Intercostales Externi: 11 pairs.

Origin:	From inferior and external surface of first 11 ribs from rib tubercles to costochondral junctions
Insertion:	Into superior and external surface of ribs 2 to 12 from rib tubercles to costochondral junctions.
Innervation:	Segmentally by ventral rami of first 11 thoracic spinal nerves (intercostal nerves) (T1 to T11).
Blood Supply:	Posterior intercostal, anterior intercostal, musculophrenic arteries and the costocervical trunk.
Action:	Elevate ribs in inspiration, support intercostal spaces in inspiration and expiration. May be active in early expiration.

Intercostales Interni: 11 pairs.

Origin:	From inferior and internal surface of first 11 ribs from the sternal border to the angle of the rib. They arise from the ridge on the medial side of the costal groove.
Insertion:	Superior border of ribs 2 to 12.
Innervation:	Segmentally by ventral rami of first 11 thoracic spinal nerves (intercostal nerves) (T1 to T11).
Blood Supply:	Posterior intercostal, anterior intercostal, musculophrenic arteries and the costocervical trunk.
Action:	Prevent the pushing out or drawing in of the intercostal spaces in inspiration and expiration. Depress the ribs in forced expiration. The portion attaching the costal cartilages of the upper four or five ribs probably help in elevation of the ribs with the external intercostals in forced inspiration.

Intercostales Intimi (Innermost Intercostals): 11 pairs.

Origin:	Similar to the internal intercostal muscles, but separated from them by the neurovascular plane containing the intercostal nerve and vessels. Originate deep to the internal intercostals from ribs 1 to 11, usually extending from the angles of the ribs posteriorly to about the costochondral junction anteriorly. May be poorly developed or absent.

Insertion:	Insert on the superior margin of ribs 2 to 12 deep to the insertion of the internal intercostal muscles.
Innervation:	Segmentally by ventral rami of first 11 thoracic spinal nerves (intercostal nerves) (T1 to T11).
Blood Supply:	Posterior intercostal, anterior intercostal, musculophrenic arteries and the costocervical trunk.
Action:	Probably act with the internal intercostal muscles. Conclusive data are lacking.

Subcostalis

Origin:	Internal surface of (usually only) lower ribs near the angles.
Insertion:	Internal surface of second or third rib below rib of origin. Variable in presence and distribution.
Innervation:	Ventral rami of thoracic spinal nerves (intercostal nerves) (T1 to T12).
Blood Supply:	Musculophrenic and posterior intercostal arteries.
Action:	Probably depress the ribs and assist the internal and innermost intercostals in expiration.

Transversus Thoracis

Origin:	Lower third of the posterior surface of the sternum and costal cartilages of lower three or four vertebrosternal ribs.
Insertion:	Inferior and internal surfaces of the costal cartilages of ribs 2, 3, 4, 5, and 6. Can vary a great deal.
Innervation:	Ventral rami of the first six thoracic spinal nerves (intercostal nerves) (T1 to T6).
Blood Supply:	Internal thoracic and anterior intercostal arteries.
Action:	Depresses the costal cartilages of the ribs to which it is attached. Probably acts in expiration.

Major Muscles of the Trunk:

The Diaphragm

Diaphragm

Origin:	Sites of attachment of the diaphragm:
Sternal Part:	attaches by 2 slips to posterior part of xiphoid process.
Costal Part:	attaches to internal surfaces of lower 6 ribs & costal cartilages. Interdigitates with transversus abdominus muscle.

Lumbar Part:	attaches to medial and lateral arcuate ligaments by two crura (singular: crus).
Right crus:	broader and longer than left. Attaches to anterolateral aspect of bodies and intervertebral disks of vertebrae L1 - L3.
Left crus:	attaches to corresponding parts of L1 and L2, shorter than right.
Median Arcuate ligament:	medial tendonous arches of both crura converge over aorta at level of thoracolumbar disk (IVD T12/L1).
Lateral Arcuate Ligament:	thickening of fascia where diaphragm overlies the quadratus lumborum muscle. Fibers run from the ligament to the central tendon of the diaphragm.
Medial Arcuate Ligament:	thickening of fascia where the diaphragm overlies the psoas major muscle. Fibers run from the ligament to the central tendon of the diaphragm.
Insertion:	Central Tendon of Diaphragm, an aponeurosis of interwoven collagenous fibers. It is located nearer the front of the thorax than the rear, so muscle fibers longer posteriorly.
Innervation:	Principal motor innervation is from the phrenic nerve, with some peripheral innervation from intercostal nerves (C3, C4, C5).
Blood Supply:	Superior and inferior phrenic arteries, musculophrenic artery, and pericardiacophrenic artery.
Action:	The essential muscle of respiration. Contraction of the diaphragm results in a lengthening of the thoracic cavity, thereby increasing the volume of the thoracic cavity, increasing negative pressure and thereby drawing air into the lungs. Contraction of the diaphragm is responsible for about two thirds of the volume of inspired air.

Major Muscles of the Trunk:

The Abdomen

Rectus Abdominus

Origin:	By two tendons, one from the pubic symphysis and one from the pubic crest.
Insertion:	By three fascicles into the cartilages of ribs 5, 6, and 7. Some fibers may attach as well to the xiphoid process and xiphicostal ligament.
Innervation:	Ventral rami of seventh to twelfth thoracic ventral rami (intercostal an subcostal nerves) (T7-T12).
Blood Supply:	Superior and inferior epigastric arteries.

Action: Flexes the vertebral column, particularly the lumbar portion. Tenses anterior abdominal wall and aids in compressing the abdominal contents. With the thorax fixed, it flexes the pelvis.

Pyramidalis

Origin: Variably present. Placed anterior to the rectus abdominus in the rectus sheath. Arises from the ventral surface of the pubis.

Insertion: Into linea alba, about midway between the pubis and the umbilicus.

Innervation: Subcostal nerve (ventral ramus of twelfth thoracic spinal nerve) (T12).

Blood Supply: Inferior epigastric artery.

Action: Tenses the linea alba, supports abdominal viscera, and is active in forced expiration.

Obliquus Externus Abdominus

Origin: Eight fleshy strips that arise from the external and inferior borders of the lower eight ribs, interdigitating with serratus anterior and latissimus dorsi.

Insertion: Ventral half of the external aspect of the iliac crest, remainder via a wide aponeurosis that helps constitute the rectus abdominus sheath, and via the sheath into the linea alba. The lower border of its aponeurosis forms the inguinal ligament.

Innervation: Ventral rami of lower six thoracic and first two lumbar spinal nerves (T7 to T12, L1, L2).

Blood Supply: Superior and inferior epigastric arteries.

Action: Compresses abdomen, supports abdominal viscera, active in forced expiration. Via actions on the ribs rotates the vertebral column to the contralateral side.

Bilateral contraction helps flex lumbar vertebral column.

Obliquus Internus Abdominus

Origin: Thoracolumbar fascia, anterior two thirds of the iliac crest deep to obliquus externus abdominus, lateral two thirds of the inguinal ligament.

Insertion: Costal cartilages of the last three ribs, and then via its aponeurosis into the rectus sheath and thus into the linea alba.

Innervation: Ventral rami of lower six thoracic and first two lumbar spinal nerves (T7 to T12, L1, L2).

Blood Supply: Superior and inferior epigastric arteries, deep circumflex iliac artery.

Action: Compresses abdomen, supports abdominal viscera, active in forced expiration. Via actions on the ribs rotates the vertebral column to the ipsilateral side.

Bilateral contraction helps flex lumbar vertebral column.

Transversus Abdominus

Origin:	Deep surfaces of the lower six costal cartilages, middle layer of the thoracolumbar fascia, anterior two-thirds of the iliac crest, deep to the obliquus internus abdominus.
Insertion:	Via the rectus sheath into the linea alba with the oblique abdominal muscles.
Innervation:	Ventral rami of the lower 6 thoracic ventral rami, iliohypogastric nerve and ilioinguinal nerve (T7 to T12, L1, L2)..
Blood Supply:	Deep circumflex iliac artery; inferior epigastric artery.
Action:	Compresses abdomen, supports abdominal viscera, active in forced expiration.

Quadratus Lumborum

Origin:	Iliolumbar ligament and posterior part of the iliac crest.
Insertion:	Inferior border of the twelfth rib, transverse processes of lumbar vertebrae 1 to 4.
Innervation:	Subcostal nerve (ventral ramus of 12th thoracic spinal nerve); ventral rami of lumbar spinal nerves 1, 2, and 3. (T12, L1, L2, L3).
Blood Supply:	Lumbar branch of the iliolumbar artery.
Action:	Depresses and fixes the twelfth rib. Acts as a muscle of inspiration by fixing the twelfth rib and thereby stabilizing the inferior attachments of the diaphragm. Laterally flexes the vertebral column when contracting unilaterally. With the pelvis fixed and contracting bilaterally, they probably help extend the lumbar vertebral column.

Upper Limb Musculature:

The Shoulder

Trapezius, Latissimus Dorsi, Levator Scapulae, Rhomboideus Minor and Rhomboideus Major

These muscles are discussed with the superficial back muscles.

Serratus Anterior

Origin:	Fleshy digitations from the external surfaces of the upper eight or nine ribs
Insertion:	Costal surface of the vertebral border of the scapula.
Innervation:	Long thoracic nerve (C5, C6, C7).
Blood Supply:	Lateral thoracic Artery.
Action:	Upward rotation and protraction of the scapula, raises ribs if scapula is fixed.

Deltoideus

- *Origin:* Lateral third of the clavicle, acromion process of the scapula, and spine of the scapula.
- *Insertion:* Deltoid tuberosity on the lateral aspect of the humerus.
- *Innervation:* Axillary nerve from the brachial plexus (C5, C6).
- *Blood Supply:* Posterior humeral circumflex artery.
- *Action:* Abduction of the arm. Posterior fibers assist with extension of the arm, anterior fibers with flexion of the arm.

Supraspinatus

- *Origin:* Supraspinatus fossa of the scapula.
- *Insertion:* Superior facet of the greater tubercle of the humerus.
- *Innervation:* Suprascapular nerve, from brachial plexus (C4, C5, C6).
- *Blood Supply:* Suprascapular artery.
- *Action:* Abduction and lateral rotation of the arm, helps keep the humeral head in contact with the glenoid fossa of the scapula. One of the "rotator cuff" muscles.

Infraspinatus

- *Origin:* Infraspinatus fossa of the scapula.
- *Insertion:* Middle facet of the greater tubercle of the humerus.
- *Innervation:* Suprascapular nerve of the brachial plexus (C5, C6).
- *Blood Supply:* Suprascapular artery.
- *Action:* Lateral rotation of the arm. One of the "rotator cuff" muscles.

Teres Minor

- *Origin:* Dorsal surface of the axillary border of the scapula.
- *Insertion:* Inferior facet of the greater tubercle of the humerus.
- *Innervation:* Axillary nerve of the brachial plexus (C5, C6).
- *Blood Supply:* Suprascapular artery.
- *Action:* Adducts and laterally rotates the arm. One of the "rotator cuff" muscles.

Teres Major

Origin:	Dorsal surface of the inferior angle of the scapula.
Insertion:	Medial lip of the bicipital groove of the humerus.
Innervation:	Lower subscapular nerve of the brachial plexus (C6, C7).
Blood Supply:	Scapular circumflex artery.
Action:	Adduction and medial rotation of the humerus.

Subscapularis

Origin:	Subscapular fossa of the ventral surface of the scapula.
Insertion:	Lesser tubercle of the humerus.
Innervation:	Upper and lower subscapular nerves of the brachial plexus (C5, C6, C7).
Blood Supply:	Scapular circumflex artery.
Action:	Medial rotation of the humerus. One of the " rotator cuff" muscles.

Upper Limb Musculature:
The Pectoral Region

Pectoralis Major

Origin:	Sternal 1/2 of clavicle (clavicular head) and lateral border of sternum to 7th rib (sternal head).
Insertion:	Lateral lip of the bicipital groove of the humerus.
Action:	Medial rotation, adduction, and flexion of the arm at the glenohumeral joint.
Innervation:	Medial and Lateral pectoral nerves (C6, C7, C8).
Blood supply:	Pectoral branches of the thoracoacromial trunk of the axillary artery.

Pectoralis Minor

Origin:	Outer surfaces of the anterior aspects of the 3rd, 4th and 5th ribs (deep to the pectoralis major).
Insertion:	Tip of the coracoid process of the scapula.
Action:	Protracts the scapula; aids the levator scapula and rhomboids in rotating the scapula downwards (depression of the glenoid).

Innervation:	Medial and lateral pectoral nerves (note: only the medial pectoral nerve pierces the muscle, the lateral sends fibers that join with the medial, but the lateral nerve itself passes superior to the pectoralis minor) (C8, T1).
Blood supply:	Pectoral branches of the thoracoacromial trunk of the axillary artery.

Subclavius

Origin:	From the junction of the first rib and first costal cartilage anterior to the costoclavicular ligament..
Insertion:	Subclavian groove on the inferior surface of the middle third of the clavicle.
Action:	Pulls the clavicle, and thus the shoulder, downward and forward.
Innervation:	Subclavian branch of the brachial plexus (C5, C6).
Blood supply:	Subclavian branch of the thoracoacromial trunk of the axillary artery.

Upper Limb Musculature: The Arm

Anterior Compartment

Coracobrachialis

Origin:	Tip of the coracoid process of the scapula.
Insertion:	Medial aspect of the humerus at midshaft.
Innervation:	Musculocutaneous nerve of the brachial plexus (C5, C6, C7).
Blood Supply:	Brachial artery.
Action:	Flexes and adducts the arm.

Brachialis

Origin:	Inferior one half to two thirds of the anterior aspect of the humerus.
Insertion:	Ulnar tuberosity and coronoid process.
Innervation:	Musculocutaneous nerve of the brachial plexus (C5, C6).
Blood Supply:	Brachial and radial recurrent arteries.
Action:	Flexion of the forearm.

Biceps Brachii

Origin: By two heads:

Long head: from supraglenoid tubercle of the scapula.

Short head: coracoid process of the scapula.

Insertion: By a common tendon into the radial tuberosity and by the bicipital aponeurosis to the common origin of the forearm flexor muscles.

Innervation: Musculocutaneous nerve of the brachial plexus (C5, C6).

Blood Supply: Brachial Artery.

Action: Supination of the forearm, flexes the forearm, flexes and adducts the arm.

Posterior Compartment

Triceps Brachii

Origin: By three heads:

Long head: from the infraglenoid tubercle of the scapula.

Lateral head: lateral aspect of the humeral shaft.

Medial head: posterior aspect of the humeral shaft, deep to the other two heads.

Insertion: Olecranon process of the ulna.

Innervation: Radial nerve from the brachial plexus (C6, C7, C8).

Blood Supply: Profunda brachii artery.

Action: Extension of the forearm; long head aids in adducting the abducted arm. Long head extends the humerus at the glenohumeral joint.

Anconeus

Origin: Dorsal aspect of lateral epicondyle of humerus, appears as an inferior extension of the triceps brachii.

Insertion: Lateral aspect of the olecranon process and superior one quarter of the dorsal aspect of the shaft of the ulna.

Innervation: Radial nerve from the brachial plexus (C7, C8, T1).

Blood Supply: Profunda brachii.

Action: Assists in extension of the forearm.

Upper Limb Musculature:

The Forearm

Anterior Compartment: Superficial Muscles

Pronator Teres

Origin:	Humeral head: Medial epicondyle of the humerus and common flexor tendon, Ulnar head: coronoid process of ulna.
Insertion:	Middle of the lateral border of the radius.
Innervation:	Median nerve from brachial plexus (C6, C7).
Blood Supply:	Anterior ulnar recurrent artery.
Action:	Pronates the forearm.

Flexor Carpi Radialis

Origin:	Medial epicondyle of humerus via common flexor tendon and the antebrachial fascia.
Insertion:	Bases of the second and third metacarpal bones.
Innervation:	Median nerve from the brachial plexus (C6, C7).
Blood Supply:	Radial Artery.
Action:	Flexes and abducts the hand at the wrist. Assists in flexing the forearm.

Palmaris Longus (Absent in 3 to 10 percent of the population.)

Origin:	Medial epicondyle of humerus via common flexor tendon.
Insertion:	Palmar aponeurosis.
Innervation:	Median nerve from the brachial plexus (C7, C8).
Blood Supply:	Posterior ulnar recurrent artery.
Action:	Flexes the wrist, cups the palm via the palmar aponeurosis.

Flexor Carpi Ulnaris

Origin:	Humeral head: Medial epicondyle of humerus via common flexor tendon.
Ulnar head:	Olecranon process and posterior border of the ulna.
Insertion:	Pisiform (which acts as a sesamoid bone in its tendon) and then continues to the base of the fifth metacarpal and to the hook of the hamate.

Innervation:	Ulnar nerve from the brachial plexus (C7, C8).
Blood Supply:	Posterior ulnar recurrent artery.
Action:	Flexes wrist, adducts hand, and assists in flexing the forearm.

Flexor Digitorum Superficialis

Origin:	Humeral head: Medial epicondyle of humerus via common flexor tendon.
Ulnar head:	From coronoid process of ulna. Radial head: from the oblique line of the radius.
Insertion:	Bases of the middle phalanges of digits two to five of the hand.
Innervation:	Median nerve from the brachial plexus (C7, C8, T1).
Blood Supply:	Ulnar and radial arteries.
Action:	Flexes the middle phalanges on the proximal phalanges, then flexes proximal phalanges on the metacarpal heads. Then aids in flexing wrist and forearm.

Anterior Compartment: Deep Muscles

Flexor Digitorum Profundus

Origin:	Proximal three quarters of the anterior surface of the ulna and the interosseousmembrane.
Insertion:	Bases of the distal phalanges of digits two to five of the hand.
Innervation:	Ulnar nerve (medial half of muscle) and median nerve (lateral half of muscle), both from the brachial plexus (C8, T1).
Blood Supply:	Ulnar and anterior interosseous arteries.
Action:	Flexes the distal phalanges on the middle phalanges after the flexor digitorum superficialis has flexed the middle phalanges. Further contraction aids in flexion of the wrist.

Flexor Pollicis Longus

Origin:	Anterior surface of the distal one quarter of the radius and the interosseous membrane.
Insertion:	Base of the distal phalanx of the thumb.
Innervation:	Anterior interosseous branch of the median nerve (from brachial plexus) (C8, T1).
Blood Supply:	Anterior interosseous artery.
Action:	Flexes the thumb.

Pronator Quadratus

Origin: Anterior surface of the distal one quarter of the ulna deep to the origin of the flexor pollicis longus.

Insertion: Distal one quarter of the anterior surface of the radius.

Innervation: Anterior interosseous branch of the median nerve (from brachial plexus) (C8, T1).

Blood Supply: Anterior interosseous artery.

Action: Pronates the forearm, helps bind the radius to the ulna.

Posterior Compartment: Superficial Muscles

Brachioradialis

Origin: Proximal two thirds of the lateral supracondylar ridge of the humerus.

Insertion: Lateral surface of the distal end of the radius.

Innervation: Radial nerve from the brachial plexus (C5, C6, C7).

Blood Supply: Radial recurrent artery.

Action: Flexes forearm after flexion is begun by brachialis and biceps brachii. May also act as a semisupinator and semipronator.

Extensor Carpi Radialis Longus

Origin: Deep to brachioradialis from the lateral supracondylar ridge of the humerus.

Insertion: Lateral side of the distal end of the radius.

Innervation: Radial nerve (C6, C7).

Blood Supply: Radial and radial recurrent arteries.

Action: Extends the wrist and abducts the hand.

Extensor Carpi Radialis Brevis

Origin: Lateral epicondyle of the humerus via the common extensor tendon.

Insertion: Base of the third metacarpal.

Innervation: Posterior interosseous branch of the radial nerve (C7, C8).

Blood Supply: Radial and radial recurrent arteries.

Action: Extends the wrist and abducts the hand.

Extensor Digitorum

Origin: Common extensor tendon and the intermuscular septum.

Insertion: Four tendons insert into the extensor expansions deep to the extensor expansion.

Innervation: Posterior interosseous branch of the radial nerve (C7, C8).

Blood Supply: Posterior interosseous artery from ulnar artery.

Action: Extends the proximal interphalangeal, distal interphalangeal and metacarpophalangeal joints of the medial four digits. Assists in extension of the wrist.

Extensor Digiti Minimi

Origin: From the common extensor tendon from the lateral epicondyle of the humerus.

Insertion: Extensor expansion of the fifth digit, often fused with the extensor digitorum.

Innervation: Posterior interosseous branch of the radial nerve (C7, C8).

Blood Supply: Posterior interosseous artery from ulnar artery.

Action: Extends the proximal interphalangeal, distal interphalangeal and metacarpophalangeal joints of the fifth digit.

Extensor Carpi Ulnaris

Origin: From the common extensor tendon from the lateral epicondyle of the humerus.

Insertion: Base of the fifth metacarpal.

Innervation: Posterior interosseous branch of the radial nerve (C7, C8).

Blood Supply: Posterior interosseous artery from ulnar artery.

Action: Extends and adducts the hand at the wrist.

Posterior Compartment: Deep Muscles

Supinator

Origin: From the common extensor tendon from the lateral epicondyle of the humerus; radial collateral and annular ligaments of the elbow, supinator crest and fossa of ulna.

Insertion: Lateral, posterior, and anterior surfaces of the proximal 1/3 of the radius.

Innervation: Deep branch of the radial nerve (C5, C6).

Blood Supply: Radial recurrent and posterior interosseous arteries.

Action: Supinates the forearm.

Abductor Pollicis Longus

Origin: Deep to supinator, proximal posterior surface of radius and ulna and the interosseous membrane.

Insertion: Base of the first metacarpal.

Innervation: Posterior interosseous branch of the radial nerve (C7, C8).

Blood Supply: Posterior interosseous artery from ulnar artery.

Action: Abducts and extends the thumb at the carpometacarpal joint.

Extensor Pollicis Brevis

Origin: Posterior surface of the radius and the interosseous membrane.

Insertion: Base of the proximal phalanx of the thumb.

Innervation: Posterior interosseous branch of the radial nerve (C7, C8).

Blood Supply: Posterior interosseous artery from ulnar artery.

Action: Extends the proximal phalanx of the thumb at the carpometacarpal joint.

Extensor Pollicis Longus

Origin: Posterior surface of the middle 1/3 of the ulna; interosseous membrane.

Insertion: Base of the distal phalanx of the thumb.

Innervation: Posterior interosseous branch of the radial nerve (C7, C8).

Blood Supply: Posterior interosseous artery from ulnar artery.

Action: Extends the thumb at the interphalangeal joint.

Extensor Indicis

Origin: Posterior surface of the proximal 1/3 of the ulna; interosseous membrane.

Insertion: Extensor expansion of digit two (index finger).

Innervation: Posterior interosseous branch of the radial nerve (C7, C8).

Blood Supply: Posterior interosseous artery from ulnar artery.

Action: Extends the second digit, assists in extending the hand at the wrist.

Upper Limb Musculature:

The Intrinsic Muscles of the Hand

Lumbricals: four in number

Origin:	From the tendons of the flexor digitorum profundus.
Insertion:	Lateral sides of the extensor expansions of digits II, III, IV, and V.
Innervation:	Lateral two: median nerve. Medial two: ulnar nerve (C8, T1).
Blood Supply:	Superficial and deep palmar arches (from radial and ulnar arteries).
Action:	Flex the digits at the metacarpophalangeal joints and extend them at the proximal and distal interphalangeal joints.

Thenar Muscles

Abductor Pollicis Brevis

Origin:	Flexor retinaculum, scaphoid and trapezium bones.
Insertion:	Lateral side of the base of the proximal phalanx of the thumb.
Innervation:	Recurrent branch of the median nerve (C8, T1).
Blood Supply:	Superficial palmar branch of the radial artery.
Action:	Abducts thumb and aids in opposition of thumb to digit V.

Flexor Pollicis Brevis

Origin:	Flexor retinaculum, tubercle of trapezium (deep to abductor pollicis brevis).
Insertion:	Lateral side of base of proximal phalanx of thumb, fuses with abductor tendon, and contains a sesamoid bone in its tendon.
Innervation:	Recurrent branch of the median nerve (C8, T1).
Blood Supply:	Superficial palmar branch of the radial artery.
Action:	Flexes thumb.

Opponens Pollicis

Origin:	Flexor retinaculum, tubercle of trapezium (deep to abductor pollicis brevis and lateral to flexor pollicis brevis).
Insertion:	Anterior and lateral side of first metacarpal.

Innervation: Recurrent branch of the median nerve (C8, T1).

Blood Supply: Superficial palmar branch of the radial artery.

Action: Opposes the thumb (draws first metacarpal anteriorly and medially) to each of the other digits.

Adductor Pollicis

Origin: Oblique head: bases of second and third metacarpals; trapezium, trapezoid and capitate bones. Transverse head: anterior surface of the shaft of the third metacarpal.

Insertion: Medial side of the base of the proximal phalanx of the thumb.

Innervation: Deep palmar branch of the ulnar nerve (C8, T1).

Blood Supply: Deep palmar branch of the ulnar artery.

Action: Adducts thumb and aids in opposition.

Hypothenar Muscles

Abductor Digiti Minimi

Origin: Pisiform bone and tendon of the flexor carpi ulnaris.

Insertion: Medial side of the base of the proximal phalanx of digit V (little finger).

Innervation: Deep branch of ulnar nerve (C8, T1).

Blood Supply: Deep palmar branch of ulnar artery, dorsal carpal branch of ulnar artery.

Action: Abducts digit V.

Flexor Digiti Minimi Brevis

Origin: Hook of hamate and flexor retinaculum.

Insertion: Fuses with tendon of abductor digiti minimi and then to base of the proximal phalanx of digit V.

Innervation: Deep branch of ulnar nerve (C8, T1).

Blood Supply: Deep palmar branch of ulnar artery, dorsal carpal branch of ulnar artery

Action: Flexes proximal phalanx of digit V.

Opponens Digiti Minimi

Origin: From the hook of the hamate and flexor retinaculum deep to short abductor and flexor muscles of digit V.

Insertion:	Medial border of fifth metacarpal.
Innervation:	Deep branch of ulnar nerve (C8, T1).
Blood Supply:	Deep palmar branch of ulnar artery, dorsal carpal branch of ulnar artery
Action:	Opposes digit V. Draws metacarpal V anteriorly and rotates it to bring palmar surface of digit V into opposition with the palmar surface of the thumb.

Palmaris Brevis

Origin:	Ulnar side of flexor retinaculum and palmar aponeurosis.
Insertion:	Skin on the medial side of the palm.
Innervation:	Superficial branch of the ulnar nerve (C8, T1).
Blood Supply:	Superficial palmar arch.
Action:	wrinkles skin on the medial side of the palm, increasing the curvature of the hollow of the palm, aids in gripping.

Interosseous Muscles

Palmar Interossei: Three in number.

Origin:	1st: ulnar side of second metacarpal; 2nd: radial side of fourth metacarpal; 3rd: radial side of fifth metacarpal.
Insertion:	1st: ulnar side of proximal phalanx of digit II; 2nd: radial side of proximal phalanx of digit IV; 3rd: radial side of proximal phalanx of digit V.
Innervation:	Deep palmar branch of ulnar nerve (C8, T1).
Blood Supply:	Deep palmar branch of ulnar artery.
Action:	Adduct digits II, IV, and V toward the midline of digit III.

Dorsal Interossei: Four in number.

Origin:	The four dorsal interossei originate by two heads, one each from the adjacent sides of the metacarpal bones.
Insertion:	1st: radial side of the proximal phalanx of digit II. 2nd: radial side of the proximal phalanx of digit III. 3rd: ulnar side of the proximal phalanx of digit III. 4th: ulnar side of the proximal phalanx of digit IV.
Innervation:	Deep palmar branch of ulnar nerve (C8, T1).
Blood Supply:	Deep palmar arch.
Action:	Abduct the digits away from the midline of digit III.

Lower Limb Musculature:
The Hip Joint Flexors

Iliopsoas

The iliopsoas is the term applied to the iliacus and psoas major muscles collectively, as they insert into the lesser trochanter of the femur by a common tendon.

Iliacus

Origin:	Superior two thirds of the iliac fossa; iliac crest; anterior sacroiliac, lumbosacral, and iliolumbar ligaments; ala of sacrum.
Insertion:	With the tendon of the psoas major into the lesser trochanter of the femur, capsule of the hip joint, and the shaft of the femur.
Innervation:	Femoral Nerve (L2, L3).
Blood Supply:	Iliac branch of the iliolumbar artery and the superior gluteal artery.
Action:	Flexes the thigh, flexes the pelvis when the thigh is fixed.

Psoas Major

Origin:	Anterior and inferior surfaces of the transverse processes, bodies, and intervertebral disks of all five lumbar vertebrae
Insertion:	After passing beneath the inguinal ligament and anterior to the capsule of the hip joint, inserts with the iliacus (as the iliopsoas muscle) into the lesser trochanter of the femur.
Innervation:	Branches of the lumbar plexus including contributions from the ventral rami of lumbar spinal nerves 2, 3, and 4 (L2, 3, 4).
Blood Supply:	Lumbar branch of the iliolumbar.
Action:	Flexes the thigh at the hip joint. If both thighs are fixed and both psoas major muscles contract together, they are important flexors of the hip as in sitting up from a supine position. While mechanically it appears to be a medial rotator of the thigh, electromyographic studies show it to be active during lateral rotation.

Psoas Minor: *The psoas minor is often absent, and is absent bilaterally in half of subjects.*

Origin:	Sides of the bodies of the twelfth thoracic and first lumbar vertebrae and the intervertebral disk between them.
Insertion:	Pectineal line and iliopectineal eminence of the innominate bone.

Innervation: Ventral ramus of the first (and sometimes second) lumbar spinal nerve (L1, L2).

Blood Supply: Lumbar branch of the iliolumbar artery.

Action: Flexes pelvis on the vertebral column, assists the iliopsoas in flexing the vertebral column on the pelvis.

Lower Limb Musculature:

The Gluteal Region

Tensor Fasciae Latae

Origin: Anterior superior iliac spine and the anterior part of the external lip of the iliac crest.

Insertion: Into the iliotibial tract, and thus into the lateral aspect of the proximal tibia.

Innervation: Superior gluteal nerve (L4, L5).

Blood Supply: Lateral femoral circumflex artery, superior gluteal artery.

Action: Abducts, medially rotates and flexes the thigh. Helps keep the knee extended and steadies the trunk on the thigh.

Gluteus Maximus

Origin: External surface of the iliac blades posterior to the posterior gluteal lines, iliac crest; dorsal surfaces of sacrum and coccyx; sacrotuberous ligament.

Insertion: Iliotibial tract, gluteal tuberosity of femur.

Innervation: Inferior gluteal nerve (L5, S1, S2).

Blood Supply: Superior and inferior gluteal arteries; first perforating branch of the profunda femoris artery.

Action: Extends thigh; laterally rotates thigh; steadies the thigh on the trunk; extends trunk; assists in raising trunk from a seated position. Active in climbing and running (forceful extension); generally inactive in quiet walking.

Gluteus Medius

Origin: Deep to gluteus maximus; from external surface of iliac blade between anterior and posterior iliac lines.

Insertion: Lateral surface of the greater trochanter of the femur.

Innervation: Superior gluteal nerve (L5, S1).

Blood Supply: Deep branch of the superior gluteal artery.

Action: Abducts thigh; medially rotates thigh when hip is extended; steadies hip joint; supports trunk in single stance and swing phases by contracting on the side in which the foot is in contact with the ground.

Gluteus Minimus

Origin: External surface of ilium between anterior and inferior iliac lines.

Insertion: Anterior surface of the greater trochanter of the femur.

Innervation: Superior gluteal nerve (L5, S1).

Blood Supply: Deep branch of the superior gluteal artery.

Action: Abducts thigh; medially rotates thigh when hip is extended; steadies hip joint; supports trunk in single stance and swing phases by contracting on the side in which the foot is in contact with the ground.

Lower Limb Musculature:
Short Rotators of the Hip

Piriformis

Origin: Anterior surface of sacrum and the sacrotuberous ligament.

Insertion: Superior border of the greater trochanter of femur.

Innervation: Ventral rami of spinal nerves S1 and S2 (S1, S2).

Blood Supply: Superior gluteal, inferior gluteal, and internal pudendal arteries.

Action: Lateral rotation of the extended thigh, abduct the flexed thigh, steadies the femur head in the acetabulum.

Obturator Internus

Origin: Pelvic surface of obturator membrane and the surrounding margins of the obturator foramen.

Insertion: Medial surface of the greater trochanter of the femur.

Innervation: Nerve to obturator internus (L5, S1).

Blood Supply: Superior gluteal and internal pudendal arteries.

Action: Lateral rotation of the extended thigh, abduct the flexed thigh, steadies the femur head in the acetabulum.

Gemellus Superior

Origin:	Ischial spine.
Insertion:	Blends with the tendon of the obturator internus, and thence on to the medial surface of the greater trochanter.
Innervation:	Nerve to obturator internus (L5, S1).
Blood Supply:	Inferior gluteal artery.
Action:	Lateral rotation of the extended thigh, steadies the femur head in the acetabulum.

Gemellus Inferior

Origin:	Ischial tuberosity.
Insertion:	Blends with the tendon of the obturator internus, and thence on to the medial surface of the greater trochanter.
Innervation:	Nerve to quadratus femoris (L5, S1).
Blood Supply:	Inferior gluteal artery.
Action:	Lateral rotation of the extended thigh, steadies the femur head in the acetabulum.

Quadratus Femoris

Origin:	Lateral border of the ischial tuberosity.
Insertion:	Quadrate tubercle of intertrochanteric crest and inferior part of crest.
Innervation:	Nerve to quadratus femoris (L5, S1).
Blood Supply:	Medial femoral circumflex artery.
Action:	Adduction and lateral rotation of the thigh, steadies the femur head in the acetabulum.

Lower Limb Musculature:

The Thigh

Anterior Compartment

Quadriceps Femoris: Composed of four parts that insert into the patella and then on to the tibial tuberosity via the patellar ligament.

Rectus Femoris

Origin: Direct head: anterior superior iliac spine. Reflected head: groove superior to the acetabulum.

Insertion: Into the superior aspect of the patella and then on to the tibial tuberosity via the patellar ligament.

Innervation: Femoral nerve (L2, L3, L4).

Blood Supply: Lateral femoral circumflex artery.

Action: Extends the leg at the knee; steadies hip joint and aids iliopsoas in flexing the thigh.

Vastus Lateralis

Origin: Greater trochanter, intertrochanteric line, lateral lip of linea aspera, capsule of hip joint, gluteal tuberosity.

Insertion: Into the superolateral aspect of the patella and then on to the tibial tuberosity via the patellar ligament.

Innervation: Femoral (L2, L3, L4).

Blood Supply: Lateral femoral circumflex artery, Lateral superior genicular artery.

Action: Extends leg at knee.

Vastus Intermedius

Origin: Anterior and lateral aspects of the shaft of the femur.

Insertion: Into the superior aspect of the patella and then on to the tibial tuberosity via the patellar ligament. Blends with the tendons of the rectus femoris and vastus lateralis and medialis.

Innervation: Femoral nerve (L2, L3, L4).

Blood Supply: Lateral femoral circumflex artery.

Action: Extends leg at knee.

Vastus Medialis

Origin: Intertrochanteric line and medial lip of the linea aspera.

Insertion: Into the superomedial aspect of the patella and then on to the tibial tuberosity via the patellar ligament.

Innervation: Femoral nerve (L2, L3, L4).

Blood Supply: Femoral artery, profunda femoris artery, superior genicular arteries.

Action: Extends leg at knee; maintains patella in the patellar groove by drawing it medially.

Articularis Genu

Origin: Inferior aspect of the anterior surface of the shaft of the femur.

Insertion: Capsule of the knee joint and walls of the suprapatellar bursa.

Innervation: Femoral nerve (L2, L3, L4).

Blood Supply: Profunda femoris artery, superior genicular arteries.

Action: Pulls the capsule of the knee joint superiorly during extension of the leg at the knee.

Sartorius

Origin: Anterior superior iliac spine and superior part of notch inferior to the spine.

Insertion: Superior part of medial surface of the tibia (pes anserinus).

Innervation: Femoral nerve (L2, L3).

Blood Supply: Femoral artery.

Action: Flexes, abducts, and laterally rotates the thigh at the hip joint, flexes the leg at the knee joint.

Posterior Compartment

Hamstring Muscles

Biceps Femoris

Origin: Long head: ischial tuberosity. Short head: lateral lip of linea aspera and lateral supracondylar line.

Insertion: Lateral side of the head of the fibula. Tendon is split by the fibular collateral ligament.

Innervation: Long head: tibial division of the sciatic nerve. Short head: common fibular (peroneal) division of the sciatic nerve (L5, S1, S2).

Blood Supply: Perforating branches of the profunda femoris artery. Popliteal artery.

Action: Flexes leg and rotates it laterally. Extends thigh (e.g., when starting to walk). Both heads flex the leg at the knee and laterally rotate the leg when it is flexed. The long head extends the thigh at the hip joint.

Semitendinosus

Origin: Ischial tuberosity.

Insertion: Medial surface of superior part of tibia (pes anserinus).

Innervation: Tibial division of sciatic nerve (L5, S1, S2).

Blood Supply:	Perforating branches of the profunda femoris artery. Popliteal artery.
Action:	Extends thigh, flexes leg, rotates lower limb medially. When thigh and leg are flexed, it can aid in extending the trunk.

Semimembranosus

Origin:	Ischial tuberosity.
Insertion:	Posterior part of medial portion of proximal tibia.
Innervation:	Tibial division of sciatic nerve (L5, S1, S2).
Blood Supply:	Perforating branches of the profunda femoris artery. Popliteal artery.
Action:	Extends thigh, flexes leg, rotates lower limb medially. When thigh and leg are flexed, it can aid in extending the trunk.

Adductor Magnus (ischial component)

The ischial component of the adductor magnus is considered a hamstring muscle. The entire muscle is described below with the adductor muscles of the thigh.

Medial Compartment

Adductor Muscles

Obturator Externus

Origin:	External margins of the obturator foramen, obturator membrane.
Insertion:	Trochanteric fossa of the femur.
Innervation:	Obturator nerve (L3, L4).
Blood Supply:	Obturator artery and medial femoral circumflex artery.
Action:	Laterally rotates thigh, adducts thigh, steadies head of femur in acetabulum.

Pectineus

Origin:	Pectineal line of pubis and pecten pubis.
Insertion:	Pectineal line of femur, between lesser trochanter and linea aspera.
Innervation:	Femoral nerve, and occasionally a branch from the obturator nerve (L2, L3).
Blood Supply:	Medial femoral circumflex artery, obturator artery.
Action:	Medially rotates, adducts, and flexes the thigh.

Adductor Longus

- *Origin:* Inferior ramus of pubis inferior to the pubic crest.
- *Insertion:* Middle third of linea aspera inferior to the pectineus.
- *Innervation:* Anterior branch of obturator nerve (L2, L3, L4).
- *Blood Supply:* Medial femoral circumflex artery, obturator artery.
- *Action:* Adducts thigh.

Adductor Brevis

- *Origin:* Inferior ramus of pubis deep to adductor longus.
- *Insertion:* Pectineal line and proximal part of linea aspera, deep to pectineus and adductor longus.
- *Innervation:* Anterior branch of obturator nerve (L2, L3, L4).
- *Blood Supply:* Medial femoral circumflex artery, obturator artery.
- *Action:* Adducts and, to a small extent, flexes the thigh.

Adductor Magnus

- *Origin:* Adductor part: inferior ramus of pubis, ramus of ischium. Hamstring part: ischial tuberosity.
- *Insertion:* Adductor part: gluteal tuberosity of femur, linea aspera, medial supracondylar line. Hamstring part: adductor tubercle of femur.
- *Innervation:* Adductor part: Posterior branch of obturator nerve. Hamstring part: tibial portion of sciatic nerve (L2, L3, L4).
- *Blood Supply:* Medial femoral circumflex artery, obturator artery, popliteal artery, perforating branches of profunda femoris artery.
- *Action:* Adducts thigh. Adductor part also flexes thigh, while the hamstring part extends the thigh.

Gracilis

- *Origin:* Inferior ramus of pubis.
- *Insertion:* Superior part of medial surface of tibia (pes anserinus).
- *Innervation:* Anterior branch of obturator nerve (L2, L3).
- *Blood Supply:* Medial femoral circumflex artery, obturator artery, profunda femoris artery.
- *Action:* Adducts thigh, medially rotates thigh, flexes leg at knee.

Lower Limb Musculature:

The Leg

Posterior Compartment: Superficial Muscles

Triceps Surae

The triceps surae is composed of the soleus and the two heads of the gastrocnemius.

Gastrocnemius

Origin:	Lateral head: lateral aspect of lateral condyle of the femur.
Insertion:	Posterior surface of calcaneus via the calcaneal ("Achilles") tendon.
Innervation:	Tibial nerve (S1, S2).
Blood Supply:	Sural arteries from the popliteal artery.
Action:	Plantarflexes foot, raises heel during walking, and flexes the knee joint.

Soleus

Origin:	Posterior aspect of the head of the fibula, superior one quarter of the posterior aspect of the fibula, from the soleal line and middle third of the medial border of the tibia, and from a tendonous arch between the tibia and fibula.
Insertion:	Posterior surface of calcaneus via the calcaneal ("Achilles") tendon.
Innervation:	Tibial nerve (S1, S2).
Blood Supply:	Posterior tibial artery, fibular (peroneal) artery, sural branches of popliteal.
Action:	Plantarflexes foot and steadies leg on foot.

Plantaris

Origin:	Inferior end of lateral supracondylar line of the femur and the oblique popliteal ligament.
Insertion:	Posterior surface of calcaneus via the calcaneal ("Achilles") tendon.
Innervation:	Tibial nerve (S1, S2).
Blood Supply:	Sural branches of the popliteal artery.
Action:	Weakly assists gastrocnemius in plantarflexing foot and flexing knee joint. May function as a sensory organ.

Appendix   Page 335

Posterior Compartment: Deep Muscles

Popliteus

Origin: Lateral surface of lateral condyle of femur and lateral meniscus of knee joint.

Insertion: Posterior surface of tibia superior to soleal line.

Innervation: Tibial nerve (L4, L5, S1).

Blood Supply: Genicular arteries from the popliteal artery.

Action: Weakly flexes knee and medially rotates tibia at beginning of flexion of the knee.

Flexor Digitorum Longus

Origin: Medial part of posterior surface of tibia inferior to soleal line. Broad aponeurosis attaches it to the fibula.

Insertion: Plantar surfaces of the bases of the distal phalanges of the lateral four digits.

Innervation: Tibial nerve (S2, S3).

Blood Supply: Posterior tibial artery.

Action: Flexes the lateral four digits and plantarflexes foot. Supports longitudinal arches of foot.

Flexor Hallucis Longus

Origin: Inferior 2/3 of the posterior surface of fibula and inferior part of interosseous membranes.

Insertion: Plantar surface of the base of the distal phalanx of great toe (hallux).

Innervation: Tibial nerve (S2, S3).

Blood Supply: Fibular (peroneal) artery.

Action: Flexes great toe at all joints and plantarflexes foot. Supports longitudinal arch of foot.

Tibialis Posterior

Origin: Interosseous membrane, posterior surface of tibia inferior to soleal line, superior two thirds of medial surface of fibula.

Insertion: Tuberosity of navicular, cuboid, and all three cuneiform bones, and plantar surfaces of the bases of the 2nd, 3rd, and 4th metatarsals.

Innervation: Tibial nerve (L4, L5).

Blood Supply: Fibular (peroneal) artery.

Action: Plantarflexes and inverts foot.

Anterior Compartment

Tibialis Anterior

Origin: Lateral aspect of the tibial plateau, superior half of lateral surface of tibia.

Insertion: Medial and inferior surfaces of the medial cuneiform bone base of the first metatarsal bone.

Innervation: Deep fibular (peroneal) nerve (L4, L5).

Blood Supply: Anterior tibial artery.

Action: Dorsiflexes and inverts foot.

Extensor Digitorum Longus

Origin: Lateral aspect of tibial plateau, superior three quarters of anterior surface of fibula, anterior surface of interosseous membrane.

Insertion: Dorsal aspect of the middle and distal phalanges of the lateral four toes through the extensor expansions.

Innervation: Deep fibular (peroneal) nerve (L5, S1).

Blood Supply: Anterior tibial artery.

Action: Extends the lateral four digits and dorsiflexes the foot.

Fibularis Tertius

Origin: Inferior 1/3 of anterior surface fibula and interosseous membrane.

Insertion: Dorsum of the base of the fifth metatarsal.

Innervation: Deep fibular (peroneal) nerve (L5, S1).

Blood Supply: Anterior tibial artery.

Action: Dorsiflexes foot and aids in eversion of foot.

Extensor Hallucis Longus

Origin: Middle part of the anterior surface of the fibula and interosseous membrane.

Insertion: Dorsal aspect of the base of the distal phalanx of the great toe.

Innervation: Deep fibular (peroneal) nerve (L5, S1).

Blood Supply: Anterior tibial artery.

Action: Extends great toe and dorsiflexes foot.

Lateral Compartment

Fibularis Longus

Origin:	Head and superior two thirds of lateral aspect of fibula.
Insertion:	Base of first metatarsal and the medial cuneiform bone.
Innervation:	Superficial fibular (peroneal) nerve (L5, S1, S2).
Blood Supply:	Anterior tibial and fibular (peroneal) arteries.
Action:	Everts foot and weakly plantarflexes foot. Also helps steady leg when standing on one foot.

Fibularis Brevis

Origin:	Inferior two thirds of lateral surface of fibula.
Insertion:	Dorsal surface of tuberosity on lateral side of base of fifth metatarsal bone.
Innervation:	Superficial fibular (peroneal) nerve (L5, S1, S2).
Blood Supply:	Fibular (peroneal) artery.
Action:	Everts and weakly plantarflexes foot.

Lower Limb Musculature:

Intrinsic Muscles of the Foot

Dorsum of Foot

Extensor Digitorum Brevis

Origin:	Anterior part of dorsal surface of calcaneus anteromedial to lateral malleolus, inferior extensor retinaculum.
Insertion:	By four tendons: one to the base of the proximal phalanx of the hallux **(extensor hallucis brevis)**, the more lateral three to the tendons of extensor digitorum longus to digits 2, 3, and 4 **(extensor digitorum brevis).**
Innervation:	Deep fibular (peroneal) nerve (L5, S1).
Blood Supply:	Dorsalis pedis.
Action:	Extends digits 1 to 4 at the metacarpophalangeal joints.

Extensor Hallucis Brevis

See preceding description of the insertion of the insertion of the extensor digitorum brevis.

Plantar Aspect of Foot: First Layer of Muscles

Abductor Hallucis

Origin:	Medial process of the calcaneal tubercle, flexor retinaculum and plantar aponeurosis.
Insertion:	Medial side of the base of the proximal phalanx of the hallux.
Innervation:	Medial plantar nerve (S2, S3).
Blood Supply:	Medial plantar artery.
Action:	Abducts and flexes hallux. Helps maintain medial arch of foot.

Flexor Digitorum Brevis

Origin:	Medial process of the calcaneal tubercle, intermuscular septum and plantar aponeurosis.
Insertion:	Both sides of the middle phalanx of the lateral four digits.
Innervation:	Medial plantar nerve (S2, S3).
Blood Supply:	Medial plantar artery.
Action:	Flexes the lateral four digits at the proximal interphalangeal and metatarsophalangeal joints. Helps maintain the longitudinal arches of the foot.

Abductor Digiti Minimi

Origin:	Medial and lateral processes of the calcaneal tuberosity, plantar aponeurosis and intermuscular septum.
Insertion:	Lateral side of the base of the proximal phalanx of digit V.
Innervation:	Lateral plantar nerve (S2, S3).
Blood Supply:	Lateral plantar artery.
Action:	Abducts and flexes the fifth digit. Helps maintain lateral longitudinal arch of foot.

Plantar Aspect of Foot: Second Layer of Muscles

Tendon of Flexor Digitorum Longus

The flexor digitorum longus muscle is described in the Posterior Compartment section..

Quadratus Plantae

Origin:	Medial surface and lateral margin of plantar surface of calcaneus.
Insertion:	Posterolateral margin of tendon of flexor digitorum longus.
Innervation:	Lateral plantar nerve (S2, S3).
Blood Supply:	Lateral plantar artery.
Action:	Assists flexor digitorum longus in flexing the lateral four digits.

Lumbricals

Origin:	From the tendons of insertion of the flexor digitorum longus.
Insertion:	Medial sides of bases of proximal phalanges of the lateral four digits and extensor expansions of the tendons of the extensor digitorum longus.
Innervation:	Medial (first) lumbrical: medial plantar nerve. Lateral three lumbricals: lateral plantar nerve (S2, S3).
Blood Supply:	Plantar metatarsal arteries.
Action:	Flex the proximal phalanges at the metatarsophalangeal joint, extend middle and distal phalanges at the proximal and distal interphalangeal joints of digits II to V.

Tendon of Flexor Hallucis Longus

The flexor digitorum longus muscle is described in the Posterior Compartment section..

Plantar Aspect of Foot: Third Layer of Muscles

Flexor Hallucis Brevis

Origin:	Plantar surfaces of cuboid bone and lateral cuneiform bone.
Insertion:	Both sides of the base of the proximal phalanx of the great toe.
Innervation:	Medial plantar nerve (S2, S3).
Blood Supply:	First plantar metatarsal artery.
Action:	Flexes the proximal phalanx of the great toe.

Adductor Hallucis

Origin: Oblique head: bases of metatarsals two to four. Transverse head: plantar ligaments of the metatarsophalangeal joints.

Insertion: Tendons of both heads insert into the lateral side of the base of the proximal phalanx of the great toe.

Innervation: Deep branch of the lateral plantar nerve (S2, S3).

Blood Supply: First plantar metatarsal artery.

Action: Adducts great toe. Assists in maintaining the transverse arch of the foot.

Flexor Digiti Minimi Brevis

Origin: Base of the fifth metatarsal bone.

Insertion: Base of the proximal phalanx of the fifth digit.

Innervation: Superficial branch of the lateral plantar nerve (S2, S3).

Blood Supply: Lateral plantar artery.

Action: Flexes the proximal phalanx of the fifth digit at the metatarsophalangeal joint.

Plantar Aspect of Foot: Fourth Layer

Plantar Interossei: Three in number.

Origin: Bases and medial sides of metatarsals three to five.

Insertion: Medial sides of bases of proximal phalanges of digits three to five.

Innervation: Lateral Plantar nerve (S2, S3).

Blood Supply: Plantar metatarsal arteries.

Action: Adduct digits three to five. Flex the metatarsophalangeal joints of digits three to five.

Dorsal Interossei: Four in number.

Origin: Adjacent sides of metatarsals one to five (two heads each).

Insertion: First: medial side of proximal phalanx of second digit. Second to fourth: lateral sides of proximal phalanges of digits two to four.

Innervation: Lateral plantar nerve (S2, S3).

Blood Supply: Dorsal metatarsal arteries.

Action: Abduct digits two to four and flex metatarsophalangeal joints of digits two to four.

Major Muscles of Mastication and of the Hyoid

Masseter

Origin: Anterior two thirds of the inferior border of the zygomatic arch and the medial surface of the zygomatic arch.

Insertion: Medial surface of mandibular ramus, at the angle and inferior and posterior part of the ramus.

Innervation: Masseteric branch of the mandibular division of the trigeminal nerve (Cranial nerve V).

Blood Supply: Superficial temporal artery, maxillary artery, facial artery.

Action: Closes the jaws by elevating the mandible.

Temporalis

Origin: Floor of the entire temporal fossa and deep surface of the temporal fascia.

Insertion: Medial surface, apex and anterior border of the coronoid process of the mandible, superior portion of the anterior border of the ramus of the mandible.

Innervation: Anterior and posterior deep temporal nerves of the mandibular division of the trigeminal nerve (Cranial Nerve V).

Blood Supply: Superficial temporal artery, maxillary artery.

Action: Elevates, closes, and retracts the mandible.

Pterygoideus Lateralis

Origin: Superior head: from infratemporal crest of cranium and lateral surface of the greater wing of the sphenoid bone. Inferior head: lateral surface of the lateral pterygoid plate of the sphenoid bone.

Insertion: Depression on the anterior part of the neck of the mandible, and anterior margin of the meniscus of the temporomandibular joint.

Innervation: Lateral pterygoid nerve from the mandibular division of the trigeminal nerve (Cranial Nerve V).

Blood Supply: Pterygoid branch of the maxillary artery.

Action: Protrudes mandible, pulls articular disc anteriorly, and assists in rotatory motions of the mandible in chewing.

Pterygoideus Medialis

Origin: Medial surface of the lateral pterygoid plate, pyramidal process of palatine bone, and tuberosity of the maxilla.

Insertion:	Inferior and posterior part of the internal aspect of the mandibular ramus and angle of the mandible.
Innervation:	Lateral pterygoid nerve from the mandibular division of the trigeminal nerve (Cranial Nerve V).
Blood Supply:	Pterygoid branches of the maxillary artery, facial artery.
Action:	Elevates the mandible, closing the jaws. Act with lateral pterygoids in protruding the mandible. Acts with the lateral pterygoids in rotatory motions of the mandible during chewing.

Styloglossus

Origin:	Anterior border of the styloid process.
Insertion:	Into the sides of the tongue.
Innervation:	Hypoglossal (Cranial Nerve XII).
Blood Supply:	Sublingual artery.
Action:	Elevation and retraction of the tongue.

Stylopharyngeus

Origin:	Medial side of the root of the styloid process.
Insertion:	Superior and posterior borders of the thyroid cartilage.
Innervation:	Glossopharyngeal nerve (Cranial Nerve IX).
Blood Supply:	Ascending pharyngeal artery.
Action:	Elevates and dilates pharynx.

Pharyngeal Constrictors

Superior

Origin:	Lower third of the posterior border of the medial pterygoid plate, alveolar process of mandible superior to the posterior end of the mylohyoid line, pterygomandibular ligament.
Insertion:	Median raphe of posterior wall of pharynx, pharyngeal tubercle of occipital bone.
Innervation:	Cranial root of accessory nerve (Cranial Nerve XI); Vagus nerve (Cranial Nerve X) through the pharyngeal plexus.
Blood Supply:	Facial and ascending pharyngeal arteries.
Action:	Contracts the pharynx during swallowing.

Middle

- *Origin:* Greater and lesser cornua of the hyoid bone.
- *Insertion:* Into median raphe of the pharynx.
- *Innervation:* Pharyngeal plexus.
- *Blood Supply:* Facial and ascending pharyngeal arteries.
- *Action:* Contracts the pharynx during swallowing.

Inferior

- *Origin:* Sides of thyroid and cricoid cartilages.
- *Insertion:* Into median raphe of the pharynx.
- *Innervation:* Pharyngeal plexus, external and recurrent laryngeal branches of the vagus (Cranial Nerve X), cranial root of accessory nerve (Cranial Nerve XI).
- *Blood Supply:* Facial and ascending pharyngeal arteries.
- *Action:* Contracts the pharynx during swallowing.

Hyoglossus

- *Origin:* Greater cornua of the hyoid bone
- *Insertion:* Sides of tongue.
- *Innervation:* Hypoglossal nerve (Cranial Nerve XII).
- *Blood Supply:* Sublingual and submental arteries.
- *Action:* Draws down sides of tongue and depresses tongue.

Mylohyoid

- *Origin:* Mylohyoid line of mandible.
- *Insertion:* Median raphe from chin to hyoid bone and onto hyoid.
- *Innervation:* Mylohyoid branch of inferior alveolar nerve from trigeminal nerve (Cranial Nerve V).
- *Blood Supply:* Lingual, facial, and inferior alveolar arteries.
- *Action:* Elevates hyoid bone and tongue, raises floor of mouth, when hyoid is fixed depresses mandible.

Omohyoid

Origin: Inferior belly: superior border of scapula and suprascapular ligament to a tendon deep to sternocleidomastoid. Superior belly: extends superiorly from this tendon.

Insertion: Inferior aspect of the body of the hyoid bone.

Innervation: Ansa cervicalis with contributions from spinal nerves C1, C2, and C3 (C1, C2, C3).

Blood Supply: Lingual and superior thyroid arteries.

Action: Fixes hyoid bone. Depresses and retracts hyoid and larynx.

Geniohyoid

Origin: Inferior genial tubercle on posterior aspect of mental symphysis of mandible.

Insertion: Anterior aspect of the body of the hyoid bone.

Innervation: First cervical spinal nerve through fibers that travel with the hypoglossal nerve (Cranial Nerve XII). (C1)

Blood Supply: Sublingual branch of lingual artery.

Action: Elevates hyoid bone and tongue.

Genioglossus

Origin: Superior genial tubercle on posterior aspect of mental symphysis of mandible.

Insertion: Lowest fibers to hyoid bone, middle fibers along inferior surface of tongue, superior fibers to tip of tongue.

Innervation: Hypoglossal nerve (Cranial Nerve XII).

Blood Supply: Sublingual and submental arteries.

Action: Posterior fibers protrude tongue, anterior fibers retract tongue, aids in depressing tongue.

Digastric

Origin: Posterior belly: from mastoid notch (digastric groove) of the temporal bone. Anterior belly: Digastric fossa on the anterior and inferior aspect of the mandible.

Insertion: Intermediate tendon between the two bellies that is affixed to the body and greater cornu of the hyoid by a ligamentous loop.

Innervation: Posterior belly: Facial nerve. Anterior belly: Mylohyoid branch of inferior alveolar nerve from trigeminal nerve (Cranial Nerve V).

Blood Supply: Posterior belly: occipital and posterior auricular arteries. Anterior belly: submental artery.

Action: Elevates and fixes hyoid bone, assists lateral pterygoid in opening the mouth by depressing the mandible.

References Cited

Bogduk, Nikola (1980) A reappraisal of the anatomy of the human erector spinae. *Journal of Anatomy* 131: 525-540.

Bustami, Faraj M. F. (1986) A new description of the lumbar erector spinae muscle in man. *Journal of Anatomy* 144: 81-91.

Greiner, T.M., M.E. Bedford, and R.A. Walker (2004) Variability in the human M. spinalisand cervicis: Frequencies and definitions. *Annals of Anatomy* 186: 185-191.